Mathematics for the Million

Lancelot Hogben has held professorial posts in the universities of Cape Town, London, Wisconsin, Birmingham and Aberdeen. He holds honorary doctorates of the universities of Wales and Birmingham and was awarded the Keith Prize and Gold Medal of the Royal Society of Edinburgh for his work in the field of mathematical genetics. He has published many books and articles in scientific journals and has travelled widely. He was the first Vice-Chancellor of the University of Guyana.

Lancelot Hogben

Mathematics for the Million

Pan Books London and Sydney

First published 1936 by George Allen & Unwin Ltd
This revised edition published 1967 by Pan Books Ltd,
Cavaye Place, London SW10 9PG
3rd printing 1978

ISBN 0 330 63129 2
Printed and bound in Great Britain by
Richard Clay (The Chaucer Press) Ltd, Bungay, Suffolk

CONTENTS

To

Jane

FOREWORD

Thirty-three years have passed since I wrote the original text of *Mathematics For The Million*. Since then much water has flowed under the bridges. I therefore welcomed the opportunity offered to me by PAN BOOKS to undertake the task of rewriting much of it, of revising the rest of previous editions and of adding new material.

In the time available, I could not have completed the task without the help of my friend Mr Terence Baylis, B.Sc., who vetted the typescript before it went to press and checked the galley proofs. My indebtedness to him is considerable.

This edition has entirely new illustrations for the execution of which I am happy to acknowledge the work of Mr David Woodcock. I have also to thank Miss Glenys Kimber for preparing the typescript.

LANCELOT HOGBEN

ILLUSTRATIONS

(between pages 328 and 329)

MATHEMATICS FOR THE MILLION

PROLOGUE

THE PARABLE OF ACHILLES AND THE TORTOISE

THERE IS a story about Diderot, the Encyclopaedist and materialist, a foremost figure in the intellectual awakening which immediately preceded the French Revolution. Diderot was staying at the Russian court, where his elegant flippancy was entertaining the nobility. Fearing that the faith of her retainers was at stake, the Tsaritsa commissioned Euler, the most distinguished mathematician of the time, to debate with Diderot in public. Diderot was informed that a mathematician had established a proof of the existence of God. He was summoned to court without being told the name of his opponent. Before the assembled court, Euler accosted him with the following pronouncement, which was uttered with due gravity: '$\frac{a + b^n}{n} = x$, donc Dieu existe, repondez!' Algebra was Arabic to Diderot. Unfortunately, he did not realize that was the trouble. Had he realized that algebra is just a language in which we describe the *sizes* of things in contrast to the ordinary languages which we use to describe the *sorts* of things in the world, he would have asked Euler to translate the first half of the sentence into French. Translated freely into English, it may be rendered: 'A number x can be got by first adding a number a to a number b multiplied by itself a certain number (n) of times, and then dividing the whole by the number of b's multiplied together. So God exists after all. What have you got to say now?' If Diderot had asked Euler to illustrate the first part of his remark for the clearer understanding of the Russian court, Euler might have replied that x is 3 when a is 1 and b is 2 and n is 3, or that x is 21 when a is 3 and b is 3 and n is 4, and so forth. Euler's troubles would have begun when the court wanted to know how the second part of the sentence follows from the first part. Like many of us, Diderot had stagefright when confronted with a sentence in size language. He left the court abruptly amid the titters of the assembly, confined himself to his chambers, demanded a safe conduct, and promptly returned to France.

In the time of Diderot the lives and happiness of individuals might still depend on holding correct beliefs about religion. Today, the lives and happiness of people depend more than most of us realize upon correct interpretation of public statistics kept by Government offices. Atomic power depends on calculations which may destroy us or may guarantee worldwide freedom from want. The costly conquest of outer space enlists immense resources of mathematical ingenuity. Without some understanding of mathematics, we lack the language in which to talk intelligently about the forces which now fashion the future of our species, if any.

We live in a welter of figures: cookery recipes, railway timetables, unemployment aggregates, fines, taxes, war debts, overtime schedules, speed limits, bowling averages, betting odds, billiard scores, calories, babies' weights, clinical temperatures, rainfall, hours of sunshine, motoring records, power indices, gas-meter readings, bank rates, freight rates, death rates, discount, interest, lotteries, wavelengths, and tyre pressures. Every night, when he winds up his watch, the modern man adjusts a scientific instrument of a precision and delicacy unimaginable to the most cunning artificers of Alexandria in its prime. So much is commonplace. What escapes our notice is that in doing these things we have learnt to use devices which presented tremendous difficulties to the most brilliant mathematicians of antiquity. Ratios, limits, acceleration, are no longer remote abstractions, dimly apprehended by the solitary genius. They are photographed upon every page of our existence.

In the course of the adventure upon which we are going to embark we shall constantly find that we have no difficulty in answering questions which tortured the minds of very clever mathematicians in ancient times. This is not because you and I are very clever people. It is because we inherit a social culture which has suffered the impact of material forces foreign to the intellectual life of the ancient world. The most brilliant intellect is a prisoner within its own social inheritance. An illustration will help to make this quite definite at the outset.

The Eleatic philosopher Zeno set all his contemporaries guessing by propounding a series of conundrums, of which the one most often quoted is the paradox of Achilles and the tortoise. Here is the problem about which the inventors of school geometry argued till they had speaker's throat and writer's cramp. Achilles runs a race with the tortoise. He runs ten times as fast as the tortoise. The tortoise has 100 yards' start. Now, says Zeno, Achilles runs 100 yards and reaches the place where the tortoise started. Meanwhile the tortoise has gone a tenth as far as Achilles, and is therefore 10 yards ahead of Achilles. Achilles runs this 10 yards. Meanwhile the tortoise has run a tenth as far as Achilles, and is therefore 1 yard in front of him. Achilles runs this 1 yard. Meanwhile the tortoise has run a tenth of a yard and is therefore a tenth of a yard in front of Achilles. Achilles runs this tenth of a yard. Meanwhile the tortoise goes a tenth of a tenth of a yard. He is now a hundredth of a yard in front of Achilles. When Achilles has caught up this hundredth of a yard, the tortoise is a thousandth of a yard in front. So, argued Zeno, Achilles is always getting nearer the tortoise, but can never quite catch him up.

You must not imagine that Zeno and all the wise men who argued the point failed to recognize that Achilles really did get past the tortoise. What troubled them was, where is the catch? You may have been asking the same question. The important point is that you did not ask it for the same reason which prompted them. What is worrying you is why they thought up funny little riddles of that sort. Indeed, what you are really concerned with is an *historical* problem. I am going to show you in a minute that the problem is not one which presents any *mathematical* difficulty to you. You know how to translate it into size language, because you inherit a social culture which is separated from theirs by the collapse of two great civilizations and by two great social revolutions. The difficulty of the ancients was not an historical difficulty. It was a mathematical difficulty. They had not evolved a size language into which this problem could be freely translated.

The Greeks were not accustomed to speed limits and passenger-luggage allowances. They found any problem involving division very much more difficult than a problem involving multiplication. They had no way of doing division to any order of accuracy, because they relied for calculation on the mechanical aid of the counting frame or abacus, shown in Fig. 7. They could not do sums on paper. For all these and other reasons which we shall meet again and again, the Greek mathematician was unable to see something that we see without taking the trouble to worry

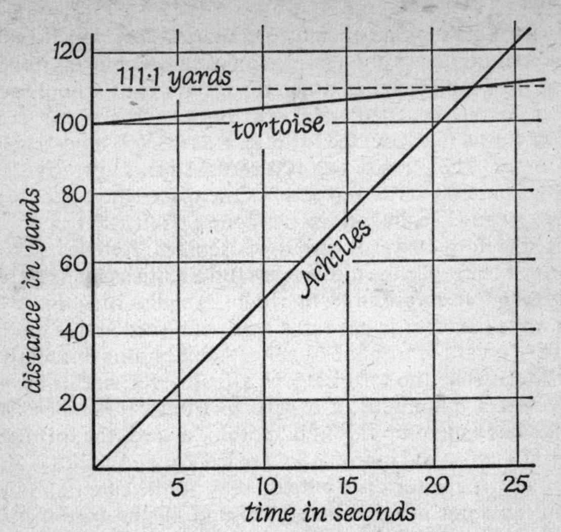

Fig. 1. The Race of Achilles and the Tortoise

Greek geometry, which was timeless, could not make it obvious that Achilles would overtake the tortoise. The new geometry of the Newtonian century put time in the picture, thereby showing when and where the two came abreast.

about whether we see it or not. If we go on piling up bigger and bigger quantities, the pile goes on growing more rapidly without any end as long as we go on adding more. If we can go on adding larger and larger quantities indefinitely without coming to a stop, it seemed to Zeno's contemporaries that we ought to be able to go on adding smaller and still smaller quantities indefinitely without reaching a limit. They thought that in one case the pile goes on for ever, growing more rapidly, and in the other it goes on for ever, growing more slowly. There was nothing in their number language to suggest that when the engine slows beyond a certain point, it chokes off.

To see this clearly, let us first put down in numbers the distance which the tortoise traverses at different stages of the race after Achilles starts. As we have described it above, the tortoise moves 10 yards in stage 1, 1 yard in stage 2, one-tenth of a yard in stage 3, one-hundredth of a yard in stage 4, etc. Suppose we had a number language like the Greeks and Romans, or the Hebrews, who used letters of the alphabet. Using the one that is familiar to us because it is still used for clocks, graveyards, and law-courts, we might write the total of all the distances the tortoise ran before Achilles caught him up like this:

$$X + I + \frac{I}{X} + \frac{I}{C} + \frac{I}{M}, \text{ and so on*}$$

We have put 'and so on' because the ancient world got into great difficulties when it had to handle numbers more than a few thousands. Apart from the fact that we have left the tail of the series to your imagination (and do not forget that the tail is most of the animal if it goes on for ever), notice another disadvantage about this script. There is absolutely nothing to suggest to you how the distances at each stage of the race are connected with one another. Today we have a number vocabulary which makes this relation perfectly evident, when we write it down as:

$$10+1+\frac{1}{10}+\frac{1}{100}+\frac{1}{1{,}000}+\frac{1}{10{,}000}+\frac{1}{100{,}000}+\frac{1}{1{,}000{,}000}, \text{ and so on}$$

In this case we put 'and so on' to save ourselves trouble, not because we have not the right number-words. These number-words were borrowed from the Hindus, who learnt to write number language after Zeno and Euclid had gone to their graves. A social revolution, the Protestant Reformation, gave us schools which made this number language the common property of mankind. A second social upheaval, the French Revolution, taught us to use a reformed spelling. Thanks to the Education Acts of the nineteenth century, this reformed spelling is part of the common fund of knowledge shared by almost every sane individual in the English-speaking world. Let us write the last total, using this reformed spelling, which we call decimal notation. That is to say:

$$10 + 1 + 0{\cdot}1 + 0{\cdot}01 + 0{\cdot}001 + 0{\cdot}0001 + 0{\cdot}00001 + 0{\cdot}000001, \text{ and so on}$$

We have only to use the reformed spelling to remind ourselves that this can be put in a more snappy form:

$$11{\cdot}111111, \text{ etc.,}$$

or still better:

$$11{\cdot}\dot{1}$$

* The Romans did not actually have the convenient method of representing proper fractions used above for illustrative purposes.

We recognize the fraction $0 \cdot \dot{1}$ as a quantity that is less than $\frac{2}{10}$ and more than $\frac{1}{10}$. If we have not forgotten the arithmetic we learned at school, we may even remember that $0 \cdot \dot{1}$ corresponds with the fraction $\frac{1}{9}$. This means that, the longer we make the sum, $0 \cdot 1 + 0 \cdot 01 + 0 \cdot 001$, etc., the nearer it gets to $\frac{1}{9}$, and it never grows bigger than $\frac{1}{9}$. The total of all the yards the tortoise moves till there is no distance between himself and Achilles makes up just $11\frac{1}{9}$ yards, and no more. You will now begin to see what was meant by saying that the riddle presents no mathematical difficulty to you. You yourself have a number language constructed so that it can take into account a possibility which mathematicians describe by a very impressive name. They call it the convergence of an infinite series to a limiting value. Put in plain words, this only means that, if you go on piling up smaller and smaller quantities as long as you can, you *may* get a pile of which the size is not made measurably larger by adding any more.

The immense difficulty which the mathematicians of the ancient world experienced when they dealt with a process of division carried on indefinitely, or with what modern mathematicians call infinite series, limits, transcendental numbers, irrational quantities, and so forth, provides an example of a great social truth borne out by the whole history of human knowledge. Fruitful intellectual activity of the cleverest people draws its strength from the common knowledge which all of us share. Beyond a certain point clever people can never transcend the limitations of the social culture they inherit. When clever people pride themselves on their own isolation, we may well wonder whether they are very clever after all. Our studies in mathematics are going to show us that whenever the culture of a people loses contact with the common life of mankind and becomes exclusively the plaything of a leisure class, it is becoming a priestcraft. It is destined to end, as does all priestcraft, in superstition. To be proud of intellectual isolation from the common life of mankind and to be disdainful of the great social task of education is as stupid as it is wicked. It is the end of progress in knowledge. No society, least of all so intricate and mechanized a society as ours, is safe in the hands of a few clever people. The mathematician and the plain man each need the other.

In such a time as ours, we may apply to mathematics the words in which Cobbett explained the uses of grammar to the working men at a time when there was no public system of free schools.

In the first of his letters on English grammar for a working boy, Cobbett wrote these words:

> 'But, to the acquiring of this branch of knowledge, my dear son, there is one motive, which, though it ought, at all times, to be strongly felt, ought, at the present time, to be so felt in an extraordinary degree. I mean that desire which every man, and especially every young man, should entertain to be able to assert with effect the rights and liberties of his country. When you come to read the history of those Laws of England by which the freedom of the people has been secured . . . you will find that tyranny has no enemy so formidable as the pen. And, while you will see with exultation the long-imprisoned, the heavily-fined, the banished William Prynne, returning to liberty, borne by the people from Southampton to London, over a road strewed with flowers: then accusing, bring to trial and to the block, the tyrants from whose hands he and his country had unjustly and cruelly suffered; while your heart and the heart of every young man in the kingdom will bound with joy at the spectacle, you ought all to bear in mind, that, without a knowledge of grammar, Mr Prynne could never have performed any of those acts by which his name has been thus preserved, and which have caused his name to be held in honour.'

Our glance at the Greek paradox of Achilles and the tortoise has suggested one sort of therapy for people who have an inbuilt fear of mathematical formulae. Cobbett's words suggest another. We can make things easier for ourselves if we think of mathematics less as an exploit of intellectual ingenuity than as an exercise in the grammar of a foreign language. Our first introduction to it may well have made this difficult. The first great mathematical treatise, that compiled by Euclid of Alexandria (about 300 BC), has cast a long shadow over the teaching of mathematics. It transmitted to posterity a mystique largely due to the Athenian philosopher Plato (about 380 BC). The leisure class of the Greek city-states played with geometry as people play with crossword puzzles and chess today. Plato taught that geometry was the highest exercise to which human leisure could be devoted. So geometry became included in European education as a part of classical scholarship, without any clear connection with the contemporary reality of measuring Drake's 'world encompassed'.

Those who taught Euclid did not understand its social use, and generations of schoolboys have studied Euclid without being told how a later geometry, which grew out of Euclid's teaching in the busy life of Alexandria, made it possible to measure the size of the world. Those measurements blew up the pagan Pantheon of star gods and blazed the trail for the Great Navigations. The revelation of how much of the surface of our world was still unexplored was the solid ground for what we call the faith of Columbus.

Plato's exaltation of mathematics as an august and mysterious ritual had its roots in dark superstitions which troubled, and fanciful puerilities which entranced, people who were living through the childhood of civilization, when even the cleverest people could not clearly distinguish the difference between saying that 13 is a 'prime' number and saying that 13 is an unlucky number. His influence on education has spread a veil of mystery over mathematics and helped to preserve the queer freemasonry of the Pythagorean brotherhoods, whose members were put to death for revealing mathematical secrets now printed in school books. It reflects no discredit on anybody if this veil of mystery makes the subject distasteful. Plato's great achievement was to invent a religion which satisfies the emotional needs of people who are out of harmony with their social environment, and just too intelligent or too individualistic to seek sanctuary in the cruder forms of animism.

The curiosity of the men who first speculated about atoms, studied the properties of the lodestone, watched the result of rubbing amber, dissected animals, and catalogued plants in the three centuries before Aristotle wrote his epitaph on Greek science, had banished personalities from natural and familiar objects. Plato placed animism beyond the reach of experimental exposure by inventing a world of 'universals'. This world of universals was the world as God knows it, the 'real' world of which our own is but the shadow. In this 'real' world symbols of speech and number are invested with the magic which departed from the bodies of beasts and the trunks of trees as soon as they were dissected and described.

His *Timaeus* is a fascinating anthology of the queer perversities to which this magic of symbolism could be pushed. Real earth, as opposed to the solid earth on which we build houses, is an equilateral triangle. Real water, as opposed to what is sometimes regarded as a beverage, is a right-angled triangle. Real fire, as opposed to fire against which you insure, is an isosceles triangle. Real air, as opposed to the air which you pump into a tyre, is a scalene triangle (Fig. 2). Lest you should find this hard to credit, read how Plato turned the geometry of the sphere into a magical

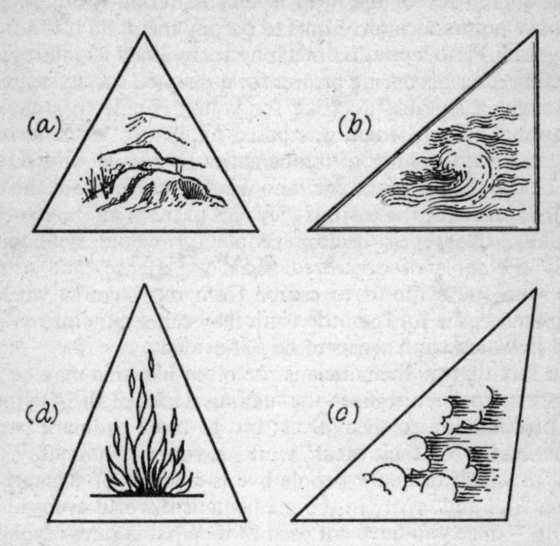

Fig. 2. Plato took Measurement out of the Geometry of his Predecessors and Reinstated the Superstitions of their Ancestors

The real world of Plato was a world of form from which matter was banished.

(*a*) An *equilateral* triangle (i.e. one of which all three sides are equal) is the elemental earth form.

(*b*) A *right-angled* triangle is the spirit of water. (To find spirit in water is the most advanced kind of magic.)

(*c*) A *scalene* triangle with no equal sides is the spirit of the air.

(*d*) An *isosceles* triangle (i.e. one of which only two sides are equal) is the elemental fire.

explanation of man's origin. God, he tells us, 'imitating the spherical shape of the universe, enclosed the two divine courses in a spherical body, that, namely, which we now term the head'. In order that the head 'might not tumble about among the deep and high places of the earth, but might be able to get out of the one and over the other', it was provided with 'the body to be a vehicle and means of locomotion, which consequently had length and was furnished with four limbs extended and jointed . . .'.

This supremacy of the head is very flattering to intellectuals who have no practical problems to occupy them. So it is not surprising that Plato's peculiar metaphysics retained its influence on education after his daring project for a planned society ceased to be thought a suitable doctrine for young people to study. An educational system which was based on Plato's teaching is apt to entrust the teaching of mathematics to people who put the head before the stomach, and who would tumble about the deep and high places of the earth if they had to teach another subject. Naturally, this repels healthy people for whom symbols are merely the tools of organized social experience, and attracts those who use symbols to escape from our shadow world in which men battle for the little truth they can secure into a 'real' world in which truth seems to be self-evident.

The fact that mathematicians are often like this may be why they are so inclined to keep the high mysteries of their Pythagorean brotherhood to themselves; but to more ordinary people, the perfection of their 'real' world savours of unreality. The world in which ordinary people live is a world of struggle and failure, trial and error. In the mathematical world everything is obvious – once you have got used to it. What is rarely explained to us is that it may have taken the human race a thousand years to see that one step in a mathematical argument is 'obvious'. How the nilometer works is obvious to you if you are a priest in the temple. If you are outside the temple, it can only become obvious through tracing out the subterranean channel which connects the temple with the river of man's social experience. Educational methods which are mixed up with priestcraft and magic have contrived to keep the rising and falling, the perpetual movement of the river from our scrutiny. So they have hidden from us the romance of what might be the greatest saga of man's struggle with the elements.

Plato, in whose school so many European teachers have grown up, did not approve of making observations and applying mathematics to arrange them and coordinate them. In one of the dialogues he makes Socrates, his master, use words which might equally well apply to many of the textbooks of mechanics which are still used.

'The starry heavens which we behold is wrought upon a visible ground and therefore, although the fairest and most perfect of visible things, must necessarily be deemed inferior far to the true motions of absolute swiftness and absolute intelligence. . . . These are to be apprehended by reason and intelligence but not by sight. . . . The spangled heavens should be used as a pattern and with a view to that higher knowledge. But the astronomer will never imagine that the proportions of night to day . . . or of the stars to these and to one another can also be eternal . . . and it is equally *absurd to take so much pains in investigating their exact truth. . . . In astronomy as in geometry we should employ problems, and let the heavens alone*, if we would approach the subject in the right way and so make the natural gift of reason to be of any use.'

This book will narrate how the grammar of measurement and counting has evolved under the pressure of man's changing social achievements, how in successive stages it has been held in check by the barriers of custom, how it has been used in charting a universe which can be commanded when its laws are obeyed, but can never be propitiated by ceremonial and sacrifice. As the outline of the story develops, one difficulty which many people experience will become less formidable. The expert in mathematics is essentially a technician. Hence mathematical books are largely packed with exercises designed to give proficiency in workmanship. This makes us discouraged because of the immense territory which we have to traverse before we can get insight into the kind of mathematics which is used in modern science and social statistics. The fact is that modern mathematics does not borrow so very much from antiquity. To be sure, every useful development in mathematics rests on the historical foundation of some earlier branch. At the same time every new branch liquidates the usefulness of clumsier tools which preceded it. Although algebra, trigonometry, the use of graphs, the calculus all depend on the rules of Greek geometry, less than a dozen from the two hundred propositions of Euclid's elements are essential to help us in understanding how to use them. The remainder are complicated ways of doing things we can do more simply when we know later branches of mathematics. For the mathematical technician these complications may provide a useful discipline. The person who wants to understand the place of mathematics in modern civilization is merely distracted and

Fig. 3. Mathematics in Everyday Life

This figure comes from Agricola's famous treatise (AD 1530) on mining technology. At that time the miners were the aristocrats of labour, and the book called attention to a host of new scientific prob-

disheartened by them. What follows is for those who have been already disheartened and distracted, and have consequently forgotten what they may have learned already, or for those who fail to see the meaning or usefulness of what they remember. So we shall begin at the very beginning.

Two views are commonly held about mathematics. One, that of Plato, is that mathematical statements represent eternal truths. Plato's doctrine was used by the German philosopher, Kant, as a stick with which to beat the materialists of his time, when revolutionary writings like those of Diderot were challenging priestcraft. Kant thought that the principles of geometry were eternal, and that they were totally independent of our sense organs. It happened that Kant wrote just before biologists discovered that we have a sense organ, part of what is called the internal ear, sensitive to the pull of gravitation. Since that discovery, the significance of which was first fully recognized by the German physicist, Ernst Mach, the geometry which Kant knew has been brought down to earth by Einstein. It no longer dwells in the sky where Plato put it. We know that geometrical statements when applied to the real world may be only approximate truths. The theory of Relativity has been very unsettling to mathematicians. In some quarters, it has become a fashion to say that mathematics is only a game. Of course, this does not tell us anything about mathematics. It tells us something only about the cultural limitations of some mathematicians. When a man says that mathematics is a game, he is making a private statement.

[*Caption continued*]

lems neglected in the slave civilizations of antiquity, when there was little co-operation between theoretical speculation and practical experience. Having measured the distance *HG*, which is the length of the stretched rope, you can get the distance you have to bore horizontally to reach the shaft, or the depth to which the shaft must be sunk, if you want to reach the horizontal boring. With a scale diagram you will see easily that the ratio of the horizontal cutting to the measured distance *HG* is the ratio of the two measurable distances $N:M$. Likewise the ratio of the shaft depth to *HG* is $O:M$. When you have done Dem. 2 in Chapter 3, this will be easier to see. The line *N* was made with a cord set horizontally by a spirit level. So it is at right angles to either of the two plumb lines. When you have read Chapters 4 and 6, you will see that the extra plumb line and the spirit level are not necessary, if you have a protractor to measure the angle at the top, and a table of sine and cosine ratios of angles.

He is telling us something about himself, his own attitude to mathematics. He is not telling us anything about the public meaning of a mathematical statement.

If mathematics is a game, there is no reason why people should play it if they do not want to. With football, it belongs to those amusements without which life would be endurable. The view which we shall explore is that mathematics is the language of size, shape and order and that it is an essential part of the equipment of an intelligent citizen to understand this language. If the rules of mathematics are rules of grammar, there is no stupidity involved when we fail to see that a mathematical truth is obvious. The rules of ordinary grammar are not obvious. They have to be learned. They are not eternal truths. They are conveniences without whose aid truths about the sorts of things in the world cannot be communicated from one person to another.

In Cobbett's memorable words, Mr Prynne would not have been able to impeach Archbishop Laud if his command of grammar had been insufficient to make himself understood. So it is with mathematics, the grammar of size. The rules of mathematics are rules to be learned. If they are formidable, they are formidable because they are unfamiliar when you first meet them – like the case forms of German adjectives or French irregular verbs. They are also formidable because in all languages there are so many rules and words to memorize before we can read newspapers or pick up radio news from foreign stations. Everybody knows that being able to chatter in several foreign languages is not a sign of great social intelligence. Neither is being able to chatter in the language of size. Real social intelligence lies in the use of a language, in applying the right words in the right context. It is important to know the language of size, because entrusting the laws of human society, social statistics, population, man's hereditary make-up, the balance of trade, to the isolated mathematician without checking his conclusions is like letting a committee of philologists manufacture the truths of human, animal, or plant anatomy from the resources of their own imaginations.

You will often hear people say that nothing is more certain than that two and two make four. The statement that two and two make four is not a mathematical statement. The mathematical statement to which people refer, correctly stated, is as follows:

$$2 + 2 = 4$$

This can be translated: ‘to 2 add 2 to get 4’. This is not necessarily a statement of something which always happens in the real world. An illustration (Fig. 4) shows that in the real world you do not always find that you have 4 when you have added 2 to 2. To say $2 + 2 = 4$ merely illustrates the meaning of the verb ‘add’, when it is used to translate the mathematical verb ‘+’. To say that $2 + 2 = 4$ is a true statement is just a grammatical convention about the verb ‘+’ and the nouns ‘2’ and ‘4’. In

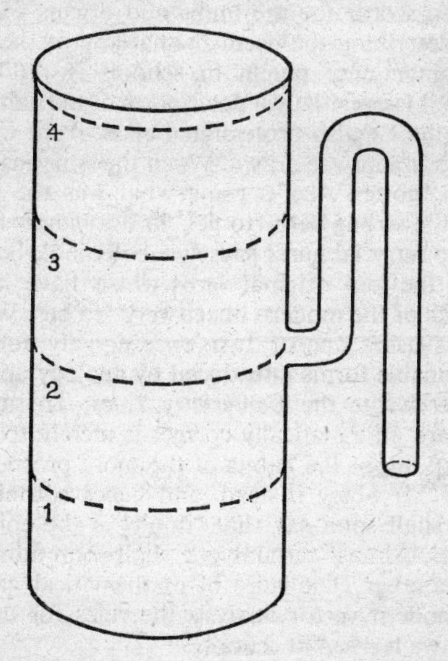

Fig. 4. In the Real World you do not always find that you have got FOUR, when you add TWO and TWO

Try filling this with water. Its laws of ‘addition’ would be: $1 \dotplus 1 = 2$; $1 \dotplus 2 = 3$; $1 \dotplus 3 = 2$; $2 \dotplus 2 = 2$, etc.

English grammar it is true in the same sense to say that the plural of ‘mouse’ is ‘mice’, or, if you prefer it, ‘add mouse to mouse to get mice’. In English grammar it is untrue to say that the plural of ‘house’ is ‘hice’. Saying ‘$2 + 2 = 2$’ may be false in precisely the same sense. A slight change in the meaning of the word ‘add’, as used to translate ‘+’, makes it a perfectly correct statement about the apparatus in Fig. 4.

We must not be surprised if we find that the rules of mathematics are not always a perfect description of how we measure the distance of a star, or count heads in a population. The rules of English grammar, as often taught, are a very imperfect description of how English is used. The people who formulated them were preoccupied with translating the bible and other classical texts. So they were over-anxious to find exact equivalents for the peculiarities of Greek and Latin. They were like the first zoologists who used words for the limbs and organs of the human body, when describing the peculiar anatomy of the insect. The English grammar once taught in schools is rather primitive zoology. Also it is essentially a description of the habits of speech prevailing in the English professional class, from which writers of books on grammar are drawn. When the American from New England says 'gotten', he is using what was the correct past participle of the strong verb 'to get' in *Mayflower* times. When the English country labourer says 'we be going', he is correctly using one of the four original verbs which have been used to make the roots of the modern mixed verb 'to be'. When he says '*yourn*', he is using one of two once equally admissible and equally fashionable forms introduced by analogy about the time when Chaucer wrote the *Canterbury Tales*. To say that 'are' and 'yours' are grammatically correct is merely to say that we have agreed to adopt the habits of the more prosperous townspeople. Since Mr Shaw is dead, and hence a topic for grammarians, we shall soon say that 'don't' is the correct way to write 'do not'. Almost certainly we shall soon admit 'it is me' as correct grammar. The rules of mathematical grammar also change. In modern vector analysis the rules for using '+' are not the rules we learned at school.

If we can unearth milestones of man's social pilgrimage in the language of everyday life, it is much more easy to do so when we study the grammar of mathematics. The language in which people describe the different *sorts* of things there are in the world is vastly more primitive and more conservative than the *size* languages which have been multiplied to cope with the increasing precision of man's control over nature. In the world which is open to public inspection, the world of inorganic and organic nature, man was not compelled to enlarge the scope of language to describe any new *sorts* of phenomena between 2000 BC and the researches of Faraday and Hertz, the father of radio. Even electric and magnetic attractions were recognized as a special sort of thing before there were any historians in the world. In the seventh century BC Thales recorded the attraction of small particles to a piece of amber (Greek 'electron') when rubbed. The Chinese already knew about the lodestone or natural magnet. Since about 1000 BC, when some men broke away from picture writing or script like the Chinese, which associates sounds with picture symbols, and first began to use an alphabet based purely on how words sound, there has only been one conspicuous invention introduced for describing the qualities of things in the world. This was made by biologists in the eighteenth century, when the confusion existing in the old herbals of medicinal plants forced them to invent an international language in which no confusion is possible. The clear description of the immense variety of organic beings has been made possible by the deliberate introduction of unfamiliar words. These words, like 'Bellis perennis', the common daisy, or 'Pulex irritans', the common flea, are taken from dead languages. Any meaning for which the biologist has no use lies buried in a social context forgotten long ago. In much the same way the North Europeans had borrowed their alphabet of sound symbols from the picture scripts, and buried the associations of distracting metaphors in the symbols used by the more sophisticated people of the ancient world.

The language of mathematics differs from that of everyday life, because it is essentially a rationally planned language. The languages of size have no place for private sentiment, either of the individual or of the nation. They are international languages like the binomial nomenclature of natural history. In dealing with the immense complexity of his social life man has not yet begun to apply inventiveness to the rational planning of ordinary language when describing different kinds of institutions and human behaviour. The language of everyday life is clogged with sentiment, and the science of human nature has not advanced so far that we can describe individual sentiment in a clear way. So constructive thought about human society is hampered by the same conservatism as embarrassed the earlier naturalists. Nowadays people do not differ about what sort of animal is meant by Cimex or Pediculus, because these words are used only by people who use them in one way. They still can and often do mean a lot of different things when they say that a mattress is infested with bugs or lice. The study of man's social life has not yet brought forth a Linnaeus. So an argument about the 'withering away of the State' may disclose a difference about the use of the dictionary when no real difference about the use of the policeman is involved. Curiously enough, people who are most sensible about the need for planning other social amenities in a reasonable way are often slow to see the need for creating a rational and international language.

The technique of measurement and counting has followed the caravans and galleys of the great trade routes. It has developed very slowly. At least four thousand years intervened between the time when men could calculate when the next eclipse would occur and the time when men could calculate how much iron is present in the sun. Between the first recorded observations of electricity produced by friction and the measurement of the attraction of an electrified body two thousand years intervened. Perhaps a longer period separates the knowledge of magnetic iron (or lodestone) and the measurement of magnetic force. Classifying things according to size has been a much harder task than recognizing the different sorts of things there are. It has been more closely related to man's social achievements than to his biological equipment. Our eyes and ears can recognize different sorts of things at a great distance. To measure things at a distance, man has had to make new sense organs for himself, like the astrolabe, the telescope, and the microphone. He has made scales which reveal differences of weight to which our hands are quite insensitive. At each stage in the evolution of the tools of measurement man has refined the tools of size language. As human inventiveness has turned from the counting of flocks and seasons to the building of temples, from the building of temples to the steering of ships into chartless seas, from seafaring plunder to machines driven by the forces of dead matter, new languages of size have sprung up in succession. Civilizations have risen and fallen. At each stage, a more primitive, less sophisticated culture breaks through the barriers of custom thought, brings fresh rules to the grammar of measurement, bearing within itself the limitation of further growth and the inevitability that it will be superseded in its turn. The history of mathematics is the mirror of civilization.

The beginnings of a size language are to be found in the priestly civilizations of Egypt and Sumeria. From these ancient civilizations we see the first-fruits of secular knowledge radiated along the inland trade routes to China and pushing out into and beyond the Mediterranean, where Semitic peoples are sending forth ships to trade in tin and dyes. The more primitive northern invaders of Greece and Asia Minor collect and absorb the secrets of the pyramid makers in cities where a priestly caste is not yet established. As the Greeks become prosperous, geometry becomes a plaything. Greek thought itself becomes corrupted with the star worship of the ancient world. At the very point when it seems almost inevitable that geometry will make way for a new language, it ceases to develop further. The scene shifts to Alexandria, the greatest centre of shipping and the mechanical arts in the ancient world. Men are thinking about how much of the world remains to be explored. Geometry is applied to the measurement of the heavens. Trigonometry takes its place. The size of the earth, the distance of the sun and moon are measured. The star gods are degraded. In the intellectual life of Alexandria, the factory of world religions, the old syncretism has lost its credibility. It may still welcome a god beyond the sky. It is losing faith in the gods within the sky.

In Alexandria, where the new language of star measurement has its beginnings, men are thinking about numbers unimaginably large compared with the numbers which the Greek intellect could grasp. Anaxagoras had shocked the court of Pericles by declaring that the sun was as immense as the mainland of Greece. Now Greece itself had sunk into insignificance beside the world of which Eratosthenes and Poseidonius had measured the circumference. The world itself sank into insignificance beside the sun as Aristarchus had measured it. Ere the dark night of monkish superstition engulfed the great cosmopolis of antiquity, men were groping for new means of calculation. The bars of the counting frame had become the bars of a cage in which the intellectual life of Alexandria was imprisoned. Men like Diophantus and Theon were using geometrical diagrams to devise crude recipes for calculation. They had almost invented the third new language of algebra. That they did not succeed was the nemesis of the social culture they inherited. In the East the Hindus had started from a much lower level. Without the incubus of an old-established vocabulary of number, they had fashioned new symbols which lent themselves to simple calculation without mechanical aids. The Moslem civilization which swept across the southern domain of the Roman Empire brought together the technique of measurement as it had evolved in the hands of the Greeks and the Alexandrians, adding the new instrument for handling numbers which was developed through the invention of the Hindu number symbols. In the hands of Arabic mathematicians like Omar Khayyám, the main features of a language of calculation took shape.

We still call it by the Arabic name, algebra. We owe algebra and the pattern of modern European poetry to a non-Aryan people who would be excluded from the vote in the Republic of South Africa.

Along the trade routes this new arithmetic is brought into Europe by Jewish scholars from the Moorish universities of Spain and by Gentile merchants trading with the Levant, some of them patronized by nobles whose outlook had been unintentionally broadened by the Crusades. Europe stands on the threshold of the great navigations. Seafarers are carrying Jewish astronomers who can use the star almanacs which Arab scholarship had prepared. The merchants are becoming rich. More than ever the world is thinking in large numbers. The new arithmetic or 'algorithm' sponsors an amazing device which was prompted by the need for more accurate tables of star measurement for use in seafaring. Logarithms were among the cultural first-fruits of the great navigations. Mathematicians are thinking in maps, in latitude and longitude. A new kind of geometry (what we call graphs in everyday speech) was an inevitable consequence. This new geometry of Descartes contains something which Greek geometry had left out. In the leisurely world of antiquity there were no clocks. In the bustling world of the great navigations mechanical clocks are displacing the ancient ceremonial functions of the priesthood as timekeepers. A geometry which could represent time and a religion in which there were no saints' days are emerging from the same social context. From this geometry of time, a group of men who were studying the mechanics of the pendulum clock and making fresh discoveries about the motion of the planets devise a new size language to measure motion. Today we call it 'the' calculus.

For the present this crude outline of the history of mathematics as a mirror of civilization, interlocking with man's common culture, his inventions, his economic arrangements, his religious beliefs, may be left at the stage which had been reached when Newton died. What has happened since has been largely the filling of gaps, the sharpening of instruments already devised. Here and there are indications of a new sort of mathematics. We begin to see possibilities of new languages of size transcending those we now use, as the calculus of movement gathered into itself all that had gone before.

Because it is good medicine to realize how long it took others, more clever than ourselves, to grasp what we now try to grasp during a half-hour period at school, we shall have to dig into a past during which the map has changed again and again. For readers who are not familiar with the geography of the Mediterranean region in antiquity, a few preliminary words about place-names will not therefore be amiss. What is now Iraq successively came under the rule of different conquerors between 3000 and 300 BC, but the temples sustained a continuous tradition of number lore throughout this period. Where books with more scholarly claims than this one speak of Sumeria, Chaldea, Babylonia, and Mesopotamia, we shall henceforth use the contemporary word which embraces them all. In doing so however, we should bear in mind that the Persia of Moslem culture in the days of Omar Khayyám embraced Iraq as well as Iran, and had its seat of government in Baghdad now the capital of the former.

The use of the adjective *Greek* also calls for comment. From about 2500 BC onwards, successive waves of people speaking kindred dialects migrated from the north into what is now Greece and the neighbouring parts of Turkey in Asia. By 1000 BC they had colonized or conquered many of the islands of the Mediterranean, including Cyprus, Crete, and Sicily. Before 500 BC they had trading ports along the coast of Asia Minor and as far west as Marseilles. What they had pre-eminently in common was their language and, after 600 BC, their alphabet borrowed largely from their Semitic trade rivals, the Phoenicians of the Old Testament when the latter had prosperous ports at Tyre and Sidon in Syria, in Africa at Carthage, and farther West in Spain.

We must therefore guard against thinking of the so-called Greeks of the ancient world as a nation with headquarters in what is now Greece. Before the conquests of Philip of Macedon and of his son Alexander, the name embraces a large number of independent city-states, several in Greece alone, often at war among themselves yet with very close cultural bonds, like those between Britain and the United States. After the death of Alexander the Great, his generals took over different parts of his ramshackle empire and his veterans settled therein. In particular, they founded dynasties in Egypt (the *Ptolemies*) and in Iraq (the *Seleucids*) on the site of the earliest temple cultures of the ancient world. There Greek culture absorbed a wealth of information which had hitherto been largely secret lore. In this setting the city of Alexandria, planned and named after the conqueror (332 BC) became pre-eminent as a centre of learning and of maritime trade in the ancient world. From Euclid onwards, its museum and library, founded in the reign of the first of the soldier dynasty, attracted the greatest scholars and inventors of the next six hundred years, and the city remained the greatest centre of learning before Rome adopted an alien faith and mythology. Greek was the medium of its teaching and writing; but the personnel was cosmopolitan with a considerable Jewish component. It is therefore misleading to speak of its great mathematicians as Greek. In this book we shall speak of them as Alexandrian, and we shall use the term Greek only for Greek-speaking individuals before the death of Alexander the Great.

Instructions for Readers of this Book

The customary way of writing a book about mathematics is to show how each step follows *logically* from the one before without telling you what use there will be in taking it. This book is written to show you how each step follows *historically* from the step before and what use it will be to you or someone else if it is taken. The first method repels many people who are intelligent and socially alive, because intelligent people are suspicious of mere logic, and people who are socially alive regard the human brain as an instrument for social activity.

Although care has been taken to see that all the logical, or, as we ought to say, the grammatical, rules are put in a continuous sequence, you must not expect that you will necessarily follow every step in the argument the first time you read it. An eminent Scottish mathematician gave a very sound piece of advice for lack of which many people have been discouraged unnecessarily. 'Every mathematical book that is worth anything,' said Chrystal, 'must be read backwards and forwards . . . the advice of a French mathematician, *allez en avant et la foi vous viendra*.'

So there are two INDISPENSABLE CAUTIONS which you must bear in mind, if you want to enjoy reading this book.

The FIRST is READ THE WHOLE BOOK THROUGH ONCE QUICKLY TO GET A BIRD'S-EYE VIEW OF THE SOCIAL INTERCONNECTIONS OF MATHEMATICS, and when you start reading it for the second time to get down to brass tacks, read each chapter through before you start working on the detailed contents.

The SECOND is ALWAYS HAVE A PEN AND PAPER, PREFERABLY SQUARED PAPER, in hand, also PENCIL AND RUBBER, when you read the text for serious study, and WORK OUT ALL THE NUMERICAL EXAMPLES AND FIGURES as you read. Almost any stationer will provide you with an exercise book of squared paper for a very small sum. WHAT YOU GET OUT OF THE BOOK DEPENDS ON YOUR CO-OPERATION IN THE SOCIAL BUSINESS OF LEARNING.

CHAPTER 1

MATHEMATICS IN REMOTE ANTIQUITY

WHEN ONE speaks of a branch of knowledge called mathematics at its most primitive level one signifies the existence of at least simple rules about calculation and measurement. The recognition of such rules presupposes that man has already at his disposal signs for numbers such as $\overline{\text{XV}}$ or 15, CXLVI or 146. To trace mathematics to its source we must therefore dig down to the primeval need for them. Most likely this was the impulse to keep track of the passage of time and, as such, to record the changing face of the heavens. For several millennia after mankind took this momentous step, star-lore was to be the pace-maker of mathematical ingenuity.

While our remote nomadic ancestors lived only by hunting and food gathering, the rising or setting position of stars on the horizon, whether they had or had not already risen when darkness fell and whether they had or had not yet done so at daybreak, were their only means of again locating a hunting ground already visited and their most reliable guide to the onset of the season at which particular game, berries, roots, eggs, shellfish, or grain would be most abundant at a particular location. As is true of all letterless communities today, members of our own species, while still nomads, must have realized that each star rises and sets a little earlier each day. They would date the routine of their journeyings by the occasion when a particular star rose or set just before dawn, and rose or set just after sunset. Before there was farming of any sort, it is likely that the older folk of the tribe had learned to reckon in lunar time (i.e. successive full or new moons) when each boy and each girl would have reached the age for the tribal initiation ceremony.

At some time not much earlier than 10,000 BC and not later than 5000 BC settled village communities of herdsmen and grain growers came into being in the fertile regions of the Middle East. Here there was a larger leisure to observe the changing face of the night sky and a greater impetus to do so in the milieu of a seasonal timetable of lambing, sowing, and reaping. Here, too, there were more favourable conditions for observing the vagaries of the sun's rising and setting positions and the varying length of its noon shadow from one rainy season to another. (See the photograph of Stonehenge in the illustrated section in the centre of this book.)

Before 3000 BC, large concentrations of population with greater division of labour made their appearance on the banks of great rivers whose flooding annually enriched the nearby soil with fresh silt.

Such were the Nile of Egypt, in Iraq the Tigris and the Euphrates, the Indus in Pakistan, and farther east the Yangtse-Kiang and Hwang-ho. From the start in Egypt and in Iraq, we find a priestly caste responsible for the custody of a ceremonial calendar, equipped with writing of a sort including number signs, and housed in buildings placed to facilitate observations on celestial bodies as signposts of the calendar round. We may confidently conclude that the recognition of the year as a unit emerged in the village communities which coalesced to make these city-states and kingdoms, and that the recognition of a lunar month of 30 days preceded that of the Egyptian year of 365 days, i.e. 12 lunar months and 5 extra days. It began when the dog star Sirius was visible on the rim of the sky just before sunrise, an event portending the nearness of the annual flooding of the sacred river. Long before barter of cattle or sheep or produce of the soil began, man had experienced the need to count the passage of time and in doing so must have faced a challenge with which the counting of cattle in one and the same plot of land did not confront him imperatively. Pigs do not fly but time does. To count time in units of any sort we therefore need a tally to aid the memory. If for this reason only, it seems to be very likely that number signs take their origin from marks chipped on stone or wood to record successive days. Indeed, all the earliest batteries of number signs in use among the priestly astronomers responsible for the custody of the calendar disclose a repetition of strokes (Fig. 6). Of this there is more than a relic in the familiar Roman numerals: I, II, III, XX, XXX, CC, for 1, 2, 3, 20, 30, 200, etc.

The next step was to group strokes in a regular way and introduce signs for such groups to expedite counting and to economize space. In taking this step, man's bodily means of handling objects dictated the pattern we signify when we speak of *ten* as the *base*

Egyptian

Babylonian

1 6 10 34 60 600

Early Greek

Ι ΙΙΙ Γ ΓΙΙΙ Δ ΔΔΓΙΙΙ

1 3 5 8 10 28

Η ΗΗΔΔΔΙΙ Χ ΧΧΗΗΗΔΓΙ

100 232 1000 2316

Fig. 6. Ancient Number Scripts

Egyptian and Iraq, about 3000 BC; Early Greek Colonies, about 450 BC.

or group number of our way of counting in tens, hundreds, thousands, etc. At the most primitive level, people use their fingers for counting or fingers and toes. A tribe of Paraguayan aborigines has names corresponding to 1 to 4, 5 (*one hand*), 10 (*two hands*), 15 (*two hands and a foot*), 20 (*two hands* and *two feet*). This recalls the special signs for 5, 50, and 500 (V, L, D) which they took over from the Etruscans, as well as for 10, 100, 1,000 (X, C, M) in the number script of the Romans. It also recalls the use of 20 as a base among the indigenous priestly custodians of the Calendar in Central America. At the time of the Spanish Conquest, the Cakchiquel Amerindians of Guatemala reckoned the passage of time in units (*ki* = 1 day), *vinaks* (20 days) *a* (20 vinaks = 400 days), and *may* (20 *a* = 8,000 days). There is a trace of this hand and foot counting in our own language, as when we speak of a *score* meaning twenty.

In the ancient priestly scripts of the Mediterranean world the numbers one to nine were actually represented by fingers. The later commercial script of the Phoenicians had a symbol for unity which could be repeated (like I, II, III in Roman script) up to nine

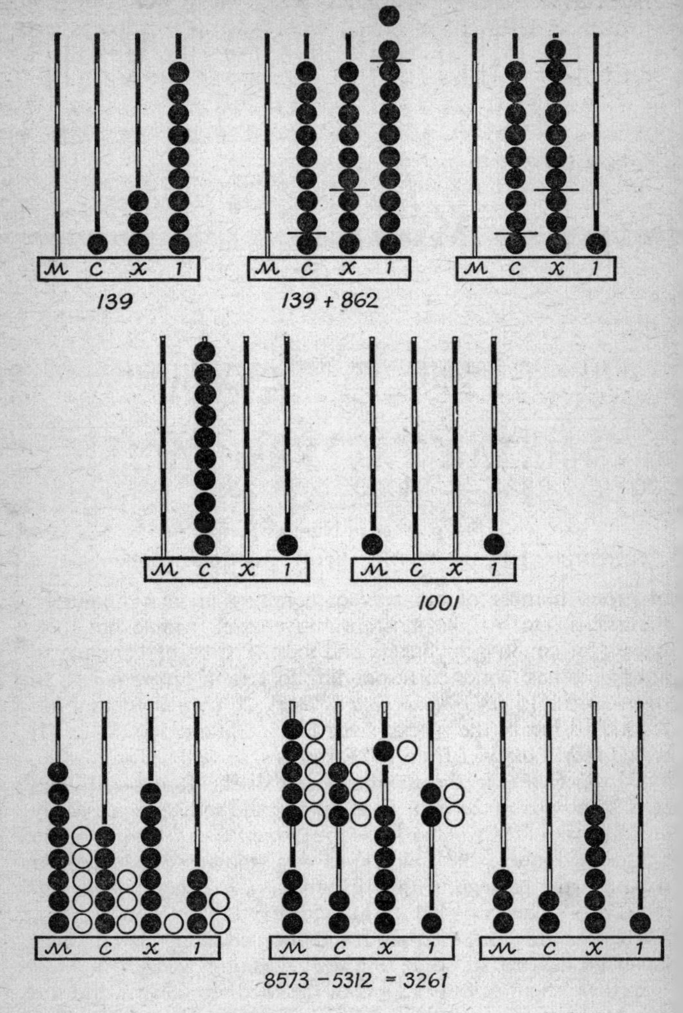

Fig. 7. A Simple Abacus
Addition and Subtraction.

times. There was a letter symbol for 10 which could be repeated (like the Roman X, XX, etc.) nine times, then another letter symbol for 100 (like the Roman C). This ancient Phoenician script, which was the basis of the numbers used by the Ionian Greeks and Etruscans, was cumbersome, but at least more rational than its successors. To make it less cumbersome, the Etruscans went back to one-hand counting, and added the symbols which in Roman numeration represent 5, 50, 500 (V, L, D). The later Greeks rejected the Ionian script, adopting and bequeathing to the Alexandrians a number system which exhausted all the letters of the alphabet, as did the Hebrew numeral system (Fig. 9). While this was compact, it had two consequences which proved to be disastrous. One, which we shall study in Chapter 4, was that it encouraged the peculiar kind of number magic called 'gematria'. The other, which is of greater importance, will be studied in Chapter 6. The introduction of a literal system for

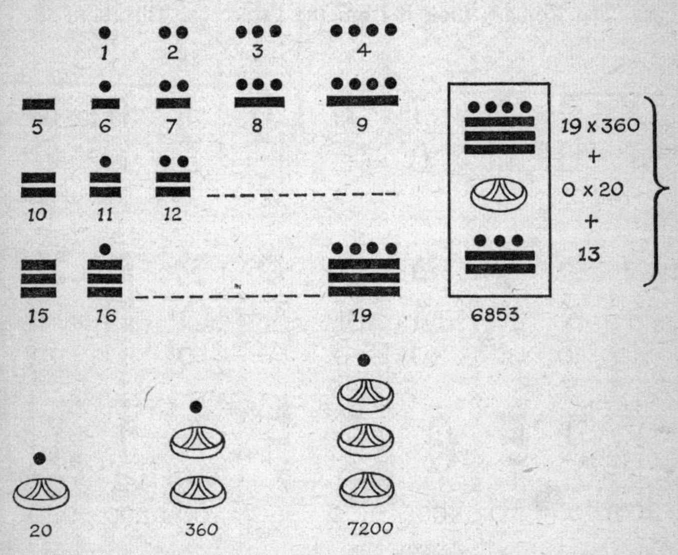

Fig. 8. The Ancient Maya Number Script

The principle of position employed in the Mayan script of which the base, alas not consistently, was twenty is as follows. The lowest group of symbols represents the units column of the abacus, that above 20s, that above 360, that above 7,200.

numbers made it impossible for the most brilliant mathematicians of Alexandria to devise simple rules of calculation without recourse to mechanical aids.

When man got beyond the stage of relying entirely upon tally sticks, representing numbers by notches, he hit on the practice of using pebbles or shells which could be rapidly discarded or used over and over again. So came the counting frame. At first it was probably a series of grooves on a flat surface. Then it was a set of upright sticks on which pierced stones, shells, or beads could be placed (Fig. 7). Finally the closed frame superseded the earlier type seen in the upper part of Fig. 99.

The counting frame or *abacus* was a very early achievement of mankind. It follows the megalithic culture routes all round the world. The Mexicans and Peruvians were using the abacus when the Spaniards got to America. The Chinese and the Egyptians already possessed the abacus several millennia before the Christian era. The Romans took it from the Etruscans. Till about the

Fig. 9. Hebrew and Attic (Late) Greek Number Scripts

beginning of the Christian era, this fixed frame remained the only instrument for calculation that mankind possessed.

To us, figures are symbols with which sums can be done. This conception of figures was completely foreign to the most advanced mathematicians of ancient Greece. The ancient number scripts were merely labels to record the result of doing work with an abacus, instead of doing work with a pen or pencil. In the whole history of mathematics there has been no more revolutionary step than the one which the Hindus made when they invented the sign '0' to stand for the empty column of the counting frame. You will see more clearly why this was important, and how it made possible simple rules of calculation, when you get to Chapter 6.

Here we can notice two things about the discovery of 'nothing'. The first is that if your base is 10, you require only nine other signs to express any number as large as you like. Your ability to represent numbers is not limited by the number of letters in your alphabet. You do not have to introduce new signs, like the Roman X, C, and M, every time you multiply by ten. The other important thing about 'nothing' you may begin to see if you look again at Fig. 7. The new number vocabulary of the Hindus allows you to add on paper in the same way as you add on the abacus. How the invention was made and how it affected the after-history of mathematics must be left for the present. The important thing to realize is that the mathematicians of classical antiquity inherited a social culture which was equipped with a number script before the need for laborious calculations was felt. So they were completely dependent upon mechanical aids which have now been banished to the nursery.

The invention of a sign for the empty column of the abacus happened only once in the Old World; but it occurred independently in the earliest indigenous civilization of the New, that of the Mayas of Guatemala and neighbouring territories (Fig. 8). The astronomer priests of the Mayan temple sites used a number script with three signs laid out in a vertical sequence of blocks of four horizontal tiers. One like a lozenge stood for zero. One of circular outline stood for a single unit, and horizontal bars for fives. A single block could accommodate at most 3 fives = 15 and 4 units above making 19 in all. The lowest block corresponded to our first position, i.e. that of 5 in 1,745, and could represent 0–19 units as our first position represents 0–9 units. The block above (our second position, 4 tens in 1,745) specified multiples of 20; the third block multiples of 360, the fourth one multiples of 7,200 (= 20 × 360), thus accommodating 19 × 7,200. Seemingly, the second position could not accommodate with 3 fives more than 2 units, so that 19 in the first position and 17 in the second represented 19 + 17(20) = 359. Otherwise, the system was consistent with the use of the base 20. But for one flaw, this system would have met the needs of rapid calculation. The anomalous place value of 360 instead of 400 is doubtless the relic of an early estimate of the year as twelve 30-day months, thus proclaiming its ancestry as tally hand-maid of the ceremonial calendar.

The first people to use number signs did so with no recognition of the fact that *counting* sheep in a flock and *estimating* the size of a field are not wholly comparable performances. If one says that a flock consists of 50 sheep one means neither more nor less than 50. What the statement that a field consists of 50 acres truly conveys depends on one's measuring device. Some of the mystification which tradition has transmitted throughout the whole of the historic record will worry us less if we clearly grasp at an early stage the distinction between *flock* numbers (e.g. how many sheep) and *field* numbers (e.g. how many acres).

When confronted with the difficulty of making whole numbers fit measurements which imperfect human beings, using imperfect sense organs, make with imperfect instruments in an imperfect and changing world, the practical man was long content to go on adding fresh divisions to his scale of measurement. You can see that this works very well up to a point by examining the following illustration. Suppose four men are asked to measure the area of an oblong field which is 300 yards broad and 427½ yards long. We will assume for the time being, as the practical man himself does assume, that it is actually possible for a field to be exactly 300 yards broad or exactly 427½ yards long. Suppose also that the first man has a rope of 100 yards, the second has a tape of 10 yards, the third has a pole of 3 yards, and the fourth a rule 1 yard long. None of them finds any difficulty with the first side. Their measures can be placed along it just 3 times, 30 times, 100 times, and 300 times respectively. The trouble begins with the side 427½ yards long. The first man finds that this is more than 4 and less than 5 times his measure, so his estimate of the area lies between that of a field of 300 × 500 square yards, or 150,000 square yards, and that of a field 300 × 400, or 120,000 square yards. The second finds that it is more than 42 and less than 43 times his measure. His estimate lies between 420 × 300 and 430 × 300 square yards. Let us make a table of all their estimates:

Measure	Lower Limit (square yards)	Upper Limit (square yards)
100 yards	300 × 400 = 120,000	300 × 500 = 150,000
10 ,,	300 × 420 = 126,000	300 × 430 = 129,000
3 ,,	300 × 426 = 127,800	300 × 429 = 128,700
1 yard	300 × 427 = 128,100	300 × 428 = 128,400

If you look at these results you will see that the upper estimate of the first crude measure is 30,000 square yards (or 25 per cent) bigger than the lower estimate of 120,000 square yards. For the last and best estimate, the upper limit is 300 square yards greater than the lower limit of 128,100 square yards. The excess is 1 in 427, or less than $\frac{1}{4}$ per cent. To put it in another way, the first estimate is 135,000 ± 15,000 square yards. The best of all the estimates is 128,250 ± 150 square yards.

Measurement is only one facet of the natural history of numbers. What we nowadays call mathematics embraces many other themes involving not merely *computation* but also *order* and *shape*. However, measurement was the first domain in which there emerged mathematical rules worthy of the title. To trace their origin we must go back to the dawn of civilization, some five thousand years ago. As we have seen, a priesthood responsible for a ceremonial calendar associated with the regulation of a seasonal economy of food production and for ritual sacrifices to propitiate the unseen then emerged as a privileged caste cut off from, and able to enrich itself by tribute exacted from, the tillers of the soil.

Thenceforth, part of the surplus wealth from the peasants provided the means of building magnificent edifices dedicated to the calendar ritual and placed to greet celestial occurrences. In Egypt, the doors of these temple observatories sometimes faced due East to greet the rising sun of the equinoxes (now March 21st and September 23rd), and sometimes more northerly or more southerly, so that the rising sun of midsummer or midwinter day cast a long shaft of light along the main corridors. Both the exaction of tribute in kind or as labour service, and the feats of architecture undertaken to construct temples or tombs such as the Pyramids for a supposedly sky-born overlord, entailed problems of practical geometry. Let us look first at the problems which faced the temple scribes responsible for the accountancy of the tax-gatherers.

According to Greek sources, the Egyptian hierarchy exacted tribute from the peasant on the basis of the area of his plot. Because the annual flooding of the Nile frequently obliterated its landmarks, there was thus a recurring need for land survey; and we may assume that the surveyor learned three lessons at an early date. One is the possibility of mapping any rectangular strip in little squares (Fig. 10), so that a strip of 50 cubits by 80 cubits

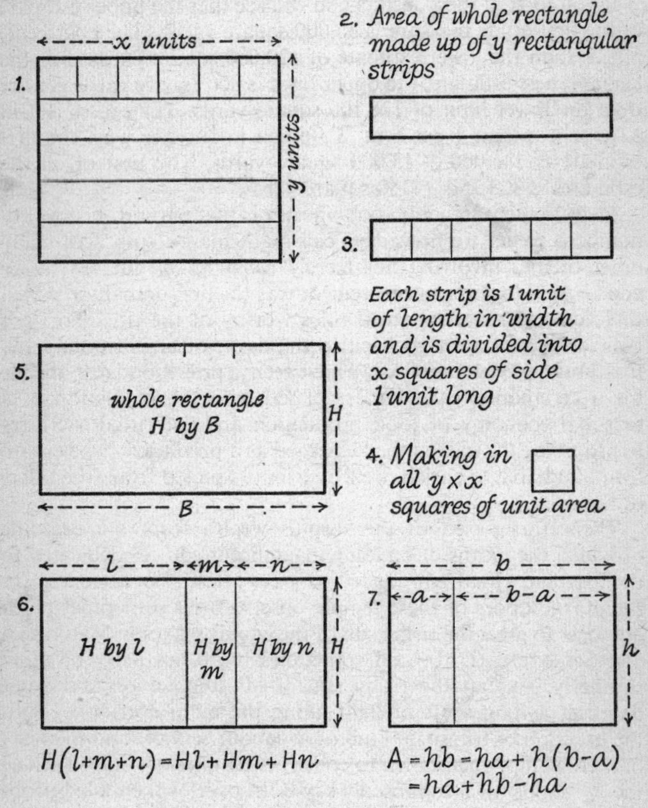

Fig. 10. The Square as Unit of Area

would be equivalent to 4,000 squares, each of whose sides is 1 cubit. A second is that a strip of any shape, if its sides are straight lines, is exactly divisible into triangles. The third is that any triangle occupies half the surface of a rectangular strip each with one side (*base*) equal and each of the same height (Fig. 11). The combination of the last two discoveries in a rule (Fig. 12) for assessing the area of any strip of land, if laid out in straight lines with a taut cord pegged down at each end, must have been a very early achievement, and one which was to exercise a decisive role in geometrical reasoning. The clue to the greater part of Greek geometry, in so far as it concerns itself with flat figures, is the quest to dissect them into equivalent triangles.

Tribute in kind raised problems of another sort. When possible, as is true of grain, flour, or wine, it is more speedy to measure produce by volume than by weight. For this, three types of measuring vessels are easiest to standardize: a box with straight

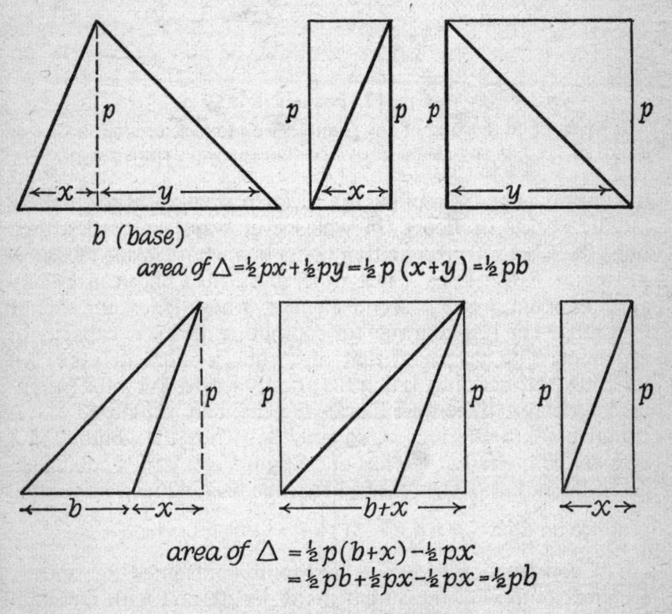

Fig. 11. Area of a Triangle
$A = \frac{1}{2}$ (Area of Rectangle of same height and base).

(*rectilinear*) edges, a hollow cylinder, and a cone. By what steps the accountant scribe of the temple civilizations of Egypt and Iraq linked the use of all three by invoking rules based on the notion of a single unit of measurement, we can merely guess. Perhaps

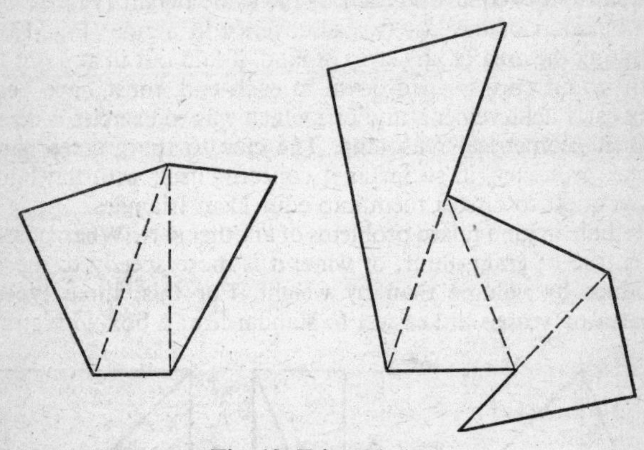

Fig. 12. Triangulation

If we can find the area of any triangle, we can measure the surface of a plot of land of any shape, provided that the walls run straight.

the notion came from piling up bricks or small discs for making mosaic panels or floors. In whatsoever way, the earliest step must have been the recognition that a box whose inside measurements are 7 feet long, 4 feet wide, and 3 feet high can exactly accommodate $7 \times 3 \times 4 = 84$ cubes, whose sides are each 1 foot long. The step to a rule for computing the cubic capacity of a cylinder and thence to that of a cone is more obscure. All we know with certainty is that the priestly scribes knew the formulae long before there were literate Greeks; and recognized that it contains the mysterious π, roughly $3\frac{1}{7}$. Thus the volume of a cylinder whose radius is r feet and height h feet is $\pi r^2 h$ cubic feet, e.g. if the radius of the base is 3 feet, the height 7 feet, its volume is approximately $\frac{22}{7} \times (3 \times 3) \times 7 = 198$ cubic feet.

The construction of temple and tomb confronted the priestly architect both with measurements of length and with measurements of direction, i.e. of angles. A very elementary problem of angular measurement is how to build a wall vertically. For this

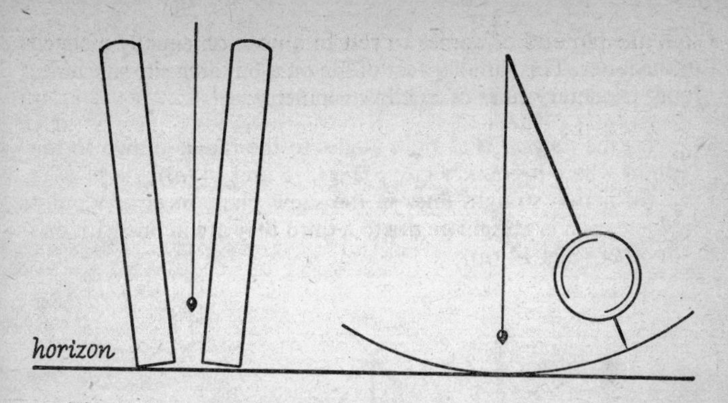

Fig. 13. Principle of Tangency Illustrated by Plumb Line and Pendulum

purpose (Fig. 13), some sort of *plumb line* is a device of great antiquity, probably of village life before the emergence of city-states. Its use puts the spotlight on more than one reason why the *right angle* (90°) was to become a fundamental unit of angular measurement. Swinging in a circular arc which grazes the flat

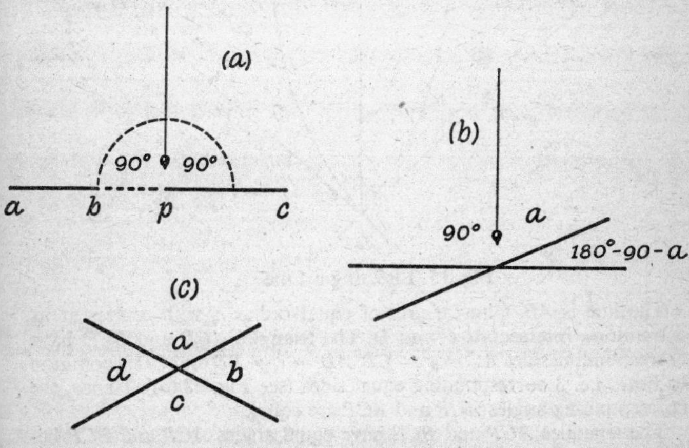

Fig. 14. Three-angle Rules

In (*c*), $a + b = 180° = b + c = c + d = d + a$. So $a + b = b + c$ and $a = c$, etc.

soil, the plumb line comes to rest in a position equally inclined to the latter. This familiar fact of life on a building site subsumes four elementary rules of Euclid's geometry:

(*a*) the *tangent* is at right angles to the radius drawn to the point where it grazes a circle (Figs. 13 and 14 (*a*));

(*b*) if two straight lines in the same plane meet at a point where each is at right angles to a third they are in line with one another (Fig. 14 (*a*));

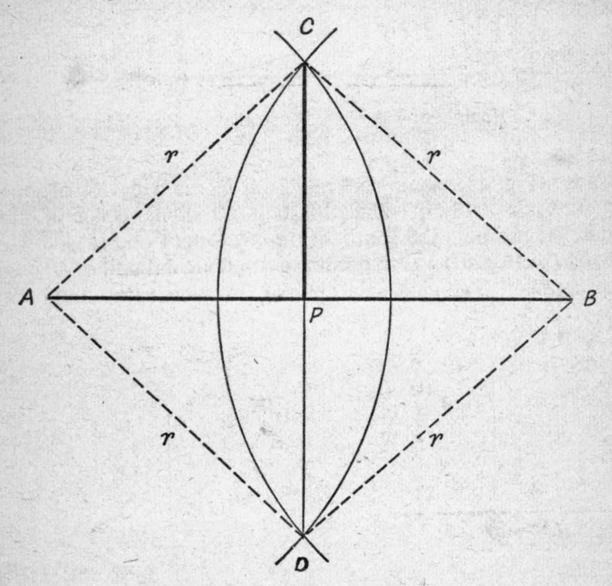

Fig. 15. Bisecting a Line

The line is *AB*. Circular arcs of equal radius *r*, with centres at its extremities, intersect at *C* and *D*. The triangles *ACD* and *BCD* have corresponding sides $AC = r = CB$, $AD = r = BD$ and *CD* is common to both, i.e. 3 corresponding equal sides (see Fig. 28 (*a*)). Hence, the corresponding angles *ACP* and *BCP* are equal.

The triangles *ACP* and *BCP* have equal angles *ACP* and *BCP* between sides of equal length, i.e. $AC = BC$ and *PC* (common to both). Hence (see Fig. 28 (*b*)), they also are equivalent, so that:

$$AP = PB \text{ and } \angle CPA = 90^\circ = \angle CPB$$

(*c*) if one straight line crosses another, the sum of the two angles it makes therewith on the same side is two right angles (Fig. 14 (*b*));

(*d*) whence, if straight lines intersect opposite angles are equal (Fig. 14 (*c*)).

One way of making a right angle horizontally, when placing two walls with a rectangular edge, is deducible from the last statement, if we bear in mind three recipes for making a unique triangle, as shown in Fig. 28. We can do so, if we bisect a line

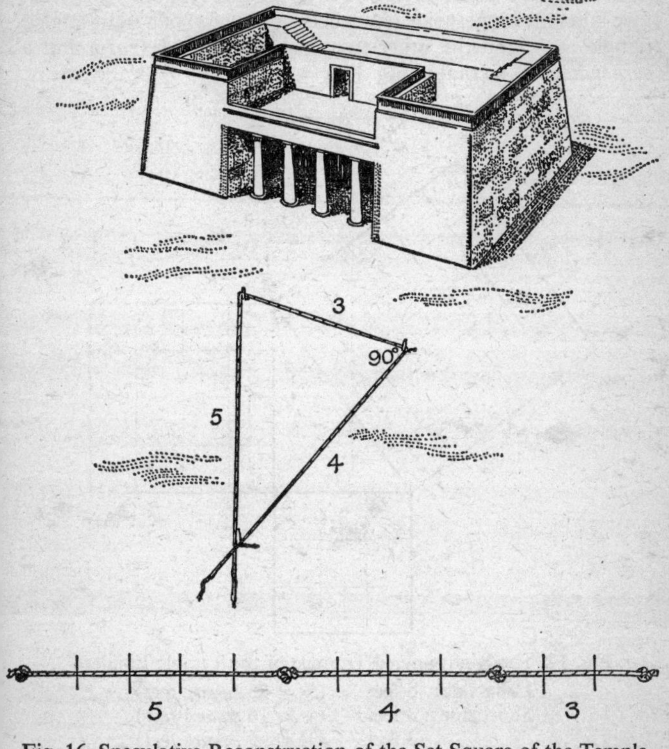

Fig. 16. Speculative Reconstruction of the Set Square of the Temple Architects

(Fig. 15 and legend) with rope and pegs on the soft soil or sand, as could doubtless the architects of the first massive buildings.

One can also make a large scale set square from a piece of cord knotted at its extremities with twelve other equidistant knots. If pegged down taut at points separated by three and four segments these with the remaining five segments form a right-angled triangle. Whether or no the temple architects used this device (Fig. 16), as one author has suggested, such a recipe for making a so-called set square was widely known among the astronomer priests of remote antiquity. As early as 2000 BC, those of Iraq recognized the 3 : 4 : 5 ratio as one example of a rule of wider application now usually referred to as the theorem of Pythagoras. If we label the longest side (so-called *hypotenuse*) of a right-angled triangle as h, and the other two respectively as b (*base*) and p (*perpendicular*), so that in Fig. 17 $h = 5, p = 4, b = 3$, the rule is:

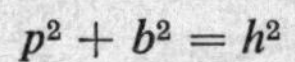

$$p^2 + b^2 = h^2$$

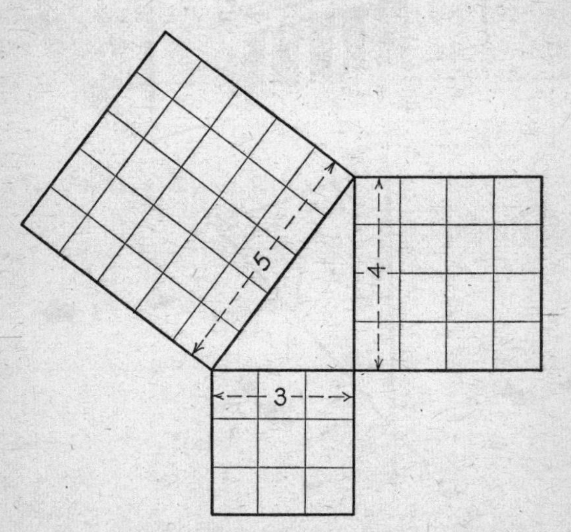

Fig. 17. The Right-angled Triangle of the Temple Builders

Long side:	5 feet	($5^2 = 25$ *square feet*)
Short sides:	4 feet	($4^2 = 16$ *square feet*)
	3 feet	($3^2 = \ \ 9$ *square feet*)

$$25 = 16 + 9 \text{ or } 5^2 = 4^2 + 3^2$$

929-2443
Mike Sheelor.

THE WORLD'S GREATEST BOOKSHOP

FOYLES

FOR BOOKS

STOCK OF OVER FOUR MILLION VOLUMES

113-119 Charing Cross Road London WC2

Open 9–6 inc. Sats. (Thurs 9–7) .

Sold by SF Dept. 2 MATHS Date 2-6-84

	£	
	1	95
£1–95		

010105 MOORE PARAGON UK LTD

37540 - 12

This bill must be produced in the event of any query regarding purchase.

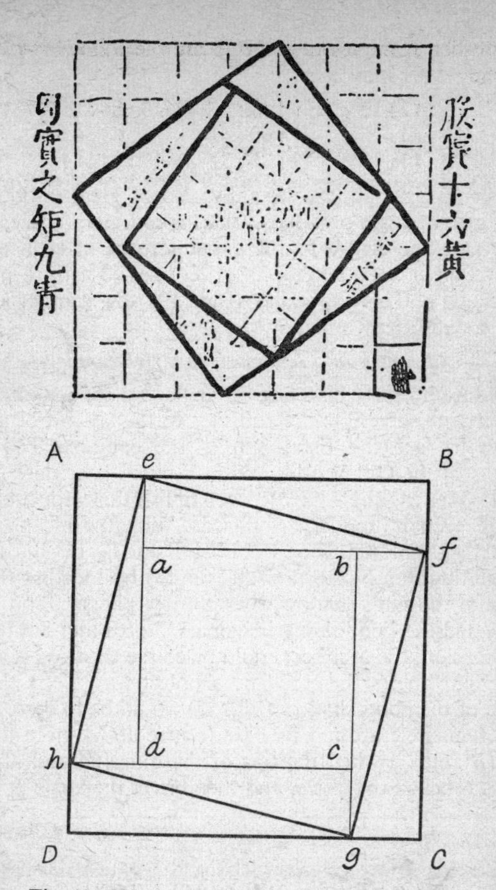

Fig. 18. Chinese Proof of Pythagoras' Theorem

The Book of *Chou Pei Suan King*, probably written about AD 40, is attributed by oral tradition to a source before the Greek geometer taught what we call the *Theorem of Pythagoras*, i.e. that the square on the longest side of a right-angled triangle is equivalent to the sum of the squares on the other two. This very early example of block printing from an ancient edition of the *Chou Pei*, as given in Smith's *History of Mathematics*, demonstrates the truth of the theorem. By joining to any right-angled triangle like the black figure *eBf* three other right-angled triangles like it, a square can be formed. Next trace four oblongs

Among other examples of the rule are the following of recipes which the reader may check

5 : 12 : 13	9 : 40 : 41	13 : 84 : 85
7 : 24 : 25	11 : 60 : 61	16 : 63 : 65
8 : 15 : 17	12 : 35 : 37	17 : 144 : 145

Another recipe, almost certainly also of great antiquity, for making a right angle on a flat surface is to trace a circle, make a line through the centre to the circumference at each end and draw two other lines joining its extremities to a third point on the boundary. This embodies (Fig. 19) the geometrical rule usually stated briefly as follows:

the angle in a semicircle is a right angle.

Procedure both for bisecting an angle and for making a right angle were indispensable prerequisites to the lay-out of a temple or tomb to face the rising sun of the equinox. The temple architects of the Old World, and at a much later date those of Central America, knew how to accomplish this with astonishing precision even by modern standards. With the means at their disposal the simplest method would be first to lay out the meridian (line pointing due North and due South) by locating the exact position of the sun's shadow when shortest, i.e. at noon. Having thus located the line joining imaginary North and South points of the horizon, the architect could place the East–West axis as a line at right angles thereto. The simplest way to fix the exact position of the noon shadow (Fig. 22) would be to erect a pillar, trace a semicircle around its base (centre at A), note the exact points (B and C) when the tips of the morning and afternoon shadows touched the circle, and then bisect the angle BAC.

[*Caption continued*]

(rectangles) like $eafB$, each of which is made up of two triangles like efB. When you have read Chapter 4 you will be able to put together the Chinese puzzle, which is much less puzzling than the proof given by Euclid. These are the steps:

$$\text{Triangle } efB = \tfrac{1}{2} \text{ rectangle } eafB = \tfrac{1}{2}\, Bf\,.\,eB$$

$$\text{Square } ABCD = \text{Square } efgh + 4 \text{ times triangle } efB$$
$$= ef^2 + 2Bf\,.\,eB$$

$$\text{Also Square } ABCD = Bf^2 + eB^2 + 2Bf\,.\,eB$$

$$\therefore\; ef^2 + 2Bf\,.\,eB = Bf^2 + eB^2 + 2Bf\,.\,eB$$

$$\therefore\; ef^2 = Bf^2 + eB^2$$

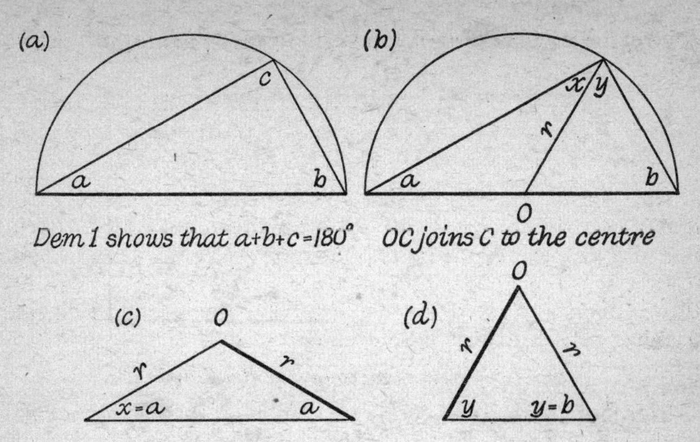

Fig. 19. The Angle in a Semicircle is a Right angle

According to tradition Thales, the father of Greek Geometry, sacrificed an ox to the gods when he was able to demonstrate why this is so. Here (*see* Dems 1 and 3 in Chapter 4) $c = x + y$, so that $c = a + b$.

$$\therefore\ a + b + c = 180° = 2a + 2b \text{ so that } 2c = 180°$$

The rule of thumb geometry of the temple precincts included recipes for making simple figures of which the right-angled triangle with two equal sides and the regular *hexagon* (i.e. a figure with 6 equal sides) are of special interest (Fig. 20). Of the former, see also Fig. 43, two angles are each equal to half a right angle (45°). The latter is divisible into six equilateral triangles, each angle of which is two-thirds of a right angle 60°. It is likely that the temple architects made use of the length of the sun's shadow to measure a height (Fig. 21). The ratio of the two would be 1 : 1 when the sunbeam made an angle of 45° with the horizon (or plumb line). Since they could undoubtedly construct an equilateral triangle and divide it into two equal right-angled triangles, it is probable that they recognized an approximate numerical value for the ratio of height to shadow when the sun's altitude (inclination to horizon) was 30° and so-called zenith distance (inclination to the plumb line) was 60° or vice versa.

If the foregoing surmise is correct, it is also likely that a very early measure of the angle was what we now call its *tangent* (see Fig. 51). For location (Fig. 24) of the heavenly bodies of the night sky by their altitude at *transit* (i.e. highest in the heavens), the astronomer-priest presumably used some other measure.

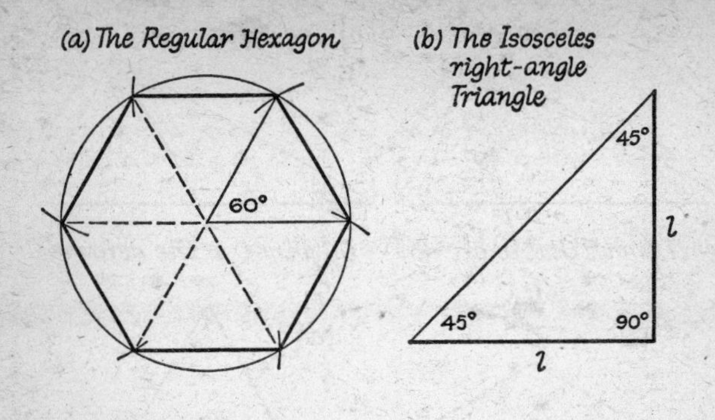

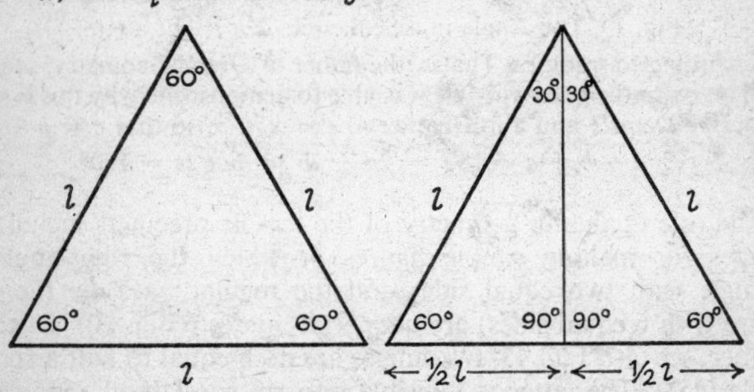

Fig. 20. Recipes for Making Angles of 30°, 45°, 60°

A regular hexagon (figure with six sides of equal length) inscribed in a circle by marking off along the boundary intersecting arcs with the same radius as the circle itself.

The division of the circle into 360° is a practice which the astronomers of Alexandria took over from the Greek-speaking colonies in contact with the temple lore of Iraq. Since twelve 30-day months was an early estimate of the year in that part of the world, the practice may have arisen from reckoning the daily course of the sun in its complete cycle in the ecliptic belt located in the night sky by the zodiacal constellations. That Chinese geometers at an early date divided the circle into 365° lends weight to this supposition.

A star map in an Egyptian tomb dated about 1100 BC testifies

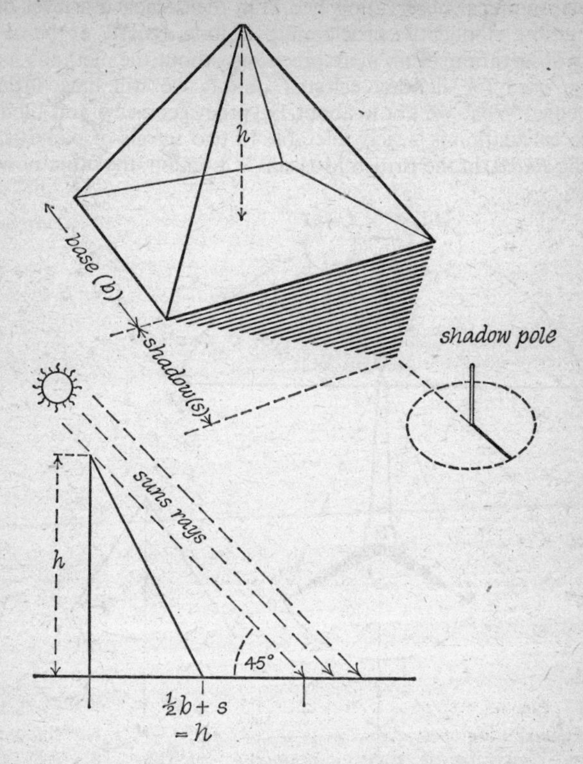

When the shadow touches the circle of radius equal to the height of the shadow-pole, the height (h) of the pyramid is got by adding the length of the shadow (s) to half the base (b)

Fig. 21

On one day of the year at Gizeh the noon sunbeam will be almost exactly at an angle of 45° to the plane of the spirit level. When the midday sun is at 45° the height of the Great Pyramid is equal to the length of the shadow and half the base.

that astronomical observation had even then reached a level of information eloquently proclaiming a long record of painstaking observation of no mere precision. About the methods its authors used for sighting celestial objects we still have little knowledge. What we know about Egyptian geometry and their aids to calculation is largely referable to two scrolls of papyrus, one (the *Rhind*) in the British Museum at London, the other now

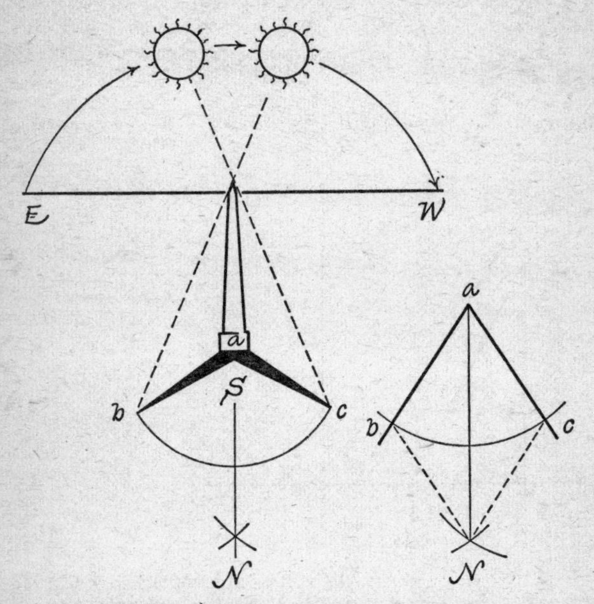

Fig. 22. Fixing the North–South Meridian

One way to fix the North–South meridian depends on the fact that the sun's noon shadow points due North in countries north of the tropics. To catch it in that position trace a circle with the pillar as centre (A) on the soil. Mark the points B and C when the forenoon and shadows just touch its boundary. Then bisect the angle BAC. The recipe for bisecting an angle is as follows. The circular arc with centre at A is of radius r. That of the circular arcs with centres at B and C are also of radius r. Hence $AB = r = AC$ and $CN = r = BN$. Thus the triangles ABN, ACN with a common side AD, having 3 corresponding equal sides, are equivalent and the corresponding angles $NAB = NAC$.

in Moscow. Their dates are respectively about 1600 and 1850 BC. The Moscow papyrus shows that the Egyptian temple libraries held the secret of a correct method of calculating the volume of a pyramid nearly 4,000 years ago. Their scribes used as a value for π (the ratio of the circumference to the diameter of a circle) 256:81, i.e. expressed in our decimal notation almost exactly 3·16, entailing an error less than 1 per cent. This may have been an empirical discovery made by measuring the boundary and width of cylinders. Not improbably (*cf.* Fig. 66) it was based on the mean perimeters (boundaries) of polygons of twelve equal sides

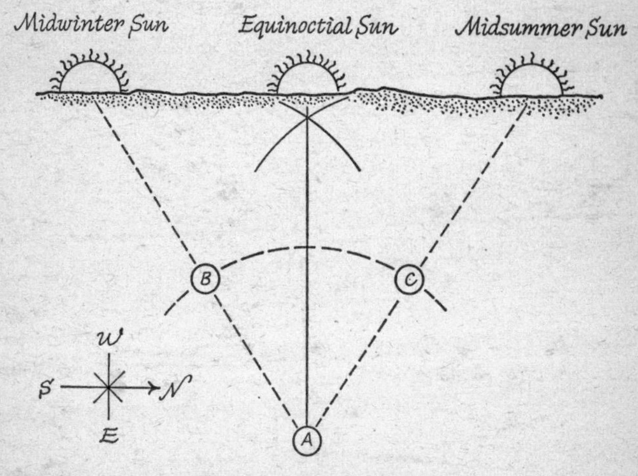

Fig. 23. Fixing the Equinox

Some early calendrical monuments suggest that the equinox was fixed by observations on the rising or setting sun of the solstices (December 21st and June 21st), when the sun rises and sets at its most extreme positions towards the South and North respectively. In this figure *A* and *B* are two poles placed in alignment with the setting sun of the winter solstice. The distance between A and *C* in line with the setting sun of the summer solstice is the same as the distance between *A* and *B*. Midway between its journey between the two extremes the sun rises and sets due East and West, and the lengths of day and night are equal. Hence, these two days (March 21st and September 23rd) are called the equinoxes. In ancient ritual, they were days of great importance. The East and West points on the horizon can be obtained by bisecting the angle *BAC*.

enclosing and enclosed by a circle of unit radius (i.e. one whose radius is the unit, e.g. foot length).

The use of π, which we need to calculate the volume of a cylinder or cone, prompts one to ask how the Egyptians handled fractions. It seems clear that they never got much beyond thinking of fractions in terms other than small units of measurement, as we might think of 1 and seven-eighteenths of a yard being 1 *yard* 1 *foot* and 2 *inches*. It is less clear why (or how) they always sought to express a fraction whose numerator is greater than two by the sum of one or more fractions with unit numerators or numerators of 2, e.g.:

$$\frac{7}{12} = \frac{1}{3} + \frac{1}{4}\ ;\ \frac{13}{45} = \frac{2}{9} + \frac{1}{18} + \frac{1}{90} \text{ or } \frac{1}{5} + \frac{2}{45} + \frac{2}{45}$$

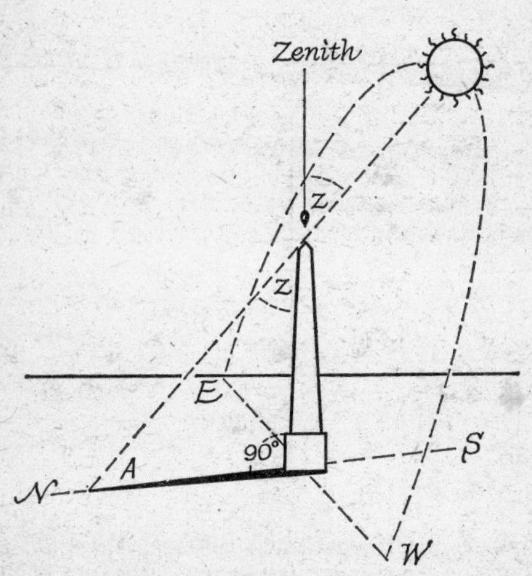

Fig. 24

Noon on the equinoxes (March 21st and September 23rd), when the sun rises due East and sets due West. The sun's shadow at noon always lies along the line which joins the North and South points of the horizon. This is also the observer's meridian of longitude which joins the North and South poles. The zenith is the name which astronomers give to the spot in the heavens directly overhead. Note that the angle (A) of the sun above the horizon (called the sun's 'altitude') and the angle (Z) which the sunbeam makes with the plumb line or vertical (called the sun's zenith distance) make up a right angle or 90°, so that $A = 90° - Z$ and $Z = 90° - A$.

Before 2000 BC, the art of calculation in the temple precincts of Iraq had reached a level much higher than that which the Egyptian priestly caste at any time attained. Oddly enough however, they remained, at the period to which the ensuing remarks refer, content with a value for π so inaccurate as 3·0, an error of over 4 per cent. Their clay tablets of 2000 BC indicate that their makers could solve by use of tables of squares and cubes problems which involve what we now call quadratic and even cubic equations. They knew examples of the Pythagorean rule for right-angled triangles other than the 5 : 4 : 3 recipe of the Egyptians. One such as 17 : 15 : 8. What is more impressive is that they had ventured into the realm of what much later generations came to call *irrational* numbers. It is not a great achievement to say: if 3^2 (three *squared*) is 9, then $\sqrt{9}$ (the *square root* of 9) is 3. This merely means if 9 is three times 3, 3 is the number which yields 9 if multiplied by itself. It was, however, a momentous feat to be first to say: let us try to find a number ($\sqrt{2}$) which is 2 when multiplied by itself. Maybe, it was more so to cite a value so good as 17 ÷ 12 in our own fractional notation. If we square this (i.e. multiply it by itself, as when we find the area of a square from the lengths of its side), the error is less than 1 in 200:

$$\left(\frac{17}{12}\right)^2 = \frac{17 \times 17}{12 \times 12} = \frac{289}{144} = 2\tfrac{1}{144}$$

Notice here that:

$$\left(\frac{16}{12}\right)^2 = \frac{256}{144} = 1\tfrac{112}{144}$$

More than two thousand years were to go by before people learned to think of $\sqrt{2}$ effortlessly both as a number which lies somewhere between (16 ÷ 12) = 1·3 and (17 ÷ 12) = $1{\cdot}41\dot{6}$, and to realize clearly that we can go on *endlessly* making the gap smaller and smaller. For instance, $\sqrt{2}$ also lies somewhere between 1·414 and 1·415. Thus:

$$\begin{aligned} 1{\cdot}415 \times 1{\cdot}415 &= 2{\cdot}002225 \\ 1{\cdot}414 \times 1{\cdot}414 &= 1{\cdot}999396 \\ \textit{Difference} &= 0{\cdot}002829 \end{aligned}$$

Today, there is nothing very irrational about $\sqrt{2}$, except the irrationality of hoping to programme a computer which will come to the end of such a process of getting nearer and nearer to its goal before the earth has become too cold for human habitation.

At a time when unusual celestial events, such as eclipses, evoked awe among the peasants, craftsmen, and small traders, the priestly astronomers had a strong incentive to exploit their credulity. About 1500 BC, the priestly astronomers of Iraq discovered an approximately 18-year cycle in which lunar and solar eclipses recur. By being, thenceforth, able to forecast them, they doubtless increased their social prestige and title to greater tribute and other privileges. The mean length of this eclipse cycle of the Iraq astronomer priests was a little over 18 years, more precisely 6585·83 days. Needless to say, it would be necessary to carry out observations over a period embracing many such cycles before reaching a reliable estimate. If, for no other reason, astronomical computations had thus to cope with numbers vastly greater than those needed to catalogue temple property or taxation. At an uncertain date before 300 BC the priestly astronomers of Iraq recognized the 25,000-year cycle in which the dates of the equinoxes complete a calendar round.

For astronomical computations, they used an earlier battery of signs for counting in multiples of 10, still retained in that sense for trade and accountancy; but they gave them new values as multiples of 60. They expressed fractions, as we continue to express fractions of a degree, by what writers in medieval Latin called minutes (*pars minutus*) and seconds (*pars secundus*). Thus, we usually write for $4{\cdot}0341\dot{6}$ degrees:

$$4^\circ\ 2'\ 3'' \text{ meaning } 4 + \frac{2}{60} + \frac{3}{60 \times 60}$$

In this system a second is less than one over three thousand, this being a small enough fraction for any calculations the mathematicians of antiquity had to handle. Moreover, the choice of 60 as a base had a peculiar merit in common with the claims of those who from time to time suggest that we ourselves should substitute 12 for 10 as the *base* (p. 34) of our number system. The advantage of the base 12 is that 12 is exactly divisible by the whole numbers (*integers*) 2, 3, 4, 6, and 12, whereas 10 is exactly divisible only by 2, 5, and 10. The number 60 is divisible by 2, 3, 4, 5, 6, 10, 12, 15, 20, 30, and 60. Thus proper fractions having any one of these numbers as its denominator in our notation is exactly expressible in minutes only. To find how many numbers are expressible in minutes and seconds, we need only

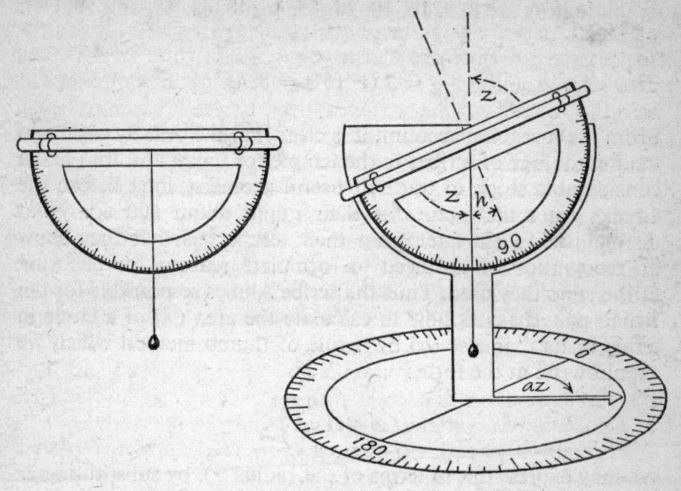

Fig. 25

A simple theodolite (*astrolabe*) for measuring the angle a star (or any other object) makes with the horizon (altitude), or the vertical (zenith distance) can be made by fixing a piece of metal tube exactly parallel to the base line of a blackboard protractor, which you can buy from any educational dealer. To the centre point of the protractor fix a cord with a heavy weight (e.g. a lump of type metal which any compositor will give you free if you ask him nicely) to act as plumb line. The division opposite the cord when the object is sighted is its zenith distance (Z), and the altitude (h) is $90° - Z$. If you mount this to move freely on an upright wooden support which revolves freely on a base with a circular scale (made by screwing two protractors on to it) and fix a pointer in line with the tube, you can measure the *azimuth* (*az*) or bearing of a star or other object (e.g. the setting sun) from the North–South meridian. To do this, fix the scale so that it reads 0° when the sighter is pointed to the noon sun or the Pole star. This was a type of instrument used to find latitude and longitude in the time of Columbus. You yourself can use it to find the latitude and longitude of your house (Chapter 3), or make an ordnance survey of your neighbourhood (Chapter 4).

ask what integers are factors of 3,600 (= 60 × 60). In the range of fractions greater than one-sixtieth, this brings in those whose denominators are 8, 9, 16, 18, 24, 25, 36, 40, 45, 48, and 50, etc., e.g.:

$$\frac{7^\circ}{48} = 7\,(1'\,15'') = 8'\,45''$$

From the foregoing account, it is clear that the priestly caste and their entourage of scribes in the temples of Egypt and Iraq had a considerable stock in trade of useful geometry long before the Greeks enter the picture as their pupils about 600 BC. What distinguishes them later from their teachers is that they show no recognition of the need to formulate reasons for believing in the rules they used. Thus the scribe Ahmes responsible for the Rhind papyrus cites how to calculate the area (A) of a circle in terms of its diameter (d) by a rule of thumb method which we should write in the form:

$$A = \left(\frac{8d}{9}\right)^2$$

We may express this in terms of the radius (r), by substituting $2r$ for d, so that

$$A = \left(\frac{16r}{9}\right)^2 = \frac{256}{81}\,r^2 = (3{\cdot}1605)\,r^2$$

This is equivalent to our formula $A = \pi r^2$, if we cite 3·1605 as the value of π, an approximation with an error less than 1 per cent from the value (3·14159 . . .) correct to five decimal places. What is noteworthy about the statement of the rule is not whether we choose to regard the error as large or great. It is that the teacher scribe feels no compunction about handing out a rule without the slightest indication of how anyone discovered it or why anyone should take it on trust.

The veneration of the Greeks by their successors is indeed due to the fact that they were first to insist explicitly on the need for *proof*. Several circumstances conspired to promote this innovation. Their ancestors who colonized and conquered so much territory around the Mediterranean came from regions free of the incubus of a highly privileged priesthood. They enriched themselves with sea-borne trade which brought them into contact with territories where architecture, star lore, and the associated arts of measurement had reached a high level of attainment long before their arrival. Prompted by a curiosity unfettered by the disposition to take all they learned by hearsay at face value, they thus brought back from their travels the knowledge of new techniques which could compel assent only in open debate with disputatious neighbours addicted to litigation.

It is also relevant that we date the birth of Greek geometry from the latter years of Thales (640–546), a wealthy merchant mariner with headquarters in the Greek colony of Miletus on the fringe of Asia minor in close propinquity to Phoenician ports. The birth of Thales, himself of Phoenician descent, happened in the dawn of a larger literacy made possible by adapting the alphabet of their Semitic trade rivals to the requirements of the Greek language whose structure was basically different from theirs. During his lifetime, the introduction from Egypt of papyrus as writing material fostered the growth of a reading public in the Greek colonies; and within twenty-five years of his death occurred the birth of Aeschylus who first institutionalized the dialogue of the tribal chorus as drama. Such use of the written word set a pattern for recording controversy about politics, religious beliefs, and other topics in an orderly and logical, or quasi-logical, way.

During the lifetime of the first of the Greek dramatists, teachers who had likewise travelled widely were able to attract pupils among the prosperous citizens. Foremost among the first of such was Pythagoras who flourished about 550 BC, and founded his school in the Greek-speaking part of South Italy. He himself left no written work, and he bound his pupils to secrecy; but Greek curiosity about the world at large, Greek partiality for disputation, and competition between rival teachers had made any attempt to monopolize knowledge impracticable within a century after his death.

Greek historians concur in the assertion that Thales gained his knowledge of, and taste for, geometry from his travels in Egypt where he came into contact with its temple lore. Pythagoras travelled not only in Egypt but also in Iraq and, it seems, farther afield. Thence he returned with interest in number-lore, including the figurate representation of numbers (Figs. 36–7) as triangles or squares. There is indeed no doubt that the raw materials of Greek mathematics were imports; but as imports they had to pass the customs of Greek incredulity. To sell his wares to citizens wealthy enough to send their sons to study under him, the traveller-teacher had to convince his customers that they were at least durable, and the demand for proof mounted with an increasing volume of public disputation. Mensuration and calculation were henceforth to be an open conspiracy.

From one point of view this was all to the good. From another, we may well have misgivings. Before Greek geometry, in the sense in which we have agreed (p. 29) to use the term *Greek*, reached its climax during the adolescence of Euclid, the notion of proof had mummified. We may indeed speak of Greek mathematics as having two phases. The first was an adventurous youth, when its exponents were still conscious of their debt to the rule of thumb, and hitherto secret, lore of the temples in Egypt and Iraq. That they did still recognize their indebtedness to their teachers of far older civilizations is abundantly clear. Thus Democritus, seemingly the first person to advance intelligible

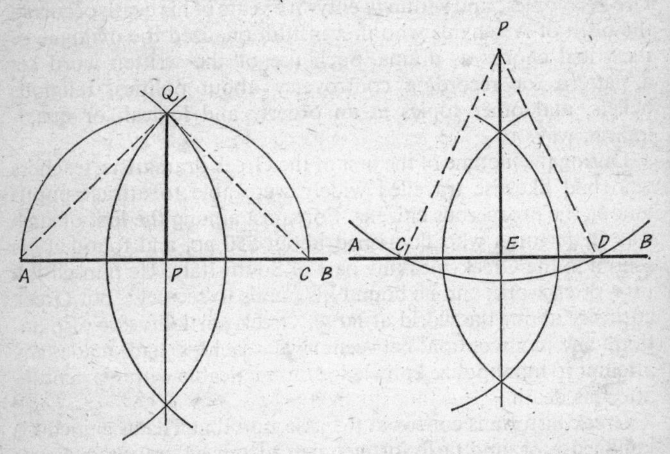

Fig. 26. Two Euclidean Constructions

To erect a perpendicular to AB at P mark off $PC = r = AC$. Bisect AC. From Fig. 15, PQ is at right angles to AC, whence also to AB. To drop a perpendicular from P on AB, first mark off $PD = R = PC$, then bisect CD. The two triangles PCE and PDE have $PD = PC$, $CE = DE$ and PE common. Hence, $CEP = DEP$.

reasons for regarding matter – including air itself – as particulate (i.e. made up of atoms), presented his credentials in the following terms:

> 'Of all my contemporaries, it is I who have traversed the greatest part of the earth, visited the most distant regions, studied climates the most diverse, countries the most varied and listened to the most men. There is no one who has surpassed me in geometrical constructions and demonstrations, no not even the geometers of Egypt among whom I passed five full years of my life.'

Democritus (about 400 BC) stands at the crossroads. His lifetime overlaps with that of Plato, who died in 389 at eighty years of age, and of that of Plato's renowned pupil Eudoxus (408–355). In the phase of Greek mathematics we associate with Plato's school we find that teachers of geometry have repudiated its roots in the practice of mankind. In becoming an end in itself it has sponsored a regimen of proof which makes no appeal to practice and has in effect banished measurement from mathematics as a respectable form of relaxation for the opulently idle. How this happened belongs to the story (Chapter 3) of Euclid's hand-out to Alexandria. It is, however, fitting to conclude this part of our story by mentioning the main features of what Plato's pupils regarded as essential to a convincing proof of what we shall continue to regard as a rule of *mensuration* such as the 3 : 4 : 5 ratio.

First, it is essential to *define* the terms used.

Secondly, it is essential to state clearly what we all agree to take for granted, e.g. $a + n = b + n$ if $a = b$.

Thirdly, it is essential to make clear, and to justify, what procedures we may invoke to define our terms or to dissect figures in order to exhibit relations between their parts.

Somewhat arbitrarily, the Platonic school restricted procedures under the third category to what we can draw with no instruments other than a ruler and a compass, i.e. straight lines and circular arcs. If we accept this limitation, we still have to show that our rules are reliable for the four main recipes of dissection:

(*a*) how to bisect a line (Fig. 15);
(*b*) how to bisect an angle (Fig. 22);

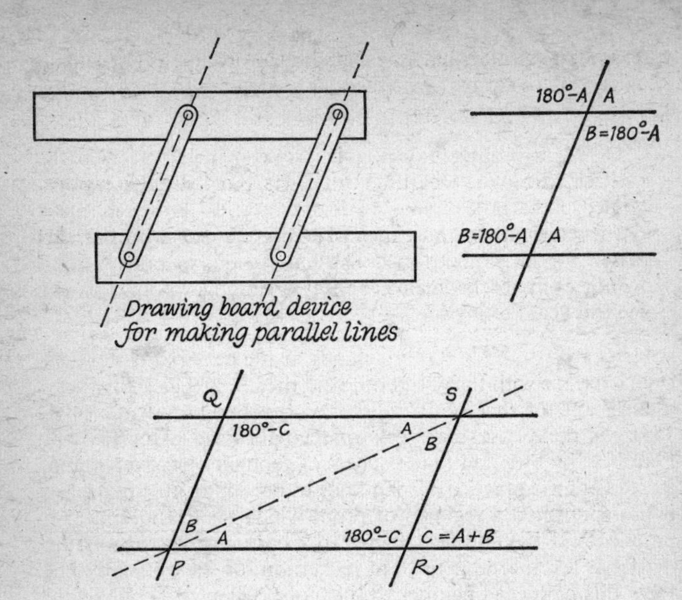

Fig. 27. Parallelism

In upper right-hand figure, A are called *corresponding*, B called *alternate* angles. In the lower figure PQ is parallel to RS and QS to PR. The triangles PQS and PRS have a common side and the angles at its extremities, see Fig. 28 (*c*), equal. They are therefore equivalent, so that $PQ = SR$ and $QS = PR$.

(*c*) how to erect a perpendicular at a particular point on a line (Fig. 26);

(*d*) how to draw through a point not on a particular line a second line perpendicular to the first line (Fig. 26).

The justification of all these procedures depends on two assumptions. The first is that the distances between the centre and every point on a circular arc are equal. This follows from the way in which we make a circle and is therefore a proper definition of what a circle is. The second depends on what we need to know to draw a particular triangle, and therefore all we need to be able to say in support of the claim that two triangles are equivalent. Any one of the following suffices (Fig. 28) to make a unique triangle (or its mirror image), whence two triangles are equivalent if our information about both of them conforms to one or other prescription:

(i) we know the length of two sides and the angle between them;
(iii) we know the length of one side and the angles at each end;
(iii) we know the length of three sides.

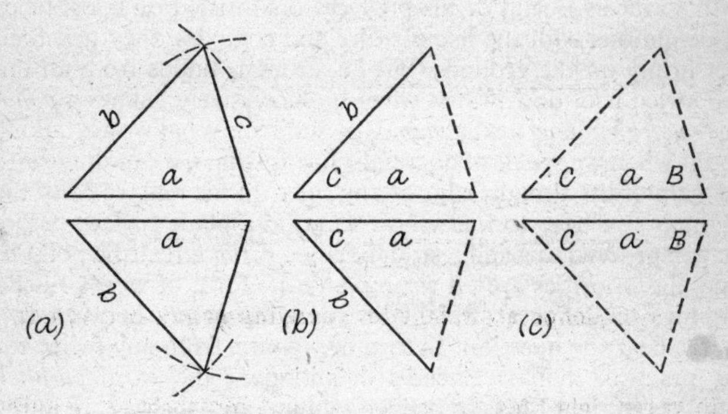

Fig. 28. How to Draw a Triangle
(*a*) Length of three sides known.
(*b*) Length of two sides and size of the angle enclosed.
(*c*) Length of one line and the two angles which the other two sides make with it known.

To illustrate the relevance of the means at our disposal to the definition of a figure, that of a square will suffice. Having drawn two lines of equal length at right angles to each other, we can complete the figure by using our compass in the following way:

(i) by tracing circular arcs of the same length as each side with their free extremities as centres;
(ii) joining the latter to the point of intersection of the arcs.

Alternatively, we may draw lines perpendicular to the extremities of the two lines first drawn. This means that we can define a square in terms of what we can do with a ruler and a compass:

(i) *either* as a figure with four equal sides of which two enclose a right angle;

(ii) *or* as a figure with two equal adjacent sides having four right angles.

What the rules of the game do not permit us to offer as a definition of a square is the assertion that a square is a figure with 4 equal sides enclosing four right angles. If our recipe for making it is (i) above, we have still to show why three remaining angles are right angles. If it is (ii) above, we have still to show why 2 remaining sides are equal to the other 2.

In so far as Euclid or his predecessors insisted on consistency of definitions with the use of ruler and compass, they had their feet firmly on the ground; but he, at least, builds from an unsupported roof downwards when he successively defines a *point*, then a *straight line*, next a *plane* (flat surface). What we are talking about when we speak of a straight line is what we can draw with the help of the straight edge of any ruler. In the real world of the engineer, we have to know how to grind a plane surface before we can proceed to grind a straight edge. From a realistic point of view the priorities are all wrong, and the form of words Euclid employs to define a straight line as lying *evenly* between two points begs the question. Even more remote from any recipe for the practical man is Euclid's definition of the word *parallel.* Parallel, straight lines, as defined by him, are 'such as are in the same plane and which, being produced ever so far both ways, do not meet'. Having defined them in this obscure way, he then inverts the proper order of definition and proof by seeking to deduce that a line crossing two parallel lines makes alternate and corresponding angles equal. The truth is that our criterion of whether two lines are parallel is that they are equally inclined to any line crossing them. The familiar drawing-board (Fig. 27) device for making two lines parallel at any required distance apart is merely a way of ensuring that they remain equally inclined to two cross pieces.

No doubt some readers will have somewhere read that we no longer believe that Euclid's geometry is necessarily true. Such statements are puzzling, unless we distinguish between two domains on which we may seek to apply it. To state the issue in the simplest terms, we may usefully narrow the field; and we may here pause to explain why. For two reasons we shall not consider Euclid's geometry of solid figures in this book. One is that better methods of dealing with what they claim to do have long since been in use. The other is that there is no convincing tie-up between Euclid's treatment of them and the definitions or methods he uses to deal with flat figures, unless we exclude any considerations relevant to orientation, e.g. the essential difference between a right-hand and a left-hand glove, or between two crystals described as optical isomers. On that understanding, we may say that the rules of Euclid's geometry of the plane do not conflict with anything the draughtsman designs on a drawing board, nor therefore with anything for which the draughtsman's design proves by experience to be a faithful model.

The last statement does not mean that Euclid's ruler and compass suffice for the construction of every curve one can draw by means of a suitable mechanical device. Even in Euclid's time, such a statement would not have been true. Whether Euclid's geometry is or is not true, in the sense that it embodies what drawing-board models correctly describe is not the bone of contention, when mathematicians consider the claims of so-called non-Euclidean geometries. Behind a smoke-screen of abstractions, what is at stake is an astronomical issue.

When we talk of outermost space, we can merely figure it as a medium in which we receive signals of light or other radiant energy. All the astronomy of antiquity rested on two assumptions: that light travels in straight lines and that light rays from any one of the fixed stars are parallel. In interstellar space, we cannot construct anything comparable with the drawing-board recipe for making a straight line; and observation alone can justify the statement that drawing-board geometry is adequate to describe faithfully how light reaches us from the Milky Way. If we substitute for Euclid's question-begging definitions of a straight line, an alternative one offered by his successor Archimedes, we do not absolve ourselves from the need to check the reliability of our model.

Archimedes defined a line as the shortest distance between two points. Since only radiations such as visible light traverse space, whatever path they do pursue is the shortest path we can know anything about. In that sense, light travels in Archimedean straight lines; but, in that sense, light does not necessarily travel in drawing-board lines. To say this should not in the least detract from the usefulness of Greek geometry at a time when astronomy and its offspring geography were the only branches of scientific knowledge which needed to enlist mathematical skill. For all

Fig. 29. The Nightly Rotation of the Stars

Diagrammatic view as we might see the sky from the Great Pyramid today in late summer. The two constellations shown, being very near the Pole, do not set below the horizon. Six months later Cassiopeia will be seen sinking after sunset and rising just before sunrise.

practical purposes which Greek astronomy embraced, drawing-board geometry was adequate, and astronomy had advanced in step with it.

Before the time of Plato, Greek astronomy had significantly outpaced what their priestly predecessor had established. In so far as the regulation of the Calendar was a major social function of the priesthood, the temples of Egypt and Iraq had long since been the repository of a considerable body of information about the rising, setting, and inclination at transit to the horizon (or plumb line) of the fixed stars, of the moon, and of the sun. We have seen that the priestly astronomers understood eclipses. They knew that lunar eclipses disclose the shadow cast by the curved rim of the earth when the latter comes between, and directly in line with, it and the sun. All this might well, and seemingly did, suggest that the earth is a circular disc. Nothing in the experience of the Temple observatory did indeed suggest the contrary. Only when Phoenician cargoes pushed out beyond the confines of the Mediterranean, and coasted northward for tin, southward for spices along the seaboard of Europe and Africa, during the three centuries before the dawn of Greek geometry, did the notion of a spherical earth emerge imperatively from common practice.

Long sea voyages in a direction north or south confronted the master mariner with an ever-growing volume of experience wholly alien to that of temple astronomers bound lifelong to the appearance of the heavens at one and the same locality. Already, before any ships had made their way beyond what the Greeks knew as the Pillars of Hercules, i.e. Straits of Gibraltar, the pilot would be familiar with the fact that one guiding star Canopus, second only to Sirius as the brightest in the firmament and visible everywhere during part of the year on the Southern boundary of the Mediterranean, would never rise north of Sicily. He would know, too, that the relative lengths (see p. 142) of the noon shadow of the sun on the equinoxes or on the solstices would be different at Marseilles and Cyrene. He had no charts other than the record of how, in the language of today, the altitude of a heavenly body at transit depends on *latitude* (Figs. 59–62).

When he did first venture beyond the confines of the Mediterranean on a northerly or southerly course, the navigator would be more than dependent on, and eager to record, the changing face of the heavens. As early as 400 BC, a master mariner Pythias, by observation of the midsummer noon shadow, cited the latitude

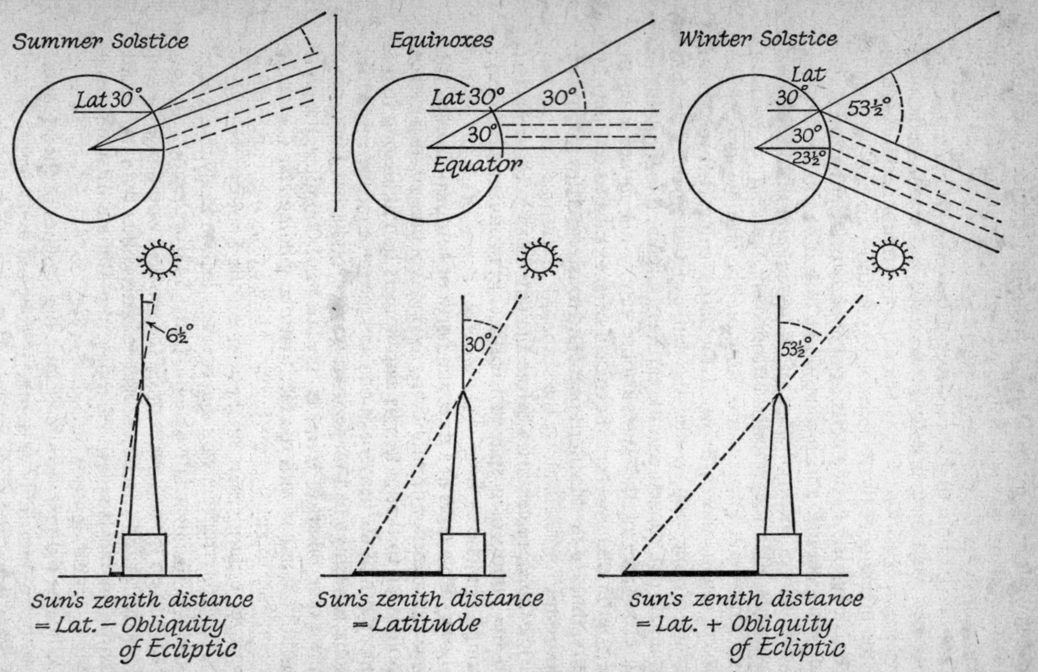

Fig. 30. Egyptian Measurement of the Obliquity of the Ecliptic from the Sun's Noon Shadow

of Marseilles correct to about a tenth of a degree in our customary reckoning; but long before that, Greek pilots and teachers had learned from the experience of their Phoenician rivals for maritime trade to think of the earth as a spherical body. Only so, can one explain why some star clusters are in some latitudes never visible, in some latitudes visible at night during only part of the year and in others throughout the whole year, as is true of Cassiopeia and the Great Bear at the latitude of London today. Fig. 29 shows that the Great Bear (*Ursa Major*) is only partly visible all the year round at the latitude of the Great Pyramid of Gizeh. If, as did the Carthaginian Hanno (about 500 BC), the sea-captain sailed southward along the African coast so far as what is now Liberia, he would first come to a place where the midsummer noon sun is directly overhead, casting no shadow. Thereafter, he would be coasting within the tropical belt, where he would encounter a phenomenon wholly unintelligible on the priestly assumption that the earth is a flat disc. During different parts of the year the noon sun's shadow would point in opposite directions.

The circumpolar stars or constellations, e.g. the two last-mentioned at the latitude of London, appear to rotate around a fixed point and all other stars in circular arcs in planes parallel to them. This fixed point (*celestial pole*) now nearly corresponds to the position of Polaris in the Little Bear (*Ursa Minor*) which rotates in a tiny circle of less than 1° from it. The angle it makes with the horizon is the same as the angle the noon sun of the equinox makes with the plumb line. Either angle (Figs. 29 and 30) tallies with what we mean by our latitude. Before the time of Euclid, the last statement was already a commonplace of Greek

---

[*Caption continued*]

At noon the sun is highest. The Pole, the earth's centre, the observer, and the sun are all in the same plane (or flat surface). On the equinoxes (March 21st and September 23rd) the sun's *zenith distance* at noon is the observer's *latitude* (30° at Memphis). If the *obliquity of the ecliptic* (i.e. the inclination of the sun's apparent *annual* track to the plane of the earth's equator) is E:

$L + E =$ sun's zenith distance on winter solstice (December 21st)
$L - E =$ sun's zenith distance on summer solstice (June 21st)

So the obliquity of the ecliptic is:

$\frac{1}{2}$ (sun's *z.d.* on December 21st – sun's *z.d.* on June 21st)

For further explanation see Figs. 59–61.

astronomy, as was the division of the world into zones of *climate*.

The astronomer priests of Egypt and Iraq knew that the sun's apparent retreat in the constellations of the zodiac (Fig. 30) is in a fixed plane (the ecliptic) inclined at a fixed angle (in our terms $23\frac{1}{2}°$) to an imaginary plane (the celestial equator) in the firmament; but they were not able to conceive the bearing of their observations on the earth's shape or temperature. To the Greeks, the fixed plane included that of the earth's equator at right angles to an axis including the celestial pole and a terrestrial one in a region of extreme cold. The ecliptic plane cut the earth surface at extreme points where the noon sun of one or other solstice was directly overhead. Such were the latitudes of the two tropics. Being able to picture the earth on the drawing board as the circular section of a sphere, a pupil of Democritus had correctly reached one challenging conclusion possibly at least a century before any ship had coasted beyond Scotland as far as Iceland. Between the North Pole and the tropic of Cancer where the sun is directly overhead at noon on Midsummer Day in the Northern hemisphere, there is a circle of latitude in which the sun is still visible at midnight on the same day. We take this on trust in our schooldays. It was one of the triumphs of scientific geography, when Greek geometry was at the peak of its usefulness.

Exercises on Chapter 1

Discoveries to Make

1. Find several circular objects, such as a dustbin lid, clock face. Measure the circumference and diameter of each and find the value of the circumference divided by the diameter in each case as accurately as you can.

The following sets of instructions relate to triangles. Notice in each instance what conclusions are suggested by them.

2. (*a*) Draw a triangle with sides 10 centimetres, 8 centimetres, and 6 centimetres. The way to do this is to draw a straight line anywhere on the paper and mark off a distance *AB* equal to 10 centimetres. Adjust your compasses so that the distance between the point and the pencil is 8 centimetres. Place the point of the compasses at *A*, and with the pencil draw an arc which will have

a radius of 8 centimetres. In the same way draw an arc with a radius of 6 centimetres with centre at *B*. Join *C*, the point where the two arcs cut, to *A* and *B*, and you will have the triangle you want.

Draw triangles with sides:

(*b*) 9 centimetres, 15 centimetres, 12 centimetres.
(*c*) 17 centimetres, 8 centimetres, 15 centimetres.

3. In all three triangles measure the angle between the two shorter sides of the triangle.

4. The Egyptian method of laying out a right angle is still in use. The following is a quotation from Bulletin Number 2 of the Ministry of Agriculture and Fisheries (1935). It is part of a set of directions for laying out a plantation of fruit trees.

> 'The easiest method of chaining a right angle is as follows: The 24th link is pegged at the point from which the right angle is to be set out, the nought end of the chain and the 96th link are pegged together, back along the base line, so that the piece of chain 0–24 is taut. If the 56th link is taken in the direction required until both the sections 24–56 and 56–96 are taut, then the point reached will be at right angles to the base line.'

If you can find a convenient space, peg out the Egyptian rope triangle and the surveyor's triangle. The link referred to is 7·9 inches long. Satisfy yourself that these are both ways of setting out a right angle.

5. (*a*) Draw a right angle. Measure off on the arms of the angle distances of 5 centimetres and 12 centimetres. Join the ends to form a triangle. Measure the other side.

(*b*) In the same way draw a right-angled triangle with sides 12 centimetres and 16 centimetres long, and measure the third side.

(*c*) Draw a right-angled triangle with sides of 7 centimetres and 24 centimetres, and measure the third side.

6. Draw a triangle with one side 2 inches long, and the angles at either end of this side equal to 30°. Draw triangles with one side 3 inches and 4 inches respectively, and the adjoining angles 30°. Measure the other sides in all three triangles.

7. Draw three triangles with one side 2 inches, 3 inches, and 4 inches respectively, and the adjoining angles 45°. Measure the sides and test the rule of Fig. 16.

8. Draw three triangles of different sizes, each with two equal sides, and measure all the angles.

9. Find out what the sum is of all the three angles added together in all the triangles you have drawn.

10. Draw two triangles and try to make their shapes different from any you have drawn before. Measure the angles and add them together in each triangle.

Tests on Triangles

1. Look back at the right-angled triangles you drew in the last section numbered 2 (*a*), (*b*), (*c*), and 5 (*a*), (*b*), (*c*). In each triangle call the longest side c, the next longest side a, and the shortest b. Find a^2, b^2, and c^2 for each triangle, and verify in each case that the following statements are true:

$$c^2 = a^2 + b^2$$
$$a^2 = c^2 - b^2$$
$$b^2 = c^2 - a^2$$

2. If in any right-angled triangle
   $c = 26$ and $a = 24$, what is b?
   If $a = 24$ and $b = 18$, was is c?
   If $c = 34$ and $b = 16$, what is a?

3. If in any triangle two angles are 45°, what is the third angle?
   If two angles are 30°, what is the third angle?
   If one angle is 30° and the other is 60°, what is the third angle?
   If one angle is 75° and the other is 15°, what is the third angle?

Things to Memorize

1. In a right-angled triangle, if c is the longest side and a and b are the other two sides:

$$c^2 = a^2 + b^2$$
$$a^2 = c^2 - b^2$$
$$b^2 = c^2 - a^2$$

2. In any triangle, if A, B, C are the angles:

$$A + B + C = 180°$$

CHAPTER 2

THE GRAMMAR OF SIZE, ORDER, AND SHAPE

AT THE quayside of any large international harbour such as Port Said or Hong Kong, the visitor can meet boys of fifteen who chatter intelligibly and fluently in half a dozen languages. Few ocean-borne passengers in transit regard this as a testimony to extraordinary intellectual endowment. It may therefore be soothing for many to whom mathematical expressions evoke a malaise comparable to being seasick, if they can learn to think of mathematics less as an exploit in reasoning than as an exercise in translating an unfamiliar script like Braille or the Morse code. In this chapter we shall therefore abandon the historical approach and deal mainly with two topics: for what sort of communication do we use this highly space-saving – now international – *written* language, and on what sort of signs do we rely. To emphasize that the aim of this chapter is to accustom the reader to approach mathematical rules as exercises in economical translation, every rule in the sign language of mathematics will have an arithmetical illustration, e.g. $3a + b = c$ means, *inter alia*, $c = 7$ if $a = 2$, $b = 1$, etc.

At the outset, we may dismiss the first of the foregoing questions briefly. In contradistinction to common speech which deals largely with the qualities of things, mathematics deals only with matters of size, order, and shape. We may take the last for granted, and defer discussion of what we mean by size and order to a later stage. First, let us consider what different sorts of signs go to the making of a mathematical statement. We may classify these as:

(i) *punctuation*;
(ii) *models*;
(iii) *Labels* (e.g. 5 or x) for *enumeration, measurement, and position in a sequence*;
(iv) signs for *relations*;
(v) signs for *operations*.

Much of this chapter will be familiar to many readers. Some may be news even to the sophisticated. The way to make the best of it is for more advanced readers to skip what has no news value till they reach something that is unfamiliar, and for others to digest carefully whatever has still news value for them.

Mathematical Punctuation. In childhood we learn the alphabet before we learn about punctuation. This has the benediction of antiquity, because punctuation did not get under way till at least four thousand years after mankind first acquired the trick of alphabetic writing. Indeed, the first simple inscriptions did not even leave a gap between words, nor use any device like STOP to separate different sentences, as we do when we draft an overseas cable. The following example shows that the use of some such device may determine whether a sequence of words does or does not make sense:

> King Charles walked and talked half an hour after his head was cut off.

In mathematical communication, it is even more important than in other sorts of writing to enlist devices to make clear what parts of a statement are or are not as separable as the head and body of King Charles. As it stands, for instance, $32 \div 8 + 8$ might mean $(32 \div 8) + 8 = 4 + 8 = 12$ or $32 \div (8 + 8) = 32 \div 16 = 2$. Two devices subserve this use:

(*a*) brackets of three sorts (. . .), [. . .], and { . . . };
(*b*) a line drawn above the connected terms, e.g.:

$$32 \div \overline{8 + 8} = 2 \textit{ and } \overline{32 \div 8} + 8 = 12$$

When it is important to distinguish several pairs or groups of numbers in a single statement (*formula* or *equation*), we may need to distinguish groups within groups. It is then clearer to use different devices chosen from the four mentioned above, e.g.:

$$5\,[32 \div (8 + 8)] = 5(32 \div \overline{8 + 8}), \text{ etc.}$$

In more complicated mathematical sentences, it may be necessary to group 4 or even more groups, so that one is enclosed within another like the layers of an onion. To unravel the sentence, however, we do not peel off the layers as we do when we peel an onion, i.e. starting first with the outermost. We have to start with the innermost, e.g.:

$$6\{20 - 5\,[32 \div (8 + 8)]\} = 6\{20 - 5\,[32 \div 16]\}$$
$$= 6\{20 - 5[2]\} = 6\{20 - 10\} = 60$$

Models. By models we here mean any *pictorial* device used to convey information about our threefold topic. Just as the earliest ingredients of the priestly writing called *hieroglyphic* were pictorial, the earliest tools of mathematical communication were either scale diagrams or the sort of patterns called *figurate* (Figs. 35–8). In spite of the considerable amount of useful (and not so useful) mathematical lore available in the Old World before the beginning of the Christian era, these were the only means available, and remained so until Moslem culture had transmitted to the West a new battery of number signs. As already mentioned and, as will become clearer at a later stage, the latter made it possible to scrap the counting frame and to enlist alphabetic letters into the formulation of simple rules such as the following –

(i) if the lengths of adjacent sides of a rectangle are a and b, its area (A) is the product of a and b, more briefly:

$$A = ab$$

e.g. $a = 7$ *yards*, $b = 3$ *yards*; $A = 21$ *square yards* (Fig. 10)

(ii) if the length of the base of a triangle is b and its vertical height is h, its area (A) is half the product of the two, more briefly:

$$A = \frac{1}{2}bh$$

e.g. $b = 10$ *yards*, $h = 15$ *yards*, $A = 75$ *square yards* (Fig. 11)

However, the mathematicians of antiquity did not rely on the pictorial method merely to frame rules of measurement about figures. They had no other means of envisaging such simple rules about manipulating numbers used for enumeration, as when we now write:

$$\frac{a}{b} \times \frac{c}{d} = \frac{ac}{bd}, \text{ e.g. } \frac{3}{5} \times \frac{4}{7} = \frac{12}{35} \text{ (Fig. 31)}$$

Although the word *geometry* literally means *earth-measurement*, much of the Platonic geometry transmitted to the boys of posterity *via* their posteriors through Euclid's best seller is merely a statement of elementary rules of calculation which we convey by letters (a, b, etc.) for measurements, operations ($+$, or 2 in a^2) and *relations* ($=$). This has left (see Fig. 32) an indelible stamp on the vocabulary of mathematics, as when we use such expressions as the *square of a* for $a \times a = a^2$, the *cube of b* for $b \times b \times b = b^3$, that we call 5 the *square root of* 25, written $\sqrt{25} = 5$ and meaning $5^2 = 25$, that we call 4 the *cube root of* 64, written $\sqrt[3]{64} = 4$ and meaning $4^3 = 64$. Examples of simple rules of computation

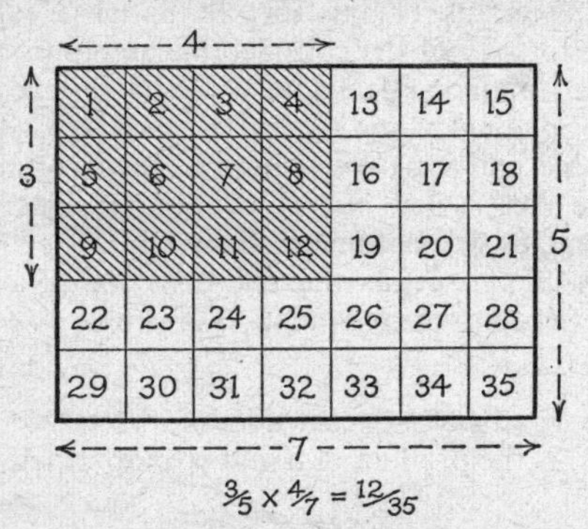

Fig. 31. Multiplication of Fractions

represented by geometrical demonstrations in Euclid's Book II are the following:

(i) $(a + b)^2 = a^2 + 2ab + b^2$ (Fig. 33)
e.g. $(3 + 4)^2 = 49 = 9 + 24 + 16$
(ii) $(a - b)^2 = a^2 - 2ab + b^2$ (Fig. 33)
e.g. $(9 - 4)^2 = 25 = 81 - 72 + 16$
(iii) $(a + b)(a - b) = a^2 - b^2$
$(7 + 4)(7 - 4) = 33 = 49 - 16$ (Fig. 34)

The last (Fig. 34) was almost certainly known to the priestly astronomers of Mesopotamia (Iraq) by 2000 BC, when the temple libraries had extensive tables of squares inscribed in the cuneiform script on clay tablets. It seems likely that they used it to perform in a roundabout way calculations we learn to do more directly when very young, and without realizing how much we owe to the civilization of India and of the Moslem World during the first millennium of the Christian era. The recipe is this. To multiply by 37×25, find the so-called *arithmetic mean* which is half the sum of the factors, i.e. $\frac{1}{2}(37 + 25) = 31$. Then check the result thus: $37 = 31 + 6$ and $25 = 31 - 6$, so that:

$$37 \times 25 = (31 + 6)(31 - 6) = 31^2 - 6^2 = 961 - 36 = 925$$

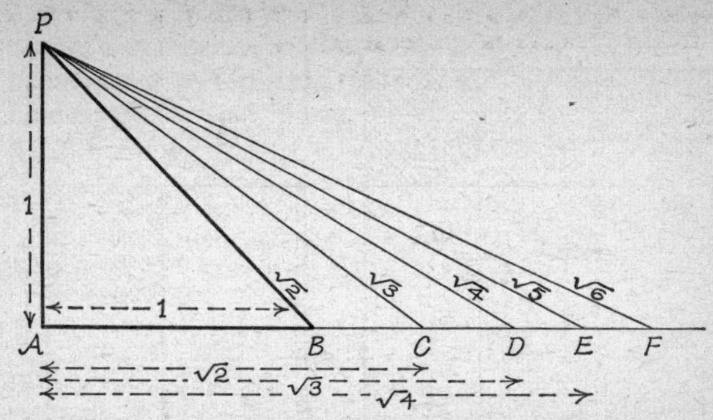

Fig. 32. Geometrical Recipe for Finding a Square Root

This depends on the so-called theorem of Pythagoras (cf. Fig. 17). Thus:

$$1^2 + 1^2 = 2 \textit{ so that } \sqrt{2} = \sqrt{1^2 + 1^2}$$
$$1^2 + (\sqrt{2})^2 = 3, \text{ etc.}$$

Similarly, to multiply 36 by 28 we can proceed thus:

$$\tfrac{1}{2}(36 + 28) = 32;\ 36 \times 28 = (32 + 4)(32 - 4) = 32^2 - 4^2 = 1024 - 16 = 1008$$

In these examples either both factors are odd (as are 37 and 25) or both are even (as are 36 and 28). The rule is a little more complicated, if one is even and one odd, e.g. 37 × 26. If we could multiply only by using clay tablets with squares of whole numbers, we should have to deal with this in one of several ways, such as the following:

$$37(26) = 37\,(25 + 1) = 37(25) + 37 = (31 + 6)\,(31 - 6) + 37$$
$$\text{and } 26(37) = 26\,(36 + 1) = 26(36) + 26 = (31 - 5)\,(31 + 5) + 26$$
$$\text{or } 37(26) = 37\,(27 - 1) = 37(27) - 37 = (32 + 5)\,(32 - 5) - 37$$
$$\text{and } 26(37) = 26\,(38 - 1) = 26(38) - 26 = (32 - 6)\,(32 + 6) - 26$$

As an exercise in translation, we may here pause to disclose why half the sum of two even or of two odd numbers is itself

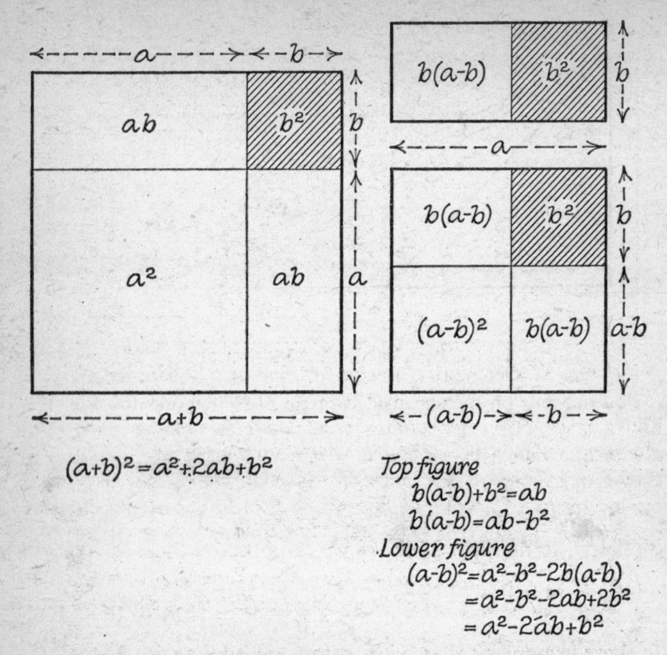

Fig. 33. Geometrical Representation of $(a + b)^2$ and $(a - b)^2$

a whole number, whereas half the sum of an even and an odd is not. If m and n are any whole numbers 1, 2, 3, etc., we can express any even number as $2m$ or $2n$ (i.e. 2, 4, 6, etc.) and any odd number as $2m - 1$ or $2n - 1$ (i.e. 1, 3, 5, etc.). The sum of two even numbers is therefore expressible as $2m + 2n$ whose half is a whole number $m + n$. If both numbers are odd, their sum is expressible as $(2m - 1) + (2n - 1) = (2m + 2n - 2)$ whose half is the whole number $(m + n - 1)$. Contrariwise, the sum of an even and of an odd number is $2n + (2m - 1)$ or $2m + (2n - 1) = 2m + 2n - 1$, whose half is $m + n - \frac{1}{2}$.

The mathematicians of antiquity limped along with the left crutch of the counting frame or with the right crutch of the sort of scale diagram we call a geometrical figure when they aspired to convey simple rules of calculation which we now learn in early childhood. Except by means of geometrical figures they

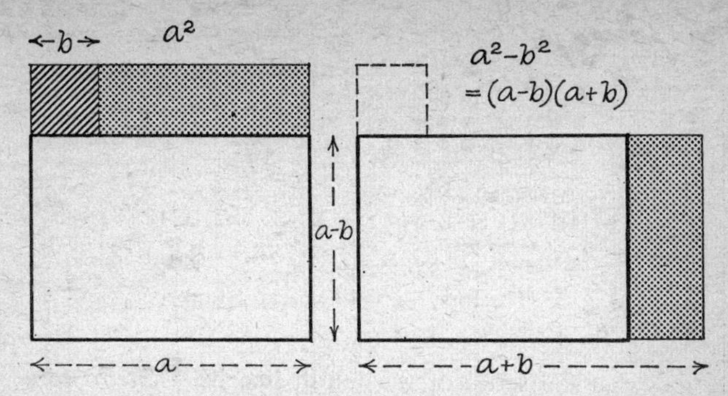

Fig. 34. Geometrical Representation of $a^2 - b^2$

had no way of representing what their successors called *incommensurables* (see below, p. 103). To represent, $\sqrt{2}$ or $\sqrt{5}$, they had to rely on the Pythagorean property (p. 48) of a right angled triangle. If one side $a = 1$ and $b = 1$, $\sqrt{2}$ or $\sqrt{3}$ etc., the diagonal d is such that (Fig. 32):

if $b = 1$: $d^2 = a^2 + b^2 = 1^2 + 1^2 = 2$, *so that* $d = \sqrt{2}$
if $b = \sqrt{2}$: $d^2 = a^2 + b^2 = 1^2 + (\sqrt{2})^2 = 3$, *so that* $d = \sqrt{3}$
if $b = \sqrt{4}$: $d^2 = a^2 + b^2 = 1^2 + (\sqrt{4})^2 = 5$, *so that* $d = \sqrt{5}$

Here we translate $d = \sqrt{2}$, etc. to mean:

(*a*) if d^2 (d multiplied by itself) is 2, we call $\sqrt{2}$ the number d;
(*b*) if d^2 (d multiplied by itself) is 3, we call $\sqrt{3}$ the number d;
(*c*) if d^2 (d multiplied by itself) is 5, we call $\sqrt{5}$ the number d;

From the same figure which exhibits $\sqrt{2}$, $\sqrt{5}$, we see how to fill in $\sqrt{6}$ as:

$$(\sqrt{6})^2 + 1^2 = (\sqrt{7})^2$$

Geometrical figures were not the only pictorial devices with which mathematicians of the pre-Christian era were familiar. Fig. 35 corresponding to the third line below shows a numerical sequence (*series*) which was part of the *mystique* of the Pythagorean brotherhoods. Such were the *triangular numbers*; why so-called will be clear from the picture. Fig. 36 also discloses why

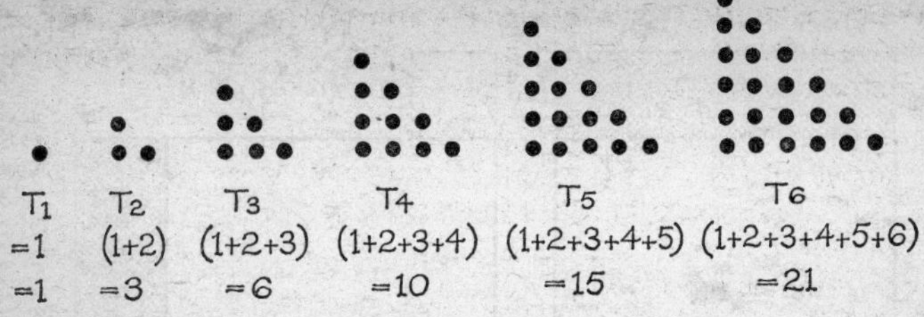

Fig. 35. The First Six Pythagorean Numbers

one may call the series of the fourth line below, *tetrahedral numbers*:

1	1	1	1	1	1	1	1	1	1 . . .
1	2	3	4	5	6	7	8	9	10 . . .
1	3	6	10	15	21	28	36	45	55 . . .
1	4	10	20	35	56	84	120	165	220 . . .

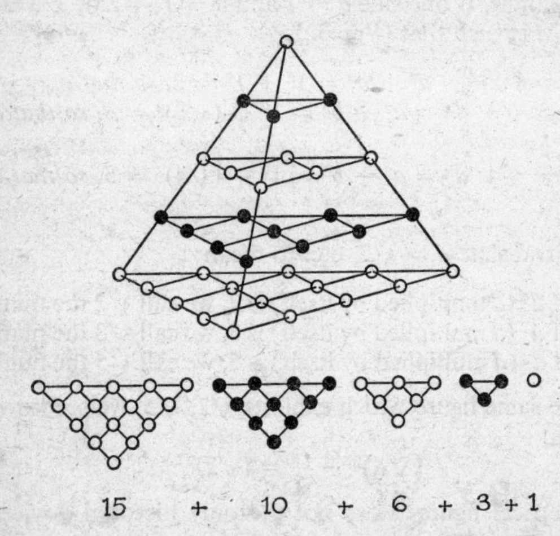

Fig. 36. The Fifth Tetrahedral Number built up from Pythagorean Triangular Numbers as we form the latter from the natural numbers, i.e.:

$1; (1 + 3) = 4; (1 + 3 + 6) = 10; (1 + 3 + 6 + 10) = 20;$
$(1 + 3 + 6 + 10 + 15) = 35$

A brief scrutiny of this table suffices to acquaint us with alternative rules of formation true of any number (so-called term) of a particular row other than the top one:

(i) it is the sum of the number to the left of it and the number above it (e.g. $6 = 5 + 1$; $21 = 15 + 6$; $56 = 35 + 21$);

(ii) it is the sum of all the numbers to the left of it in the row above it and of the number corresponding to it in the same row (e.g. in line 4, $35 = 1 + 3 + 6 + 10 + 15$).

Before we can make a rule which is more *economical*, i.e. both time-saving and space-saving, to generate such numbers without adding term to term, we shall need to label the order in which each term occurs in its sequence. First let us look at another family of series built up from its top row in the same way (see Figs. 37 and 38):

2	2	2	2	2	2	2	2	2	2 . . .
1	3	5	7	9	11	13	15	17	19 . . .
1	4	9	16	25	36	49	64	81	100 . . .
1	5	14	30	55	91	140	204	285	385 . . .

Here the *odd* numbers replace the so-called *natural* numbers (1, 2, 3, 4, etc.) of the previous table in the second row, the *square* numbers replace the triangular ones in the third and the *pyramidal* numbers replace the *tetrahedral* ones in the fourth. A simple formula to calculate the pyramidal numbers effortlessly had a small pay-off during comparatively recent times when armies used cannon balls and stacked them in pyramidal heaps ready for the quarter-master to check the size of the stock.

It may well be that the fascination of the so-called triangular numbers for an early secret society of mathematicians had some tie-up with what land surveyors call *triangulation*. One can dissect (Fig. 12) any *rectilinear* figure, i.e. one whose boundary is made up of straight lines, into triangles whose area is measurable. In the same way we can split into triangular numbers any sequence of numbers pictured by flat figurate patterns as, for instance:

Squares	Triangles
1	1
4	3 + 1
9	6 + 3
16	10 + 6
25	15 + 10
36	21 + 15
etc.	etc.

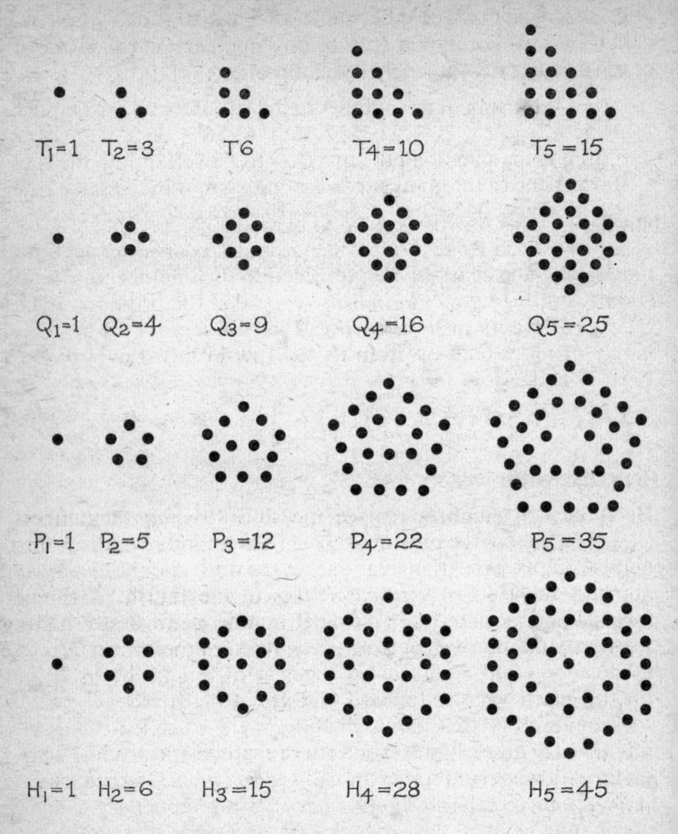

Fig. 37. A Family of Plane Figurates built up like the Triangular Numbers

(cf. Fig. 38 for the squares Q_r)

Before we take leave of figurate numbers for the time being, let us look at the second line (natural and odd numbers) of the last two tables. We may there label the *sum* of the first 6 *natural* numbers by the symbol S_6, so that:

$$S_6 = 1 + 2 + 3 + 4 + 5 + 6$$
$$S_6 = 6 + 5 + 4 + 3 + 2 + 1$$
$$2 \,.\, S_6 = (6 + 1) + (5 + 2) + (4 + 3) + (3 + 4) + (2 + 5) + (1 + 6)$$
$$= 7 + 7 + 7 + 7 + 7 + 7$$
$$2 \,.\, S_6 = 6\,(7) \text{ and } S_6 = 3\,(7) = \frac{6}{2}(1 + 6)$$

Now let us write for the first 8 odd numbers Z_8, so that:

$$Z_8 = 1 + 3 + 5 + 7 + 9 + 11 + 13 + 15$$
$$\therefore Z_8 = 15 + 13 + 11 + 9 + 7 + 5 + 3 + 1$$
$$\therefore 2 \,.\, Z_8 = (15 + 1) + (13 + 3) + (11 + 5) + (9 + 7) + (7 + 9) + (5 + 11) + (3 + 13) + (1 + 15)$$
$$= 16 + 16 + 16 + 16 + 16 + 16 + 16 + 16$$
$$\therefore 2 \,.\, Z_8 = 8\,(16) \text{ and } Z_8 = 4\,(16) = \frac{8}{2}(1 + 15) = 8^2$$

Here, the reader will see that:

(i) in each sequence, any number after the first differs from its predecessor by an equal amount: 1 in the case of the *natural*; 2 in the case of the *odd* numbers;

(ii) if we denote by f the first (here 1 in each case) and by l the last (here 6 or 15) S_n of the first n numbers is obtainable without adding term to term by recourse to the formula:

$$S = \frac{n}{2}(f + l)$$

If we had to put the last formula in words, as would Archimedes or Omar Khayyám, the equivalent would be somewhat like the following assertion:

to get the sum of an afore-mentioned sequence of terms from first to last inclusive we multiply one half their number by the sum of the first and the last of them.

Thus our formula clearly takes up much less space on paper than any suitable form of words. Also, it shows us how to save time. We can now write down promptly the sum of the first hundred natural numbers as:

$$\frac{100}{2}(1 + 100) = 50\,(101) = 5{,}050$$

The same formula is not so convenient for finding the sum of the first 100 odd numbers, but the third line of our second table suggests that (and Fig. 38 shows why) it is $(100)^2 = 10{,}000$; and a stop-watch is not necessary to vindicate once more the claim that the language of mathematics conveys rules which are *time-saving* as well as *space-saving*. For the record, we call a sequence built up by adding the same number d to each member (*term*) in turn an *arithmetic series*. Thus the following are seven successive terms of an arithmetic series in which $d = 5$:

$$-8, -3, 2, 7, 12, 17, 22$$

According to our formula above (as the reader can check):

$$S_n = \frac{7}{2}(-8 + 22) = \frac{7\,(14)}{2} = 49$$

A third kind of pictorial language used in mathematics has its origin in the construction of star maps and later of earth maps in the last two centuries before the beginning of the Christian era. Such is co-ordinate geometry for which the slang word is *graphs*. It did not come into its own before the century of Newton, whereafter it led to many discoveries. Unlike Euclid's geometry, it can bring the measurement of *time* into the picture. For instance, it exposes (Fig. 1) why and when Achilles (pp. 12–14) caught up with (and passed) the tortoise.

Numbers. When we have discussed what we mean by *size*, *order*, and *operations*, there will be much more to mention about the use of numbers and different ways in which we distinguish them. Here we shall distinguish between two sorts, each of which crops up in all three domains. We may liken one to *proper* nouns, e.g. Napoleon the Great or Joan of Arc. Such are: 0·7 and 3,264. We may liken the other to *collective* nouns, e.g. emperors and saints. Such are a or b, N or M, x or y. However, we must not think of a letter of the alphabet necessarily as a collective noun. In antiquity, Jews, Greeks, and Romans used them as proper nouns, e.g. $L = 50$, $C = 100$, $D = 500$, $M = 1{,}000$ in the Roman system. Moreover, and for reasons which will transpire, we still use e and the Greek letter π for numbers expressible by infinite series (p. 14) employing no number signs other than the ten so-called Hindu–Arabic *cyphers*, i.e. 0, 1, 2, 3, . . . 9.

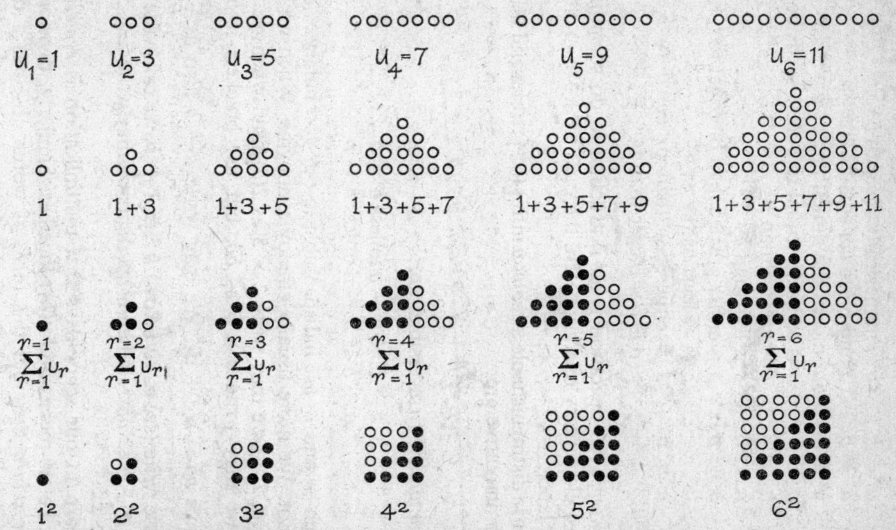

$u_1=1$	$u_2=3$	$u_3=5$	$u_4=7$	$u_5=9$	$u_6=11$
1	$1+3$	$1+3+5$	$1+3+5+7$	$1+3+5+7+9$	$1+3+5+7+9+11$
$\sum_{r=1}^{r=1} u_r$	$\sum_{r=1}^{r=2} u_r$	$\sum_{r=1}^{r=3} u_r$	$\sum_{r=1}^{r=4} u_r$	$\sum_{r=1}^{r=5} u_r$	$\sum_{r=1}^{r=6} u_r$
1^2	2^2	3^2	4^2	5^2	6^2

Fig. 38. Sum of the Squares of the first n Odd Numbers

Relations. Instead of attempting a verbal definition of what one means by relations in general, a short glossary will tell the reader all the signs we shall use for relations in this book:

$a = b$	a is numerically *equal* to b
$a \neq b$	a and b are *not* equal
$a^2 \equiv a \times a$	a^2 means the same thing as $a \times a$
$a \simeq b$	a is as nearly equal to b as it needs be
$a > b$	a is greater than b
$a < b$	a is less than b
$a \geqslant b$	a is either greater than or equal to b, i.e. a is at least as big as b
$a \leqslant b$	a is either less than or equal to b, i.e. a is not bigger than b

Operations. An operation is an instruction to do something. The most elementary operations are the ones we learn in childhood to label: $+$, $-$, $\times$, $\div$.

They are distinguishable as *commutative* (reversible) and otherwise. Commutative are:

$$a + b \equiv b + a \qquad a \times b \equiv b \times a$$

Non-commutative (irreversible) are:

$$a - b \neq b - a \text{ unless } b = a$$
$$a \div b \neq b \div a \text{ unless } b = a$$

Though $+$ and $-$ are indispensable to our vocabulary $\times$ and $\div$ are not. We more usually employ synonyms. When we use digits (e.g. 2, 5) we may write 2×5 as 2(5); and when we use letters, we may write $a \times b$ as ab, but use brackets if $b = c + d$ thus $ab = a(c + d)$. We may write $a \div b$ also as $\frac{a}{b}$ or $a : b$. If we write (see p. 90 below) $b^{-1} \equiv 1 \div b$, we can write on one line $a \div b \equiv ab^{-1}$. When using digits we may write 2×5 in the form 2.5.*

We learn in our schooldays that multiplication is *repetitive addition*, i.e. ab means add a(bs) together. Similarly, division is *repetitive subtraction*; but this form of words obscures a difference. If a and b are whole numbers (*integers*) ab is a whole number, e.g. $28 \times 3 = 84$. In short, the operation comes to an end; but the operation ab^{-1} may lead to no exact result in this

* See Footnote on p. 186 of Chapter 4.

sense. Thus 28 and 3 are integers, but $28 \div 3 \equiv 28 . 3^{-1} = 9.3$ is not an integer.

Just as there is a relation between addition and multiplication or between subtraction and division, there is an instructive relation between addition and subtraction and between multiplication and division. One conveys it by saying that one member of the pair is the *inverse* operation. In the sign language of mathematics we may express this as below:

(i) $(a + b) = c \equiv a = (c - b) \equiv b = (c - a)$

$$\text{e.g. } (3 + 4) = 7 \equiv 3 = (7 - 4)$$

(ii) $ab = c \equiv a = c \div a$

$$\text{e.g. } 3 . 4 = 12 \equiv 3 = 12 \div 4$$

From the last, we get the *diagonal rule* for cross multiplication of fractions:

$$\frac{a}{b} = \frac{c}{d} \equiv ad = bc$$

$$\text{e.g. } \frac{5}{3} = \frac{35}{21} \equiv 5(21) = 3(35)$$

When dealing with statements set out as equations, i.e. linked by the relation $=$, we can add to, subtract from or multiply by the same number both sides without making the equality sign invalid, i.e.:

$$a = b \equiv (a + c) = (b + c) \equiv (a - c) = (b - c) \equiv ac = bc$$

The above are true even if $c = 0$, so that for any number N including $N = 0$:

$$N + 0 = N \text{ and } N(0) = 0$$

The corresponding rule of *cancellation*, involving the operation of division, is true only if c is not zero, i.e.:

$$\frac{a}{b} = \frac{ac}{bc} \text{ if } c \neq 0,$$

$$\text{e.g. } \frac{5}{7} = \frac{15}{21} = \frac{3 . 5}{3 . 7}$$

This restriction is necessary because $0 \div 0$ is not a nameable entity. Corresponding to the statement $a(b + c) \equiv ab + ac$, we may say (if $a \neq 0$) that $(b + c) \div a \equiv (b \div a) + (c \div a)$, whence by using the rule of cancellation, we get the *addition and subtraction rules for division*:

$$\frac{a}{c} + \frac{b}{d} \equiv \frac{ad}{cd} + \frac{bc}{cd} \equiv \frac{ad + bc}{cd}$$

$$\frac{a}{c} - \frac{b}{d} \equiv \frac{ad}{cd} - \frac{bc}{cd} \equiv \frac{ad - bc}{cd}$$

The multiplication rule for fractions is as follows. On the understanding that neither b nor d is zero, we put $a = pb$ and $c = qd$, so that by the rule of cancellation:

$$\left(\frac{a}{b}\right)\left(\frac{c}{d}\right) = pq = \frac{pq \,.\, bd}{bd} = \frac{pb \,.\, qd}{bd} = \frac{ac}{bd}$$

At an early stage in this book, we shall need a sign for an operation which we can perform on a sequence of numbers connected by a *rule of succession*, as are 1, 4, 7, 10, 13, . . . , or 1, 3, 9, 27, 81, . . .; but we cannot fully see the point of it until (p. 105) we are also clear about the use of numbers to label the order in which the items (so-called *terms*) of such sequences occur.

First, let us notice that one uses, in many branches of mathematics, digits or letters which stand for them, to indicate how often we are to perform an operation. For instance, we write 3^4 for $3 \times 3 \times 3 \times 3$; and most of us think of this as four threes multiplied together. It is better to think of it as: *performing the operation of multiplying unity by* 3 *four times*, i.e.

$$3^4 = 1 \times 3 \times 3 \times 3 \times 3;\ 3^3 = 1 \times 3 \times 3 \times 3;\ 3^2 = 1 \times 3 \times 3, \textit{ so that } 3^1 = 1 \times 3 (= 3) \textit{ and } 3^0 = 1$$

(one multiplied by 3 *no* times). A common convention in higher mathematics is to indicate the so-called *inverse* operation by a minus sign, so that 3^{-4} represents *dividing* unity by 3 four times. We shall later find that this way of interpreting the so-called powers of a number is very instructive in more than one context. To remind ourselves let us make an abridged table as below.

$$n^1 = 1 \,.\, n \qquad n^{-1} = 1 \,.\, \frac{1}{n}$$

$$n^2 = 1 \,.\, n \,.\, n \qquad n^{-2} = 1 \,.\, \frac{1}{n} \,.\, \frac{1}{n}$$

$$n^3 = 1 \,.\, n \,.\, n \,.\, n \qquad n^{-3} = 1 \,.\, \frac{1}{n} \,.\, \frac{1}{n} \,.\, \frac{1}{n}$$

$$n^4 = 1 \,.\, n \,.\, n \,.\, n \,.\, n \qquad n^{-4} = 1 \,.\, \frac{1}{n} \,.\, \frac{1}{n} \,.\, \frac{1}{n} \,.\, \frac{1}{n}$$

$$n^5 = 1 . n . n . n . n . n \qquad n^{-5} = 1 . \frac{1}{n} . \frac{1}{n} . \frac{1}{n} . \frac{1}{n} . \frac{1}{n}$$

$$n^0 = 1$$

The use of $-a$ to label an operation in n^{-a} is consistent with our notion of subtraction in so far as we subtract -1 from a when we divide n^a by n, so that $n^a \div n = n^{a-1}$, e.g.:

$$100{,}000 \div 100 = 10^5 \div 10^2 = 10^3 = 1000$$

However, this would have been very difficult at a time when the use of the abacus, with or without the aid of pictorial models, circumscribed the art of calculation for all practical purposes. It is therefore fitting that we should here pause to consider more carefully what we mean by subtraction.

The use of the Minus Sign. When we write $a - b$, meaning subtract b from a, we can visualize the operation denoted by the minus sign either as removing b balls from an abacus set of a or as laying off b units along a line of length a units, so that the segment is of length $(a - b)$ units. In which way we do so, we have to assume two things:

(i) $a > 0$, so that, in what we shall later regard as a somewhat misleading description, a is a so-called *positive* number;

(ii) $a \geqslant b$, i.e. a is greater than b, or in the limiting case $a = b$ and $(a - b) = 0$.

This primitive use of the minus sign for the operation of subtraction implies what we call *balancing an equation*, i.e. changing $+$ to $-$, or vice versa when we transfer an item from one side to the other. One of the axioms of Euclid's geometry is the self-evident statement that if a and c are equal, $a + n = c + n$, e.g. if $a = 3 = c$ and $n = 5$, $a + n = 8 = c + n$. Similarly, $a = c$ means the same thing as $a - n = c - n$, e.g. if $a = 7 = c$ and $n = 4$, $(a - n) = 3 = (c - n)$. If we combine the last two statements, we see that:

$$(a - b) = c \equiv (a - b) + b = c + b \text{ so that } (a - b) = c \equiv a = c + b$$

Similarly,

$$(a + b) = c, \text{ if } a = (c - b)$$

Though we can picture $(a - b)$ as a segment of a line or as an operation on the abacus only if $a \geqslant b$, we can explore what it might otherwise mean by considering a double entry account in which a and b are entries respectively on the credit and debit side, so that if $a < b$ there is a deficit. Numerically, we can call the deficit c, meaning $a + c = b$ or $a = b - c$. It does not affect the deficit or credit balance, if we add equal amounts to both sides so that:

$$(a - b) = (b - c) - b = -c$$

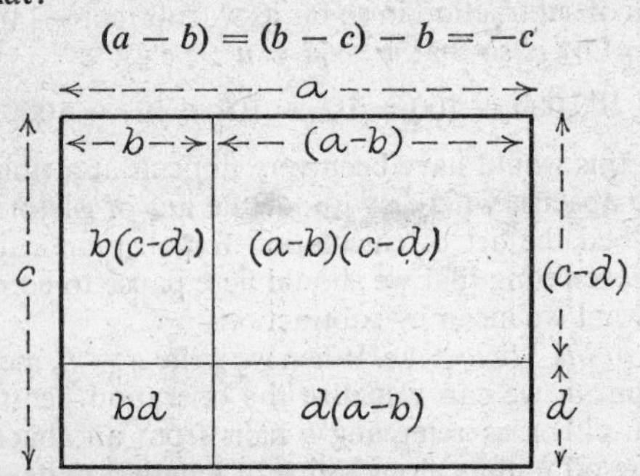

Fig. 39. Geometric Representation of the Law of Signs

On the assumption that $a > b$ and $c > d$:

$$(a - b)(c - d) + bd + b(c - d) + d(a - b) = ac$$
$$(a - b)(c - d) + bd + bc - bd + ad - bd = ac$$
$$(a - b)(c - d) + bc + ad - bd = ac$$

Take away bc and ad from each side and add bd to each, then

$$(a - b)(c - d) = ac - ad - bc + bd$$

This gives an intelligible meaning to $-c$, e.g. -10 dollars mean 10 dollars in the red; and this is quite consistent with the use of the minus sign on the thermometer scale. If we say that the air temperature dropped during the night from 5° by 8° at dawn, we mean that the dawn temperature was $(5 - 8) = -3°$. However, neither temperature scale nor double entry book-keeping necessarily confers an intelligible meaning on what we learn at school as the *rules of signs*:

$$(+a) \times (+b) \equiv +ab$$
$$(+a) \times (-b) \text{ or } (-a) \times (+b) \equiv -ab$$
$$(-a) \times (-b) \equiv +ab$$

The rules for division are consistent with this, i.e.:

$$(+a) \div (+b) \equiv +(a \div b)$$
$$(+a) \div (-b) \text{ or } (-a) \div (+b) \equiv -(a \div b)$$
$$(-a) \div (-b) \equiv +(a \div b)$$

So long as $(a + b)$, $(a - b)$, $(c + d)$, and $(c - d)$ are each greater than zero we can represent them by line segments and make diagrams in which all the products (Fig. 39) refer to rectangular areas, as below:

$$(a + b)(c + d) = ac + ad + bc + bd$$
$$(a + b)(c - d) = ac - ad + bc - bd$$
$$(a - b)(c + d) = ac + ad - bc - bd$$
$$(a - b)(c - d) = ac - ad - bc + bd$$

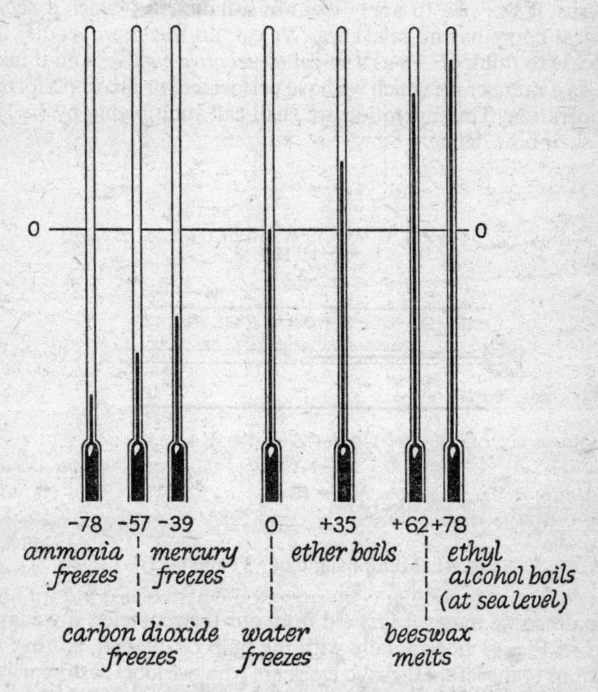

Fig. 40. So-called Negative Numbers in the World's Work

However, diagrams of this sort are relevant to the rules of signs only with the reservation stated. They cannot justify as an isolated statement $(-a) \times (-b) = ab$. Nor does the analogy of making a balance sheet or thermometer scale help us to do so.

Clearly we cannot meaningfully speak of multiplying one overdraft by another, an overdraft by a credit balance or one credit balance by another. Nor can we speak intelligibly of multiplying 5° below zero by 8° above. We can merely talk of multiplying the overdraft or credit balance, or readings on the negative or positive part of the temperature scale by a number which is *signless*. For we know what we mean by writing $7(-3°) = -27°$; but we do not obviously mean anything by writing $(-3°)\ (-7°)$. In short, we have to seek some other way of interpreting our laws of signs, if we wish to apply them in situations to which a geometrical figure has no relevance. We can do this more readily if we cease to think of $-n$ as a so-called *negative number*, and think of it as a number on which we have performed, or are to perform an operation. This operation we shall call multiplying by (-1) and shall now define.

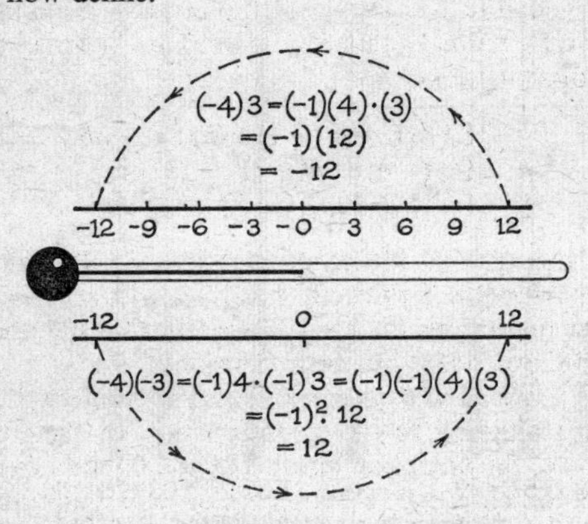

Fig. 41. Multiplying by (-1) and $(-1)^2$

To do so we may get first aid from our thermometer, if we lay it, as in Fig. 41 horizontally with the bulb on the left, so that t divisions to the left of the zero mark are what we label customarily as $-t°$. We can then picture $-t$ as the result of an *anti-clockwise* rotation of t scale divisions in the positive part of the scale through an angle of 180°; and if we do this twice we get back to t divisions in the positive domain.* We can then think of $-t$ as

* In this context, the direction clockwise or anti-clockwise is irrelevant. It is however relevant when we seek a meaning for $\sqrt{-1}$.

both a number t and an operation (-1) performed (or to perform) on it. We can write this as $(-1)t$; and can speak figuratively of the operation as multiplying by (-1). If we perform this operation twice, we may write the double operation $(-1)(-1)$ as $(-1)^2$, and this corresponds to a rotation through 360°, so that

$$-t \equiv (-1) \,.\, t \quad \textit{and} \quad (-1)^2 \,.\, t \equiv t$$

Now there is no objection to writing $n(-t) = -nt$, meaning lay off t scale divisions successively n times in the negative part of the scale; but we shall now regard $(-n) \,.\, t$, in which t means t divisions in the other part, as the double operation of multiplying first by n, then by (-1), i.e. $(-1)(n \,.\, t)$. We can do this in either order, i.e. *either* first lay off t scale divisions successively n times in the positive part then rotate through 180°, *or* first rotate through 180° t divisions in the positive domain then lay off successively n times in their new domain. We may thus write our rules of signs as below:

$$(-a)(b) \equiv (-1)\,a\,(b) \equiv -ab$$
$$(a)(-b) \equiv a(-1)b \equiv -ab$$
$$(-a)(-b) \equiv (-1)^2\,ab \equiv ab$$

We shall later see that the mysterious $\mathrm{i} \equiv \sqrt{-1}$, which mathematicians of Newton's time still regarded with misgivings, need be no more mysterious than -1, if we regard it like the latter as an instruction to rotate our scale divisions.

Division, Ratio, and Proportion. To some readers, it may well have been reassuring to have confirmation of their suspicions that the use of the operation labelled by the minus sign may have a meaning which Euclid's geometry or manipulations on the abacus cannot convey. Let us now consider the non-commutative operation we call *division.* Today, it seems to be immaterial to distinguish between $27 \div 4$, the ratio $27 : 4$ and the so-called improper fraction $\frac{27}{4}$. In short, it matters to us as little as whether we write $a \times b$ as ab or $a(b)$. This was by no means so in antiquity. One can divide on the abacus one whole number a exactly by another whole number b, only if a is the product of b and a third whole number m.

In practice today, we now regard statements about numbers as exact, *if we can state exact limits within which they lie*; and our decimal notation for fractions enables us to do this for division involving two numbers neither of which is exact in Plato's sense.

Consider the meaning of $a \div b$ when a and b are the following numbers:

$$1{\cdot}125 < a < 1{\cdot}126; \quad 2{\cdot}666 < b < 2{\cdot}667$$

In this situation $a \div b$ may mean:

$$1{\cdot}125 \div 2{\cdot}666;\ 1{\cdot}125 \div 2{\cdot}667;\ 1{\cdot}126 \div 2{\cdot}666;\ 1{\cdot}126 \div 2{\cdot}667$$

Of these the smallest fraction is one with the largest possible denominator (2·667) and the smallest possible numerator (1·125). The largest fraction is the one with the largest possible numerator (1·126) and the smallest possible denominator (2·666), so that

$$\frac{1{\cdot}125}{2{\cdot}667} < \frac{a}{b} < \frac{1{\cdot}126}{2{\cdot}666}$$

In short, a $\div$ b lies between 0·4218 and 0·4223. This is as exact a type of statement as we can ever hope to make about the ratio of two *measurements*.

If we are to probe more deeply into what we mean by a ratio, it is worth while at the outset to be clear about what we mean by *proportion*. In common speech, we use the word proportion in two different senses. When we say that the proportion of youths in a co-educational college is 60 per cent we mean that the number of one part of a collection, i.e. *male students*, expressed as a fraction of the whole collection, i.e. *all students*, is three-fifths; but when we say that petrol consumption is *directly proportional* to distance covered we are not expressing one part of a collection as a fraction of another. We are expressing a relation between measurements in different systems of units, e.g. gallons and miles. The statement implies the assumption that we can pair off *corresponding* values in the two sets, so that m_a (e.g. 70) miles correspond to a consumption of g_a (e.g. 2) gallons and that m_b (e.g. 175) miles correspond to a consumption g_b (e.g. 5) gallons. The word *directly* implies, in addition, that every value of the set m is a fixed multiple (called the *proportionality constant*) of the corresponding value in the g set. For instance, our consumption may be 35 miles to the gallon, in which event $m = 35\ g$ and the constant of proportionality is 35.

If we say that x is directly proportional to y when $x = Cy$ for corresponding pairs x, y, the numerical value of the constant C depends on the particular system of units (e.g. *miles* and *gallons* or *kilometres* and *litres*) employed. We can, however, express what we mean in more general terms without reference to a numerical constant which ties us down to particular units. Since we may write $m_a = 35g_a$ and $m_b = 35g_b$

$$\frac{m_a}{m_b} = \frac{35\ g_a}{35\ g_b}, \textit{ so that } \frac{m_a}{m_b} = \frac{g_a}{g_b}$$

In other words, we say that one quantity (q) is *directly proportional* to another (Q), if the *ratio* of any pair ($q_1 : q_2$) of values the first may have is numerically equal to the ratio of the corresponding pair ($Q_1 : Q_2$) of the second. In thus stating what we mean, both quantities on the same side of the equation are in the same system of units, here either both in miles or both in gallons. So long as we stick to one or the other, such a definition does not restrict us to what units we choose, e.g. whether we measure distance in miles or kilometres and consumption in gallons or litres.

The volume (v) of air in a tyre depends on the pressure (p) in a different way. We can convey it by saying that the product pv for corresponding values of pressure and volume is a fixed number K for the particular units (e.g. cubic feet and pounds per square foot) employed. We then speak of pressure and volume as *inversely* (in contradistinction to *directly*) proportional for the following reason which illustrates the diagonal rule. For corresponding pairs of each we may write:

$$p_a \,.\, v_a = K = p_b \,.\, v_b, \textit{ so that } \frac{p_a}{p_b} = \frac{v_b}{v_a}$$

Here the ratio of any pair of values of pressure is equal to the *reciprocal* of the ratio of corresponding values of volume; and this statement of what we mean by inverse proportion is independent of the units we use if we use them consistently. Notice *en passant*, the following useful convention. In the formula connecting two so-called *variables* (x and y), which we can pair off like miles and gallons or pressure and volume, it is customary to express by *capital letters* what quantities (*Constants*) have a fixed numerical value if we use the same system of units, as when we write for direct proportion $x = Cy$ or for inverse proportion $xy = K$.

The two foregoing ways of defining what we mean by *direct proportion* disclose the gist of the crisis which led to the divorce of Greek geometry from the world's work. For reasons which we shall examine later, perfectionists inspired by Plato's teaching refused to countenance the notion that the ratio of two so-called *incommensurables* (e.g. $\sqrt{2}$ or $\sqrt{5}$) could be an honest-to-God number; but they were not squeamish about saying that the ratio of one pair could be equal to that of another. In accordance with this doctrine, elaborated by Plato's pupil Eudoxus, and taken over lock, stock, and barrel by Euclid, one could say of two circles that the ratios of their boundaries ($p_1 : p_2$) and the ratios of

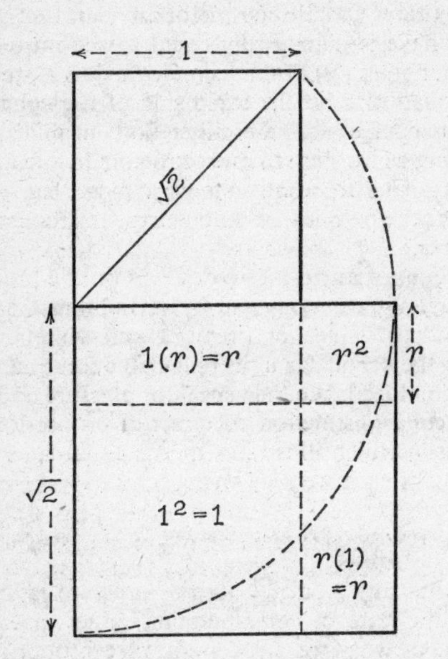

Fig. 42. Geometrical Representation of $(1 + r)^2 = 2$

their corresponding diameters ($d_1 : d_2$) were equal. Contrariwise, Euclid would not concede that the notion of a ratio is reducible to the operation of division. We need not here dwell on so subtle a distinction. What may well concern us more is that a rule expressed in this way does not give us any recipe for computing the length of the boundary of a circle whose diameter we already know. If we wish to do this we have to bring the *proportionality constant* into the picture. Since *perimeter* is the Greek equivalent of *boundary-measurement* and π is the Greek equivalent of Roman *P*, it has long been customary to call this constant π (see p. 60). So an alternative statement of the rule is $p = \pi d$. To say that Archimedes and his Alexandrian successors reinstated measurement in the arid wilderness of Euclid's system, i.e. put back *number* therein, means that Archimedes recognized that we can calculate to *any required level of precision* a proportionality constant which makes this formula meet all the requirements of imperfect people with imperfect measuring instruments.

Two Sorts of Size. To say that mathematics is the language of size, order, and shape signifies that we know what we mean by the last three nouns. We may here take *shape* for granted; but the words *size* and *order* each invite cross examination. By size, we may mean the outcome of either *counting* or *measuring*. The distinction is vital. The result of counting the sheep in a field is unique, if correct. The result of measuring the side of the field is not. If we call the number of sheep n_s, the statement, $n_s = 25$ is either correct or false. If we label the length one side of the field (in yards) as L_f, the statement $L_f = 25$ cannot be wholly correct; and is not indeed a precise statement of any observation we could make.

We have already (p. 40) recognized this distinction; but it is worthy of second thoughts in terms of what one may call a *matching* process. When we know that there are 25 sheep in the field, we can imagine them in a row and place a turnip beside each of them. If no more turnips remain when we have done so, we know that there are also 25 turnips in the field. The matching process involved in measuring the length of its size is not like this. Although the scale divisions of our yardstick may be spaced as equally as any manufacturer can make them, no manufacturer can guarantee perfect equality of gaps and perfectly equal thickness of the marks which bound them. Aside from the margin of uncertainty inherent in the limitations of any manufacturing process, measurement also involves a handicap inherent in the limitations of our own *powers of observation*.

Consider the measurement of the length (L) of a piece of foolscap on which I am writing at the moment. I take a foot rule whose scale divisions have a gap (as nearly as possible) of one-tenth of an inch. I place the beginning of the scale (as nearly as possible) in line with one edge and find that the opposite extremity is nearer to the 109th scale division beyond the initial (zero) mark than to the 108th or 110th. Accordingly, I may say in common parlance that its length is 10·9 inches; but it would not be strictly true to write $L = 10{\cdot}9$. What I can legitimately convey is that it lies *either* in the half interval between 10·85 inches and 10·9 inches *or* in the half interval between 10·9 inches and 10·95 inches. In fact, I may mean either of the first two sentences below combined in the third:

(i) $10{\cdot}85 < L < 10{\cdot}9$; (ii) $10{\cdot}9 < L < 10{\cdot}95$;
(iii) $10{\cdot}85 < L < 10{\cdot}95$.

In more general terms, if d is the *size* (itself an approximation) of the gap between scale divisions, nd the nearest match, and L a length, what we mean when we loosely say $L = nd$ is at best one of the three following:

$$(n - \tfrac{1}{2})d < L < nd; \quad nd < L < (n + \tfrac{1}{2})d;$$
$$(n - \tfrac{1}{2})d < L < (n + \tfrac{1}{2})d$$

Allied to this important practical distinction between how we use numbers as labels for counting and measuring is another, as when we label them *rational* or otherwise. One calls numbers *rational*, if one can express them as the ratio between two *nameable* whole numbers (so-called *integers*), i.e. the numbers we use when we count things. Thus any of the following is rational:

$$\frac{51}{1}(= 51); \quad \frac{52}{4}(= 13); \quad \frac{53}{8}(= 6{\cdot}625); \quad \frac{51}{13} = 3{\cdot}\dot{9}2307\dot{6}$$

As we learned in our schooldays, the last of the four above means:

3·923076 923076 923076 923076 *and so on endlessly*

In short, so-called rational numbers include the *integers* (like the first of the four above) and *vulgar fractions* of which we may distinguish three sorts:

(i) reducible to an integer;
(ii) expressible as a *terminating* decimal fraction;
(iii) expressible by a *non-terminating* fraction of which one or more figures repeat themselves *periodically without end.*

The so-called period (repeating portion) of (iii) may involve one figure, a sequence of two figures, or a sequence of any finite (i.e. nameable) whole number of figures as below:

$$3{\cdot}1\dot{8} = \frac{287}{90}; \quad 3{\cdot}\dot{1}\dot{8} = \frac{315}{99}; \quad 3{\cdot}12\dot{8}3\dot{9} = \frac{312527}{99999}$$

Our Hindu heritage of decimal fractions makes it clear that $3{\cdot}12\dot{3}3\dot{9}$ . . . could never exceed 3·1284 as the result of adding more cyphers after 9 and could attain the value last named only if every subsequent cypher added end without end happened to be 9. Furthermore, $3{\cdot}12\dot{8}3\dot{9}\ldots$ can exactly equal 3·12839 only if all the figures following endlessly are zeros. In the shorthand of mathematics:

$$3{\cdot}12840 \geqslant 3{\cdot}12\dot{8}3\dot{9}\ldots \geqslant 3{\cdot}12839$$

We can therefore conceive the existence of numbers which we can *locate between limits as small as we like without any prospect of defining them exactly*. We may speak of them as *non-terminating* decimal fractions which are *non-periodic* and in traditional jargon as *non-rational*.

One cannot construct a machine which can produce an endless sequence of digits. From the point of view of the computer, the distinction between such non-rational numbers is therefore immaterial. It is, however, one of great historical significance. It bogged down the pioneers of Greek geometry at a time when the notion of a *limit*, e.g. that a series (pp. 12–14) can *choke off*, a simple rule of thumb for division without recourse to a counting frame was as alien to Greek number lore as our familiar device of representing fractions in decimal notation. The crisis which shook the first secular mathematicians of the ancient world was the discovery that the square root of 2 is not expressible in terms of the ratio of any pair of finite (*namable*) numbers. How they hit on this will come later, first let us look at the issue in a practical way. We know that the square ($1{\cdot}5 \times 1{\cdot}5$) of 1·5 is 2·25 and that the square ($1{\cdot}4 \times 1{\cdot}4$) of 1·4 is 1·96. So the number which yields 2 when squared lies between 1·5 and 1·4, i.e. $1{\cdot}4 < \sqrt{2} < 1{\cdot}5$. Proceeding by trial and error the hard way, we find:

$$1{\cdot}41 < \sqrt{2} < 1{\cdot}42;\quad 1{\cdot}410 < \sqrt{2} < 1{\cdot}415;\quad 1{\cdot}414 < \sqrt{2} < 1{\cdot}415 \text{ etc.}$$

The mere fact that we can go on, till we are too tired to continue, without pinning down $\sqrt{2}$ as a recurring decimal fraction does not prove that we could never do so. Let us then look at it from the viewpoint of Dr Plato's select Academy for Young Gentlemen. Their dilemma arose in this way. By application of the earliest recipe for making a set square (see p. 48), we can make an oblong whose sides are as accurately as possible (in Biblical units) 3 and 4 cubits with a diagonal (as accurately as possible) of 5 cubits. If we apply (Figs. 32 and 42) the same rule to the construction of a square whose sides are 1 cubit long, the diagonal (d) should be $\sqrt{2}$ cubits, since $d^2 = 1^2 + 1^2 = 2$. Greek arithmetic had contrived no way of giving precision to the smallest convenient limits within which $\sqrt{2}$ lies.

The fact that we cannot express the square root of 2 in the simplest fractional form, i.e. as a ratio of two *namable* whole numbers, is not so easy to see as some pundits assume. So we shall look at the Greek dilemma in easy stages. First, we have to reckon with the fact that any such ratio, if not already reduced to that of two integers (i.e. whole numbers) without a single common factor, is reducible to that form, e.g.:

$$\frac{33}{24} = \frac{11}{8}; \quad \frac{50}{35} = \frac{10}{7}; \quad \frac{119}{77} = \frac{17}{11}$$

When so reduced, it cannot be the ratio of two even numbers, all of which have two as a factor. Since (p. 167) ancient writers thought of them respectively as *male* and *female*, we shall call M and m odd and F even numbers. Our reduced ratio can therefore be one of only three sorts, here illustrated by ratios whose square is near to two:

$$\frac{M}{F}, \text{ e.g. } \frac{11}{8}; \quad \frac{F}{M}, \text{ e.g. } \frac{10}{7}; \quad \frac{M}{m}, \text{ e.g. } \frac{17}{11}$$

Let us examine each possibility on its own merits:

(i) If the first ($M : F$) is the square root of 2, so that $(M : F)^2 = 2$, it is also true that $M^2 = 2F^2$. Since an odd number does not contain 2 as a factor, its square cannot. So this cannot be true.

(ii) The third could be true, only if $M^2 = 2m^2$, in which event 2 would be a factor of M, and M could not be an odd number. So this cannot be true.

(iii) The remaining possibility means that $F^2 = 2M^2$. Now the square of an even number must contain 4 as a factor, in which event our equation could be true only if M^2 contained 2 as a factor. If M is odd, it cannot. So the only remaining possibility is also inconsistent with the possibility that the square root of two is expressible as a fraction of the usual sort.

Had Plato's contemporaries been familiar with the distinction between unending series which do or do not choke off, or more simply with what our use of decimal fractions exposes, the argument might have taken a different turn. Whether a number is even or odd depends on its last digit. For instance, 5943, 4945, 5947, 5949 are odd, 5940, 5942, 5944, 5946, 5948 are even. If we go on endlessly adding digits to a and b in the fraction $a \div b$, we do not reach a point at which we can say what is the last digit in either; and surely this is one thing we mean by saying that a non-terminating and non-repeating decimal fraction is not expressible as a vulgar fraction whose numerator and denominator we can pin down as nameable numbers.

The disclosure that we cannot represent the diagonal of a square of unit area (sides each one unit) by what they could think of as an exact number led geometers in the Platonic tradition to speak of it as *incommensurable* with the side. In real life, however, the sides themselves are incommensurable with one another. For we have seen that the only legitimate statement we can make about the length (L or l) of a side is of the form $L_1 < L < L_2$ or $l_1 < l < l_2$ in which L_1, l_1 are lower and L_2, l_2 are upper limits. If so, a ratio $R(= L : l)$ will be least when L is least and l greatest. It will be greatest when L is greatest and l is least. So we may say that:

$$R_1 < R < R_2, \textit{ if } R_1 = (L_1 \div l_2) \quad \textit{and} \quad R_2 = (L_2 \div l_1)$$

Professional mathematicians distinguish between two categories of non-rational numbers: (i) the *irrational* numbers such as the square root of 2 or the fifth root of 4; (ii) *transcendental* numbers such as π, the ratio of the circumference to the diameter of a circle. Many readers will recall that this is roughly 22 : 7. More precisely, i.e. with a margin of uncertainty less than 1 in 30 million:

$$3{\cdot}1415926 < \pi < 3{\cdot}1415927$$

With a margin of uncertainty less than 1 in a million the square roots of 2 and 10 are expressible as:

$$1{\cdot}414213 < \sqrt{2} < 1{\cdot}414214$$
$$3{\cdot}162277 < \sqrt{10} < 3{\cdot}162278$$

In terms of a programme for an electronic brain, the distinction last named is of meagre (if any conceivable) interest. All we can hope to do is to frame an instruction for specifying any non-rational number with as much precision as we require (e.g. $\sqrt{2} \simeq 1{\cdot}4142$ or $\pi \simeq 3{\cdot}1416$). With that end in view, we may proceed in one of three ways. One is iteration, i.e. successive approximation as already illustrated for $\sqrt{2}$ above (p. 102). A second is the use of an *infinite* (i.e. endless) series the sum of whose terms beyond a specifiable number can never exceed a specifiable value. Such a series is obtainable (p. 416) for $\sqrt{2}$ by means of the *Binomial Theorem*. One for π (which the reader will here also have to take on trust, see p. 485) is:

$$\tfrac{1}{4}\pi = 1 - \tfrac{1}{3} + \tfrac{1}{5} - \tfrac{1}{7} + \tfrac{1}{9} - \tfrac{1}{11} + \tfrac{1}{13} \ldots, \text{ etc.}$$

There is a third recipe which generates the square root of any integer, a value for π and a value for another very important transcendental number e (p. 422). The last named is defined as the infinite series:

$$e = 1 + \frac{1}{1} + \frac{1}{2\,.\,1} + \frac{1}{3\,.\,2\,.\,1} + \frac{1}{4\,.\,3\,.\,2\,.\,1} + \frac{1}{5\,.\,4\,.\,3\,.\,2\,.\,1}, \text{ etc.}$$

It is not difficult to see that the sum of this series cannot exceed a namable limit. The denominators of the eleventh and twelfth terms are respectively $10 . 9 . 8 \ldots 3 . 2 . 1$ and $11 . 10 . 9 \ldots 3 . 2 . 1$. Thus every term after the eleventh is less than one-tenth of its successor, and therefore less than a decimal fraction represented by adding the first 11 terms with successive units in the twelth decimal place and thereafter. With no error in the seventh place, it is expressible as:

$$2{\cdot}71828182 < e < 2{\cdot}71828183$$

Our third way of representing a so-called incommensurable is one which is of an earlier vintage than the second. In so far as it concerns $\sqrt{2}$, it is as follows. Since $2^2 = 4$ and $1^2 = 1$ we know that $1 < \sqrt{2} < 2$. So we write (Fig. 42) $(1 + r) = \sqrt{2}$ in which event $(1 + r)^2 = 2 = 1 + 2r + r^2$ and $2r + r^2 = 1$, whence $r(2 + r) = 1$, so that

$$r = \frac{1}{2 + r}$$

We can go on replacing r on the right side indefinitely thus:

$$r = \cfrac{1}{2 + \cfrac{1}{2 + r}} \qquad\qquad r = \cfrac{1}{2 + \cfrac{1}{2 + \cfrac{1}{2 + r}}}$$

It makes less and less difference to the left hand r if we knock out r on the right as the number of tiers increases; and the square of

$1 + r$ based on successive estimates obtained in this way alternately exceeds or falls short of 2. The first three are

$$\tfrac{1}{2} = 0{\cdot}5; \quad \cfrac{1}{2 + \cfrac{1}{2}} = 0{\cdot}4; \quad \cfrac{1}{2 + \cfrac{1}{2 + \cfrac{1}{2}}} = 0{\cdot}41\dot{6}$$

These yield $\sqrt{2} \simeq 1{\cdot}5$, $\sqrt{2} \simeq 1{\cdot}40$, $\sqrt{2} \simeq 1{\cdot}41\dot{6}$. The next approximation derived by the same method is $\sqrt{2} \simeq \frac{41}{29}$. The first four approximations so obtained when squared thus yield:

$$2\tfrac{1}{4};\ 1\tfrac{24}{25};\ 2\tfrac{1}{144};\ 1\tfrac{840}{841}$$

The last differs from 2 by little more than one in a thousand. The reader may apply the method to the determination of $\sqrt{4}$ by putting $\sqrt{4} = (3 - r)$, so that $r(6 - r) = 5$. The estimate gets nearer and nearer to 2, in this case its exact value, if we proceed as above. In short, the square root of any integer whether rational or otherwise is expressible as a *continued fraction.* To do so, we may proceed in either of two ways. We suppose that N lies between n^2 and $m^2 = (n + 1)^2$. If it is nearer to n^2, we write $(n + r)^2 = N$; but it is better to express it in the form $(m - r)^2 = N$ if N lies nearer to m^2. For instance, $\sqrt{5}$ and $\sqrt{8}$ both lie between 2 and 3 since $2^2 = 4$, $(2{\cdot}5)^2 = 6{\cdot}25$, and $3^2 = 9$. Here 2 is nearer $\sqrt{5}$ but 3 is nearer to $\sqrt{8}$, and it is most convenient to write:

$$(2 + r)^2 = 5 \textit{ so that } 4 + 4r + r^2 = 5 \textit{ and } r(4 + r) = 1$$
$$(3 - r)^2 = 8 \textit{ so that } 9 - 6r + r^2 = 8 \textit{ and } r(6 - r) = 1$$

Rank and Order. Besides using number symbols to enumerate things, to represent measurements and to specify an operation (e.g. n^3), we use them to record an orderly succession of events as in a calendar. We have seen that this was almost certainly the earliest use of numbers. It is also one which plays a pivotal role in the large tract of mathematics which deals with so-called *series.* A series is a number sequence in which a single rule relates each term (member) to one or more of its predecessors. The natural numbers arranged in order of size constitute a series for which the simplest conceivable rule connects each term to its immediate predecessor, e.g.:

$$1;\ 2 = (1 + 1);\ 3 = (2 + 1);\ 4 = (3 + 1);\ 5 = (4 + 1), \text{etc.}$$

One familiar series is the decimal fraction $1\cdot\dot{1}$, which we have met in connection with the Achilles–tortoise paradox (p. 12). To express as snappily as possible a rule which connects successive members (*terms*) of such a series, we use a whole number to label each, and speak of such an integer as its *rank*. Though it may be convenient to call the initial term the term of rank 1, this is not essential. All that matters is to label the successor of a term whose rank is r as the term of rank $(r + 1)$ and, excluding the initial term, the predecessor of a term whose rank is r as the term of rank $(r - 1)$. We can then use a *subscript* notation which labels as t_r the *t*erm whose *r*ank is r, e.g. in the sequence below $t_3 = 10^{-3}$ is the *t*erm of *r*ank 3:

rank (r) =	1	2	3	4	5 . . .
terms (t_r) =	0·1	0·01	0·001	0·0001	0·00001 . . .
	$\frac{1}{10}$	$\frac{1}{10^2}$	$\frac{1}{10^3}$	$\frac{1}{10^4}$	$\frac{1}{10^5}$ . . .

When we label the terms in this way, we recognize alternative ways of stating the rule. If we label the term of rank r as t_r, we may write:

$$t_r = \frac{1}{10} \cdot t_{r-1} \quad or \quad t_r = \frac{1}{10^r}$$

There is nothing sacred about this way of labelling such a series. Consider another formed by a comparable rule.

(r) =	1	2	3	4	5 . . .
term (t_r) =	10	100	1000	10000	100000 . . .
	10^1	10^2	10^3	10^4	10^5 . . .

Here alternatively, we can write:

$$t_{r-1} = \frac{1}{10} t_r \quad or \quad t_r = 10^r$$

Actually, we can bring the series whose sum is $0\cdot0\dot{1}$ into the same pattern if we write:

r =	−1	−2	−3	−4 . . .
t_r =	0·1	0·01	0·001	0·00001 . . .
	10^{-1}	10^{-2}	10^{-3}	10^{-4} . . .

Below we see that each term is one-tenth of its predecessor (on the left), if each rank is one less than its predecessor:

$r =$	4	3	2	1	0	-1	-2	$-3\ldots$
$t_r =$	10,000	1000	100	10	1	$\frac{1}{10}$	$\frac{1}{100}$	$\frac{1}{1000}\ldots$
	10^4	10^3	10^2	10^1	10^0	10^{-1}	10^{-2}	$10^{-3}\ldots$

Alternatively, each term is ten times its predecessor on the left, if we arrange the terms in ascending order, so that each rank is one more than its predecessor; as below:

$r =$	-3	-2	-1	0	1	2	3	4
$t_r =$	0·001	0·01	0·1	1	10	100	1000	10,000
	10^{-3}	10^{-2}	10^{-1}	10^0	10^1	10^2	10^3	10^4

Similarly, with another so-called *geometric* series in which each term is *twice* its predecessor, we may lay out the pattern thus:

$r =$	-3	-2	-1	0	1	2	3	$4\ldots$
$t_r =$	$\frac{1}{8}$	$\frac{1}{4}$	$\frac{1}{2}$	1	2	4	8	$16\ldots$
	$\frac{1}{2^3}$	$\frac{1}{2^2}$	$\frac{1}{2}$	1	2	4	8	$16\ldots$
	2^{-3}	2^{-2}	2^{-1}	2^0	2^1	2^2	2^3	$2^4\ldots$

If we then say that $b^0 = 1, b^{-1} = \frac{1}{b}, b^{-2} = \frac{1}{b^2}$ etc., we are *not making statements which require proof*, we are merely stating *what labels we shall use when talking about a sequence of numbers connected by the rule that each term when arranged in ascending order is* b *times its predecessor*. We have chosen this way of labelling them not because it is the only legitimate way, but because it makes the rule most explicit, i.e. $t_r = b^r$.

Let us look at another series for which the rule of formation is not so easy to see. The so-called *triangular numbers* of Fig. 35 run:

$r =$ 1	2	3	4	5	6 . . ., etc.
$t_r =$ 1	3	6	10	15	21 . . ., etc.

Here each term is made by adding its rank to its predecessor, a fact recorded by the sentence:

$$t_r = t_{r-1} + r, \qquad \text{e.g. } t_6 = t_5 + 6 = 15 + 6 = 21$$

This rule is incomplete, unless we already know one term and its rank, e.g. $t_r = 1$ if $r = 1$. By trial and error, you may notice that the terms conform to the following pattern:

$r =$	1	2	3	4	5	6 . . ., etc.
$t_r =$	1	3	6	10	15	21 . . ., etc.
	$\frac{1(2)}{2}$	$\frac{2(3)}{2}$	$\frac{3(4)}{2}$	$\frac{4(5)}{2}$	$\frac{5(6)}{2}$	$\frac{6(7)}{2}$

We can now discern a likely rule as follows:

$$t_r = \frac{r(r+1)}{2}$$

How can we be sure that it is always true? To answer this, recall that:

$$t_r = t_{r-1} + r$$

whence

$$t_{r+1} = t_r + (r+1)$$

If the rule is true therefore:

$$t_{r+1} = \frac{r(r+1)}{2} + (r+1) = \frac{r^2 + r + 2(r+1)}{2} = \frac{r^2 + 3r + 2}{2}$$

$$= \frac{(r+1)(r+2)}{2}$$

If we write $r + 1$ as n, this is equivalent to writing

$$t_n = \frac{n(n+1)}{2}$$

So the same rule holds good, if we substitute for one value (r) of the rank the value ($r + 1$) of its *successor*; and the fact that it does so, illustrates one recipe for obtaining reliable rules to calculate any term of a series without the labour of building it up step by step. The last qualification is necessary because we can speak of an assemblage of numbers as a series only when arranged in a way which exhibits a relation between any term and one (or more) of its predecessors; and in what follows we assume that such a relation exists between each term and its immediate predecessor, e.g. $t_r = 2t_{r-1}$ if $t_r = 2^r$ and $t_r = t_{r-1} + r$ if $t_r = \frac{1}{2}r(r+1)$ as in the foregoing example. The recipe (known as *mathematical induction*) is as follows:

If the formula for the $(r + 1)$th term is obtainable by the substitution of $r + 1$ for r in what seems to be a good enough formula for the rth term, it will be true for all *subsequent* terms if it is also true for the *initial* term.

That this method is of much wider scope than it might seem to be at first sight is because the sum of any number of terms of a series is itself a series. We have seen (p. 86) that this is true of the sum (S_r) of the first r odd numbers; and may use the method of induction to justify the formula. Accordingly, we set out the terms as below:

$r = 1$	2	3	4	5 . . .
$n_r = 1$	3	5	7	9 . . .
$S_r = 1$	4	9	16	25 . . .
$= 1$	2^2	3^2	4^2	5^2 . . .

The terms of the second line (odd numbers) are obtainable from the first by the formula $n_r = 2r - 1$. The terms of the next row are connected with those of the one above by the relation $S_r = S_{r-1} + n_r$ (e.g. $S_5 = S_4 + n_5 = 16 + 9 = 25$). Since every term n_r in the second row is expressible in terms of any term (r) of the first row, every term in the third is connected with its predecessor in a definable way, i.e.:

$$S_r = S_{r-1} + (2r - 1) \text{ or } S_{r+1} = S_r + 2(r + 1) - 1 = S_r + (2r + 1)$$

Now the fourth row above suggests (Fig. 38) that any term of the third row is expressible as r^2; and this is evidently true of the first term. All that remains to show is that we can derive $S_{r+1} = (r + 1)^2$ from what we already know,

i.e. $S_{r+1} = S_r + (2r + 1)$. If $S_r = r^2$, therefore:

$$S_{r+1} = r^2 + (2r + 1) = r^2 + 2r + 1 = (r + 1)^2.$$

Thus the rule is true for the $(r + 1)$ term, if true for the rth. Since it is true for the first, it is true for the second. If true for the second, it must be true for the third and so on.

When using this device to find a formula for any term of rank r in an ordered sequence of numbers, it is often most convenient to label the *initial* term as the term of rank 0 rather than as that of rank 1; but when we do so we need to bear in mind: (*a*) that the term of rank r is the last of $(r + 1)$ terms; (*b*) the same formula does not apply to both. If we label the first odd number as the odd number of rank 1, the formula for the odd number of any rank r is $(2r - 1)$, e.g. the sixth odd number is 11. If we label the first odd number as that of rank 0, the formula is $(2r + 1)$. In this case the latter procedure has no advantage; but below is a series for which the general formula is less cumbersome, if we label the initial terms as in the bottom row:

$r = 1$	2	3	4	5	6...
$t_r = 3$	7	11	15	19	23...
$r = 0$	1	2	3	4	5...

The above is a so-called *arithmetic* series. If we label the initial term as of rank 1, the formula for the rth term is $3 + 4(r - 1)$. If we label it as 0, the formula is $3 + 4r$. A comparable simplification results from labelling the initial term of a so-called *geometric* series as that of rank 0, e.g. the following example:

$r = 1$	2	3	4	5	6...
$t_r = 3$	6	12	24	48	96...
$r = 0$	1	2	3	4	5...

For the above, the two formulae are:

$t_r = 3(2^{r-1})$ if the initial term is of rank 1

$t_r = 3(2^r)$ if the initial term is of rank 0

Exercises on Chapter 2

Discoveries

1. On the assumption that $x > a > b$ make diagrams like Fig. 39 to illustrate the following rules:

$(x + a)(x + b) = x^2 + (a + b)x + ab$; e.g. $x^2 + 7x + 12 = (x + 4)(x + 3)$

$(x + a)(x - b) = x^2 + (a - b)x - ab$; e.g. $x^2 + 2x - 15 = (x + 5)(x - 3)$

$(x - a)(x + b) = x^2 + (b - a)x - ab$; e.g. $x^2 - 3x - 40 = (x - 8)(x + 5)$

$(x - a)(x - b) = x^2 - (a + b)x + ab$; e.g. $x^2 - 10x + 16 = (x - 8)(x - 2)$

Make up other numerical examples and check, e.g.:

$$(x - 5)(x + 2) = x(x + 2) - 5(x + 2) = x^2 + 2x - 5x - 10 = x^2 - 3x - 10$$

2. Make figures to illustrate the statements:

$$(x + y + z)^2 = x^2 + y^2 + z^2 + 2xy + 2xz + 2yz$$

$$(g + f)(a + b + c + d) = g(a + b + c + d) + f(a + b + c + d)$$

$$(g + f)(a + b + c + d) = ga + gb + gc + gd + fa + fb + fc + fd$$

Check the former by putting $x = 2$, $y = 4$, $z = 7$, etc.

3. Find the sum of all the numbers:

(*a*) From 7 to 21.
(*b*) From 9 to 29.
(*c*) From 1 to 100.

Check the results by addition.

4. Find by formula and by direct addition the sum of the following sets of numbers:

(*a*) 3, 7, 11, 15, 19, 23, 27, 31, 35.
(*b*) 5, 14, 23, 32, 41, 50.
(*c*) 7, $5\frac{1}{2}$, 4, $2\frac{1}{2}$, 1, $-\frac{1}{2}$.

5. Draw an angle of 30°. At any point in one arm of the angle draw a line at right angles to it meeting the other arm. You thus have a right-angled triangle with one angle equal to 30°. The sides of a right-angled triangle have special names. The longest side, that opposite the right angle, is called the hypotenuse. The other sides are named with reference to one of the other angles. At present we are interested in the angle of 30°, so we call the side opposite to it the perpendicular, and the other short side the base. Draw several right-angled triangles with one angle of 30°, each a different size and in different positions on the paper, and see that you can recognize the perpendicular, base, and hypotenuse immediately in any position.

6. In each triangle measure the following ratios:

$$\frac{\text{perpendicular}}{\text{hypotenuse}}, \quad \frac{\text{base}}{\text{hypotenuse}}, \quad \frac{\text{perpendicular}}{\text{base}}$$

7. In the same way draw several triangles with an angle of 60° and several with an angle of 45°. Measure the ratios in each triangle. These ratios have been given names. If A is any angle the hypotenuse and base are the lines which enclose it, and the perpendicular is the side *opposite A*. The ratio $\frac{\text{perpendicular}}{\text{hypotenuse}}$ is

then called the SINE of the angle A. The ratio $\frac{\text{base}}{\text{hypotenuse}}$ is called the COSINE of A. The ratio $\frac{\text{perpendicular}}{\text{base}}$ is called the TANGENT of A. These ratios are written for brevity: sin A, cos A and tan A. By drawing right-angled triangles of various sizes you will see that each of these ratios has a fixed value for any particular angle.

8. By drawing two or three right-angled triangles and taking the mean of your measurements make a table of the sines, cosines, and tangents of 15°, 30°, 45°, 60°, 75°.

Find out by looking at the triangles you have drawn why:

(*a*) $\sin (90° - A) = \cos A$.
(*b*) $\cos (90° - A) = \sin A$.
(*c*) $\tan A = \frac{\sin A}{\cos A}$.

9. Draw a circle with radius of 1 inch. Draw two radii enclosing an angle of 15°. One way of measuring an angle is by the ratio of the arc bounding it to the radius of the circle. This is called the number of *radians*, so that when the arc equals the radius, the angle is *one radian*. You have already found an approximate value for the ratio of the circumference to the diameter of a circle, and have learnt to call this ratio π. If you look at your diagram, you will see that you can get 24 angles of 15° into the circle, and that therefore the arc bounding it is $\frac{1}{24}$ of the circumference. So the arc will be $\frac{\pi}{24}$ times the diameter, and as the radius is half the diameter it will be $\frac{\pi}{12}$ times the radius. We have drawn a circle of unit radius, so we see that an angle of 15° can also be described as an angle of $\frac{\pi}{12}$ radians.

10. In the same way find the values of angles of 30°, 60°, 90°, and 180° in radians.

11. How many degrees are there in a radian (take $\pi = 3\frac{1}{7}$)? What fraction of a radian is a degree?

Tests on Measurement

1. Suppose you want to measure accurately the length of a garden plot. To get first an approximate measure you would pace it out. Distances of this sort are reckoned by surveyors in chains and links. A chain is 66 feet, and a link is $\frac{1}{100}$ part of a chain, that is 7·92 inches. A natural pace is considered to be about $2\frac{2}{3}$ feet, so that the surveyor's rule is to multiply the number of paces by four and point off two figures from the right-hand end to stand for links. So if the length of a fence is 120 paces this would be taken as being 4 chains and 80 links ($4 \times 120 = 480$). You thus obtain the required distance in chains and links.

(*a*) Imagine you have paced the length of your garden plot and found it to be exactly 80 paces. According to the surveyor's rule it should measure 3 chains 20 links, or 211 feet 2·4 inches. Now mark one of your own paces with chalk on the floor and measure it with a foot rule. How would you correct your estimate of the length of the garden plot on account of the difference between your pace and the standard pace? N.B. – If you are not fully-grown, your pace will have to be rather a long one, or you will be a long way out.

(*b*) Land surveyors measure distances along a straight line with a Gunter's chain. This is a chain 66 feet long, divided up into 100 links. Each link is 7·92 inches long. Within what limits would you expect the measurement of your garden by this chain to lie? Remember that what you know about the length so far is that it is 211 feet 2·4 inches corrected by the difference between your pace and the standard pace.

(*c*) A linen tape 60 feet long, divided into feet and inches, is sometimes used. Such a tape was once found to have shrunk to 55 feet 4 inches after being used on wet grass. If you were so unlucky as to use a tape in such a state and later discovered the amount of shrinkage, how would you correct your measurements?

2. The division marks on a dressmaker's tape may be as much as $\frac{5}{8}$ of a millimetre thick. A millimetre is about $\frac{1}{25}$ of an inch. A careless person might take his measurements from the outsides of the marks, or from the insides of the marks. Measuring in this way, what could be the highest and lowest estimate of the length of:

(*a*) the side of a curtain about 6 feet long;

(*b*) a space of about $\frac{1}{2}$ inch between 2 buttons on a baby's frock?

What percentage of the mean is the difference between two pairs of measurements in (*a*)?

Tabulation

3. By the method of approximation given in this chapter, make a table of the square roots of all the numbers from 1 to 20 inclusive correct to three decimal places.

Tabulate 2^n from $n = 1$ (when $2^n = 2$) to $n = 12$ (when $2^n =$ 4,096). Do the same with 3^n from $n = 1$ to $n = 10$. Use these results to tabulate $(1\frac{1}{2})^n$ and $(\frac{2}{3})^n$ from $n = 1$ to $n = 8$, correct to three decimal places.

Translation into Size Language

4. Translate the following into mathematical language:

(*a*) Multiply twice the length added to twice the breadth (measured in yards) by the price of fencing per yard to find the cost of fencing a plot.

(*b*) Take one spoonful of tea per person and add one spoonful for the pot to get the amount of tea required to make a pot for a party (call the number of people n).

(*c*) If you know that the weight of a crate containing n eggs is W and the weight of the empty crate is w, then you must subtract w from W and divide the result by n to get the average weight of one egg.

(*d*) Multiply the base by the perpendicular height and divide by 2 to get the area of a triangle.

(*e*) Write down the rule for finding the amount if a sum of money (£a) is left to accumulate for n years at r per cent simple interest.

Algebraic Manipulation

5. When learning how to use symbols, it is very useful to check all your work arithmetically, as in the following example—

Simplify: $a + 2b + 3c + 4a + 5c + 6b$

In algebra to simplify an expression means to change it to a convenient form for doing further work with it. We can simplify this expression by adding together all the a's, all the b's, and all the c's. We then have

$$a + 2b + 3c + 4a + 5c + 6b = 5a + 8b + 8c$$

To check our work by arithmetic, put $a = 1$, $b = 2$, and $c = 3$. Then

$$\begin{aligned} a + 2b + 3c + 4a + 5c + 6b &= 1 + 2 \times 2 + 3 \times 3 + 4 \times 1 + 5 \times 3 + 6 \times 2 \\ &= 1 + 4 + 9 + 4 + 15 + 12 \\ &= 45 \end{aligned}$$

Also

$$\begin{aligned} 5a + 8b + 8c &= 5 \times 1 + 8 \times 2 + 8 \times 3 \\ &= 5 + 16 + 24 \\ &= 45 \end{aligned}$$

Check your results in this way in the following tests and whenever in doubt if you have dealt correctly with an algebraic expression—

Simplify:

(*a*) $x(x + 2y) + y(x + y)$.

(*b*) $(x + 2y + 3z) + (y + 3x + 5z) + (2z + 3y + 2x)$.

(*c*) $(a + 1)(a + 2) + (a + 2)(a + 3) + (a + 3)(a + 1) + 1$.

(*d*) $(x - 1)^2 - (x - 2)^2$.

(*e*) $a^2 - ab - (b^2 + ab)$.

(*f*) $(zx)(xy) + (xy)(yz) + (yz)(zx)$.

(*g*) $(2ab)(3a^2b^3)$.

(*h*) $(x^3)^2 + (x^2)^3$.

(*i*) $(a - b)(a + 2b) - (a + 2x)(a + x) - (a - 2x + 2b)(a - x - b)$.

(*j*) $\dfrac{2x^4y^5}{4x^3y^2}$.

(*k*) $\dfrac{(3ab^2)^3}{9a^2b^5}$.

(*l*) $\dfrac{2ab}{3c} \times \dfrac{4cd}{8b}$.

6. Satisfy yourself that the statements (*a*) x is inversely proportional to y when z is kept the same; and (*b*) x is directly proportional to z when y is kept the same are both contained in the single equation:

$$xy = kz$$

This is very important to remember. (*Clue*: replace the quantity kept the same by some other letter, e.g. C or c to represent a 'constant'.)

Simple Equations

7. In each case make sure that the value you have found for x satisfies the equation.

(*a*) $3x + 7 = 43$

(*b*) $2x - 3 = 21$

(*c*) $17 = x + 3$

(*d*) $3(x + 5) + 1 = 31$

(*e*) $2(3x - 1) + 3 = 13$

(*f*) $x + 5 = 3x - 7$

(*g*) $4(x + 2) = x + 17$

(*h*) $\frac{x}{4} = \frac{1}{8}$

(*i*) $\frac{x + 2}{5} = \frac{x - 1}{2}$

(*j*) $\frac{x}{2} - \frac{x}{3} + 7 = \frac{5x}{6} - 5$

(*k*) $\frac{3}{x} = 3$

(*l*) $4 + \frac{15}{x} = 7$

(*m*) $- 2x - 5 + 12x - 3 - 4 = 8$

Find x in terms of a and b in the following:

(*n*) $x - a = 2x - 7a$

(*o*) $2(x - a) = x + b$

(*p*) $a(a - x) = 2ab - b(x + b)$

Simple Problems Involving an Equation

Check all your results.

8. Divide £540 between A and B so that A gets £30 more than B. (Call B's share x. A's will be $x + 30$.)

9. Divide £627 between A and B and C so that A gets twice as much as B and three times as much as C. (Call C's share x.)

10. Tom walks at a rate of 4 miles per hour, and Dick at a rate of 3 miles per hour. If Dick has half an hour's start, how long will it be before Tom catches up with him? (Call time from Dick's start till he is caught up x).

11. A page of print contains 1,200 words if in large type, and 1,500 in small type. If an article of 30,000 words is to occupy 22 pages, how many pages must be in small type?

12. Automobile A does 30 miles to a gallon of petrol and 500 miles to a gallon of oil. Automobile B does 40 miles to a gallon of petrol and 400 miles to a gallon of oil. If oil costs as much as petrol, which will be the cheaper automobile to run?

13. A 60-foot row of early peas yields 12 pecks. An 80-foot row of maincrop peas yields 18 pecks. If maincrop peas fetch 1*s*. 4*d*. a peck, what price must early peas fetch to make them equally profitable?

14. Express as continued fractions $\sqrt{3}$, $\sqrt{5}$, $\sqrt{6}$, and $\sqrt{10}$. Obtain the first four approximations for each and check the error by squaring.

Things to Memorize

1.
$$(a + b)(a + b) = a^2 + 2ab + b^2$$
$$(a - b)(a - b) = a^2 - 2ab + b^2$$
$$(a + b)(a - b) = a^2 - b^2$$

2. If A is any angle:

$$\sin A = \frac{\text{perpendicular}}{\text{hypotenuse}} \qquad \sin(90° - A) = \cos A$$

$$\cos A = \frac{\text{base}}{\text{hypotenuse}} \qquad \cos(90° - A) = \sin A$$

$$\tan A = \frac{\text{perpendicular}}{\text{base}} \qquad \tan A = \frac{\sin A}{\cos A}$$

3. If $x \propto z$, y being fixed and $x \propto \frac{1}{y}$, z being fixed, $zy = kz$.

4.
$$- = - \div + \qquad + = + \times +$$
$$- = + \div - \qquad - = + \times -$$
$$+ = - \div - \qquad + = - \times -$$

CHAPTER 3

EUCLID AS A SPRINGBOARD

IN OUR Prologue, we have done justice to one reason why knowledge of the history of mathematics or science can be helpful to many of us. There is another reason very relevant to the fact that the image of Euclid has fallen from its pedestal in the school or college curriculum since half-way into our own century. To some extent, any sort of education, perpetuates traditions which have ceased to be relevant to contemporary conditions. When we understand why this and that is no longer relevant, we can save ourselves unnecessary time and effort by taking short-cuts which our forefathers could not foresee.

Since we do not know the birthplace of the first of the great teachers associated with the cosmopolitan University of Alexandria, Euclid may well have been an Egyptian or a Jew: but like others of the school, he wrote in Greek his thirteen books composed about 300 BC. His lifetime overlaps that of Aristotle, possibly also that of Plato, the influence of whose school is manifest in all his mathematical writings. The thirteen books, not all of which survive, encompass all the achievements of Greek mathematics. Only seven of them deal with geometrical figures. The first four books contain 115 proofs, the sixth 33, the eleventh and twelfth 70. With two exceptions, those of the last two books deal with solid figures.

In the time of Newton (AD 1642–1727) Euclid's treatises were the foundations of all mathematical teaching in the western world. It was still an essential part of the teaching of mathematics in the majority of schools and colleges of the western world at the beginning of our own century. This is no longer true; and we may be glad of it for two reasons. Like Plato, Euclid was a thoroughly bad educationist. To them, there was 'no royal road' to mathematical proficiency; and they had no compelling incentive to make the journey as painless as possible. In short, Euclid's method of stating a case is forbidding. From a modern viewpoint it is also excessively long-winded. For instance, we do not reach the rationale of the most important discovery, that about right-angled triangles illustrated by the 3 : 4 : 5 ratio (Fig. 17), till we have digested 40 other proofs of Book 1.

What most helps to make his method of teaching forbidding is that he was still bogged down by the notion of incommensurables (p. 98), and therefore unwilling to present a rule as a valid recipe for drawing-board measurement. Instead of talking about equal lengths, he prefers to talk about equal lines. Instead of talking about equal areas, he prefers to talk about equivalent figures (triangles, circles, rectangles, etc.). The whole of his fifth book is devoted to what we ourselves may well regard as a remarkably pedantic exposition of the trick by which his predecessor Eudoxus sought to define equality of ratios without bringing number into the picture and without conceding the possibility of representing the so-called incommensurables by devices such as we have examined in connection with the square root of 2 or as we have foreshadowed to get a good enough value for π.

As we have now seen, the notion of ratio and of an unending process of division presented understandable difficulties to geometers who relied on the abacus and used numeral signs which do not fulfil the requirements of rapid calculation. This may explain Euclid's reluctance to introduce the notion of ratio at an early stage. Unnecessary circumlocution also results from the adoption of a definition (p. 66) of parallel straight lines having no firm foundation in fact. If, as did Euclid's successor Archimedes (about 240 BC), we reinstate measurement, define parallelism in drawing-board terms, and waste no time over Euclidean results we can more readily derive by later methods, very few of the rules of Euclid's geometry suffice to pinpoint all we need for the understanding of any mathematical discoveries before the American Revolution or the Battle of Waterloo. Most readers will already have made their acquaintance at some time, but will not have realized their role in a development which was one of the supreme achievements of the Alexandrians from the time of Hipparchus (about 150 BC) to that of Ptolemy (about AD 130).

Long before Euclid systematized Greek geometry in accordance with Plato's philosophy, his predecessors had made every discovery relevant to the build up of a table of so-called *trigonometrical ratios* to meet the needs of the surveyor and of the astronomer. In this chapter, we shall take advantage of hindsight to set forth the indispensable rules of Greek geometry as signposts towards this goal. If much of it is familiar to the reader, the viewpoint may not be. First, a few words about definitions will not be amiss. When Euclid speaks of two triangles as being *equal in all respects*, he means that if we are free to move one of them on the page and to turn it from left to right or vice versa, we can superimpose one on the other so that corresponding sides and corresponding angles coincide exactly. By corresponding angles ABC and DEF we then mean angles respectively enclosed by sides $AB = DE$ and $BC = EF$. By corresponding sides $AB = DE$, etc., we mean sides whose extremities lie between the angles $ABC = DEF$ and $BAC = EDF$. Modern textbooks usually call triangles *congruent* if they are in this sense equal in all respects. In what follows we shall call them *equivalent*. For brevity, we may write $\angle ABC$ to specify the angle whose tip is B in contradistinction to $\triangle ABC$ the triangle whose vertices are A, B, and C.

Let us now summarize under eight headings what we may take for granted:

(i) We need not set forth as a proof two rules which follow from the definition of a right angle:

ANGLE RULE ONE (Fig. 14)

If one straight line meets another at a point on the latter the two angles it makes therewith are equal to two right angles (180°).

ANGLE RULE TWO (Fig. 14)

If two straight lines cross one another opposite angles are equal.

(ii) If we define (see p. 64) parallel straight lines by the way in which we make them, i.e. as straight lines equally inclined to a cross-piece, we may assume:

PARALLEL RULE ONE (Fig. 27)

Straight lines are parallel if equally inclined to any straight line (transversal) which crosses them.

PARALLEL RULE TWO (Fig. 27)

Any transversal to a pair of parallel lines make equal *alternate angles* with them.

(iii) We may take each of three recipes for constructing a unique triangle (or its mirror image) as a sufficient criterion of when two triangles are equivalent, i.e. that we know:

TRIANGLE RULE ONE (Fig. 28)

The lengths of two sides and the angle between them.

TRIANGLE RULE TWO (Fig. 28)

The length of one side and the two angles at its extremities.

TRIANGLE RULE THREE (Fig. 28)

The lengths of all three sides.

(iv) From the way in which we draw a circle we shall assume that the distances between the centre and any points on its *perimeter* are the same, whence with the help of (iii) we can:

(*a*) bisect a straight line (Fig. 15);
(*b*) bisect an angle (Fig. 22);
(*c*) erect a perpendicular at any point on a straight line (Fig. 26);
(*d*) drop a perpendicular on a straight line from any point outside it (Fig. 26).

(v) We define a chord of a circle as a straight line joining any two points on the perimeter of a circle. By using (iii) with what we may assume (p. 64) about a circle, we see that straight lines joining the extremities of equal chords to the centre of a circle enclose equal angles.

(vi) Areas of segments of a circle cut off by two radii are equal if the radii enclose equal angles at the centre. A diameter is a chord which divides a circle into segments of equal area, and its length is twice that of the radius.

(vii) We here define a parallelogram as a 4-sided figure with two pairs of parallel sides and a rectangle as a parallelogram of which *one* angle is a right angle. Anything else we say about them calls for proof (Fig. 27).

(viii) The area of a rectangle (Fig. 10) is equivalent to the product of the lengths of adjacent sides and that of a triangle (Fig. 11) is equivalent to half that of a rectangle of the same base and height.

We are now ready to condense in seven rules all that we shall need to know about the seven books of Euclid mentioned above. Textbooks usually call them theorems. Here we shall follow the lead of the first atomist by calling them *demonstrations*.

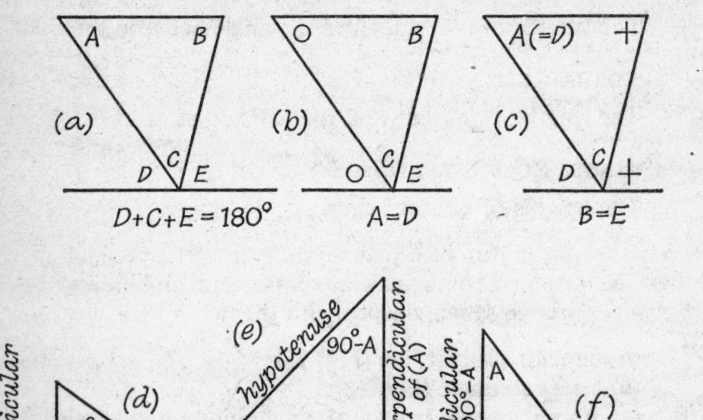

Fig. 43. Construction for Demonstration 1.
If $D + C + E = 180°$; $A = D$ and $B = E$, then: $A + B + C = 180°$.

Demonstration 1

The three angles of a triangle add up to two right angles.

To demonstrate this, all you have to do is to tilt one corner of a triangle on a straight edge till one side is parallel to it. All the relevant facts are in Fig. 43.

From this rule follows an important peculiarity of any right-angled triangle. If we label the right angle as *C* and its others as *A* and *B*:

$$A + B + 90° = 180°, \text{ so that } A + B = 180° - 90° = 90°$$
$$\therefore B = 90° - A \text{ and } A = 90° - B$$

Thus we know the third angle if we know one of the other two, e.g. if $A = 45°$, $B = 45°$; if $A = 30°$, $B = 60°$, and vice versa. As we see from (i) and (ii) in Fig. 44, this means that:

(i) all right-angled triangles with the same corner angle (i.e. either angle less than 90°) are *equiangular*, see below;

(ii) right-angled triangles which we can place so that two pairs of adjacent sides are in line are likewise equiangular.

We shall need later to have labels for the sides of a right-angled triangle as in the lower half of Fig. 43. One calls the longest side the *hypotenuse*. The side opposite the angle A is the *perpendicular* of A, and the remaining side the *base* of A. It follows that the perpendicular of $(90° - A)$ is the base of A and vice versa.

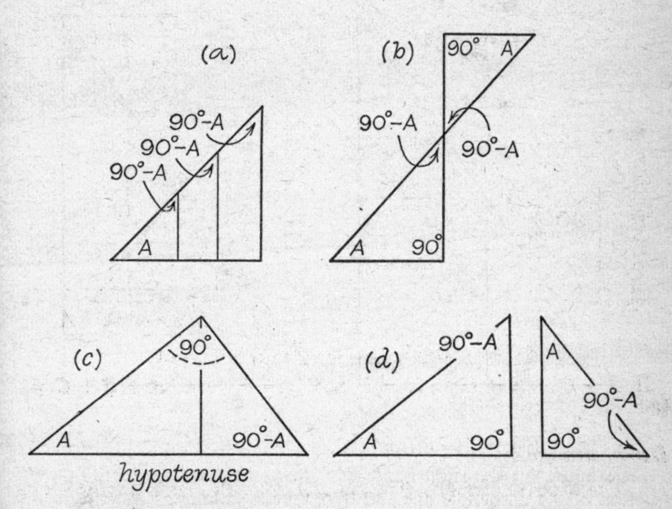

Fig. 44. Some Properties of Right-angled Triangles.

One defines two triangles as *similar* or *equiangular* if their corresponding angles are equal. From (*c*) and (*d*) of Fig. 44, we see that the following rule of dissection is true for all right-angled triangles:

The perpendicular dropped from the right angle to the hypotenuse divides a right-angled triangle into two other right-angled triangles each equiangular with it.

Having defined a parallelogram and a rectangle as in (vii) on p. 121, we may now justify what else we can say about them. Either diagonal divides a parallelogram into triangles having in common one side (Fig. 27) the diagonal and the angles at its extremities equal in virtue of the fact that a transversal, by (ii) p. 120, makes equal alternate angles with parallel lines. It follows that the two triangles are equivalent, whence the third angle of each is equal. Thus opposite angles of a parallelogram are equal, and if one is a right angle the other three must be right angles. If we define a square as a rectangle with two equal adjacent sides, the same construction shows that all four sides are equal:

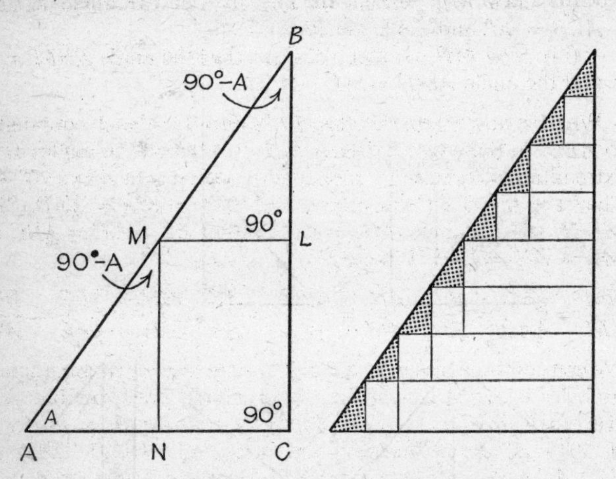

Fig. 45. Construction for Demonstration 2.

The figure on the right-hand side shows how we can continue bisection of the perpendicular and hypotenuse as long as we care to do so to generate smaller and smaller similar triangles.

Demonstration 2

The ratios of the lengths of corresponding sides of equiangular right-angled triangles are equal.

This demonstration is the gist of what Euclid's pupils reached in Book VI of his curriculum for affluent sophomores. Actually, it is true of all triangles that the ratio of corresponding sides is the same, if they are equiangular. Since we can split any triangle into two right-angled ones, the reader should be able to puzzle out why this must be true, if we can also say the same about triangles with a right angle. The exercise is not compulsory. To make the jump from Athens to Alexandria by putting back measurement into geometry, it is necessary only to name the fixed ratios of the surveyor and astronomer of posterity as given below.

To demonstrate the rule, as stated above (Fig. 45):

(i) bisect the hypotenuse of the right-angled triangle ABC, in which $\angle C = 90°$ and $\angle B = 90° - A$, so that $AM = BM$;

(ii) draw ML parallel to AC, so that the angle $MLB = ACB = 90°$ and the angle $BML = A$;

(iii) draw MN parallel to LC, so that the angle $ANM = 90°$ and the angle $AMN = 90° - A$.

We have now three triangles: BLM and AMN each equiangular to ABC. In these two triangles $MB = AM$ and the angles at the extremities of these sides are equal respectively to A and $90° - A$. They are therefore equivalent, and $BM = MA = \frac{1}{2}AB$. Since $MLCN$ is a rectangle, $MN = LC$, so that $BL = LC = \frac{1}{2}BC$ and $ML = AN = \frac{1}{2}AC$. Whence:

$$\frac{NM}{AM} = \frac{\frac{1}{2}BC}{\frac{1}{2}AB} = \frac{BC}{AB};\ \frac{AN}{AM} = \frac{\frac{1}{2}AC}{\frac{1}{2}AB} = \frac{AC}{AB};\ \frac{MN}{AN} = \frac{\frac{1}{2}BC}{\frac{1}{2}AC} = \frac{BC}{AC}$$

We can continue bisecting AB indefinitely thereby making smaller triangles equiangular with AB by a similar construction. However small we make AB, AC, BC in the triangle whose angles are A, $90° - A$, and $90°$ each of the ratios $BC : AB$, $AC : AB$, and $BC : AC$ has therefore a fixed value depending only on A.

The practical importance of this conclusion depends on the fact that we can make a scale diagram of any figure which we can dissect into right-angled triangles. When we have done so, the measurement of one angle A and one side of each right-angled triangle suffices for the reconstruction of all lengths we have not measured or are not accessible.

Our picture (Fig. 46) of one way in which Thales may have measured the height of the Great Pyramid at Gizeh shows a very simple application of the principle. To make the best use of it, we shall later need names for the ratios mentioned above. The current ones are as follows:

sine A (pronounced *sign A*) $\equiv$ sin A

$$= \frac{\text{length of perpendicular to } A}{\text{hypotenuse}}$$

$$\text{cosine } A \text{ (pronounced } KO\text{-}sign\ A) \equiv \cos A = \frac{\text{length of base to } A}{\text{hypotenuse}}$$

$$\text{tangent } A \equiv \tan A = \frac{\text{length of perpendicular to } A}{\text{length of base to } A} = \frac{\sin A}{\cos A}$$

As Fig. 43 shows, *base to* $90° - A =$ *perpendicular to* A, and vice versa, so that:

$$\sin A = \cos (90° - A) \textit{ and } \cos A = \sin (90° - A)$$

By means of this rule, we can divide a line into any number of equal parts, e.g. we can *trisect* one as in Fig. 47. A more import-

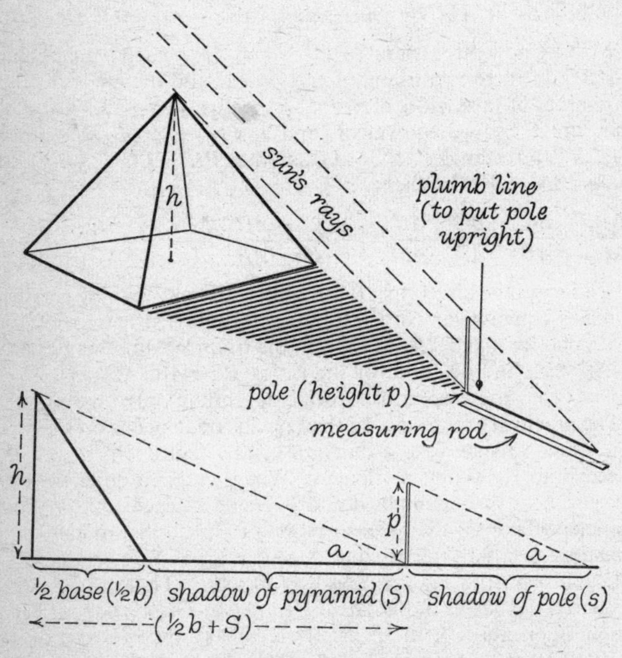

Fig. 46. One way in which Thales may have measured the height of the Great Pyramid at Gizeh by measuring the length of the Noon Shadow:

$$H \div (\tfrac{1}{2}b + S) = \tan A = p \div s$$

$$\therefore\ H = p(\tfrac{1}{2}b + S) \div s$$

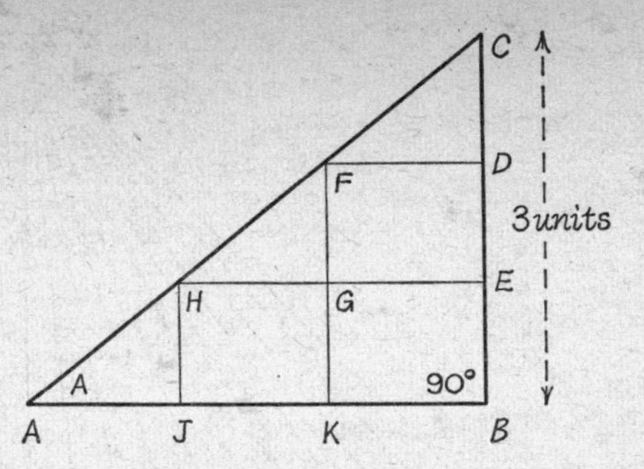

Fig. 47. Trisection of a Line

To AB, the line to trisect, erect the perpendicular BC of length 3 Units, so that $CD = DE = EB = 1$. Draw FD and HE parallel to AB, also FK and HJ parallel to CB. Then

$$FD = \frac{1}{\tan A} = KB;\ HG = \frac{1}{\tan A} = JK;\ AJ = \frac{1}{\tan A}$$

$$\therefore\ AJ = JK = KB$$

ant application of the recognition that these ratios are fixed quantities is a construction for making a scale of which each division of length d units is in a fixed ratio to that of another with division D units apart. Fig. 48 shows the recipe and Fig. 49 an important application, the *vernier*, which revolutionized precision of measurements in Newton's time.

Demonstration 3

If two sides of a triangle are of equal length the angles opposite to them are equal, and if two angles of a triangle are equal the sides opposite to them are of equal length.

(i) The first half of this assertion implies that $AB = l = AC$ in (*a*) of Fig. 50. If we bisect the angle A ($= BAC$) by the straight line AP, we have two triangles in one of which AP and AB enclose the angle $\frac{1}{2}A$ and in the other AP and AC ($= AB$) enclose the angle $\frac{1}{2}A$. These triangles having two equivalent sides, and the included angle are equivalent. Hence the corresponding angles ABC and ACB opposite, respectively, AC and AB are equal.

(ii) The second half of the statement implies that $\angle ABC = \angle ACB$, as in (ii) of Fig. 50. If we bisect the angle BAC as before

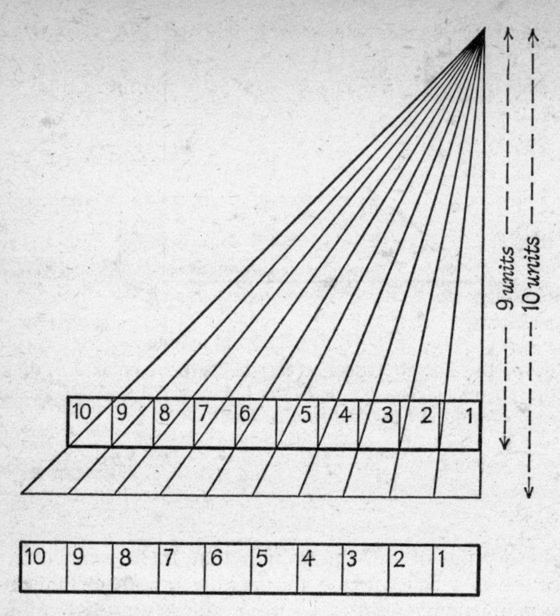

Fig. 48. Recipe for Making One Scale Nine-tenths as Long as Another

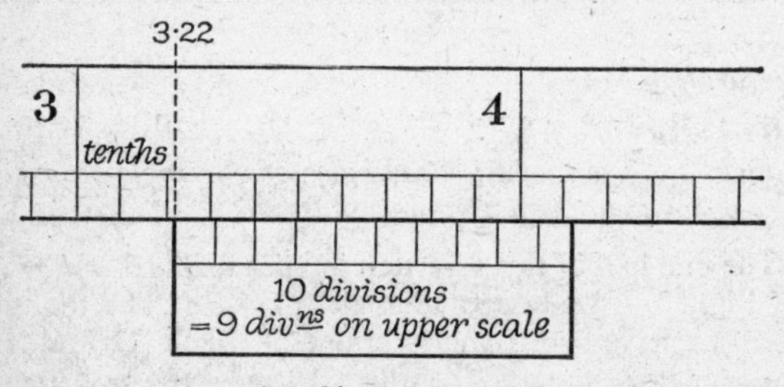

Fig. 49. The Vernier

Invented early in the seventeenth century (AD) by Pierre Vernier, a French mathematician, the Vernier Scale is an auxiliary movable ruler device, which permits great accuracy. On the lower ruler, we see a

by the straight line AP we see that $\angle ABP = \angle ABC$ and $\angle ACP = \angle ACB$, so that

$$\angle ABP + \angle \tfrac{1}{2}A + \angle APB = 180° = \angle ACP + \angle \tfrac{1}{2}A + \angle APC$$

$$\therefore \angle ABC + \angle \tfrac{1}{2}A + \angle APB = 180° = \angle ABC + \angle \tfrac{1}{2}A + \angle APC$$

$$\therefore \angle APB = 90° = \angle APC$$

The two triangles ABP and ACP, having one side of equal length and three equal angles are equivalent. Hence the corresponding sides AB and AC are of equal length.

It is customary to speak of a triangle with two equal sides as an *isosceles* triangle and one with three equal sides and (by our new rule) three equal angles as an equilateral triangle. From Dem. 1, we see that (Fig. 20):

(i) if one angle (A) of a right-angled triangle is 45°, the other is 45°, whence the two sides enclosing the right angle are of equal length; (ii) each angle of an equilateral triangle is 60°; and by bisection of any one of them we can divide it into two triangles whose angles are 30°, 60°, and 90°.

We can now take our first step (Fig. 51) towards tabulating *trigonometrical ratios*, i.e.

$$\tan 45° = \frac{l}{l} = 1; \quad \cos 60° = \frac{\frac{1}{2}l}{l} = \tfrac{1}{2}; \quad \sin 30° = \frac{\frac{1}{2}l}{l} = \tfrac{1}{2}$$

[*Caption continued*]

distance equivalent to nine divisions on the upper scale divided into ten parts. To measure an object the end of which is marked by the thin line between 3·2 and 3·3 in the figure, set the beginning of the Vernier Scale at this level and look for the first division which exactly coincides with a division on the upper scale. In the figure this is the second division and the correct measurement is 3·22. The theory of the device is as follows. If x is some fraction of a division on the upper scale to be ascertained, the correct measurement is $3{\cdot}2 + a$. On the lower scale, the first whole number a of the smaller divisions coincident with the upper differs from a division on the upper scale by the distance x. Now 1 division on the lower scale is $\frac{9}{10}$ of a division on the upper. Hence

$$(\tfrac{9}{10}a) + x = a \quad \textit{so that} \quad x = \tfrac{1}{10}a$$

If a is 2, x = 2-tenths of a scale division on the upper scale. If a division on the upper scale is 0·1, $x = 0{\cdot}02$.

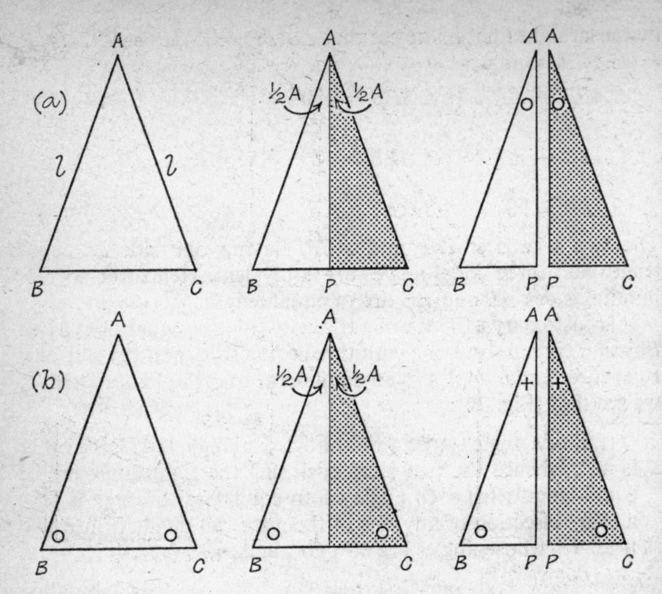

Fig. 50. The Pons Asinorum
We may judge how forbidding was the teaching of Euclid's geometry only a century ago from the fact that Dem. 2 was still a stumbling block to many pupils, whence called the *Bridge of Asses*.

Tradition credits Thales with the following rule which follows from Dems. 1 and 3. If we connect by straight lines the extremities of the diameter (Fig. 19) of a circle with any point on the perimeter, the figure so found is a right-angled triangle. It is usual to state this rule by saying that *the angle in a semicircle is a right angle*. According to the source of the story, Thales sacrificed an ox to the gods, when he found a satisfactory proof of the truth of this assertion. As we have seen (p. 50), it is an eminently useful recipe for making a set square.

We can usefully apply the fact that *tan* $45° = 1$ in several ways, e.g.:

(i) when the shadow of an upright pole is equal to its height, that of a wall or of a vertical tower, if we add to it half their width respectively, is also equal to their heights (see Fig. 21);

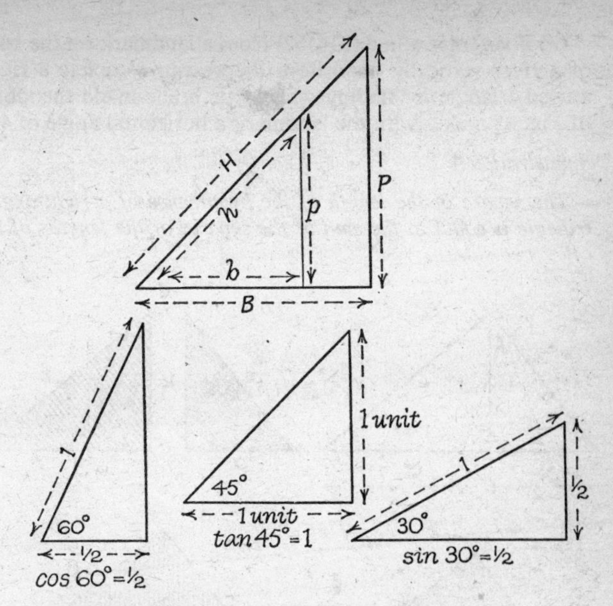

Fig. 51. First Step to Construction of a Trigonometrical Table

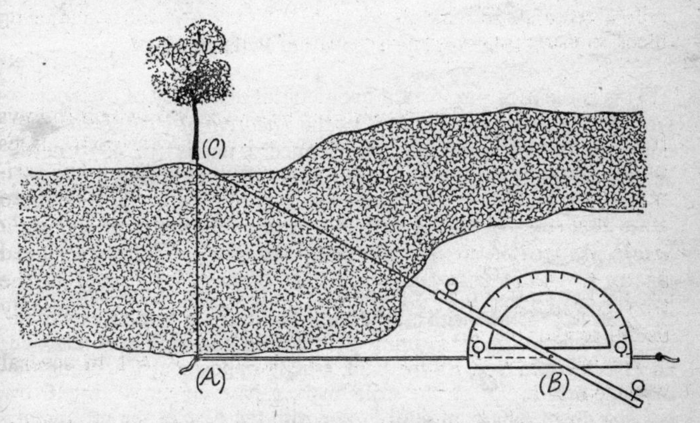

Fig. 52. Measuring the Width of a River

A simple instrument (see Fig. 25) may be made by nailing a strip of wood to the centre of a protractor so that it can revolve freely. Screw

(ii) if we trace a line (Fig. 52) from a landmark on the bank of a river vertically opposite to it, we can complete a right-angled triangle by finding where our home-made theodolite of Fig. 25 makes with the landmark a horizontal angle of 45°.

Demonstration 4

The square of the length of the hypotenuse of a right-angled triangle is equal to the sum of the squares of the lengths of the other two sides.

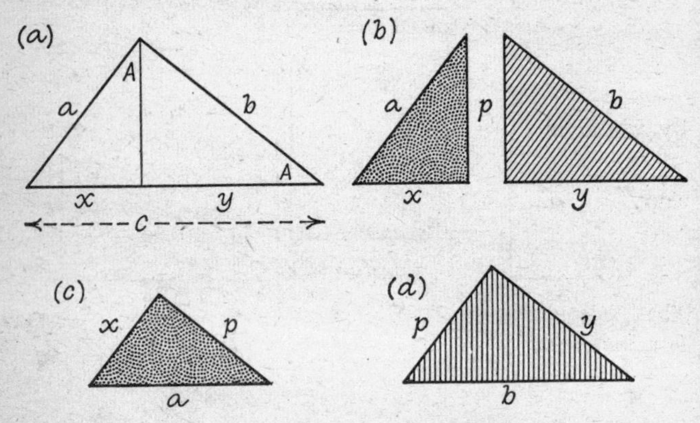

Fig. 53. The Theorem of Pythagoras

This is the most important geometrical discovery of the ancient world. One usually refers to it as the *Theorem of Pythagoras*. The converse, which need not here concern us, is (p. 48) one of several recipes for making a right angle. Fig. 18 shows a Chinese

---

[*Caption continued*]

into the two ends of this strip and into the two ends of the base line of the protractor screws with eyes (such as are used for holding extensible curtain rods) to use for sighting. Standing on one bank at A select an an object, e.g. a tree, on the bank exactly opposite at C. Setting the movable arm at 90° on the scale, make a base line at 90° to *AC* by pegging down a piece of cord in line with the base of the protractor. Walk along this line to *B*, where *C* is seen to be exactly at 30° to *AB*. Measure *AB*. Then *ABC* is a right-angled triangle in which $AC = \frac{1}{2}BC$ and $AB = (\sqrt{3}/2)BC$. So *AC*, the width of the stream, is $AB \div \sqrt{3}$.

demonstration of its truth current at least in the first century of the Christian era. Here we use the dissection exhibited in Fig. 44 on p. 123. This divides the right-angled triangle to the left in the top row of Fig. 53 into two triangles equiangular with it, and placed in (*c*)–(*d*), so that each lies like the composite triangle with its hypotenuse facing the bottom of the page and corresponding sides or angles are easy to recognize at a glance. Dem. 2 then tells us:

$$\frac{a}{c} = \frac{x}{a}, \text{ so that } a^2 = cx$$

$$\frac{b}{c} = \frac{y}{b}, \text{ so that } b^2 = cy.$$

By combining these results:

$$a^2 + b^2 = cx + cy = c(x + y)$$

since

$$(x + y) = c,\ c\,(x + y) = c^2 \text{ and}$$

$$a^2 + b^2 = c^2$$

The last equation expresses what we set out to prove. We may note also that

$$\frac{p}{x} = \frac{y}{p}, \quad \text{so that } p^2 = xy \quad \text{and } p = \sqrt{xy}$$

In the last expression we call p the *geometric mean* of x and y. Thus if $x = 3$ and $y = 27$, $p^2 = 27 \times 3 = 81$ and $p = 9$. So x, p, y are terms of the geometric series $3;\ 3^2 = 9;\ 3^3 = 27$. We shall recall this when we come to logarithms.

We are now ready to take a big stride forward. We first recall the isosceles right-angled triangle of Fig. 51. If the two equal sides are of length 1 unit, we may write for that of the hypotenuse (h):

$$h^2 = 1^2 + 1^2 = 1 + 1 = 2$$

$$\therefore h = \sqrt{2}$$

Whence (Fig. 54):

$$\sin 45° = \frac{1}{\sqrt{2}} = \cos 45°.$$

Next recall a triangle (Fig. 51) whose angles are $A = 30°$, $B = 60°$, $C = 90°$. If we derive this by splitting an equilateral triangle whose sides are of length 1 unit, we may write for the hypotenuse of

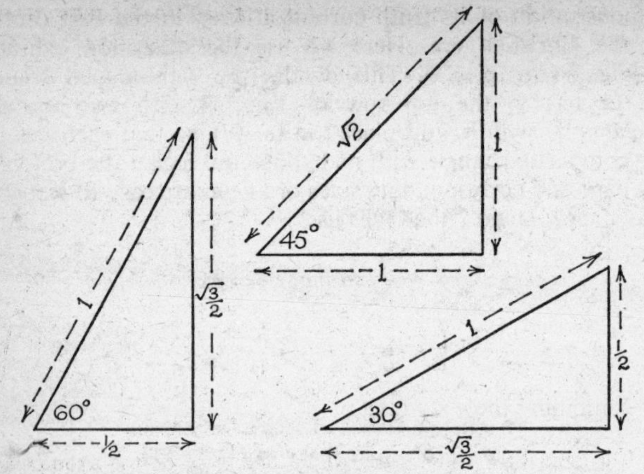

Fig. 54. Second Step to Construction of a Trigonometrical Table

$\tan 45° = 1$; $\sin 45° = 1/\sqrt{2}$; $\cos 45° = 1/\sqrt{2}$
$\tan 60° = \sqrt{3}$; $\sin 60° = \sqrt{3}/2$; $\cos 60° = \frac{1}{2}$
$\tan 30° = 1/\sqrt{3}$; $\sin 30° = \frac{1}{2}$; $\cos 30° = \sqrt{3}/2$

$\triangle ABC$, $h = 1$, and one of the remaining sides $b = \frac{1}{2}$. Then if p is the third side $h^2 = b^2 + p^2$, so that $p^2 = h^2 - b^2$ and

$$p^2 = 1^2 - (\tfrac{1}{2})^2 = 1 - \tfrac{1}{4} = \tfrac{3}{4}$$

$$\therefore \quad p = \frac{\sqrt{3}}{2}$$

We thus arrive at the result (Fig. 54):

$$\sin 60° = \frac{\sqrt{3}}{2} = \cos 30°.$$

Similarly (Fig. 54), we derive:

$$\tan 30° = \frac{1}{\sqrt{3}} \quad \textit{and} \quad \tan 60° = \sqrt{3}$$

Before we make our first miniature table of trigonometrical ratios, we may note that:

(i) as the angle A (Fig. 55) of a right-angled triangle becomes nearer and nearer to 90°, p becomes nearer and nearer to h, so that, when $A = 90°$, $p = h = 1$;

(ii) as A becomes nearer and nearer to zero, b becomes nearer and nearer to h, so that when $A = 0$, $b = h = 1$.

Thus we can add to our list:

$$\sin 90° = 1 = \cos 0°$$
$$\sin 0° = 0 = \cos 90°$$

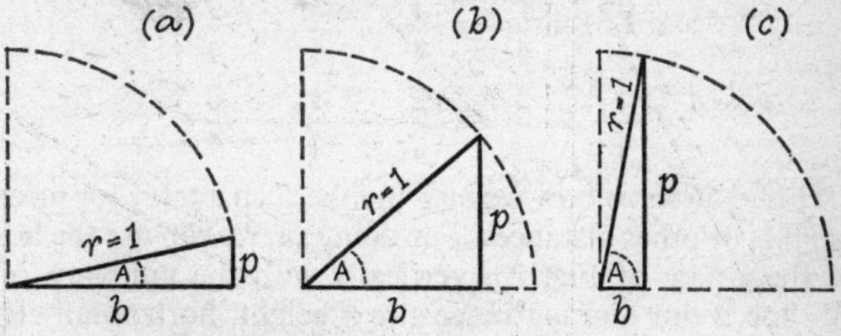

Fig. 55. Ratios of Small Angles, etc.

The circle here drawn has radius 1 unit long. So the hypotenuse (r) is in each case 1 unit ($r = 1$).

$$\sin A = \frac{p}{r} = p, \textit{ and } \cos A = \frac{b}{r} = b$$

One important rule which follows from Dem. 4 is the following. Since $h^2 = p^2 + b^2$ if *sin* $A = p \div h$ and *cos* $A = b \div h$:

$$\frac{h^2}{h^2} = \frac{p^2}{h^2} + \frac{b^2}{h^2}, \textit{ so that } 1 = (\sin A)^2 + (\cos A)^2$$

It is usual to write for brevity:

$$\sin^2 A \equiv (\sin A)^2; \quad \cos^2 A \equiv (\cos A)^2; \quad \tan^2 A \equiv (\tan A)^2$$

This shorthand is unfortunate, because it suggests that $sin^{-1} A$ means something very different (see p. 376) from what it does. However, nearly all books in circulation use it, and our rule accordingly takes the form:

$$\sin^2 A + \cos^2 A = 1$$

$$\therefore \quad \sin A = \sqrt{1 - \cos^2 A} \quad \textit{and} \quad \cos A = \sqrt{1 - \sin^2 A}$$

This means that we can convert a table of sines into one of cosines or vice versa. Test the rule on the foregoing table, e.g.:

$$\sin 30° = \sqrt{1 - \cos^2 30°} = \sqrt{1 - \tfrac{3}{4}} = \tfrac{1}{2}$$

$$\cos 60° = \sqrt{1 - \sin^2 30°} = \sqrt{1 - \tfrac{1}{4}} = \frac{\sqrt{3}}{2}$$

If we can convert a table of sines into a table of cosines or vice versa, either suffices to make a table of tangents, since *tan* $A =$ *sin* $A \div$ *cos* A. The formula last given helps us to remember the figures in our first table if we write them thus:

$A°$	0°	30°	45°	60°	90°
$\sin^2 A$	$\frac{0}{4}$	$\frac{1}{4}$	$\frac{2}{4}$	$\frac{3}{4}$	$\frac{4}{4}$
$\cos^2 A$	$\frac{4}{4}$	$\frac{3}{4}$	$\frac{2}{4}$	$\frac{1}{4}$	$\frac{0}{4}$

Figs. 21 and 56 show how we may invoke such a table for measuring heights or other distances. For doing so, we can use the home-made theodolite of Fig. 25, vertically with the guidance of its plumb line if our aim is to measure a height, horizontally (with the aid of a spirit level), if we wish to measure a width. In either

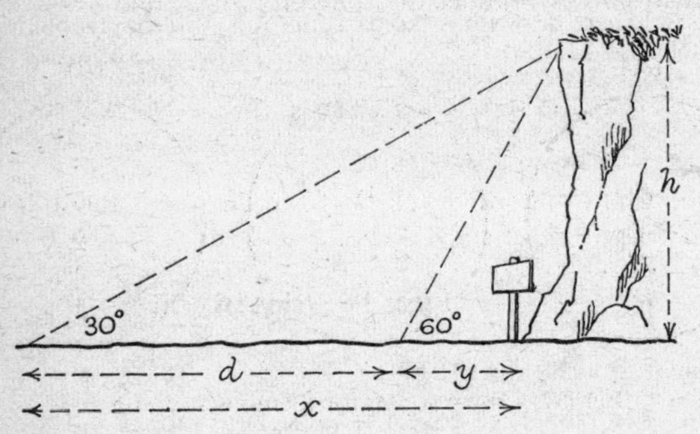

Fig. 56. Measuring the height of a cliff when trespassers near the bottom will be prosecuted

As drawn, you cannot find x or y, but you can measure out $d = (x - y)$.

$$\therefore\ x - d = y$$

$$\therefore\ \frac{h}{x} = \frac{1}{\sqrt{3}} \quad \text{or} \quad h\,.\sqrt{3} = x : \frac{h}{y} = \sqrt{3} \quad \text{or} \quad y = \frac{h}{\sqrt{3}}$$

$$h\,.\sqrt{3} - d = \frac{h}{\sqrt{3}}$$

Multiplying both sides by $\sqrt{3}$, we get: $3h - d\sqrt{3} = h$

$$\therefore\ 2h = d\sqrt{3} \quad \textit{and} \quad h = \frac{\sqrt{3}}{2}\,.\,d$$

Measuring the distance of the moon from the earth is essentially like this.

case, we assume that we can find places at a measurable distance along a base line from which we can sight a particular landmark (e.g. the top of a cliff or a tree on a river bank) at a pair of angles such as 30° and 60° or 90° and 45°. Our pictorial representation from the measurement of the height of a cliff makes the base line at ground level. In practice it would be at eye level. The reader may make a sketch to show how one can correct for this source of error. If the cliff is a high one, it will be trivial.

In real life, it may not be possible, and will commonly be at best laborious, to find at a measurable distance, two places from which we can get sight lines at a pair of angles cited in our table, as illustrated in the two pictures last mentioned. Before we take leave of Euclid as a springboard from which to build up a source of surveying, we shall need to be able to extend our table to cope with any situation in which we can sight an object in angular measure at two extremities of a measured base line.

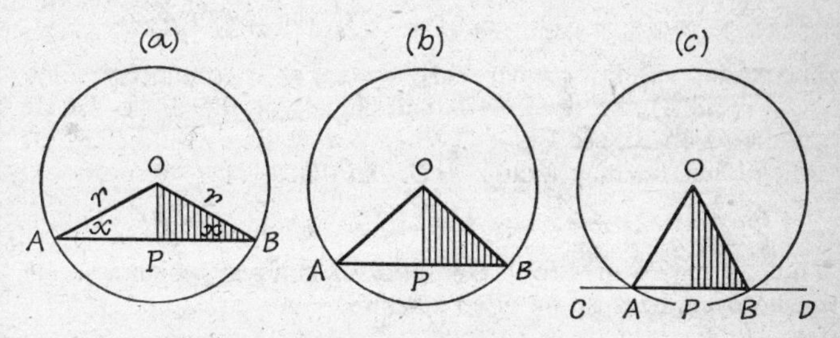

Fig. 57. Construction for Demonstration 5

Demonstration 5

> *The straight line (tangent) grazing a circle is at right angles to the straight line which joins the centre to the spot where it touches the boundary.*

Fig. 13 shows an informal way of demonstrating that this is so. When the plumb line swings through an arc of a circle from whose centre it hangs it comes to rest where the circle grazes the plane of the horizon. For a more formal demonstration (Fig. 57) we may join the mid-point P of any chord (AB) of a circle of radius r to the centre O. We have then two triangles with sides $OA = r = OB$, $AP = PB$ and OP common. By *Triangle Rule Three* they are thus equivalent, and $OPA = 90° = OPB$ (since $OPA +$

$OPB = 180°$). We now extend the line APB to any points C and D beyond the boundary of the circle, and notice that we can bring OB and OA closer and closer to OP till $\angle AOB = 0°$. P is then the grazing point, CD the tangent, and OP perpendicular to it.

The reader may here ask: why do we use the same word (tangent) both for a line which grazes (Latin *tangere* = to touch, as in *tangible*) a circle and for the ratio defined on p. 126? The answer is instructive. As we shall see later, it helps us to enlarge and to simplify what we mean by the trigonometrical ratios. Fig. 58 discloses the relevant facts. If OR is the radius of a circle and PR is at right angles (i.e. is *tangent* to it) at R and $\angle POR = A$, our original definition is that *tan* $A = PR \div OR$. Now this ratio (by *Dem.* 2) does not depend on the absolute length of OR. It is dependent *only* on how large (or small) A is. Consequently, we can make our circle one of *unit radius*, i.e. $OR = 1 = OS$, and then see that:

$$\sin A = SQ; \quad \cos A = OQ; \quad \tan A = PR$$

The reader will now see why early writers on trigonometry called SQ (= *sin* A) the *semichord* (half-chord) of the angle $2A$. If unable to do so, see Fig. 84. When not dealing with a circle of unit radius, we may write $r = OR$ so that:

$$\sin A = SQ \div r; \quad \cos A = OQ \div r; \quad \tan A = PR \div r$$

Thus we may express SQ, OQ, and PR in a way which is *worth memorizing*, because we often need it:

$$SQ = r \,.\, \sin A; \quad OQ = r \,.\, \cos A; \quad PR = r \,.\, \tan A$$

The practical applications of this demonstration are many. We shall consider two of them after (p. 214) we have found how to

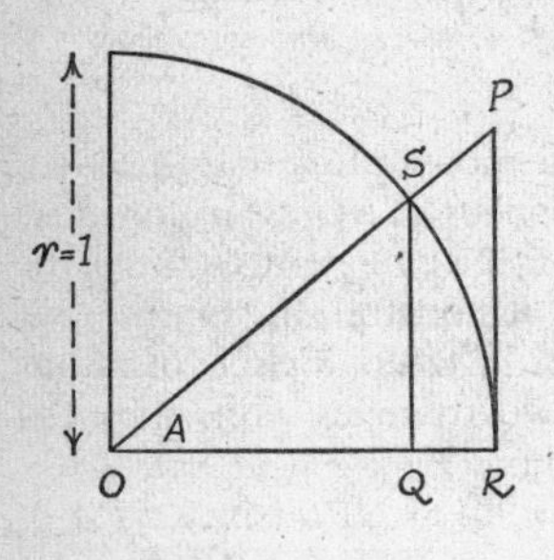

Fig. 58. A Useful Way of Defining the Trigonometrical Ratios

In the circle of unit radius here shown: $OS = 1 = OR$, and:

$$\sin A = SQ \div OS = SQ$$
$$\cos A = OQ \div OS = OQ$$
$$\tan A = PR \div OR = PR$$

This explains why early writers on trigonometry spoke of $PR \div OR$ as the *tangent* of A, and why they spoke of $SQ \div OS$ as the *semichord* of A.

express the boundary of a circle, hence the circumference of the earth, if we also know the length of its radius. It is, indeed, a kingpin of scientific geography. On the assumption that light travels in straight lines it signifies (Fig. 59) that:

(i) *the straight line joining the observer to a point on the horizon boundary is at right angles to the line which joins the observer to the earth's centre, i.e. to a continuation of the plumb line;*

(ii) *the zenith* (*p.* 56), *the observer's plumb line and the earth's centre are in one and the same straight line;*

(iii) *When a celestial body, e.g. the sun at noon, is at transit, i.e. highest above the horizon, it is in the same plane as the plumb line and the earth's centre.*

The last three statements contain all the essential information we require in order to bring the navigator's notion of latitude (and of longitude) into line with drawing-board design on the assumption that the earth's shape is spherical. As we have seen (p. 71) this belief which was a prominent feature of the teaching of the Phoenician Pythagoras, is of great antiquity. It arises inescapably from the experience of seafaring people. Even before ships tracked north or south along the coast-line beyond the Pillars of Hercules, the mariner in port was accustomed to see a ship appear or disappear piecemeal above or below the limit of vision and the mariner at sea could be familiar with the gradual appearance or disappearance of a mountaintop as his ship came nearer to land or left it farther behind. A flat earth surface could not account for such facts. How experience of long distance contributed to the notion of a spherical earth divided into zones of latitude we have also seen. Let us recall briefly the essential data.

Finding Your Latitude (Figs. 60 and 61)

The stars appear to revolve in circles upwards from the east and downwards to the west, about an axis which passes through a point which is called the celestial pole. Nowadays we explain this by saying that the earth revolves around an axis which passes through its centre, its poles, and the celestial pole, in the opposite direction to the apparent motion of the heavenly bodies. Most stars set below the horizon and are only visible at night during part of the year, but the stars very near the pole, like those in the

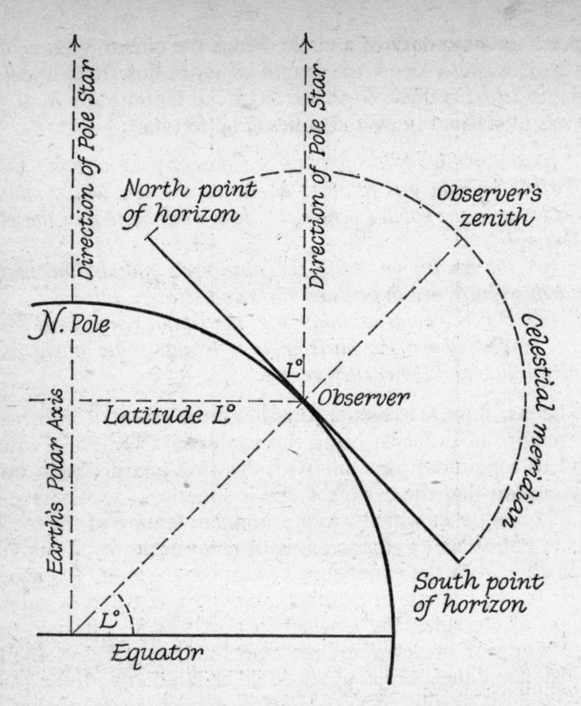

Fig. 59. A Star at Meridian Transit lies in the same plane as the Observer, the Earth's Poles, the Zenith, the South and North Points of the Horizon, and the Earth's Centre

A straight line which does not lie on some particular flat slab of space (so-called plane) can cut it only once. A straight line which passes through more than one point on a plane therefore lies on the same plane as the points through which it passes. The plane bounded by the observer's great circle of longitude and the earth's axis includes as points the earth centre, observer, and the earth's poles. The line joining zenith and observer passes also through the earth centre, i.e. through more than one point on this plane. Hence it lies wholly on the plane. The North and South points of the horizon are simply points where lines from the earth centre through the meridian of longitude pierce the horizon plane. They must also lie in the same plane as these lines. So South and North points, zenith and observer are all in the same plane with the earth centre and poles. A circle can cut a plane in which it does not itself lie only at two points. The circle drawn through the North and

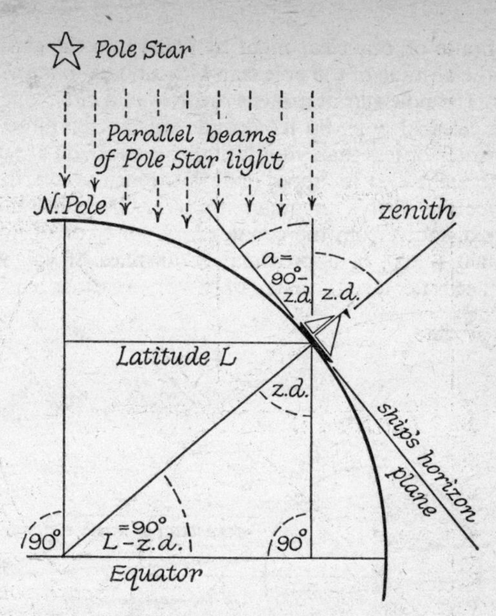

Fig. 60. Latitude from Pole Star

The altitude (horizon angle) of the pole star, if located exactly at the Celestial Pole, is the observer's latitude, both being equivalent to 90° – zenith distance of pole star.

constellation of the Great Bear, the Little Bear, Lyra, Draco, and Cassiopeia, in Great Britain never sink below the horizon, being seen most of the night below the pole at some seasons and above it at others. One star, the *pole star*, is so near the celestial pole that it always seems to be in the same place. It lies almost exactly in line with the earth's north pole and the earth's centre. Since star beams are parallel, the rays which reach us from the pole star are parallel to the earth's axis. You will see from Fig. 60 that the latitude of a place is the angle ('altitude') which the celestial pole makes with the horizon. So you can get the latitude

---

[*Caption continued*]

South points and the zenith passes through more than two points in the same plane, and therefore lies wholly on it. So any point on this imaginary circle (the celestial meridian) is also on it.

of your house on any clear night by going into the garden and sighting the altitude of the pole star with a home-made astrolabe (Fig. 25). The pole star at present revolves in a circle one degree from the celestial pole. So its altitude will not be more than a degree greater or less than your latitude, even if you are unlucky enough to sight it at its upper or lower transit across the meridian. Since the earth's circumference is 25,000 miles, this gives you your distance from the equator with an error of not more than 25,000 ÷ 360, or approximately 70 miles. If you want to be really accurate, take the mean of two observations made at the

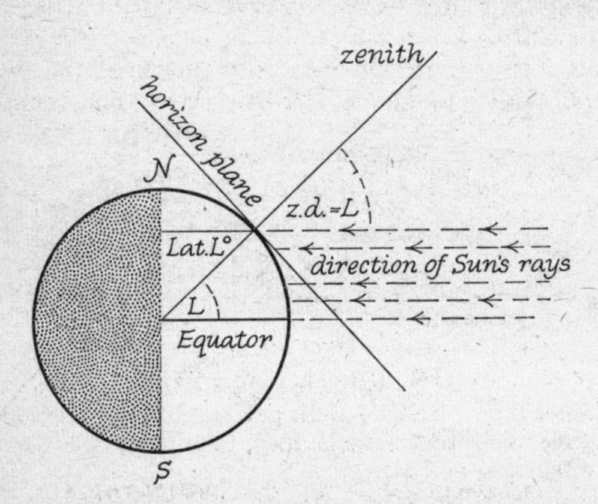

Fig. 61. Latitude by the Equinoctial Noon Shadow

On March 21st and September 23rd day and night are of equal length throughout the world. So the sun lies directly above the equator. At noon the sun always lies over the line joining the North and South points of the horizon, i.e. the observer's meridian of longitude. So the sun, the poles, observer, zenith, and earth centre are all in the same flat slab (or 'plane') of space. Since the edges of a sunbeam are parallel, the sun's *z.d.* at noon on the equinoxes is the observer's latitude.

same time of night, one six months after the other, when the pole star will be just as much above the celestial pole as it was previously below it, or vice versa.

At the same time you may like to know how to find your longitude (Fig. 62). Nowadays, this is a very simple matter because ships have accurate clocks which can keep Greenwich time over a long voyage, and most of us can tune in for Greenwich time on the radio. Noon is the time when the sun is exactly above the meridian at its highest point in the heavens. If our sundial registers noon an hour after it is noon at Greenwich, the sun has to travel, as the ancients would say, 15° farther west, or our earth has to rotate eastwards through 15° on its axis, between the times of the two noons. We are therefore 15° W. of Greenwich. The peoples of antiquity discovered that time recorded by the shadow clock does not synchronize in different places through observing when an eclipse occurred, or when a planet passed behind the moon's disc. The Babylonians had hour-glasses of sand, and could observe the time which elapsed between noon on a particular day and the beginning or end of an eclipse or occulation. Before chronometers were invented this was the chief way of finding longitude. If at one place a lunar eclipse was

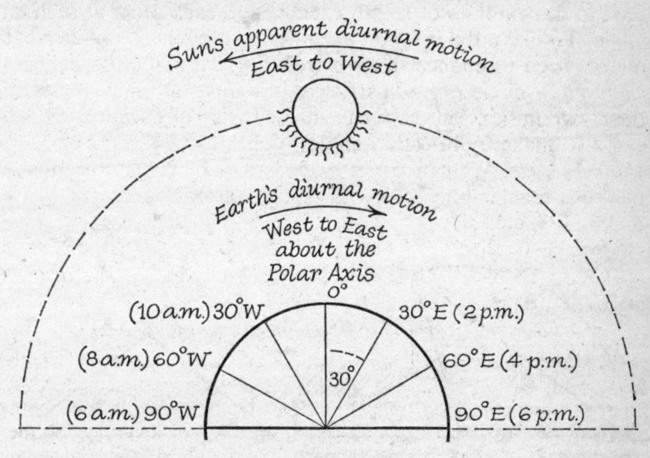

Fig. 62. Longitude

At noon the sun lies directly over the line joining the north and south points of the horizon, i.e. the meridian of longitude on which you are located. In the figure, it is directly above the Greenwich meridian, and it is therefore noon at Greenwich. If you are 30° East of Greenwich the earth has rotated through 30° since your sundial registered noon. It has therefore made one-twelfth of its twenty-four-hourly revolution, so that it is now two o'clock by the sundial. If you are 60° West the earth has still to rotate 60° before the sun will be over your meridian, i.e. one-sixth of its twenty-four-hourly rotation; so your sundial will register 8 am.

seen to begin at 8 hours after local noon and at another $9\frac{1}{2}$ hours after local noon, noon in the second place occurred $1\frac{1}{2}$ hours earlier than noon at the first. So the second was $1\frac{1}{2} \times 15°$ $= 22\frac{1}{2}°$ East of the first. The construction of maps based on latitude and longitude was never achieved by the Greeks. It was done when Greek geometry was transferred to the great shipping centre of the classical world, Alexandria.

One application of the principle of tangency is a recipe for making a regular polygon enclosing a circle. A regular polygon, including equilateral triangles with three and squares with four, has n equal sides. It is therefore divisible into n triangles each having two sides (see Fig. 67) equal to the radius of the enclosed or enclosing circle and vertices equal to $360° \div n$. If we can make such an angle (e.g. $60°$ if $n = 6$ and $45°$ if $n = 8$), we proceed to draw n lines of length r inclined to each other at such an angle. To make the polygon *enclosed* by a circle of radius r, we merely need to connect their extremities. To make the polygon *enclosing* a circle of radius r, we draw lines at right angles to their extremities. Since we can make angles of 60° and 45°, we can also make by bisection angles of 30°, 15°, $7\frac{1}{2}°$, etc., or $22\frac{1}{2}°$, $11\frac{1}{4}°$, $5\frac{5}{8}°$, etc. We can therefore construct by rule and compass methods regular polygons with 6, 12, 24, 48, 96, etc., or 8, 16, 32, 64, 128, etc., sides.

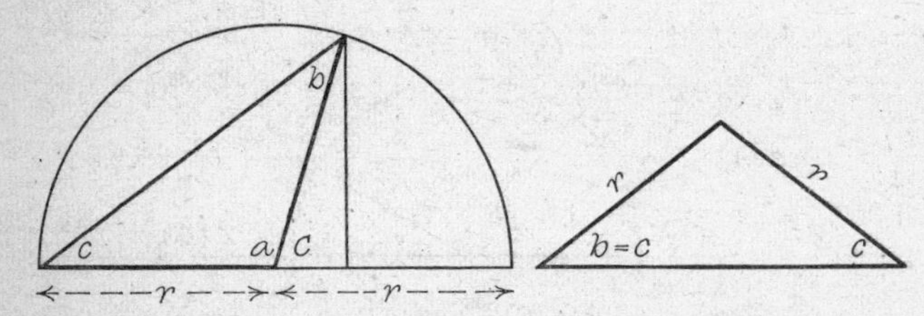

Fig. 63. Construction for Demonstration 6

Demonstration 6

The angle enclosed by straight lines from the extremities of a chord to a point on the perimeter of a circle is half as large as the angle enclosed by straight lines joining the extremities of the chord to the centre.

Such is the way in which we learn this key rule as it comes down to us from the ancients. In fact, all that is important is to see what it means if we divide the circle into two parts as in Fig. 63. There the larger triangle has two equal sides (r being the radius of the circle). We have to show that the angle (c) formed by joining a point to the extremity of the diameter is half the angle (C) formed by joining the point to the centre of the circle.

By Dem. 3: $b = c$ (because $r = r$)
By Dem. 1: $a + b + c = 180° = a + 2c$
By ANGLE RULE ONE (p. 120) $a + C = 180°$

$$a + 2c = 180° = a + C,$$
$$\textit{so that } 2c = C \text{ and } c = \tfrac{1}{2}C$$

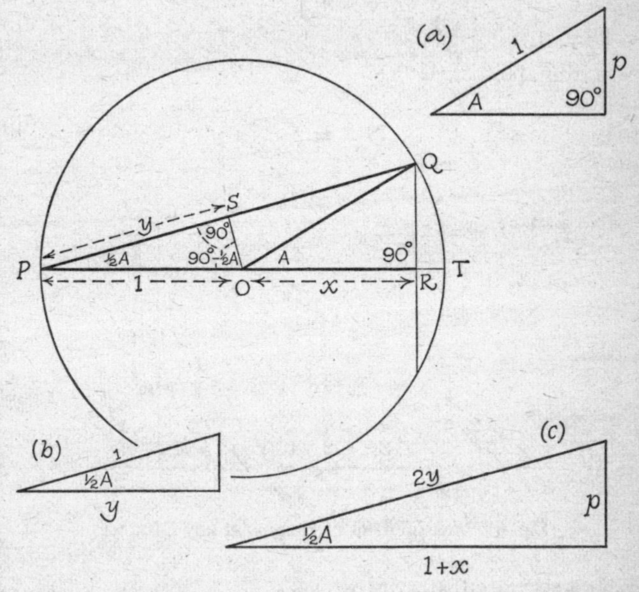

Fig. 64. Third Step to Construction of a Trigonometrical Table

At first sight this is not very formidable; but it leads to a remarkable result which we can expose more easily if (as on p. 138) we exhibit (Fig. 64) the same construction in a circle whose radius is 1 unit, labelling $C = A$ and $c = \frac{1}{2}A$. The only additional dissection is that the triangle POQ is cut up into two right-angled triangles by OS at right angles to PQ. The two right-angled triangles are equivalent by TRIANGLE RULE TWO because

$$SO = SO$$
$$\text{The enclosed } \angle SOP = 90^\circ - \tfrac{1}{2}A = \angle SOQ$$
$$OP = 1 = OQ$$

So if PS is y units long, $PS = \frac{1}{2}PQ$ and $PQ = 2y$. Since the radius of the circle is 1 unit long, $OQ = OP = 1$. The figure shows

$$\cos A = \frac{x}{OQ} = x \quad \dots\dots\dots \quad \text{(i)}$$

$$\cos \tfrac{1}{2}A = \frac{PR}{PQ} = \frac{1+x}{2y} \quad \dots\dots\dots \quad \text{(ii)}$$

Also from the triangle POS

$$\cos \tfrac{1}{2}A = \frac{y}{PO} = y \quad \dots\dots\dots \quad \text{(iii)}$$

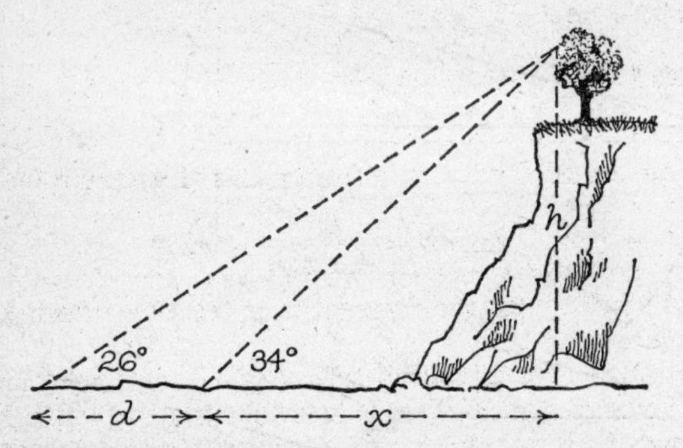

Fig. 65. Measurement of Heights at any Distance

Combining (i), (ii), and (iii), we have

$$\cos \tfrac{1}{2}A = \frac{1 + \cos A}{2 \cos \frac{1}{2}A}$$
$$2(\cos \tfrac{1}{2}A)^2 = 1 + \cos A$$
$$(\cos \tfrac{1}{2}A)^2 = \tfrac{1}{2}(1 + \cos A)$$
$$\cos \tfrac{1}{2}A = \sqrt{\tfrac{1}{2}(1 + \cos A)} \quad \dots\dots \quad (a)$$

Before going farther, test this. You already know that cos $60° = 0{\cdot}5$ and cos $30° = \frac{1}{2}(\sqrt{3})$. From the new rule

$$\cos 30° = \cos \tfrac{1}{2}(60°) = \sqrt{\tfrac{1}{2}(1 + \cos 60°)} = \sqrt{\tfrac{1}{2}(1{\cdot}5)} = \sqrt{\tfrac{3}{4}}$$

$$\text{i.e. } \cos 30° = \tfrac{1}{2}(\sqrt{3})$$

We get the sine rule for half angles in the same way. Thus

$$\sin A = p$$

$$\sin \tfrac{1}{2}A = \frac{p}{2y} = \frac{\sin A}{2 \cos \tfrac{1}{2}A} \quad \dots\dots \text{(iv)}$$

You can check the sine rule for half angles first by using what you already know, namely, that sin $60° = \frac{1}{2}(\sqrt{3})$, sin $30° = \frac{1}{2}$, and cos $30° = \frac{1}{2}(\sqrt{3})$:

$$\sin 30° = \sin \tfrac{1}{2}(60°) = \frac{\sin 60°}{2 \cos 30°} = \frac{(\frac{1}{2}\sqrt{3})}{2(\frac{1}{2}\sqrt{3})} = \tfrac{1}{2}$$

Similarly, you can check both rules against the result obtained with the scale diagram of Fig. 84, thus:

$$\sin 15° = \sin \tfrac{1}{2}(30°) = \frac{\frac{1}{2}}{2 \cos 15°}$$

$$\cos 15° = \cos \tfrac{1}{2}(30°) = \sqrt{\tfrac{1}{2}(1 + \cos 30°)} = \sqrt{\tfrac{1}{2}(1 + 0{\cdot}866)}$$

$$\text{i.e. } \cos 15° = \sqrt{0{\cdot}933} = 0{\cdot}966$$

You can get the last step from tables of square roots. So

$$\sin 15° = \frac{0{\cdot}5}{2(0{\cdot}966)} = 0{\cdot}259$$

This differs from the value we got from the scale drawing by less than one per cent.

If you are now convinced that the half-angle rules are good ones, you can make a table of sines like the one which Hipparchus made at Alexandria about 150 BC, with the added advantage that we have more accurate tables of square roots and a decimal system of fractions. We have got cos $15° = 0{\cdot}966$. So sin $(90° - 15°)$ or sin $75° = 0{\cdot}966$.

We have also got sin $15° = 0{\cdot}259$. So cos $(90° - 15°)$ or cos $75° = 0{\cdot}259$. We next get cos $7\frac{1}{2}°$, using tables of square roots as before:

$$\cos 7\tfrac{1}{2}° = \cos \tfrac{1}{2}(15)° = \sqrt{\tfrac{1}{2}(1{\cdot}966)} = \sqrt{0{\cdot}983} = 0{\cdot}991$$

$$\sin 7\tfrac{1}{2}° = \frac{\sin 15°}{2 \cos 7\frac{1}{2}°} = \frac{0{\cdot}259}{2(0{\cdot}991)} = 0{\cdot}131$$

This gives us

$$\cos 7\tfrac{1}{2}^\circ = 0{\cdot}991 = \sin 82\tfrac{1}{2}^\circ$$
$$\sin 7\tfrac{1}{2}^\circ = 0{\cdot}131 = \cos 82\tfrac{1}{2}^\circ$$

From $\cos 75^\circ = 0{\cdot}259$ and $\sin 75^\circ = 0{\cdot}966$:

$$\cos 37\tfrac{1}{2}^\circ = \cos \tfrac{1}{2}(75)^\circ = \sqrt{\tfrac{1}{2}(1{\cdot}259)} = \sqrt{0{\cdot}629} = 0{\cdot}793$$

$$\sin 37\tfrac{1}{2}^\circ = \sin \tfrac{1}{2}(75)^\circ = \frac{0{\cdot}966}{2(0{\cdot}793)} = 0{\cdot}609$$

This gives us

$$\cos 37\tfrac{1}{2}^\circ = 0{\cdot}793 = \sin 52\tfrac{1}{2}^\circ$$
$$\sin 37\tfrac{1}{2}^\circ = 0{\cdot}609 = \cos 52\tfrac{1}{2}^\circ$$

From $\cos 45^\circ = 0{\cdot}707 = \sin 45^\circ$:

$$\cos 22\tfrac{1}{2}^\circ = \cos \tfrac{1}{2}(45)^\circ = \sqrt{\tfrac{1}{2}(1{\cdot}707)} = \sqrt{0{\cdot}853} = 0{\cdot}924$$

$$\sin 22\tfrac{1}{2}^\circ = \frac{0{\cdot}707}{2(0{\cdot}924)} = 0{\cdot}383$$

This gives us:

$$\cos 22\tfrac{1}{2}^\circ = 0{\cdot}924 = \sin 67\tfrac{1}{2}^\circ$$
$$\sin 22\tfrac{1}{2}^\circ = 0{\cdot}383 = \cos 67\tfrac{1}{2}^\circ$$

We can now tabulate the foregoing results, calculating the fourth column in the following table from $\tan A = \dfrac{\sin A}{\cos A}$:

TABLE OF TRIGONOMETRICAL RATIOS IN INTERVALS OF $7\frac{1}{2}^\circ$

Angle (A°)	sin A	cos A	tan A
90	1·000	0·000	∞
$82\frac{1}{2}$	0·991	0·131	7·56
75	0·966	0·259	3·73
$67\frac{1}{2}$	0·924	0·383	2·41
60	0·866	0·500	1·73
$52\frac{1}{2}$	0·793	0·609	1·30
45	0·707	0·707	1·00
$37\frac{1}{2}$	0·609	0·793	0·77
30	0·500	0·866	0·58
$22\frac{1}{2}$	0·383	0·924	0·41
15	0·259	0·966	0·27
$7\frac{1}{2}$	0·131	0·991	0·13
0	0·000	1·000	0·00

Of course, if you like to do so, you can go on making the interval smaller: $\frac{1}{2}(7\frac{1}{2}) = 3\frac{3}{4}$; $\frac{1}{2}(3\frac{3}{4}) = 1\frac{7}{8}$; $\frac{1}{2}(1\frac{7}{8}) = \frac{15}{16}$, etc. Hipparchus, who published, as far as we know, the first table of sines, did not go farther than we have gone. If you have actually checked all the arithmetic so far, you will have grasped how a table of sines, etc., can be built up, and you will not be likely to forget it quickly.

Let us now stop to ask what advantage we have gained. To begin with, we have made scientific map-making and surveying on a scale adequate for long-range geographical work possible. The method used for the cliff illustrated in Fig. 56 presupposes, first, that you can get near enough to find the angles you want (30° and 60°), and secondly, that you have time to walk about finding exactly where the cliff has an elevation of 30° and 60° from the ground. The first is often impracticable, and the second is an unnecessary nuisance. With tables of sufficiently small intervals, all you have to do is to sight the angle of a distant object, then walk *any* measured distance away from it in a direct line, and measure a second angle. In Fig. 65 the measured distance is d, and the angles are 34° and 26°. To make a map we can get all we want, namely, the height of the hill and the horizontal distance to the spot vertically below it. Calling the height h and the horizontal distance from the nearest place of observation x, we look up the tables and find

$$\tan 34^\circ = 0{\cdot}674 \quad \tan 26^\circ = 0{\cdot}488$$

We may suppose that the measured distance d is 64 yards. You will then see at once that

$$\frac{h}{x} = \tan 34^\circ \quad \text{or} \quad x = \frac{h}{0{\cdot}674} \quad . \quad . \quad . \quad . \quad \text{(i)}$$

Also $$\frac{h}{x+64} = \tan 26^\circ \quad \text{or} \quad h = 0{\cdot}488(x + 64) \quad . \quad . \quad \text{(ii)}$$

Combining (i) and (ii), we get

$$h = 0{\cdot}488\left(\frac{h}{0{\cdot}674} + 64\right) = \frac{0{\cdot}488}{0{\cdot}674}h + 64(0{\cdot}488)$$

So that $$h - \frac{0{\cdot}488}{0{\cdot}674}h = 64(0{\cdot}488) \quad \text{or} \quad h\left(1 - \frac{0{\cdot}488}{0{\cdot}674}\right) = 31{\cdot}232$$

Hence $$0{\cdot}276h = 31{\cdot}232$$

$$h \simeq \frac{31{\cdot}232}{0{\cdot}276} = \underline{113{\cdot}16 \text{ yards or } 339\tfrac{1}{2} \text{ feet}}$$

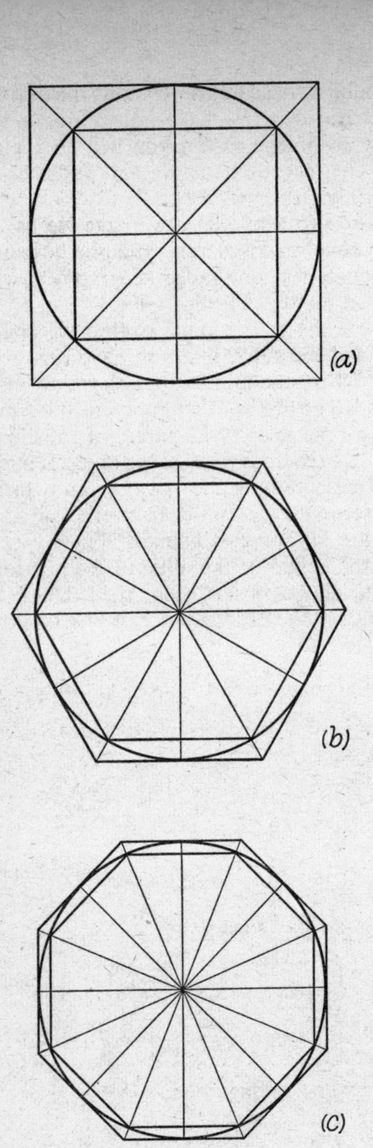

Fig. 66. Enclosed and Enclosing Regular Polygons (Dem. 7)

Once we have found h, we can also get x at once from (i), i.e.

$$x = \frac{113{\cdot}16}{0{\cdot}674} \simeq 168 \text{ yards.}$$

Try again with four-decimal values of tan 34° and tan 26°. You will get the better values $h = 112{\cdot}76$ yards, $x = 167{\cdot}2$ yards. This tells you how far your previous results can be trusted.

Demonstration 7

The ratio of the perimeters of two regular polygons with the same number of sides is equal to the ratio of the radii of circles enclosing them and to the ratio of circles enclosed by them.

This is the gist of the most important theorem in Euclid's twelfth Book. It epitomizes the whole mensuration of the circle. As already mentioned, however (p. 98), Euclid does not set out

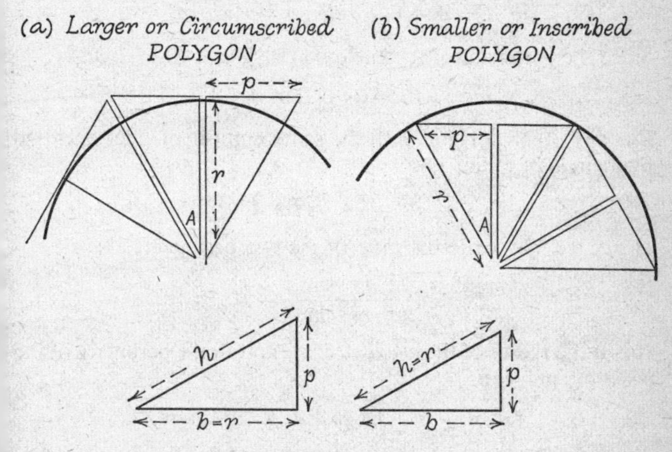

Fig. 67. How Archimedes found a value for π

The angle $A = \dfrac{360°}{2n}$ when the polygon has n equivalent sides.

$$\frac{p}{r} = \tan A = \tan \frac{360°}{2n} \qquad \frac{p}{r} = \sin A = \sin \frac{360°}{2n}$$

$$\therefore\ p = r \,.\tan \frac{360°}{2n} \qquad \therefore\ p = r \,.\sin \frac{360°}{2n}$$

his proof in a way which leads his affluent and persevering pupils to any numerical result. We shall therefore examine the issue from the point of view of his successor Archimedes, who put number back into geometry.

As stated on p. 51, a *regular* polygon of n sides is a figure whose perimeter (boundary) consists of n equal straight lines, e.g. an equilateral triangle, a square, a regular hexagon,* etc. Every such figure (Fig. 66) is divisible into n equal isosceles triangles and hence into $2n$ equal right-angled triangles of which the angles at the common centre are each equal to $A = (360 \div 2n)$ degrees. Thereby (Fig. 67), we divide the perimeter (P) into $2n$ equal segments, each of which we may speak of as the side (p) *perpendicular* to A. The radius of the *enclosing* circle is the *hypotenuse* (h) of each of the $2n$ right-angled triangles and the radius of the *enclosed* circle is the *base* (b).

If r_1 is the radius of the *enclosing* circle:

$$\sin A = p \div r_1, \textit{ so that } p = r_1 \sin A$$

Since the perimeter of the enclosed polygon is $P_1 = 2np$,

$$P_1 = 2n \,.\, r_1 \sin A$$

For any other polygon with the same number of sides enclosed by a circle of radius r_2

$$P_2 = 2n \,.\, r_2 \sin A$$

Hence we may write the ratio of the two perimeters as:

$$\frac{P_1}{P_2} = \frac{2n \,.\, r_1 \,.\, \sin A}{2n \,.\, r_2 \,.\, \sin A} = \frac{r_1}{r_2}$$

If r_1 is the radius of the *enclosed* circle and P_1 the perimeter of the *enclosing* polygon:

$$\tan A = p \div r_1, \textit{ so that } p = r_1 \,.\, \tan A$$

* The rule and construction of a regular hexagon (Fig. 20) follows from the fact that all three angles of an equilateral triangle are 60°, so that six placed together at one of the apexes of each make a polygon with a central angle of 6 × 60 = 360°, six equal spokes which constitute the equal radii of a circle, and six external sides each of length likewise equal to its radius. Hence, one can construct it by starting at any point on the perimeter of a circle and marking off successive cords of length equal to the radius.

Whence, if P_2 and r_2 are respectively the perimeter and radius of a second enclosing polygon and enclosed circle:

$$\frac{P_1}{P_2} = \frac{2n \,.\, r_1 \tan A}{2n \,.\, r_2 \tan A} = \frac{r_1}{r_2}$$

This is as far as Euclid deigns to take us, albeit by a different route. Let us press forward with Archimedes. We need then set no limit to the number (n) of sides a regular polygon can have. We can construct with rule and compass one of which n is any recognizable multiple of 2 or 3, e.g. 6, 12, 24, 48, 96, etc., or 4, 8, 16, 32, 64, 128, etc. The greater we make n the smaller becomes the difference between the perimeters of polygons bounding and bounded by a circle of fixed radius r and diameter $d = 2r$. In short, the two eventually become indistinguishable when:

$$2n \,.\, r \,.\, \sin \frac{360°}{2n} = 2n \,.\, r \,.\, \tan \frac{360°}{2n} \text{ or } n \,.\, d \,.\, \sin \frac{360°}{2n} = n \,.\, d \,.\, \tan \frac{360°}{2n}$$

Here we make our first encounter with a great mathematical discovery. As we increase n, both *sin A* and *tan A* become nearer and nearer to zero; but the products $n \,.\, \sin A$ and $n \,.\, \tan A$ approach a limiting value at which both are the same. This limiting value is what we call π, and we can now define it as a number which lies in a gap which we can make as small as we care to by making n bigger and bigger, so that

$$n \,.\, \sin \frac{360°}{2n} < \pi < n \,.\, \tan \frac{360°}{n}$$

For instance, if a polygon has 24 sides so that $2n = 48$ and $360 \div 2n = 7\frac{1}{2}°$

$$24 \sin 7\tfrac{1}{2}° < \pi < 24 \tan 7\tfrac{1}{2}°$$

From the tables (correct to 4 decimal places) at the end of this book, we get: *sin* $7\frac{1}{2}° = 0{\cdot}1305$ and $\tan 7\frac{1}{2}° = 0{\cdot}1317$, so that:

$$24\,(0{\cdot}1305) < \pi < 24\,(0{\cdot}1317)$$

$$\therefore 3{\cdot}1320 < \pi < 3{\cdot}1608$$

By the same reasoning, it is clear that we may diminish the intervening gap until the enclosing and enclosed regular n-sided polygons merge into one another and coalesce. We may thus speak of the perimeter (P) of a circle of diameter d ($= 2r$) as:

$$P = \pi d = 2\pi r$$

Area of a Circle. To find the area of a circle most conveniently, we consider it as enclosed between an *inscribed* regular polygon of n sides ($n = 8$ in Fig. 68) and an *escribed* (i.e. outer) one of $\frac{1}{2}n$ sides (4 in Fig. 68). We dissect the inner polygon into n equivalent triangles whose apical angles are each $A = 360° \div n$. We dissect the outer one into $2\,(\frac{1}{2}n) = n$ equivalent triangles whose apical angles are also $A = 360° \div n$. The bases of each are of length R (the radius of the circle), their heights being respectively $R \sin A$ and $R \tan A$. So their areas are respectively:

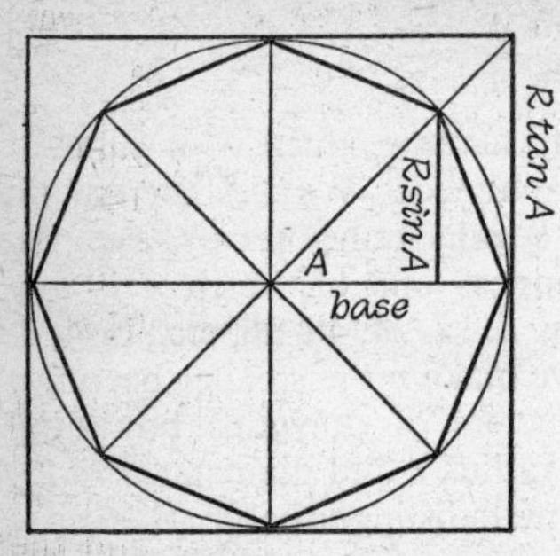

Fig. 68. Square Enclosing and Octagon Enclosed by a Circle

$$\tfrac{1}{2}R \,.\, R \sin A = \tfrac{1}{2}R^2 \sin A \textit{ and } \tfrac{1}{2}R \,.\, R \tan A = \tfrac{1}{2}R^2 \tan A$$

The total areas of the two polygons are thus

$$\tfrac{1}{2}nR^2 \sin A = \tfrac{1}{2}nR^2 \sin \frac{360°}{n}$$

$$\tfrac{1}{2}nR^2 \tan A = \tfrac{1}{2}nR^2 \tan \frac{360°}{n}$$

If we write $\frac{1}{2}n = N$, we thus obtain for the area S of the circle.

$$N \,.\, R^2 \sin \frac{360°}{2N} < S < N \,.\, R^2 \tan \frac{360°}{2N}$$

When N becomes unmeasurably great, we may write:

$$N \,.\, \sin \frac{360°}{2N} \simeq \pi \simeq N \,.\, \tan \frac{360°}{2N}$$

For the area (S) of the circle of radius R.

$$S = \pi R^2$$

We are now ready to make use of the table (p. 148) based on the half-angle formula for trigonometrical ratios to obtain a few successive approximations for the numerical value of π.

Number of Sides (n)	$n.\sin\frac{360°}{2n}$	$n.\tan\frac{360°}{2n}$	Mean π	Error Per cent
3	2·598	5·196	3·90	24
4	2·828	4·000	3·41	8·5
6	3·000	3·464	3·23	2·8
8	3·062	3·314	3·19	1·5
12	3·106	3·215	3·16	0·6
18	3·125	3·173	3·150	0·3
36	3·139	3·150	3·144	0·07

Both in the derivation of a formula for the perimeter and for that of the area of a circle, we have encountered one guise in which one may meet the notion of a *limit*, that of a product in which we conceive one factor (n) to increase endlessly and the other (*sin A* or *tan A*) to diminish endlessly. In short, the former is free to become immeasurably large, the latter (concurrently) immeasurably small. The outcome is a *finite* number, one we can put a name to, while its factors are infinitely (i.e. namelessly) large and small. The recognition of such a possibility must have been beset by immense difficulty to the founding fathers of Greek geometry wholly dependent on the abacus for performing computations and equipped with no very explicit way of representing fractions. For us, the difficulty should be less formidable, as the following example shows.

$$(5{\cdot}0)\,(0{\cdot}1) = 0{\cdot}5 = (50{\cdot}0)\,(0{\cdot}01) = (500{\cdot}0)\,(0{\cdot}001)$$
$$= (5{,}000{,}000{,}000{\cdot}0)\,(0{\cdot}000{,}000{,}0001), \text{ etc.}$$

Here we have stopped, when there are 9 zeros to the left of the decimal point of one factor and 9 to the right of the decimal point of the other; but there is no reason why we should not insert 99 or 999 or 9999 zeros, and so on endlessly, without affecting the finite numerical value of the product, i.e. 0·5.

By recourse to the sign ∞ used for *infininty*, i.e. the immeasurably large, the customary way of writing π as a limit in terms consistent with our method of deriving it above is

$$\underset{n\to\infty}{\mathfrak{L}t}\; n\,.\sin\frac{360}{2n} = \pi = \underset{n\to\infty}{\mathfrak{L}t}\; n\,.\tan\frac{360}{2n}$$

Let us now return to the world's work. With the aid of this demonstration (and that of Dem. 5) we can now understand one way of calculating the circumference of the earth and how to

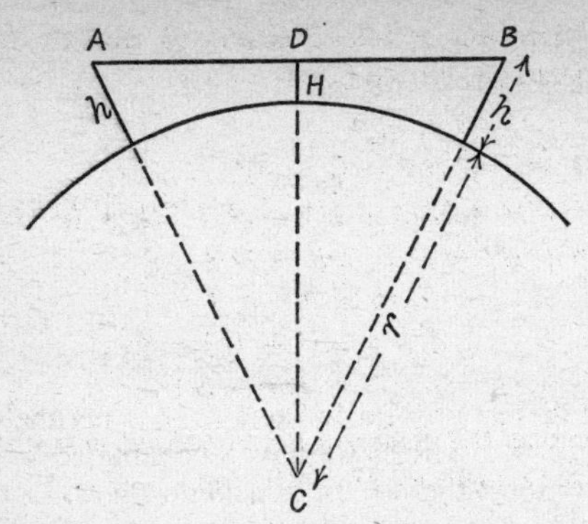

Fig. 69. Canal Method of Measuring the Earth's Circumference

determine the distance at which an object, such as a ship or mountain, ceases to be visible on the horizon.

Earth's Circumference

A. R. Wallace, associated with Darwin in the great evolutionary controversy, started life as a surveyor, and suggested a very simple method of measuring the radius or circumference of the earth. It was the outcome of a wager to debunk the credentials of a very vocal contemporary evangelist of the Flat Earth school. Two sticks (Fig. 69), the upper ends (A and B) of which are separated by a measured distance AB in a *straight* canal, are driven in so that they stand upright at the same height h above the level of the water. Exactly midway between them, a third stick is driven in so that its upper end D is in the line of sight with A and B. Since the earth's surface and, therefore, that of the water in the canal, is really curved, the height H of D above the level of the water will be a little less than h. If we measure h, H, and BD accurately we can find the radius of the earth by applying Dems. 6 and 8. Since

$$AC = (r + h) = BC$$

the $\triangle ABC$ is an isosceles triangle in which

$$AD = \tfrac{1}{2}AB = DB$$

So CD is at right angles to AB (Dem. 6), and the $\triangle DBC$ is a right-angled triangle. Hence (Dem. 8)

$$DB^2 + DC^2 = BC^2$$
$$DB^2 + (r + H)^2 = (r + h)^2$$
$$\therefore \quad DB^2 + r^2 + 2rH + H^2 = r^2 + 2rh + h^2$$
$$\therefore \quad DB^2 + H^2 - h^2 = 2rh - 2rH$$
$$= 2r(h - H)$$
$$\therefore \quad r = \frac{DB^2 + H^2 - h^2}{2(h - H)}$$

Since the distance DB is very long compared with the height of the sticks we may neglect $(H^2 - h^2)$, and

$$r = \tfrac{1}{2}DB^2 \div (h - H)$$
$$= \tfrac{1}{8}AB^2 \div (h - H)$$

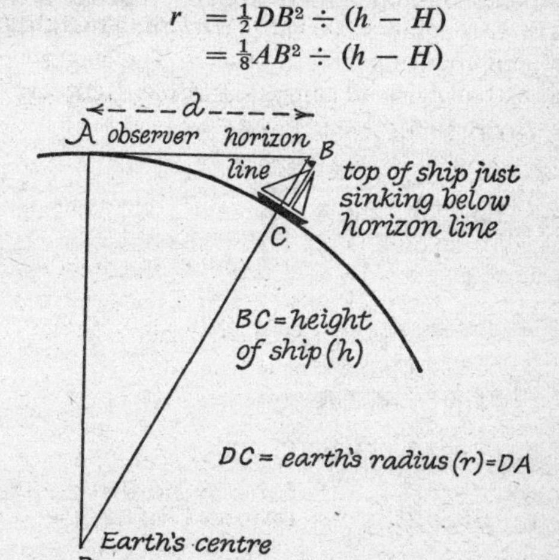

Fig. 70. The Visible Limits of the Horizon.

Distance of the Horizon

In Fig. 70 the observer is at A and BC is the distant object (e.g. mountain or ship), of which he can only just see the topmost point B, the rest being below his horizon line AB. Since light travels in straight lines, this is the line which goes through B and just grazes the circumference of the earth at A. So $\angle BAD$ is a right angle. Applying Dem. 8.

$$
\begin{aligned}
AB^2 + AD^2 &= DB^2 \\
&= (DC + CB)^2 \\
&= DC^2 + 2DC \,.\, CB + CB^2
\end{aligned}
$$

Since AD and DC are both earth radii, $AD = r = DC$

$$\therefore AB^2 + AD^2 = AD^2 + 2DC \,.\, CB + CB^2$$

$$\therefore AB^2 = 2DC \,.\, CB + CB^2$$

Calling AB (the distance of the object as it vanishes below the horizon) d, and BC (its height if it were fully visible) h, we have

$$
\begin{aligned}
d^2 &= 2rh + h^2 \\
&= h(2r + h)
\end{aligned}
$$

Since the highest mountains are about 5 miles and the radius of the earth is roughly 4,000 miles, $(r + h)$ cannot differ from r by more than about one in a thousand. The height h of a ship is of course extremely small compared with r. Hence we may put $(2r + h) = 2r$, so that

$$d^2 = 2hr$$

This shows how far away a mountain 2,000 feet high must be when it just dips below the sea line, if the observer's eye is level with the sea.

$$
\begin{aligned}
d^2 &= 2 \times \tfrac{2000}{5280} \times 4{,}000 \text{ (square miles)} \\
&= \tfrac{100000}{33}
\end{aligned}
$$

$$
\begin{aligned}
\therefore \quad d &\sqrt{100{,}000 \div 33} \\
&= 55 \text{ miles (approximately)}
\end{aligned}
$$

Discoveries and Tests on Chapter 3

1. Two straight lines cross one another making the four angles A, B, C, D. Make diagrams putting in the other three angles when A is (i) 30°, (ii) 60°, (iii) 45°.

2. A triangle has three sides, lengths a, b, c, opposite the angles A, B, C respectively. Continue a beyond C to E, draw the figure and find what $\angle ACE$ is when (i) $A = 30°$, $B = 45°$; (ii) $A = 45°$, $B = 75°$. If $\angle ACE$ is called the 'exterior angle' at C, what is the general rule connecting an exterior angle of a triangle with the two opposite interior angles?

3. Draw an equilateral triangle with sides 1 unit in length. Draw a perpendicular from one corner to the opposite side. Express the

area of the triangle in terms of (*a*) sin 60°, (*b*) cos 30°. If the length of a side is *a* units, what will the area be?

4. Draw an isosceles triangle with one angle of 120°. If the equal sides are of unit length, find an expression for the area of the triangle. What is the area if the equal sides are *a* units in length?

5. Draw diagrams to illustrate the following statements:

$$(2a + 3b)^2 = 4a^2 + 12ab + 9b^2$$
$$(3a - 2b)^2 = 9a^2 - 12ab + 4b^2$$
$$(2a + 3b)(3a - 2b) = 6a^2 + 5ab - 6b^2$$
$$(2a + 3b)(2a - 3b) = 4a^2 - 9b^2$$

6. In the last chapter you found out how to write down $(a + b)^2$, $(a - b)^2$. These identities can be used for squaring many different types of expressions. For example:

$$\left(\boxed{x + y} + 1\right)^2 = \boxed{x + y}^2 + 2 . 1 . \boxed{x + y} + 1^2$$
$$= x^2 + 2xy + y^2 + 2x + 2y + 1$$

This is more usually written:

$$\{(x + y) + 1\}^2 = x^2 + 2xy + y^2 + 2x + 2y + 1$$

Notice that when two sets of brackets are used, one inside the other, they are of different shapes to avoid confusion. In this way write down the values of the following:

$$(x + y + 2)^2 \qquad (x + 1)^2$$
$$(x + y - 2)^2 \qquad (x - 1)^2$$
$$(2a^2 + 3y^2)^2 \qquad (4a - 5b)^2$$
$$(x^2 + y^2)^2 \qquad (xy - 1)^2$$
$$(x^2 - y^2)^2$$

7. By reversing the process you can write down the square roots of any expression of the form

$$a^2 \pm 2ab + b^2$$

Write down the square roots of the following:

$$9x^2 + 42xy + 49y^2 \qquad a^2 + 6a + 9$$
$$4a^2 - 20ab + 25b^2 \qquad x^2 - 2x + 1$$
$$16a^2 - 72ab + 81b^2 \qquad x^2 + 2x + 1$$
$$x^2 + 24xy + 144y^2$$

8. Using the identity $(a + b)(a - b) = a^2 - b^2$, find the values of the following:

$$(x + 1)(x - 1) \qquad (x + 3)(x - 3)$$
$$(ab + 1)(ab - 1) \qquad (a^2 - b^2)(a^2 + b^2)$$
$$(x + y - 2)(x + y + 2)$$

9. It is very useful to be able to split up a complicated expression into factors. You have already seen how to find the factors of expressions like $a^2 + 2ab + b^2$. The identity $a^2 - b^2 = (a - b)(a + b)$ can be used to find the factors of any expression which is the difference of two squares.

$$\text{e.g. } 64x^4 - 81y^2 = (8x^2)^2 - (9y)^2$$
$$= (8x^2 - 9y)(8x^2 + 9y)$$

In this way write down the factors of:

$$x^2 - 1 \qquad a^2 - (b + c)^2$$
$$(a + b)^2 - c^2 \qquad (x + y)^2 - 1$$
$$a^2 - (b - c)^2 \qquad x^8 - y^8$$
$$a^4 - b^4 \qquad a^2 + 2ab + b^2 - 1$$
$$81 - x^2 \qquad x^2 + 2xy + y^2 - 2^2$$
$$(x + 2)^2 - (x - 1)^2$$

10. Write down the value of the third angle of a triangle when the other two angles have the following values:

(i) 15°, 75° (ii) 30°, 90°
(iii) 49°, 81° (iv) 110°, 60°
(v) 90°, 12°

11. If you refer to Fig. 13 you will see what is meant by the zenith distance (z.d.) and altitude (a) of a heavenly body. Explain why $a = 90° -$ z.d. and z.d. $= 90° - a$.

12. If the altitude of the pole star is 30° at Memphis, 41° at New York, and $51\frac{1}{2}$° at London, what is its zenith distance at these places?

13. The star Sirius is $106\frac{1}{2}$° measured along the meridian from the pole star. Draw diagrams to show its position relative to the pole star when on the meridian at each of the three places mentioned in the last paragraph. What is its zenith distance and altitude in each case?

14. Draw four right-angled triangles in which one angle is 10°, 30°, 45°, 75° respectively. In each triangle drop a perpendicular

from the right-angled corner to the hypotenuse. Into what angles does the perpendicular divide the right angle in each case?

15. A ladder leaning against a vertical wall makes an angle of 30° with the wall. The foot of the ladder is 3 feet from the wall. How high up the wall does the ladder reach and how long is the ladder?

16. A wardrobe 5 feet high stands in an attic in which the roof slopes down to the floor. If the wardrobe cannot be put nearer the wall than 2 feet, what is the slope of the roof?

17. A thatched roof has a slope of 60°. It ends 15 feet above the ground. In building an extension the roof can be continued until it is 6 feet from the ground. How wide can the extension be?

18. At noon a telegraph pole known to be 17 feet high cast a shadow 205 inches long. What was the sun's approximate zenith distance? (Use the table of tangents.)

19. At noon, when the sun's zenith distance was 45°, the shadow of a lamp-post just reached to the base of a 12-foot ladder whose top touched the top of the lamp-post. How much longer was the shadow of the lamp-post later in the day when the sun's zenith distance was 60°? (Draw a figure. No calculation needed.)

20. The shadow of a vertical pole 3 feet 6 inches long was found to be 5 feet at four o'clock in the afternoon. At the same time the shadow of a cliff with the sun directly behind it was 60 yards. How high was the cliff?

21. A surveyor wants to measure the width of a river he cannot cross. There is a conspicuous object P on the opposite bank. From a point A on his own side to the left of P, he finds that the angle between his bank and the direction of P is 30°. From another point B to the right on the surveyor's side, the direction of P is 45°. He then measures AB and finds it to be 60 feet. Draw a diagram of this, and find the width of the river. (Hint. Find the relations between the perpendicular from P to AB in terms of the segments of AB and add the segments.)

22. A halfpenny (diameter 1 inch) placed at a distance of 3 yards from the eye will just obscure the disc of the sun or moon. Taking the distance of the sun as 93 million miles, find its diameter. Taking the diameter of the moon as 2,160 miles, find its distance.

23. If $\sin A = \cos 60°$, what is A?
If $\sin A = \cos 45°$, what is A?
If $\cos A = \sin 15°$, what is A?
If $\cos A = \sin 8°$, what is A?

24. If $\sin x = \frac{\sqrt{3}}{2}$ and $\cos x = \frac{1}{2}$, what is $\tan x$?

If $\sin x = 0{\cdot}4$ and $\cos x = 0{\cdot}9$, what is $\tan x$?
If $\cos x = 0{\cdot}8$ and $\sin x = 0{\cdot}6$, what is $\tan x$?
If $\sin x = 0{\cdot}8$ and $\cos x = 0{\cdot}6$, what is $\tan x$?

25. Use tables of squares or square roots to find the third side in the right-angled triangles whose other sides are:

(*a*) 17 feet, 5 feet.
(*b*) 3 inches, 4 inches.
(*c*) 1 centimetre, 12 centimetres.

How many different possible values are there for the third side in each triangle?

26. Make two different geometrical constructions with careful scale diagrams to tabulate the squares of the whole numbers from 1 to 7.

27. Make geometrical constructions to find the arithmetic and geometrical mean of 2 and 8, 1 and 9, 4 and 16.

28. What is a star's zenith distance when it is just grazing the horizon? When it is directly over the meridian, Canopus, next to Sirius the brightest star, is $7\frac{1}{2}°$ above the south point of the horizon in the neighbourhood of the Great Pyramid (Lat. 30°). What is the angle between Canopus and the pole star? On the assumption that the angle between any two stars when they lie over the meridian is fixed, what is the most northerly latitude at which Canopus can be seen at all?

29. If the sun is directly over the Tropic of Cancer (Lat. $23\frac{1}{2}°$ N) on June 21st, show by the aid of a diagram like that of Figs. 61 and 62 what are its altitude and zenith distance at New York (Lat. 41° N) when it is over the meridian (i.e. at noon). What is the most southerly latitude at which the sun can be seen at midnight on that day?

30. What is the zenith distance of the pole star at New York (Lat. 41° N) and London ($51\frac{1}{2}°$ N), and the altitude of the noon sun on September 23rd?

31. In a Devonshire village the shadow of a telegraph pole was shortest at the time when the radio programme gave Greenwich time as 12.14 pm on December 25th. What was its longitude?

32. By dividing a polygon of x equivalent sides into x equivalent triangles, show that the angle between any two sides is the fraction $\frac{2x - 4}{x}$ of a right angle.

33. What is the height of a lighthouse if its light can be seen at a distance of 12 miles?

34. From a ship's masthead 60 feet above sea level it is just possible to see the top of a cliff 100 feet high. How far is the ship from the cliff?

35. At noon on a certain day the shadows of two vertical poles *A*, *B*, each 5 feet high, are 3 feet 3 inches and 3 feet $1\frac{1}{2}$ inches respectively. If *A* is 69 miles north of *B*, what is the radius of the earth?

36. If a square is drawn outside a circle of 1-inch radius so that its sides just touch the circle, show that the length of its boundary is 8 tan 45°. If it is drawn inside the circle so that the corners of the square lie on the circumference, show that the length of the boundary is 8 sin 45°. Similarly show that the boundary of a circumscribed hexagon is 12 tan 30°, and of an inscribed hexagon 12 sin 30°. What would you expect the boundaries to be of a circumscribed and inscribed octagon (8 sides) and dodecagon (12 sides)?

37. Calculate the numerical values of the boundaries of a square, hexagon, octagon, and dodecagon, both circumscribed and inscribed.

Tabulate your results to show between what values π lies, using the tables of sines and tangents.

38. For a circle of unit radius, show that the area of the circumscribed square is 4 tan 45°, and of the inscribed square 4 sin 45° cos 45°. What are the areas of the circumscribed and inscribed hexagon? Give a general expression for the area of circumscribed figures with *n* equivalent sides, noticing that in the case of a square the area is 4 tan $\frac{360}{8}$°.

39. Since the area of a circle of unit radius is π ($\pi r^2 = \pi$ when $r = 1$), use the general expressions you have just obtained to find the limits between which π lies, taking π to lie between the areas of circumscribed and inscribed figures with 180 equal sides.

40. If the radius of the earth is taken to be 3,960 miles, what is the distance between two places with the same longitude, separated by one degree of latitude?

41. What is the distance apart of two places on the equator separated by one degree of longitude?

42. A ship after sailing 200 miles due West finds that her longitude has altered by 5°. What is her latitude? Use the table on p. 624–5.

43. On Midsummer Day the sun is directly above the Tropic of

Cancer (Lat. $23\frac{1}{2}°$ N). On Midwinter Day it is directly above the Tropic of Capricorn (Lat. $23\frac{1}{2}°$ S). Make a diagram to show at what angles the noon's sun is inclined to the horizon at London (Lat. $51\frac{1}{2}°$ N) on June 21st and December 21st.

44. Show by diagrams that the noon shadow always points north at New York (Lat. 41° N).

45. How would you know by watching the noon shadow throughout the year whether you were:

(*a*) North of Latitude $66\frac{1}{2}°$ N?
(*b*) Between Latitude $66\frac{1}{2}°$ N and $23\frac{1}{2}°$ N?
(*c*) Between Latitude $23\frac{1}{2}°$ N and the equator?
(*d*) Exactly at the North Pole?
(*e*) Exactly on the Arctic Circle?
(*f*) Exactly on the Tropic of Cancer?
(*g*) Exactly on the equator?

46. At what latitude will the sun's noon shadow be equal to the height of the shadow pole on (*a*) June 21st, (*b*) March 21st, (*c*) December 21st?

47. A ship's chronometer on September 23rd recorded Greenwich time as 10.44 am when the sun crossed the meridian at an angle of 56° above the *northern* horizon. What port was it approaching? (Use a map.)

48. If New York is on the meridian of Longitude 74° W and Moscow on $37\frac{2}{3}°$ E, what will be the local time in New York and Moscow when Greenwich time is 9.0 pm?

49. Using Dem. 9 and the definition of a circle as a figure in which every point on the boundary is equidistant from a fixed point called the centre, show that the centre is also the point where lines drawn at right angles to the mid-point of any two chords of a circle cross one another.

50. How would you use this, if you wished to make the base of a home-made theodolite like the one in Fig. 12 from the circular top of a second-hand three-legged stool, or to pierce the centre of a circular tin?

51. If one side BC of a triangle ABC is extended to a point D, show that $\angle ACD = \angle CAB + \angle ABC$. When two observers at B and C sight an object at A, $\angle CAB$ is called its *parallax* with reference to the two observers. Explain by Dem. 5 why the parallax of an object is the difference between its elevations at B and C, if A has the same azimuth for both observers.

Things to Memorize

1. In a triangle which has a side b opposite $\angle B$, a side a opposite $\angle A$, a side c opposite $\angle C$, and a perpendicular height h from B to b:

(i) Area $= \frac{1}{2}hb$ (ii) $A + B + C = 180°$

If $B = 90°$

(i) $C = 90° - A$
$A = 90° - C$

(ii) $b^2 = c^2 + a^2$

(iii) $\sin A = \dfrac{a}{b} = \cos C$

$\cos A = \dfrac{c}{b} = \sin C$

$\tan A = \dfrac{a}{c} = \dfrac{1}{\tan C}$

2. In a circle of radius r (diameter d):

Boundary $= 2\pi r$ (or πd)
Area $= \pi r^2$

3. Two triangles are equivalent:

(i) If all three sides are equivalent.
(ii) If two sides and the enclosed angle of one triangle are equivalent to two sides and the enclosed angle of the other.
(iii) If one side and the angles the other two sides make with it in one triangle are equivalent to one side and the two corresponding angles in the other.

4.

Angle ($A°$)	$\sin A$	$\cos A$	$\tan A$
90	1	0	∞
60	$\frac{\sqrt{3}}{2}$	$\frac{1}{2}$	$\sqrt{3}$
45	$\frac{1}{\sqrt{2}}$	$\frac{1}{\sqrt{2}}$	1
30	$\frac{1}{2}$	$\frac{\sqrt{3}}{2}$	$\frac{1}{\sqrt{3}}$
0	0	1	0

5. $\cos^2 A + \sin^2 A = 1$; $\sin A = \sqrt{1 - \cos^2 A}$; $\cos A = \sqrt{1 - \sin^2 A}$

6. $\cos \tfrac{1}{2}A = \sqrt{\tfrac{1}{2}(1 + \cos A)}$

$$\sin \tfrac{1}{2}A = \frac{\sin A}{2 \cos \tfrac{1}{2}A} = \sqrt{\tfrac{1}{2}(1 - \cos A)}$$

CHAPTER 4

NUMBER LORE IN ANTIQUITY

TODAY, we commonly use the word *arithmetic* for the art of calculation without recourse to mechanical or pictorial aids on which mathematicians before the Christian era relied. The word is Greek, and, as used in the Greek-speaking world of antiquity, it referred to a classification of numbers much of which is traceable to superstitious beliefs, some of them beliefs which Pythagoras probably picked up in his wanderings east of the regions in contact with Asiatic Greek colonies. Though there is still controversy about the date of the earliest Chinese mathematical treatises, in particular the *Book of Permutations*, some of the number lore incorporated in the rigmarole of the Pythagorean brotherhood may have come into the West through the trans-Asiatic caravan routes which have transmitted to us from the same source, *inter alia*: sugar, silk, paper, block printing, gunpowder, kites, coal, and lock gates for canals.

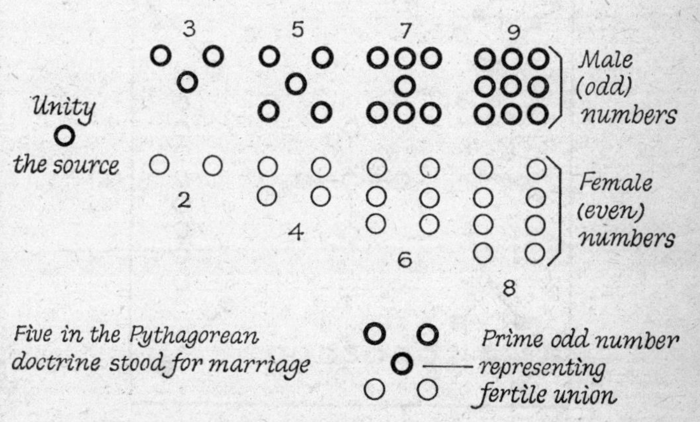

Fig. 71. Phallic Symbolism in Ancient Number Lore

The *Book of Permutations* sets forth in figurate form (Fig. 72) what may have been the first representation of a so-called magic square. This is a n by n (e.g. 3 by 3 or 4 by 4) lay out of numbers of which the sum of column, row and diagonal entries are identical, e.g. the following which you can lay out for yourself:

First row: $4 + 9 + 2 = 15$; second: $3 + 5 + 7 = 15$; third: $8 + 1 + 6 = 15$.

First column: $4 + 3 + 8 = 15$; second: $9 + 5 + 1 = 15$; third: $2 + 7 + 6 = 15$.

Diagonals: $4 + 5 + 6 = 15$ and $2 + 5 + 8 = 15$.

In the Chinese figurate representation of the entries of this square, white circles make up odd and black circles make up even numbers respectively distinguished as *male* and *female* numbers, a classification which throws a lurid light (Fig. 71) on the peculiar significance which the disciples of Pythagoras attached to the number 5. To the Pythagoreans, who invested both numbers and geometrical figures with moral (or immoral) qualities, we may trace some of the obscurantism of Plato's teaching, as referred to in the Prologue.

One, being regarded as the source of all numbers rather than itself a number, stood for reason, two for opinion, four for justice,

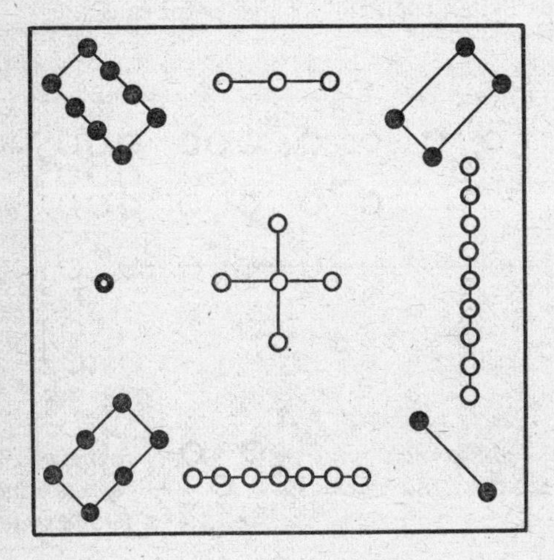

Fig. 72. Chinese Figurate Representation of a MAGIC SQUARE

five for marriage, because it is formed by the union of the first male number 3 and the first female number 2 (Fig. 71). In the properties of five lay the secret of colour, in six the secret of cold, in seven the secret of health, in eight the secret of love, i.e. three (potency) added to five (marriage). The six-faced solid figure held the secret of the earth. The pyramid held the secret of fire (later the *logos spermatikos* of the Stoics, the light which lighteth every man). The twelve-faced solid held the secret of the heavens. The sphere was the most *perfect* figure. The distances of the stars were supposed to form a harmonic number series like the lengths of the wires of early stringed instruments, whence the 'harmony of the spheres'. Numbers were put into classes of bright and obliging or dull and discontented boys and girls. There were *perfect* numbers of which all the whole-number factors add up to the number itself. The first of these is 6, of which all the divisors are 1, 2, and 3 ($1 + 2 + 3 = 6$). The second is 28, of which all the factors are 1, 2, 4, 7, and 14 ($1 + 2 + 4 + 7 + 14 = 28$). The neo-Pythagorean Nichomachus of Alexandria spent a great deal of time hunting for the next two, which are 496 and 8,128. There is not another till we get to 33,550,336. In a fruitless effort to get so far Nichomachus discovered that 'the good and the beautiful are rare and easily counted, but the ugly and bad are prolific'. There were also *amicable* numbers. Asked what a friend was, Pythagoras replied, 'One who is the other I. Such are 220 and 284.' Being interpreted, this means that all the divisors of 284 (1, 2, 4, 71, and 142) add up to 220, and all the divisors of 220 (1, 2, 4, 5, 10, 11, 20, 22, 44, 55, and 110) add up to 284. You can amuse yourself like the audiences of Pythagoras by trying to find others.

Another class of good omen was that of triangular numbers (Fig. 35). A story about these shows you how different was mathematics as cultivated by the Pythagorean Brotherhood from the mathematics which the Greek merchants and craftsmen could use as Thales used his knowledge to measure how far a ship at sea was from port. In Lucian's dialogue a merchant asks Pythagoras what he can teach him. 'I will teach you to count,' says Pythagoras. 'I know that already,' replies the merchant. 'How do you count?' asks the philosopher. The merchant begins, 'One, two three, four. . . .' 'Stop,' cries Pythagoras, 'what you take to be four is ten, or a perfect triangle, and our symbol.' Seemingly, the audiences of Pythagoras wanted charades. He gave them brighter and better charades. It was a short step to the queer ceremonial prayers which his pupils offered to magic numbers. 'Bless us, divine number, who generatest Gods and men, O holy *tetraktys*, that containest the root and source of the eternally flowing creation.' Such was the incantation of his pupils to the number four.

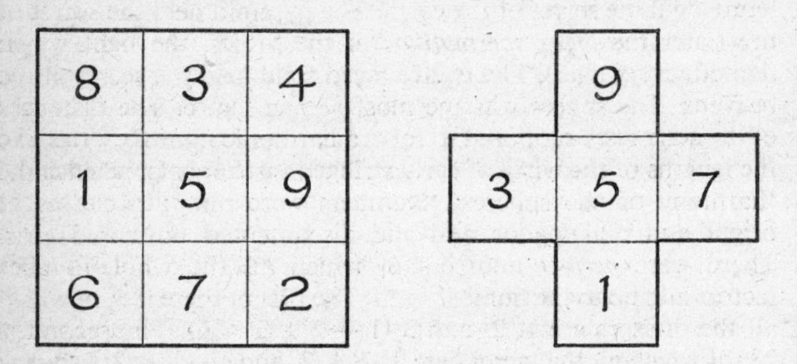

Fig. 73. Magic Squares

On left: the Chinese square of Fig. 72 in our notation.

On right: Cabalistic use of the middle horizontal and vertical rows concealing the name of Jehovah.

Doubtless one circumstance which fostered the persistence of number magic in the ancient world was a very backward step which the Hebrews and later mainland Greeks took (Fig. 9). The priestly number scripts and those of the Phoenicians embody the repetitive principle (p. 34). As such they were lucid but too bulky to find favour with merchants and seamen. Both the Jews and the Greeks eventually settled for a much more economical way of *recording* numbers, i.e. if the only criterion of economy is for the signs to occupy the minimum of space on the writing surface. They conscripted every letter of the alphabet to represent a number, and represented numbers exclusively by such letters. If the change was thrifty in one sense, it was anything but economical in another. It successfully stifled the impulse to enlist number signs in the service of ready computation. This we shall come to later. Another consequence, more relevant to our present theme, is that it encouraged the species of magic called *gematria*.

Gematria is the name for the quaint superstitions which arose in connection with the use of alphabet letters for numbers of the Hebrews and Greeks. In those days, when men were first learning to use signs for numbers, they found themselves entangled in a bog of confusion by first attempts to invent symbols which took up less space than the earlier hieroglyphic forms. When each letter came to represent a number, each word had its characteristic number formed by adding all the separate numbers alternatively represented by its letters. When the numbers of two words were the same there were dark forebodings of hidden mysteries for hosts of commentators (Fig. 73). The superiority of Achilles over Hector was due to the fact that the letters of Achilles add up to 1,276, whereas the name of Hector was only equivalent to 1,225. In Hebrew *Eliasar* adds up to 318. The Hebrew saga tells us that Abraham drove out 318 slaves when he rescued Eliasar. Gematria linked up stars and planets and portents in the astrological writings of the therosophists and the astrologers of the middle ages. A Latin proverb shown in Fig. 74 illustrates an analogous game.

Plato's obscure number, which was 'lord of better and worse births', stimulated a great deal of useless intellectual effort among the Platonists. The number of the Beast in the book of Revelation gave later investigators prolific opportunities for practice in this branch of arithmetic. So did the book of Daniel to which Newton devoted the intellectual efforts of his declining years. Peter

S	A	T	O	R
A	R	E	P	O
T	E	N	E	T
O	P	E	R	A
R	O	T	A	S

Sator arepo tenet opera rotas

Fig. 74. A Verbal Magic Square

Bungus, a Catholic theologian, wrote a book of 700 pages to show that the number 666 of the Beast was a cryptogram for the name of Martin Luther. The Protestants, who sponsored the new mercantile arithmetic, were much better at this line of propaganda. Luther replied by interpreting it as a prophecy of the duration of the papal regime, which was happily approaching its predestined end. Stifel, a convert of Luther, and the first European mathematician to use the signs, $+$, $-$, and $\sqrt{}$ in a book on algebra, traced his conversion to the discovery that 666 refers to Pope Leo X. When written in full Leo X is LEO DECIMVS. The simplicity of the proof merits repetition. Stifel first saw that E, O, and S are not numbers in Roman script. So their inclusion is merely an oversight. The number letters arrange themselves with little help to give MDCLVI, i.e. 1,656. This is $666 + 990$. It is only fair, argued Stifel, to add in X the alternative way of writing DECIMVS. This gives us $666 + 1{,}000$. The Latin equivalent of 1,000 is M, the initial letter of *mysterium*. Hence the apocalyptic reference to the 'mystery' of the Beast. Napier, who is now famous for his logarithms, attached equal importance to his own method of

1	15	14	4
12	6	7	9
8	10	11	5
13	3	2	16

Fig. 75. A 4×4 Medieval Magic Square

identifying the Pope as antichrist. This was no more absurd than St Augustine's claim that: 'God made all things in six days because this number is perfect.' So also is 28, the number of days in February, if not in a leap year.

Before we take leave of *gematria* and therewith the *Magic*

Square, the reader may wish to know a recipe for making one. A general formula for the 3 by 3 square is as in the grid below. Here a, b, c are integers, of which $a > b + c$ to insure that no entries are negatives. Also $2b$ must not be equal to c. Otherwise, the same entry will occur more than once. The sum of the terms in rows, columns, and diagonals is $3a$. The cells contain the integers from 1 to 9 inclusive, as in the early Chinese example, if $a = 5$, $b = 3$, and $c = 1$. Others are:

$a = 6$ $b = 3, c = 2$; $b = 3, c = 1$
$a = 7$ $b = 3, c = 2$; $b = 4, c = 2$; $b = 3, c = 1$; $b = 4, c = 1$
$a = 8$ $b = 6, c = 1$; $b = 5, c = 2$; $b = 5, c = 1$
$b = 4, c = 3$; $b = 4, c = 1$
$b = 3, c = 2$; $b = 3, c = 1$

$a + c$	$a + b - c$	$a - b$
$a - b - c$	a	$a + b + c$
$a + b$	$a - b + c$	$a - c$

One class of numbers which attracted early interest is the *Primes*. These are odd numbers and numbers which are not expressible as the product of two whole numbers, other than unity and themselves, e.g. the only integers which yield 17 when multiplied are 17 and 1. Figuratively (Fig. 71), we cannot represent them by equivalent rows of bars, dots, or circles. Thus, 3, 5, 7, 11, and 13 are primes. The odd numbers 9, 15, 21, and 25 are not prime numbers. The recognition of this class of numbers was not a very useful discovery except in so far as it simplified finding square roots before modern methods were discovered.

To get all the primes between 1 and 100 you first leave out all the even numbers (which are divisible by 2) and all numbers ending in 5 or 0 in our notation (because these are divisible by 5), except, of course, 2 and 5 themselves. This leaves:

1 2 3 5 7 9 11 13 17 19
21 23 27 29 31 33 37 39
41 43 47 49 51 53 57 59
61 63 67 69 71 73 77 79
81 83 87 89 91 93 97 99

Now you must throw out all the numbers (other than 3 or 7) which are exactly divisible by 3 or 7. This leaves:

1	2	3	5	7	11	13	17	19
23	29	31	37	41	43	47		
53	59	61	67	71	73	79		
83	89	97						

We have already rejected all numbers divisible by 9, since these are all divisible by 3, and all numbers divisible by 6, 8, and 10, since these are all divisible by 2. Numbers up to 100, divisible by 11 or a higher number are also divisible by one of the first ten numbers, since higher multiples of 11 are greater than 100. So all those left are prime.

The use of primes for finding square roots depends upon an important rule which you will meet again and again. It is illustrated by the following examples:

$$\sqrt{4 \times 9} = \sqrt{36} = 6 = 2 \times 3 = \sqrt{4} \times \sqrt{9}$$
$$\sqrt{4 \times 16} = \sqrt{64} = 8 = 2 \times 4 = \sqrt{4} \times \sqrt{16}$$
$$\sqrt{4 \times 25} = \sqrt{100} = 10 = 2 \times 5 = \sqrt{4} \times \sqrt{25}$$
$$\sqrt{9 \times 16} = \sqrt{144} = 12 = 3 \times 4 = \sqrt{9} \times \sqrt{16}$$
$$\sqrt{4 \times 49} = \sqrt{196} = 14 = 2 \times 7 = \sqrt{4} \times \sqrt{49}$$
$$\sqrt{9 \times 25} = \sqrt{225} = 15 = 3 \times 5 = \sqrt{9} \times \sqrt{25}$$
$$\sqrt{9 \times 49} = \sqrt{441} = 21 = 3 \times 7 = \sqrt{9} \times \sqrt{49}$$

These examples illustrate the rule:

$$\sqrt{ab} = \sqrt{a} \times \sqrt{b} \quad \text{or} \quad (ab)^{\frac{1}{2}} = a^{\frac{1}{2}} . b^{\frac{1}{2}}$$

So we may put:

$$\sqrt{6} = \sqrt{2} . \sqrt{3}$$
$$\sqrt{8} = \sqrt{4} . \sqrt{2} = 2\sqrt{2}$$
$$\sqrt{12} = \sqrt{4} . \sqrt{3} = 2\sqrt{3}$$
$$\sqrt{18} = \sqrt{9} . \sqrt{2} = 3\sqrt{2}$$
$$\sqrt{24} = \sqrt{4} . \sqrt{6} = 2\sqrt{2} . \sqrt{3}$$

In other words, if we know $\sqrt{2}$ and $\sqrt{3}$, we can get the square root of any number formed by multiplying twos and threes, such as 32, 48, 72, 96. If we also know $\sqrt{5}$, we can get all square roots of numbers formed by multiplying fives with twos and threes, e.g. 10, 15, 30, 40, 45, 50, 60. You can test this as follows:

If we take $\sqrt{2} = 1{\cdot}414$ and $\sqrt{3} = 1{\cdot}732$, then $\sqrt{6} = 1{\cdot}414 \times 1{\cdot}732 = 2{\cdot}449$ correct to three decimal places. By multiplying out, we get:

1·414	1·732	2·449
1·414	1·732	2·449
1·414	1·732	4·898
0·5656	1·2124	0·9796
0·01414	0·05196	0·09796
0·005656	0·003464	0·022041
1·999396	2·999824	5·997601

The error, as we should expect, is greater in the third product, because we have taken only the result of multiplying $\sqrt{2}$ and $\sqrt{3}$ correct to three decimals, and are multiplying the errors in the values we gave to them. The final result is less than one in two thousand (0·0024 in 6·0) out.

This rule is one which we shall use often in later chapters. You must be able to recognize it when it involves fractions, e.g.

$$\sqrt{\frac{a}{b}} = \sqrt{a \times \frac{1}{b}} = \sqrt{a} \times \sqrt{\frac{1}{b}} = \frac{\sqrt{a}}{\sqrt{b}}$$

$$\text{e.g.} \quad \sqrt{\frac{3}{4}} = \frac{\sqrt{3}}{2}$$

Notice how we can use it when cancelling expressions, e.g.

$$\frac{3}{\sqrt{3}} = \frac{\sqrt{3} \times \sqrt{3}}{\sqrt{3}} = \sqrt{3} \quad \text{and} \quad \frac{1}{\sqrt{2}} = \frac{\sqrt{2}}{2}$$

Also notice its use in the following sentences, which we shall meet when we come to find a value for π:

$$\sqrt{1 - \left(\frac{2}{3}\right)^2} = \sqrt{1 - \frac{2^2}{3^2}}$$

$$= \sqrt{\frac{3^2 - 2^2}{3^2}} = \tfrac{1}{3}\sqrt{3^2 - 2^2} = \frac{\sqrt{5}}{3}$$

The triangular numbers of the Pythagoreans had a later payoff. Even Diophantus, most notable of the latter day mathematicians of Alexandria wrote (about AD 250) a treatise on figurate numbers with no prevision of their use by the Persian Omar Khayyám (about AD 1100) as a key to the so-called *Binomial Theorem* (p. 279) or later by James Bernoulli (1713), who was one of the Founding Fathers of the mathematical treatment of Choice and of Chance. Here, as in our dealings with Euclid

(Chapter 3), we shall take advantage of hindsight to examine them as a way of initiation into an increasingly important aspect of size language in later generations, i.e. *how to pack up meaningfully as much information as possible in the smallest possible space*. One might choose other themes to illustrate how one can do so; but figurate numbers keep our feet on the firm ground of simple arithmetic in a picturesque landscape.

They will also give us opportunities both to understand the difference between *discovery* and *proof*, and to understand why our teachers sometimes put the cart before the horse. In much of what we learn at school, or may learn at college, proof comes first, discovery second, or, more often, not at all. In the history of growing knowledge, discovery comes first. It is the proper task of proof to decide how far, and in what circumstances, a discovery is a reliable rule of conduct. If we do not here take much trouble to cite proofs for the defence of discovery, it is because the reader has now (p. 108) one tool – *mathematical induction* for providing proof.

What follows should help the reader to make discoveries.

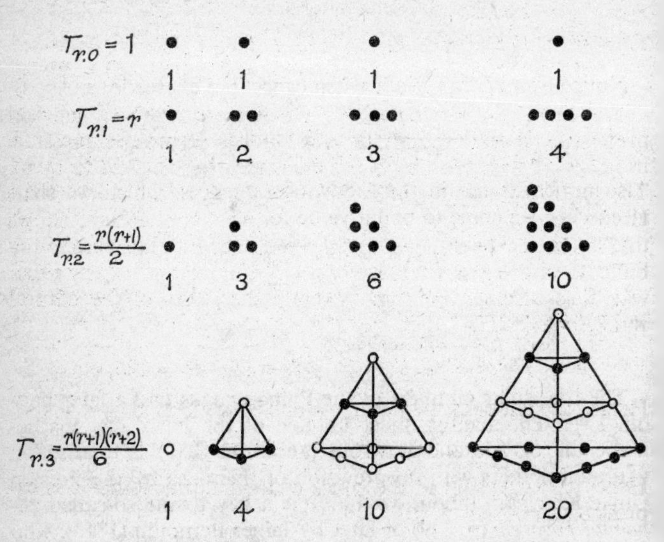

Fig. 76. Dimensions of Figurate Numbers – Triangular and Tetrahedral Numbers

It also provides what to many readers will be a novel introduction to the art of labelling symbols used in mathematical reasoning in a way which helps the user to recall what he or she is reading, writing, or talking about. If we call the price of cotton x and the price of wheat y, it is easy to forget which is which. It is easy to remember which is which if we label the price of cotton as p_c and the price of wheat as p_w. Some readers may find it best to skim through the rest of this chapter quickly, returning to it after reading the first three-quarters of Chapter 7.

Figurate Numbers also offer us an opportunity to study how we can express rules for generating series in the most economical form. We have already (p. 107) seen how to generate the *triangular* numbers (T_r) from the *natural* numbers ($N_r = 1, 2, 3, 4, 5, \ldots$) by addition in accordance with the generation of the natural numbers from the units ($U_r = 1$), as shown below:

$$U_r = \quad 1 \qquad\qquad 1 \qquad\qquad 1 \qquad\qquad 1$$
$$N_r = (0 + 1) = 1; (1 + 1) = 2; (2 + 1) = 3; (3 + 1) = 4$$
$$1 \qquad\qquad 1 \qquad\qquad \ldots$$
$$(4 + 1) = 5; (5 + 1) = 6 \ldots$$

$$T_r = (0 + 1) = 1; (1 + 2) = 3; (3 + 3) = 6; (6 + 4) = 10;$$
$$(10 + 5) = 15; (15 + 6) = 21 \ldots$$

The series T_r, to which the Pythagoreans attached magical properties, is one example of vast families of series expressible in figurate form. Each such has an interesting relation to them. The following example of a discovery made in the story of the Hindu contribution early in the so-called Christian era, shows that the use of the triangular numbers can give us the clues to the build up of a series which we cannot easily picture in any other way. For instance, we may generate the cubes of the natural numbers as:

$$1^3 = 1; \quad 2^3 = 8; \quad 3^3 = 27; \quad 4^3 = 64; \quad 5^3 = 125, \text{ etc.}$$

Hence, the cubes of the natural numbers are:

$$1; \quad 8; \quad 27; \quad 64; \quad 125, \text{ etc.}$$

Hence also, the sum of the cubes of the natural numbers is expressible as below:

$$0 + 1 = 1; \quad 1 + 8 = 9; \quad 9 + 27 = 36; \quad 36 + 64 = 100; \quad 100 + 125 = 225, \text{ etc.}$$

A series for the sum of the cubes of the natural numbers with its generating series of the natural numbers, labelled in accordance with the first term (term 1) as the term of rank 1, therefore takes shape thus:

$r = 1$	2	3	4	5 . . .
$S_r = 1$	9	36	100	225 . . .
$= 1$	3^2	6^2	10^2	15^2 . . .

The items of the last line are squares of the triangular numbers whose general term is

$$\frac{r\,(r+1)}{2}, \text{ so that its square is } \frac{r^2\,(r+1)^2}{4}$$

Having this clue, we see that:

$$S_{r+1} = \frac{r^2\,(r+1)^2}{4} + (r+1)^3 = \frac{(r+1)^2\,(r^2+4r+4)}{4}$$

$$\therefore\ S_{r+1} = \frac{(r+1)^2\,(r+2)^2}{4}$$

We have therefore shown that the formula for the $(r + 1)$*th* term is true, if the formula for the rth is true. To complete the acid test of induction (p. 108), all that remains is to show that the formula is true for the initial term (i.e. unity):

$$S_1 = \frac{r^2(r+1)^2}{4} = \frac{1\,(1+1)^2}{4}$$

As a check on this formula, let us ask what is the sum of the cubes of the first seven natural numbers, i.e.:

$$S_7 = 1^3 + 2^3 + 3^3 + 4^3 + 5^3 + 6^3 + 7^3$$
$$= 1 + 8 + 27 + 64 + 125 + 216 + 343 = 784$$

The formula (*less laboriously*) yields the same result:

$$S_7 = \frac{7^2\,(8^2)}{4} = 49 \times 16 = 784$$

As represented above, the generating series of the triangular numbers is the series of natural numbers (N_r) starting with unity (*rank* 1); and the law of formation is:

$$T_{r+1} = T_r + N_{r+1}$$

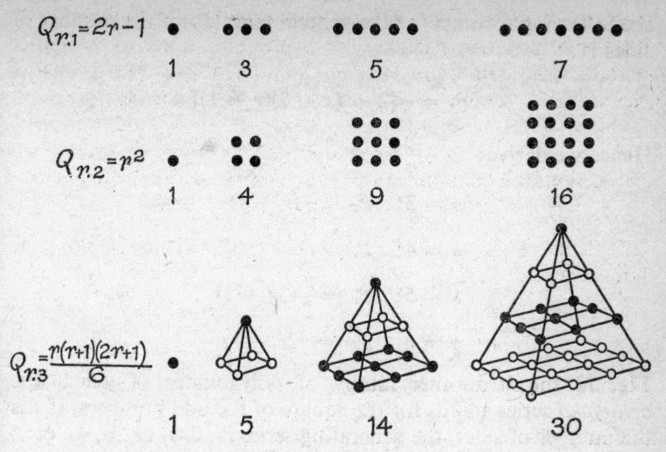

Fig. 77. Dimensions of Figurate Numbers – Squares and Pyramidal Numbers

The natural numbers constitute an arithmetic series, i.e. one of which any two consecutive terms differ by the same amount d, the so-called common difference. If the series is 1, 2, 3, 4, etc., $d = 1$. If $d = 2$, the corresponding series with the same initial terms consists of the odd numbers 1, 3, 5, 7, etc. If $d = 3$, it is the sequence 1, 4, 7, 10, etc. With the same law of formation as that of the triangular numbers, we can thus generate successive series as follows:

Square Numbers ($d = 2$)

1 3 5 7 9 11 13 . . .
1 4 9 16 25 36 49 . . .

Pentagonal Numbers ($d = 3$)

1 4 7 10 13 16 19 . . .
1 5 12 22 35 51 70 . . .

Hexagonal Numbers ($d = 4$)

1 5 9 13 17 21 25 . . .
1 6 15 28 45 66 91 . . .

There is no limit to the number of such series which we can represent (Fig. 37) figuratively in this way. The reader may check the following formula for the general term (P_r) if the number of sides is s:

$$P_r = \frac{r}{2}[2 + (s - 2)(r - 1)]$$

Hence we derive:

$$s = 3; \quad P_r = \frac{r}{2}(r + 1)$$

$$s = 4; \quad P_r = r^2$$

$$s = 5; \quad P_r = \frac{r}{2}(3r - 1)$$

$$s = 6; \quad P_r = r(2r - 1)$$

Fig. 78 shows another family of polygonates of which the *octagonal* series stands for the square of the odd numbers. If s is the number of sides, the generating series is: 0, s, $2s$, $3s$, $4s$, etc.; and the reader will detect the law of formation from below:

0	s	$2s$	$3s$	$4s$	$5s$	$6s \ldots$
1	$1 + s$	$1 + 3s$	$1 + 6s$	$1 + 10s$	$1 + 15s$	$1 + 21s \ldots$

So that if $s = 8$, the build up is:

0	8	16	24	32	40 ...
1	9	25	49	81	121 ...

In the countless families of series which we can represent by plane figures which disclose their formulae, the same series may have more than one figurate representation. As an example of how we may use the Pythagorean triangular numbers to dissect them, and thereby to suggest the formula for the general term, i.e. the term of rank r, let us look at the pentagonal numbers of Fig. 37, represented by $\frac{1}{2}r(3r - 1)$. Below (to the left) we see the first six terms laid out vertically:

$$\begin{aligned} 1 &= 1 &&= 1 + 0 \\ 5 &= 3 + 2 &&= 3 + 2(1) \\ 12 &= 6 + 6 &&= 6 + 2(3) \\ 22 &= 10 + 12 &&= 10 + 2(6) \\ 35 &= 15 + 20 &&= 15 + 2(10) \\ 51 &= 21 + 30 &&= 21 + 2(15) \end{aligned}$$

If we call the general term P_r and T_r the triangular number of rank r, the above suggests:

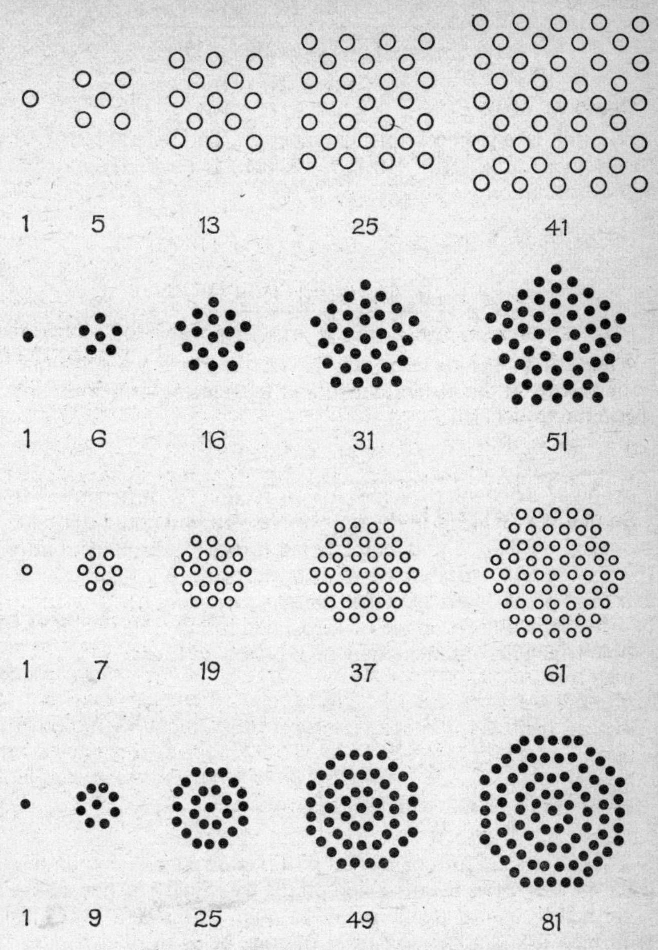

Fig. 78. A Family of Polygonal Numbers

$$P_r = T_r + 2 \,.\, T_{r-1}$$

$$= \tfrac{1}{2}r(r+1) + \frac{2(r-1)r}{2} = \frac{(r^2+r) + (2r^2-2r)}{2}$$

$$= \frac{r}{2}(3r-1)$$

We still have to check this by induction. The general term (g_r) of the generating series (1, 4, 7, 10, etc.) is $(3r - 2)$; and the law of formation is:

$$P_{r+1} = \mathrm{P}_r + g_{r+1} = \frac{r}{2}(3r-1) + (3r+1)$$

$$= \frac{(r+1)\,[3(r+1)-1]}{2}$$

Thus the $(r + 1)$th term tallies with the rth if we substitute $(r + 1)$ for r in the general term; and the general term is true of the first ($P_1 = 1$) since

$$P_1 = \tfrac{1}{2}(3-1) = 1$$

Another triangular recipe for dissection is instructive. It yielded a big dividend in the time of Newton, but could not take shape till the Hindu and Chinese had learned to bring *zero* into the picture. All series which we can represent in a figurate way can form what one may call a *vanishing triangle*, i.e. a triangle of which the apex is made of zeros. Vanishing triangles of the simple triangular numbers may be represented thus:

$$\begin{array}{ccccccccccc}
 & & & & & 0 & & & & & \\
 & & & & 0 & & 0 & & & & \\
 & & & 0 & & 0 & & 0 & & & \\
 & & 1 & & 1 & & 1 & & 1 & & \\
 & 2 & & 3 & & 4 & & 5 & & 6 & \\
1 & & 3 & & 6 & & 10 & & 15 & & 21
\end{array}
\quad \textit{or briefly} \quad
\begin{array}{ccccccc}
 & & & 0 & & & \\
 & & 1 & & 1 & & \\
 & 2 & & 3 & & 4 & \\
1 & & 3 & & 6 & & 10
\end{array}$$

Such triangles are formed by first putting down a certain number of terms of a series in succession on the base line. The line above is formed by *subtracting* adjacent base line terms. The next one is formed in a similar way from the one below it. This is continued, till we get nothing but zeros. The reason why series of the triangular family vanish is easy to see. The successive series of triangular numbers are based on adding adjoining terms of the parent series. They all have the same ancestry. Thus the parent

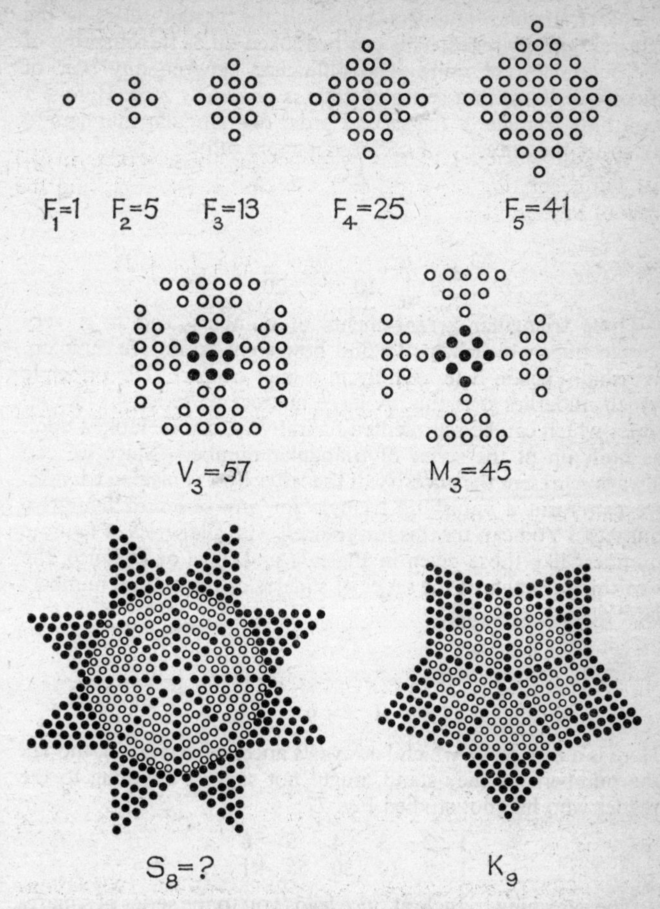

Fig. 79. Finding a Formula

You will find the answers after *Things to Remember* at the end of this Chapter.

series of the second order of triangular numbers is the series of simple triangular numbers of which the parent series is the natural number series. This can be looked on as the offspring of the succession of units. The difference between any pair of successive terms in a series of units is obviously zero. Here is a vanishing triangle of the second order of triangular numbers to illustrate the ancestry of such series more fully:

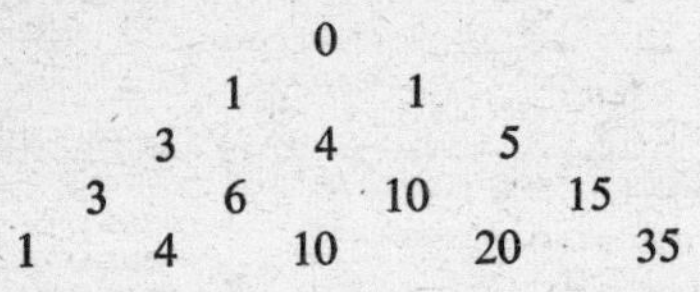

These triangular arrangements of numbers lead to a very simple trick which helps to find how some series are built up. We shall explain it more fully in a later chapter. The principle which underlies it is this. A large number of series, in fact all series which can be represented figuratively, can be looked upon as built up of the series of triangular numbers. Since we can always represent the ancestry of the latter by a vanishing triangle, we can form a vanishing triangle for any series of figurative numbers. You can try this for yourself with the series of figurate numbers like those given in Figs. 37 and 78. For instance, the vanishing triangle of the series of squares of the natural numbers is:

0
2 2
3 5 7
1 4 9 16

Here is a new series which betrays its ancestry very soon, though the numbers, as they stand, might not suggest anything to the reader who had not studied Fig. 77:

1 2 3 4 5 6 . . .
1 5 14 30 55 91 . . .

The vanishing triangle at once leads you to the series of squares of the natural numbers:

0
2 2
5 7 9
4 9 16 25
1 5 14 30 55

Thus the nature of the series is:

$$1^2, (1^2 + 2^2), (1^2 + 2^2 + 3^2), (1^2 + 2^2 + 3^2 + 4^2), \text{etc.}$$

A Fourth Dimension. During the twenties, when Einstein's theory of gravitation was a daily topic of conversation, the Press (and people who should know better) made much ado about a *fourth dimension.* Figurate figures provide us with a readily understandable clue to what professional mathematicians mean by this term, when they use it.

For reasons which will be more clear to the reader after scanning Chapter 9, if not clear enough already, one speaks of a flat figure (such as a square) as 2-dimensional and a solid figure (such as a cube) as 3-dimensional. In the same way, we may speak of a straight line, which Euclid defines as having length but no breadth, as 1-dimensional and of a point, which Euclid defines as having only position, as 0-dimensional. Fig. 76 should make it sufficiently clear that we can speak of the sequence of units as 0-dimensional, the sequence of natural numbers as 1-dimensional, the sequence of triangular numbers as 2-dimensional, and the sequence of tetrahedral numbers as 3-dimensional. To make clear what we can legitimately mean by a 4-dimensional sequence, or one of d-dimensions when $d > 3$, we may label up the terms of rank r in this family of series (Fig. 76) as follows:

$F_{r.0}$	*Units*
$F_{r.1}$	*natural* numbers
$F_{r.2}$	*triangular* numbers
$F_{r.3}$	*tetrahedral* numbers

We get the fifth natural number, by adding to the fourth a unit, the fifth triangular number by adding to the fourth the fifth natural number, and the fifth tetrahedral number by adding to the fourth, the fifth triangular number. In short:

$$F_{5.3} = F_{4.3} + F_{5.2}; \quad F_{5.2} = F_{4.2} + F_{5.1}; \quad F_{5.1} = F_{4.1} + F_{4.0}$$

Thus the same law of formation applies to $F_{r.1}$, $F_{r.2}$, $F_{r.3}$, $F_{r.4}$, namely:

$$F_{r.d} = F_{(r-1).d} + F_{r.(d-1)}$$

To leave no doubt about the meaning of our symbols, let us

lay out as below the figures of Fig. 76 in a table which shows that:

$$F_{7.3} = F_{6.3} + F_{7.2} \quad \textit{means} \quad F_{7.3} = 56 + 28 = 84$$

		Rank (r)							
		0	1	2	3	4	5	6	7...
	0	1	1	1	1	1	1	1	1...
Dimension (d)	1	0	1	2	3	4	5	6	7...
	2	0	1	3	6	10	15	21	28...
	3	0	1	4	10	20	35	56	84...

This table includes all we have represented by figures in a space of three dimensions; but there is no reason why we should not continue to generate new series by applying the same law of formation, e.g.:

$$F_{7.4} = F_{6.4} + F_{7.3} \textit{ or } F_{7.5} = F_{6.5} + F_{7.4}$$

We may thus go on adding rows indefinitely according to the same rule, e.g.:

d								
3	0	1	4	10	20	35	56	84...
4	0	1	5	15	35	70	126	210...
5	0	1	6	21	56	126	252	462...

We have already found a general formula for $F_{r.2}$ (the *triangular* numbers) and the reader can try out the method of triangular dissection to get a clue to that of the tetrahedral numbers, i.e.:

$$F_{r.3} = \frac{r\,(r+1)\,(r+2)}{6}, \text{ e.g. } F_{7.3} = \frac{7\,.\,8\,.\,9}{6} = 84$$

We may then ask: can we formulate a more general rule for $F_{r.d}$ regardless of whether $d = 0, 1, 2, 3$, etc. Let us tabulate* what

* It is unfortunate that British does not conform to American usage with respect to the position of the dot in a decimal fraction, e.g. one-quarter. It is customary for British school books to write the dot somewhat above the line, i.e. 0·25. In the United States, it is customary to place it on the line as in 0.25. No ambiguity arises if we use the first convention and also write (1) (2) (3) = 6 as 1 . 2 . 3 = 6. Nor does it arise, if we use the second and write 1 · 2 · 3 = 6. The writer trusts that the reader used to writing a quarter as 0.25 will infer from context when the dot at line level signifies the operation of multiplication, i.e. 1 . 2 . 3 . 4 . 5 = 120.

we know so far for brevity using 1 . 2 for 1 × 2, 1 . 2 . 3 for 1 × 2 × 3, 1 . 2 . 3 . 4 for 1 × 2 × 3 × 4, etc.:

$$F_{r.0} = 1 \qquad = r^0$$

$$F_{r.1} = r \qquad = \frac{r}{1}$$

$$F_{r.2} = \frac{r(r+1)}{2} \qquad = \frac{r(r+1)}{1.2}$$

$$F_{r.3} = \frac{r(r+1)\ (r+2)}{6} = \frac{r(r+1)(r+2)}{1.2.3}$$

This suggests that

$$F_{r.4} = \frac{r(r+1)(r+2)(r+3)}{1.2.3.4}$$

$$F_{r.5} = \frac{r(r+1)(r+2)(r+3)(r+4)}{1.2.3.4.5}$$

By reference to the last table shown, the reader can check that this is so. It is evident that there is here a meaningful pattern, and equally clear that an expression to convey what it is may become very unwieldy, when d is large. This raises a new question: how can we adapt our battery of signs for operations to convey such a rule as compactly as possible. Let us start our quest by recalling (left below) our definition of n^r on p. 90 and place on the right a comparable convention for expressing a product such as $n(n + 1)(n + 2)(n + 3)$, etc.:

$n^0 = 1$	$n^{[0]} = 1$
$n^1 = 1 . n$	$n^{[1]} = 1 . n$
$n^2 = 1 . n . n$	$n^{[2]} = 1 . n\,(n + 1)$
$n^3 = 1 . n . n . n$	$n^{[3]} = 1 . n\,(n + 1)(n + 2)$
$n^4 = 1 . n . n . n . n$	$n^{[4]} = 1 . n\,(n + 1)(n + 2)(n + 3)$
$n^5 = 1 . n . n . n . n . n$	$n^{[5]} = 1 . n\,(n + 1)(n + 2)(n + 3)(n + 4)$

Notice first that the brackets in $n^{[r]}$ are square. Later, we shall give a different meaning to $n^{(r)}$ of which the brackets are curved. Notice also that the last factor on the right of $n^{[5]}$ is $(n + 4) = (n + 5 - 1)$ and more generally that of $n^{[r]}$ is $(n + r - 1)$. So we may now write:

$$n^{[r]} = 1 . n . (n + 1)(n + 2(n + 3) \ldots (n + r - 1)$$

Let us now examine the meaning which this shorthand gives to $1^{[r]}$, usually written $r!$ and called *factorial* r:

$$
\begin{array}{lll}
1^{[0]} = 1 = 0! & = 1 & = 1 \\
1^{[1]} = 1 \,.\, 1 = 1! & = 1 & = 1 \\
1^{[2]} = 1 \,.\, 1 \,.\, 2 = 2! & = 1 \,.\, 2 & = 2 \\
1^{[3]} = 1 \,.\, 1 \,.\, 2 \,.\, 3 = 3! & = 1 \,.\, 2 \,.\, 3 & = 6 \\
1^{[4]} = 1 \,.\, 1 \,.\, 2 \,.\, 3 \,.\, 4 = 4! & = 1 \,.\, 2 \,.\, 3 \,.\, 4 & = 24 \\
1^{[5]} = 1 \,.\, 1 \,.\, 2 \,.\, 3 \,.\, 4 \,.\, 5 = 5! & = 1 \,.\, 2 \,.\, 3 \,.\, 4 \,.\, 5 & = 120
\end{array}
$$

We can now write:

$$F_{r.0} = \frac{r^{[0]}}{0!} = 1$$

$$F_{r.1} = \frac{r^{[1]}}{1!} = r$$

$$F_{r.2} = \frac{r^{[2]}}{2!} = \frac{r(r+1)}{2}$$

$$F_{r.3} = \frac{r^{[3]}}{3!} = \frac{r(r+1)(r+2)}{6}$$

$$F_{r.4} = \frac{r^{[4]}}{4!} = \frac{r(r+1)(r+2)(r+3)}{24}$$

In short, we can now express for this family of figurates any term of rank r and dimension d the common pattern tersely in the form:

$$F_{r.d} = \frac{r^{[d]}}{d!}$$

We may now add a new entry to our dictionary of signs for operations, the pattern being:

$$5_{[3]} \equiv \frac{5^{[3]}}{3!}; \quad 7_{[8]} \equiv \frac{7^{[8]}}{8!}, \text{ etc.}$$

More generally:

$$n_{[r]} \equiv \frac{n^{[r]}}{r!} \quad \textit{and} \quad F_{r.d} = r_{[d]} \equiv \frac{r^{[d]}}{d!}$$

We can then write even more compactly than above:

$$F_{r.6} = \frac{r(r+1)(r+2)(r+3)(r+4)(r+5)}{6!} \equiv r_{[6]}$$

Let us now seek for a more explicit way of stating the law of formation of this family of figurates. By scrutiny of the table, we see that:

$F_{5.1} = 1 + 1 + 1 + 1 + 1 = 5$ *Sum* of first five terms of dimension 0.

$F_{5.2} = 1 + 2 + 3 + 4 + 5 = 15$ *Sum* of first five terms of dimension 1.

$F_{5.3} = 1 + 3 + 6 + 10 + 15 = 35$ *Sum* of first five terms of dimension 2.

$F_{5.4} = 1 + 4 + 10 + 20 + 35 = 70$ *Sum* of first five terms of dimension 3.

In words, the rule is: to get the term of rank r and dimension d add up all the terms inclusive from 0 to rank r and dimension $(d - 1)$. To get this into the most economical form we need a shorthand for summation. This will be our next task.

Saving More Space. To do justice to the limitless families of series about which we have been talking is difficult unless we can exhibit their relations in a more compact form than hitherto. What is common to all of them is that the law of formation involves in the last resort a process of addition. Our next concern in quest of a space-saving device may therefore be to equip our battery of signs for operations with a new one to convey a process of summation. With no trappings, it is the Greek capital letter (pronounced *Sigma*) for *s*, i.e. the initial letter of *sum*, and we may illustrate it thus:

$$\sum_{r=0}^{r=6} 10^r \equiv 1 + 10 + 100 + 1{,}000 + 10{,}000 + 100{,}000 + 1{,}000{,}000$$
$$\equiv 10^0 + 10^1 + 10^2 + 10^3 + 10^4 + 10^5 + 10^6$$

$$\sum_{r=0}^{r=5} (4 + 3r) \equiv 4 + 7 + 10 + 13 + 16 + 19$$
$$\equiv 4 + (4 + 3) + (4 + 6) + (4 + 9) + (4 + 12) + (4 + 15)$$

In the wider language of size and order our definition is:

$$\sum_{r=m}^{r=n} t_r \equiv t_m + t_{m+1} + t_{m+2} \ldots + t_{n-1} + t_n$$

The left-hand operation is translatable verbally as follows:

> Add up all the terms from the term (t_m) of rank m to the term (t_n) of rank n inclusive.

Our new sign thus allows us to pack up an enormous amount of information in very little space. For instance:

$$(1 + 2 + 3 + 4 + 5 + 6 + 7 + 8) = 36$$
$$\equiv \sum_{r=1}^{r=8} r = \frac{8(8 + 1)}{2}$$

When we are dealing with an unending (so-called infinite) series of which we label the initial term as the term of rank 0, we write it thus:

$$\sum_{r=0}^{r=\infty} 10^{-r} = 1 . \dot{1} = \frac{10}{9} = 1 + 0{\cdot}1 + 0{\cdot}01 + 0{\cdot}001 + 0{\cdot}0001 \ldots$$
and so on

Here $r = \infty$ does not mean that r is a namable number. It merely means: *make* r *as big as conceivable and more so.*

As an exercise in how to construct a mathematical sentence in the most economical way, let us now look at two series of which successive terms are *numerically* alike, but different because terms of the second have alternately positive and negative signs. The first is:

$$S_a = 1 + \frac{1}{2} + \frac{1}{3} + \frac{1}{4} + \frac{1}{5} + \frac{1}{6} + \frac{1}{7} + \frac{1}{8} + \frac{1}{9}, \textit{ and so on without end}$$

$$= 1^{-1} + 2^{-1} + 3^{-1} + 4^{-1} + 5^{-1} \ldots = \sum_{r=1}^{r=\infty} r^{-1}$$

We can express this alternatively as the sum of two series:

$$S_a = (1^{-1} + 3^{-1} + 5^{-1} + 7^{-1} \ldots) + (2^{-1} + 4^{-1} + 6^{-1} + 8^{-1} \ldots)$$

Here the terms of the first series are the reciprocals of the odd numbers, i.e. $\frac{1}{2r - 1}$, if we write the initial term as the term of rank 1 or $\frac{1}{(2r + 1)}$, if we write the initial term as the term of rank 0. The terms of the second series are the reciprocals of the even numbers for which we may write $\frac{1}{2r}$, the initial term being 0 if we label it as the term of rank 0. Alternatively, we may write:

$$S_a = \sum_{r=1}^{r=\infty} (2r-1)^{-1} + \sum_{r=1}^{r=\infty} (2r)^{-1}$$

or

$$S_a = \sum_{r=0}^{r=\infty} (2r+1)^{-1} + \sum_{r=0}^{r=\infty} (2r)^{-1}$$

Let us now look at the following series:

$$S_b = 1 - \frac{1}{2} + \frac{1}{3} - \frac{1}{4} + \frac{1}{5} - \frac{1}{6} + \frac{1}{7} - \frac{1}{8} \ldots, \textit{ and so on without end}$$

$$= (1 + \frac{1}{3} + \frac{1}{5} + \frac{1}{7} + \frac{1}{9} \ldots) - (\frac{1}{2} + \frac{1}{4} + \frac{1}{6} + \frac{1}{8} \ldots)$$

$$= \sum_{r=0}^{r=\infty} (2r+1)^{-1} - \sum_{r=0}^{r=\infty} (2r)^{-1}$$

There is, however, another and more economical way of expressing S_b, if we apply the sign rule as below:

$$(-1)^0 = 1; (-1)^1 = -1; (-1)^2 = 1; (-1)^3 = -1; (-1)^4 = 1; (-1)^5 = -1, \text{ etc.}$$

Hence if $r = 1, 2, 3$, etc.

$$(-1)^{r-1} = (-1)^0 = 1; (-1)^{r-1} = (-1)^1 = -1; (-1)^{r-1} = (-1)^2 = 1, \text{ etc.}$$

Thus we can write:

$$S_b = (-1)^0 \cdot \frac{1}{1} + (-1)^1 \cdot \frac{1}{2} + (-1)^2 \cdot \frac{1}{3} + (-1)^3 \cdot \frac{1}{4} \ldots, \text{ etc.}$$

$$= \sum_{r=1}^{r=\infty} (-1)^{r-1} \cdot r^{-1}$$

Alternatively, if we label the initial term (unity) as the term of rank 0, our layout is

$r =$	0	1	2	3	4	5
$t_r =$	1	-2^{-1}	3^{-1}	-4^{-1}	5^{-1}	-6^{-1}

$(0+1)^{-1}; -(1+1)^{-1}; (2+1)^{-1}; -(3+1)^{-1}; (4+1)^{-1}; -(5+1)^{-1}$, etc.

The expression for the sum of the terms is then:

$$S_b = \sum_{r=0}^{r=\infty} (-1)^r (r+1)^{-1}$$

We can represent the sum of the series below alternatively as:

$$At_0 + At_1 + At_2 + At_3 \ldots At_n = A(t_0 + t_1 + t_2 + t_3 \ldots t_n)$$

$$\therefore \sum_{r=0}^{r=n} A \,.\, t_r = A \,.\, \sum_{r=0}^{r=n} t_r \quad \textit{and} \quad \sum_{r=1}^{r=n+1} \mathrm{A} \,.\, t_{r-1} = A \,.\, \sum_{r=1}^{r=n+1} t_{r-1}$$

It may not be obvious to the reader that a succession of units is a series of which the so-called *general term* (i.e. formula from which we can derive every term from its rank) is $t_r = r^0$. Here the sum of n terms is:

$$\sum_{r=1}^{r=n} 1 = n = \sum_{r=0}^{r=n-1} 1 \quad \textit{and} \quad A \sum_{r=1}^{r=n} 1 = A \,.\, n = A \sum_{r=0}^{r=n-1} 1$$

If we wish to sum a series from any term of rank a up to any term of rank b inclusive, the lay-out is:

$$r = 0, 1, 2, 3, \ldots, a-1, a, a+1, \ldots, b$$

$$t_r = t_0, t_1, t_2, t_3, \ldots, t_{a+1}, t_a, t_{a-1}, \ldots t_b$$

$$\therefore \sum_{r=0}^{r=a-1} t_r + \sum_{r=a}^{r=b} t_r = \sum_{r=0}^{r=b} t_r$$

$$\therefore \sum_{r=a}^{r=b} t_r = \sum_{r=0}^{r=b} t_r - \sum_{r=0}^{r=a-1} t_r$$

An important rule for use in connection with the summation sign is the following: if the general term of the series is the sum of several terms, we can represent it as the combined sum of the separate series. Think of the series whose general term is $t_r = (r+1)^2 = r^2 + 2r + 1$. So that:

$$t_0 = 0 + 0 + 1 = 1;\ t_1 = 1 + 2 + 1 = 4;\ t_2 = 4 + 4 + 1 = 9;\ t_3 = 9 + 6 + 1 = 16$$

We can collect our terms up to rank 3 as follows:

$$(0 + 1 + 4 + 9) + 2(0 + 1 + 2 + 3) + (1 + 1 + 1 + 1)$$

In short:

$$\sum_{r=0}^{r=n} (r+1)^2 = \sum_{r=0}^{r=n} (r^2 + 2r + 1) = \sum_{r=0}^{r=n} r^2 + 2\sum_{r=0}^{r=n} r + \sum_{r=0}^{r=n} 1$$

If we apply this procedure to the sum of the first n odd numbers, and label 1 as the term of rank 1, so that the general term is $t_r = (2r - 1)$, we can split the sum into the sequence of natural numbers (whose sum is the corresponding triangular number) and a series of units thus:

$$\sum_{r=1}^{r=n} (2r-1) = 2\sum_{r=1}^{r=n} r - \sum_{r=1}^{r=n} 1 = 2\frac{n(n+1)}{2} - n$$

$$= n^2 + n - n = n^2$$

We may now state the law of formation from its predecessor (and hence eventually from the sequence of units of one row in our table of p. 179 based on Fig. 77) as:

$$\sum_{n=1}^{n=r} F_{n\,.\,(d-1)} = F_{r\,.\,d}$$

Now we have seen, without conclusive proof, that we may express $F_{r\,.\,d}$ in terms of r and d. To prove the truth of the formula conclusively by induction, the reader will need to test it first for r when d remains constant, then for d when r is constant. Here we shall take it for granted and exhibit an application. We may then write

$$\sum_{n=1}^{n=r} n_{[d-1]} = r_{[d]}$$

Hence we may write:

$$\sum_{n=1}^{n=r} n_{[2]} = r_{[3]}$$

In this formula:

$$n_{[2]} = \tfrac{1}{2}n(n+1) = \tfrac{1}{2}n^2 + \tfrac{1}{2}n$$

$$\therefore \tfrac{1}{2}\sum_{n=1}^{n=r} n^2 + \tfrac{1}{2}\sum_{n=1}^{n=r} n = \frac{r(r+1)(r+2)}{6}$$

$$= \frac{r^3 + 3r^2 + 2r}{6}$$

Here, for the second term on the left, we write in accordance with the law of formation of the triangular numbers:

$$\sum_{n=1}^{n=r} n = \frac{r(r+1)}{2} = \frac{r^2 + r}{2}$$

$$\therefore \tfrac{1}{2}\sum_{n=1}^{n=r} n^2 = \frac{r^3 + 3r^2 + 2r}{6} - \frac{r^2 + r}{4}$$

$$\therefore \tfrac{1}{2}\sum_{n=1}^{n=r} n^2 = \frac{2r^3 + 6r^2 + 4r}{12} - \frac{3r^2 + 3r}{12}$$

$$= \frac{2r^3 + 3r^2 + r}{12} = \frac{r(r+1)(2r+1)}{6}$$

$$\therefore \quad \sum_{n=1}^{n=r} n^2 = \frac{r(r+1)(2r+1)}{6}$$

We have thus obtained an expression for the sum of the *squares* of the first r natural numbers, in fact, for the third dimensional terms of which those of one dimension are the odd numbers in accordance with the additive law of formation:

1	3	5	7	9	11	13	15	17	19
1	4	9	16	25	36	49	64	81	100
1	5	14	30	55	91	140	204	285	385

In agreement with the above, the sum of the squares of the first ten natural numbers is according to our formula:

$$\frac{10\,(10 + 1)\,(20 + 1)}{6} = \frac{10 \,.\, 11 \,.\, 21}{6} = 5 \,.\, 11 \,.\, 7 = 385.$$

By the same method the reader may derive the formula already given on p. 178 for the sum of the *cubes* of the first r natural numbers.

Binomial Coefficients: We are now ready to construct a family of numbers, usually known as binomial coefficients and sometimes wrongly called Pascal's triangle after the seventeenth-century author of a tract on the topic, himself the first to formulate the notion of mathematical probability. As a first step let us look at a very simple problem of *choice*; in how many ways can we choose (*regardless of order*) from six things 1, 2, 3, up to six different ones? Let us label them as *ABCDEF*. There is only one way of choosing *none* at a time; and six ways of *choosing* 1 at a time. We shall write these results as: ${}^6C_0 = 1$ *and* ${}^6C_1 = 6$. If we set out our results in sequence we have the following way of choosing 2, 3 etc., and can label them accordingly as ${}^6C_2$, ${}^6C_3$, etc.:

AB, AC, AD, AE, AF, BC, BD, BE, BF, CD, CE, CF, DE, DF, EF

$$\therefore {}^6C_2 = 15$$

ABC ABD ABE ABF ACD ACE ACF ADE ADF AEF
BCD BCE BCF BDE BDF BEF CDE CDF CEF DEF

$$\therefore {}^6C_3 = 20$$

ABCD ABCE ABCF ABDE ABDF ABEF ACDE ACDF
ACEF ADEF BCDE BCDF BCEF BDEF CDEF

$$\therefore {}^6C_4 = 15$$

ABCDE ABCDF ABCEF ABDEF ACDEF BCDEF

$$\therefore {}^6C_5 = 6$$

Needless to say, there is only *one* way of choosing (regardless of order) six different things out of a heap of six, i.e. ${}^6C_6 = 1$. So we may tabulate these results as below:

${}^6C_0$	${}^6C_1$	${}^6C_2$	${}^6C_3$	${}^6C_4$	${}^6C_5$	${}^6C_6$
1	6	15	20	15	6	1

To get further light on the build up of such series, let us return to the family of which the Pythagorean triangular numbers represent the 2-dimensional series, viz.:

1	1	1	1	1	1	1	1	1
1	2	3	4	5	6	7	8	9
1	3	6	10	15	21	28	36	45
1	4	10	20	35	56	84	120	165
1	5	15	35	70	126	210	330	495
1	6	21	56	126	252	—	—	—
1	7	28	84	210	—	—	—	—

Now let us tilt through 45° clockwise as much of the table as we care to do, so that it looks like this:

1

1 1

1 2 1

1 3 3 1

1 4 6 4 1

1 5 10 10 5 1

1 6 15 20 15 6 1

Here the bottom row corresponds to the series: ${}^6C_0$, ${}^6C_1$, ${}^6C_2$, etc. By combining letters as for ${}^6C_r$, the reader will be able to ascertain that we may write the terms of the triangle:

$$
\begin{array}{ccccccccc}
 & & & & {}^0C_0 & & & & \\
 & & & {}^1C_0 & & {}^1C_1 & & & \\
 & & {}^2C_0 & & {}^2C_1 & & {}^2C_2 & & \\
 & {}^3C_0 & & {}^3C_1 & & {}^3C_2 & & {}^3C_3 & \\
{}^4C_0 & & {}^4C_1 & & {}^4C_2 & & {}^4C_3 & & {}^4C_4 \\
 & & \text{etc.} & & \text{etc.} & & \text{etc.} & &
\end{array}
$$

To tie up the meaning of these signs in terms of the figurate entries, it is best to push the layers, so that the first term of each row starts in the same vertical column. The result takes shape thus:

$$
\begin{array}{lllllll}
1 & & & & & & \\
1 & 1 & & & & & \\
1 & 2 & 1 & & & & \\
1 & 3 & 3 & 1 & & & \\
1 & 4 & 6 & 4 & 1 & & \\
1 & 5 & 10 & 10 & 5 & 1 & \\
1 & 6 & 15 & 20 & 15 & 6 & 1
\end{array}
\qquad
\begin{array}{lllllll}
F_{1.0} & & & & & & \\
F_{1.1} & F_{2.0} & & & & & \\
F_{1.2} & F_{2.1} & F_{3.0} & & & & \\
F_{1.3} & F_{2.2} & F_{3.1} & F_{4.0} & & & \\
F_{1.4} & F_{2.3} & F_{3.2} & F_{4.1} & F_{5.0} & & \\
F_{1.5} & F_{2.4} & F_{3.3} & F_{4.2} & F_{5.1} & F_{6.0} & \\
F_{1.6} & F_{2.5} & F_{3.4} & F_{4.3} & F_{5.2} & F_{6.1} & F_{7.0}
\end{array}
$$

In the signs we have defined for brevity to express $F_{r.d}$ we can now write the table as below:

$$
\begin{array}{lllllll}
\frac{1^{[0]}}{0!} & & & & & & \\[6pt]
\frac{1^{[1]}}{1!} & \frac{2^{[0]}}{0!} & & & & & \\[6pt]
\frac{1^{[2]}}{2!} & \frac{2^{[1]}}{1!} & \frac{3^{[0]}}{0!} & & & & \\[6pt]
\frac{1^{[3]}}{3!} & \frac{2^{[2]}}{2!} & \frac{3^{[1]}}{1!} & \frac{4^{[0]}}{0!} & & & \\[6pt]
\frac{1^{[4]}}{4!} & \frac{2^{[3]}}{3!} & \frac{3^{[2]}}{2!} & \frac{4^{[1]}}{1!} & \frac{5^{[0]}}{0!} & & \\[6pt]
\frac{1^{[5]}}{5!} & \frac{2^{[4]}}{4!} & \frac{3^{[3]}}{3!} & \frac{4^{[2]}}{2!} & \frac{5^{[1]}}{1!} & \frac{6^{[0]}}{0!} & \\[6pt]
\frac{1^{[6]}}{6!} & \frac{2^{[5]}}{5!} & \frac{3^{[4]}}{4!} & \frac{4^{[3]}}{3!} & \frac{5^{[2]}}{2!} & \frac{6^{[1]}}{1!} & \frac{7^{[0]}}{0!}
\end{array}
$$

We can now discern a pattern in our bottom row, if we set forth the terms vertically:

$$\frac{1^{[6]}}{6!} = \frac{1.1.2.3.4.5.6}{1.2.3.4.5.6} = 1 = \frac{6.5.4.3.2.1}{1.2.3.4.5.6}$$

$$\frac{2^{[5]}}{5!} = \frac{1.2.3.4.5.6}{1.2.3.4.5} = 6 = \frac{6.5.4.3.2}{1.2.3.4.5}$$

$$\frac{3^{[4]}}{4!} = \frac{1.3.4.5.6}{1.2.3.4} = 15 = \frac{6.5.4.3}{1.2.3.4}$$

$$\frac{4^{[3]}}{3!} = \frac{1.4.5.6}{1.2.3} = 20 = \frac{6.5.4}{1.2.3}$$

$$\frac{5^{[2]}}{2!} = \frac{1.5.6}{1.2} = 15 = \frac{6.5}{1.2}$$

$$\frac{6^{[1]}}{1!} = \frac{1.6}{1} = 6 = \frac{6}{1}$$

$$\frac{7^{[0]}}{0!} = \frac{1}{1} = 1 = 1$$

On the left, the pattern ties up recognizably with that of our table of the family of Pythagorean figurates; but it has been usual during the past two centuries to express the general term in the form:

$$^{n}C_r = \frac{n(n-1)(n-2)\ldots(n-r+1)}{r!}$$

A shorthand in common use for the numerator of this fraction is the following:

$$n^{(0)} = 1$$
$$n^{(1)} = 1.n$$
$$n^{(2)} = 1.n(n-1)$$
$$n^{(3)} = 1.n(n-1)(n-2)$$
$$n^{(4)} = 1.n(n-1)(n-2)(n-3)$$
$$n^{(5)} = 1.n(n-1)(n-2)(n-3)(n-4)$$
$$n^{(r)} = 1.n(n-1)(n-2)(n-3)\ldots(n-r+1)$$

In this shorthand:

$$n^{(n)} = n! = 1^{[n]}$$

$$n^{[r]} = (n+r-1)^{(r)} \quad \textit{and} \quad (n-r+1)^{[r]} = n^{(r)}$$

We may also write for brevity:

$$\frac{n^{(r)}}{r!} = n_{(r)} = {}^nC_r$$

If the reader asks why the more conventional $n^{(r)}$ rather than the author's square bracket notation introduced for representation of figurate numbers has gained favour, the answer is that it plays a more useful role in the *Theory of Choice and Chance.* This will be the topic of a later chapter. The reader need therefore experience no sense of defeat if the next few paragraphs fail to register. Their aim is to *indicate* (*rather than to justify*) some reasons why $n^{(r)}$ and $n_{(r)}$, like $n!$, turn out to be such useful devices when discussing different ways of choosing things.

One customarily speaks of ${}^nC_r$ as the number of *combinations of n different things taken r at a time* on the assumption that the order of choice is irrelevant. If so, each of the following is a single combination of three letters: *ABC*, *ACB*, *BAC*, *BCA*, *CAB*, *CBA*; but if our concern is with the different ways (linear *permutations*) in which we can arrange three different letters *in a row* from the set of six *ABCDEF*, one sometimes writes this as ${}^6P_3$. As we shall see later (Chapter 12), but the reader may discover from Fig. 80:

$${}^nP_r = n^{(r)}$$

The number of such arrangements of 6 letters taken all at a time is $6^{(6)}, = 6 . 5 . 4 . 3 . 2 . 1 = 1 . 2 . 3 . 4 . 5 . 6 = 1^{[6]} = 6! = 720$.

In the *Theory of Choice*, $n_{(r)}$ defined as above has more than one meaning. If r of the n letters are identical and the remainder though not like them are also identical, $n_{(r)}$ is the number of *distinguishable* arrangements which we can make if we use all of them, e.g. from *AABBBB* we can make:

AABBBB	*ABABBB*	*ABBABB*	*ABBBAB*	*ABBBBA*
BAABBB	*BABABB*	*BABBAB*	*BABBBA*	*BBAABB*
BBABAB	*BBABBA*	*BBBAAB*	*BBBABA*	*BBBBAA*

It is customary to write this as ${}^6P_{2.4}$, and in general:

$${}^nP_{r\,.\,(n-r)} = n_{(r)}$$

The reader may ask: why write ${}^nP_{r\,.\,(n-r)}$ rather than ${}^nP_{(n-r)\,.\,r}$?

First Place (one at a time) five ways	Second Place (two at a time) 5 . 4 = 20 ways	Third Place (three at a time) 5 . 4 . 3 = 60 ways			Fourth Place (four at a time) 5 . 4 . 3 . 2 = 120 ways		
A	*AB*	*ABC*	*ABD*	*ABE*	*ABCD*	*ABDC*	*ABEC*
					ABCE	*ABDE*	*ABED*
	AC	*ACB*	*ACD*	*ACE*	*ACBD*	*ACDB*	*ACEB*
					ACBE	*ACDE*	*ACED*
	AD	*ADB*	*ADC*	*ADE*	*ADBC*	*ADCB*	*ADEB*
					ADBE	*ADCE*	*ADEC*
	AE	*AEB*	*AEC*	*AED*	*AEBC*	*AECB*	*AEDB*
					AEBD	*AECD*	*AEDC*
B	*BA*	*BAC*	*BAD*	*BAE*	*BACD*	*BADC*	*BAEC*
					BACE	*BADE*	*BAED*
	BC	*BCA*	*BCD*	*BCE*	*BCAD*	*BCDA*	*BCEA*
					BCAE	*BCDE*	*BCED*
	BD	*BDA*	*BDC*	*BDE*	*BDAC*	*BDCA*	*BDEA*
					BDAE	*BDCE*	*BDEC*
	BE	*BEA*	*BEC*	*BED*	*BEAC*	*BECA*	*BEDA*
					BEAD	*BECD*	*BEDC*
C	*CA*	*CAB*	*CAD*	*CAE*	*CABD*	*CADB*	*CAEB*
					CABE	*CADE*	*CAED*
	CB	*CBA*	*CBD*	*CBE*	*CBAD*	*CBDA*	*CBEA*
					CBAE	*CBDE*	*CBED*
	CD	*CDA*	*CDB*	*CDE*	*CDAB*	*CDBA*	*CDEA*
					CDAE	*CDBE*	*CDEB*
	CE	*CEA*	*CEB*	*CED*	*CEAB*	*CEBA*	*CEDA*
					CEAD	*CEBD*	*CEDB*
D	*DA*	*DAB*	*DAC*	*DAE*	*DABC*	*DACB*	*DAEB*
					DABE	*DACE*	*DAEC*
	DB	*DBA*	*DBC*	*DBE*	*DBAC*	*DBCA*	*DBEA*
					DBAE	*DBCE*	*DBEC*
	DC	*DCA*	*DCB*	*DCE*	*DCAB*	*DCBA*	*DCEA*
					DCAE	*DCBE*	*DCEB*
	DE	*DEA*	*DEB*	*DEC*	*DEAB*	*DEBA*	*DECA*
					DEAC	*DEBC*	*DECB*
E	*EA*	*EAB*	*EAC*	*EAD*	*EABC*	*EACB*	*EADB*
					EABD	*EACD*	*EADC*
	EB	*EBA*	*EBC*	*EBD*	*EBAC*	*EBCA*	*EBDA*
					EBAD	*EBCD*	*EBDC*
	EC	*ECA*	*ECB*	*ECD*	*ECAB*	*ECBA*	*ECDA*
					ECAD	*ECBD*	*ECDB*
	ED	*EDA*	*EDB*	*EDC*	*EDAB*	*EDBA*	*EDCA*
					EDAC	*EDBC*	*EDCB*

Fig. 80. Linear Permutations

Here are shown from five different items (*ABCDE*), the five ways (${}^5P_1$) of taking one, $5 \times 4 = 20$ ways (${}^5P_2$) of taking two and so on. For each way of choosing four, there is only one way of choosing five, so that ${}^5P_4 = {}^5P_5 = 5 . 4 . 3 . 2 . 1$. More generally:

$${}^nP_r = n(n-1)(n-2)\ldots(n-r+1) = n^{(r)} \text{ and } {}^nP_n = n^{(n)} = n!$$

Either would be appropriate. There is no ambiguity in this formula, since

$$\frac{n^{(r)}}{r!} = \frac{n(n-1)(n-2)\ldots(n-r+1)}{(1.2.3.4\ldots r)}$$

$$= \frac{n(n-1)(n-2)\ldots(n-r+1)}{(1.2.3.4\ldots r)} \times \frac{(n-r)(n-r-1)\ldots 4.3.2.1}{1.2.3.4\ldots(n-r-1)(n-r)}$$

$$\therefore \frac{n^{(r)}}{r!} = \frac{n!}{r!(n-r)!} \quad \text{and} \quad \frac{n^{(n-r)}}{(n-r)!} = \frac{n!}{r!(n-r)!}$$

$$\therefore {}^nC_r = {}^nC_{n-r} \text{ and } {}^nP_{r.(n-r)} = {}^nP_{(n-r).r}$$

In this introductory excursion into the art of *translating* the language of size, order, and shape, we have made the acquaintance of a trick (so-called *subscript notation*) rarely, if ever, employed in very elementary books too often hag-ridden with x or y, a, or b, etc., which suggest nothing about what they stand for at a verbal level. When we write t_r for the *t*(erm) of *r*(ank) r or S_r for the *s*(um) of all the terms up to and including that of *r*(ank) r, we are not likely to forget what our signs stand for. Many of the pitfalls of the beginner result from failure to keep track of the meaning of signs which do not suggest the entities to which they refer, and we can sidestep them by making our signs suggestive in more ways than one. For instance, if our task is a computation involving the cost of peaches and pineapples, we may use p_e for the price of the former and p_i for that of the latter. Alternatively, since pineapples are larger than peaches, P for the price of a pineapple and p for that of a peach.

Exercises on Chapter 4

1. Given that

$$\sqrt{2} = 1{\cdot}4142$$
$$\sqrt{3} = 1{\cdot}7321$$
$$\sqrt{5} = 2{\cdot}2361$$

find correct to three places of decimals $\sqrt{27}$, $\sqrt{18}$, $\sqrt{12}$, $\sqrt{24}$, $\sqrt{10}$, and $\sqrt{30}$.

2. If the hypotenuse of a right-angled triangle is 1 unit of length, find the third side when the second side is $\frac{3}{4}$, $\frac{2}{3}$, and $\frac{4}{5}$ of the hypotenuse.

3. Make up any arithmetical series of five terms. Call the sum

S. Write the series backwards underneath so that the last term comes under the first term, etc. Add the two together, thus getting $2S$. Verify arithmetically that the sum of an arithmetical progression (AP) of n terms when the first term is f and the last term is l is $\frac{n}{2}(f + l)$.

4. Repeat this with another arithmetical series.

5. Write down the abstract number series $f, f + d, f + 2d$, etc. If there are n terms, express l in terms of f, n, and d. Express f in terms of n, l, and d. Express the last term but two in terms of (*a*) n, f, and d, and (*b*) n, l, and d. In this way build up the general rule without using proper numbers.

6. Find the fifth term, the tenth term, and the sum of ten terms in the following arithmetical progressions. First apply the formulae you have found, then check your results by writing out the first ten terms and adding them up.

$1 \quad 3 \quad 5 \ldots \qquad -6 \quad -2 \quad +2 \ldots$
$1 \quad 4 \quad 7 \ldots \qquad a \quad +0 \quad -a \quad -2a \ldots$
$5 \quad 10 \quad 15 \ldots \qquad 3 \quad +\frac{4}{3} \quad -\frac{1}{3} \ldots$
$\frac{1}{2} \quad 1 \quad 1\frac{1}{2} \ldots$

7. Find the nth term and the sum of n terms in the foregoing arithmetical progressions. Find the first term and the differences between two successive terms in the AP in which the sixth term is 13 and the twelfth is 25.

8. Find the sum of the first n natural numbers (1, 2, 3, . . .).

9. Find four numbers between 6 and 15 such that together with 6 and 15 they form six terms of an AP.

10. Find three terms between 1 and 3, which together with 1 and 3 form five terms of an AP.

11. Show how to insert n numbers between f and l so that together they form an AP of $n + 2$ terms, with first term f and last term l.

This is sometimes called inserting n arithmetic means between f and l. It is rather a stupid name, but it is a useful thing to know. It enables you, for instance, to find a given number of points at equal distances apart on a straight line.

12. Build up the formula for the sum of a geometrical series of n terms, f being the first term, fr the second term, fr^2 the third term, and so on.

Show that it is $\frac{f(r^n - 1)}{r - 1}$.

13. Find the fifth term and the sum of five terms in the following geometrical series:

$$1, 2, 4, \ldots$$
$$0{\cdot}9, 0{\cdot}81, 0{\cdot}729, \ldots$$
$$\tfrac{3}{4}, \tfrac{3}{8}, \tfrac{3}{16}, \ldots$$
$$x^5, a \,.\, x^4, a^2x^3, \ldots$$
$$1, 3, 3^2, \ldots$$

Check your results arithmetically.

14. Find the nth term and the sum of n terms in the foregoing geometrical series.

15. Insert two numbers between 5 and 625 so that the four terms form a geometrical progression. This is sometimes called inserting two geometric means between 5 and 625.

16. Insert three geometric means between $\frac{1}{3}$ and $\frac{16}{243}$.

17. Find the formula for inserting n geometric means between two numbers of which the first is f and the second is l.

18. Make up a geometrical series in which the first term is f and r (of the formula in example 12) is a fraction less than 1. Write out ten terms of the series. What would you expect the last term to be if you went on for ever?

19. Remembering that if you can make a quantity as small as you like you can ignore it compared with quantities which have a definite value, see whether you can show, by examining the formula in example 12, that the sum of a diminishing geometrical series which goes on for ever (first term a, common ratio r) cannot be larger than $\dfrac{a}{1-r}$.

20. You can write any recurring decimal in the following way:

$$0{\cdot}666\ldots \qquad = 0{\cdot}6 + 0{\cdot}06 + 0{\cdot}006\ldots$$

Use the result of the last example to express the following recurring decimals as proper fractions:

$$0{\cdot}\dot{6}$$
$$0{\cdot}252525\ldots$$
$$0{\cdot}791791791791\ldots$$

21. Find a formula for the sum of the first n terms of the series: a, ar, ar^2, ar^3, etc.

22. Find the nth term:

(*a*) in the series of hexagonal numbers, shown in Fig. 78, and the alternative hexagonals 1, 6, 15, 28, 45, etc.;
(*b*) in the series of stellate numbers (Fig. 79).

Check your formula by drawing figures of the third and fifth numbers.

23. Find by experiment (diagram) and formula how many different arrangements there are of the four aces in a pack of cards, the four aces and four kings, all three kinds of picture cards, all cards lower than 6 (excluding aces).

24. Using a diagram to check the formula, find how many different sets, irrespective of order (i.e. combinations of cards), can be obtained by taking three cards, any one of which may be either a king or a queen, from a full pack, or four cards, any one of which may be a king, queen, or jack, or five cards, any one of which may be a king, queen, jack, or ace.

25. How many different peals can be rung with six bells, all different, using all the bells once only?

26. How many different scores can be obtained from: (*a*) three; (*b*) five tosses of a die?

27. A committee consists of chairman, secretary, treasurer, and four ordinary members. In how many different ways can they sit at a straight table on the platform behind the speaker if: (*a*) no places are reserved; (*b*) the middle seat is reserved for the chairman; (*c*) the secretary and the treasurer must sit on either side of the chairman, who is occupying the middle seat; and (*d*) the secretary sits on the right and the treasurer on the left of the chairman occupying the middle seat.

28. A bag contains six coloured balls (all different). How many different pairs of two colours can be drawn from it if (*a*) each pair is replaced, and if (*b*) each pair is left out, when withdrawn?

29. Test the formula:

$$^nC_0 + {}^nC_1 + {}^nC_2 + {}^nC_3 \ldots {}^nC_n = \sum_{r=0}^{r=n} {}^nC_r = 2^n$$

Things to Memorize

1. $n^{(r)} = 1 \,.\, n(n-1)(n-2) \ldots (n-r+1) \,.\, and\ n^{(0)} = 1$

2. If ${}^{n}C_{r}$ stands for the number of combinations of n different things taken r at a time (none taken more than once)

$$ {}^{n}C_{r} = \frac{n(n-1)(n-2)\ldots(n-r+1)}{r!} = \frac{n!}{(n-r)!r!} = n_{(r)} $$

$$ {}^{n}C_{r} = {}^{n}C_{n-r} $$

3. ${}^{n}C_{0} = 1 = 0! = n^{(0)}$

$$ {}^{n}C_{0} = {}^{n}C_{n} $$

4. If ${}^{n}P_{r}$ is the number of linear permutations (i.e. arrangements in a straight line) of r different objects taken from n different objects (none taken more than once),

$$ {}^{n}P_{r} = n^{(r)} \quad \text{and} \quad {}^{n}P_{n} = n! $$

Answers to Problems on Fig. 79

$F_r = 2r^2 - 2r + 1$; $V_r = r(5r + 4)$; $M_r = 5r^2$; $S_r = 1 + 8r(r-1)$; $K_r = 1 + 5r(r-1)$.

CHAPTER 5

THE RISE AND DECLINE OF THE ALEXANDRIAN CULTURE

FOR SIX centuries between the death of Alexander the Great and the adoption of Christianity as the official creed of the crumbling Roman Empire, the city founded by, and named after, the Conqueror of Egypt, the Middle East, and Persia was the greatest centre of commerce, and a beacon of learning, in the Western World. From its foundation it was cosmopolitan – part Greek, part Jewish, part Egyptian, with a sprinkling of persons of Phoenician and Persian descent. Its intelligentsia had lively contacts with Greek-speaking settlements of Alexander's veterans on the temple sites of Iraq. Though Greek was the medium of its secular culture, its character owed much to what its personnel absorbed from people of other speech communities, and much to an unwritten body of geographical knowledge common to those whose parents had participated in overland campaigns which extended the conquests of Alexander and of his father Philip from Macedonia to India in the course of half a dozen years.

The outlook of its thinkers had little in common with the intellectual aloofness of the Athenian circles in which Plato's influence was predominant; and its so-called Museum, with a library at one time said to be stocked with three-quarters of a million volumes, was far more akin to our ideas of a University than were the Athenian Schools associated with the names of Plato and Aristotle. To be sure, its first teacher of renown adhered to the Athenian tradition of geometry; but Euclid himself anticipated the more workmanlike temper of his successors in the composition of treatises on optics. His use of geometry to interpret the properties of mirrors left an enduring impression on the astronomical advances which have made Alexandria most famous. Apart from the phenomenon of refraction first studied by Claudius Ptolemaios (about AD 150), Euclid's optics subsumes all theoretical knowledge of the propagation of light available until the time of Leonardo da Vinci (about AD 1500).

The same may be said of the mechanics of one of his immediate successors, Archimedes (about 225 BC), who established both the principle of the lever and the principle of hydrostatic buoyancy. Archimedes was pre-eminent both as a mathematician and as an engineer. Tradition credits him with the invention of the pumping device known as the *Archimedean Screw*, with the design of catapults for defence of his native city, and with the construction of enormous concave mirrors capable of concentrating solar energy sufficient to set on fire the ships of an invading navy. None the less, the range of his mathematical contributions, to only a few of which we can refer in this chapter, was formidable.

At a later date, probably contemporary with Caesar's landing which reputedly resulted in the destruction of the first Alexandrian library by fire, Heron, a mathematician of less, but of no mean, stature, has to his credit the invention of a steam-driven bellows which was the still-born predecessor of the turbine. He also applied his knowledge of trigonometry to prescribe how to dig a tunnel through a mountain by simultaneous excavation at each end. Among other memorable inventions of the Alexandrian age were water-clocks of remarkable complexity and no little service to the astronomers. Archimedes himself invented a wheel-driven model of the celestial sphere to reproduce the apparent diurnal motion of the fixed stars.

It was, indeed, in the field of astronomy and its sister science of geography that Alexandrian mathematics left its most lasting influence on everyday life. So far as we know, the most enterprising mathematicians of the Greek mainland made no serious attempt to measure either the size of the earth or its distance from celestial bodies. Alexandrian mathematicians did so with as much success as anyone else before transatlantic navigation by Columbus and Cabot gave a new impetus to astronomical and geographical study. The story begins with a calculation by Eratosthenes (about 250 BC), librarian of Alexandria during the period between the death of Euclid and the birth of Archimedes.

To understand his method, we need to recall only four elementary considerations which give us a clue to Euclid's interest in optics: (*a*) rays of light coming from a great distance appear to be parallel, a familiar experience of life in ancient times; (*b*) a line crossing two parallel lines makes correspond-

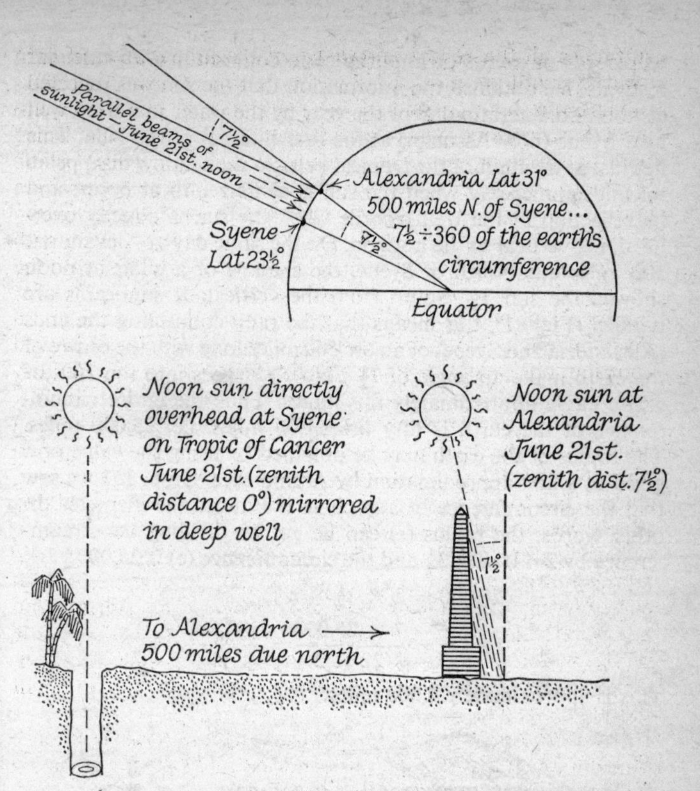

Fig. 81. How Eratosthenes Measured the Earth's Circumference

Note that at noon the sun lies directly over the observer's meridian of longitude. Syene and Alexandria have nearly the same longitude. So the sun, the two places, and the earth's centre may be drawn on the same flat slab of space.

ing angles equivalent (PARALLEL RULE ONE, Fig. 27); (*c*) when a heavenly body is directly overhead (*at its zenith*) a line joining the heavenly body to the observer passes through the centre of the earth (Fig. 59); (*d*) at noon the sun lies above some point on the observer's meridian of longitude (Fig. 24). Eratosthenes was librarian of Alexandria. As such, he had access to records of events of importance in connection with calendar festivals. He obtained the information that the sun was reflected at noon on a certain day of the year by the water in a deep well near Syene (now Assouan) at the first cataract of the Nile. This lies just at the limit of the tropical belt. So the shadow disappears on midsummer day when the sun is at its zenith at noon; and its reflection in the well records when the sun is directly overhead, i.e. vertical to the horizon. On the same day at Alexandria, 500 miles due North of Syene, the shadow of a pillar at noon showed the sun $7\frac{1}{2}°$ South from the vertical. If sunbeams are parallel (Fig. 81), this means that the radii connecting the ends (Alexandria and Syene) of an arc 500 miles long with the centre of the earth make an angle of $7\frac{1}{2}°$. Now $7\frac{1}{2}°$ goes into the 360° of entire circle approximately fifty times. Thus the entire circumference of the earth is fifty times 500 miles, i.e. 25,000 miles. The radius of the earth may be obtained by using the value of π given as a first approximation by Archimedes. On p. 153 we saw that the circumference of any circle is π times the diameter. In other words, the radius (r) can be got by dividing the circumference by 2π. If $\pi = 3\frac{1}{7}$ and the circumference (c) is 25,000:

$$25{,}000 = 2 \times 3\tfrac{1}{7} \times r$$

$$\text{i.e.} \quad r = \frac{7 \times 25{,}000}{22 \times 2} = \frac{175{,}000}{44}$$

This is approximately 4,000 miles.

Fig. 82. Ptolemy's Map of the World, about AD 150

This is one of three plane projections of the Old World as then known, made by Ptolemy.

There is a beautiful simplicity about the method which Eratosthenes used. It invokes no mathematical principles which had not been current in the Greek-speaking world two centuries before his time; and its importance to posterity lies less in any direct impetus to theoretical inquiry than to the fact that it provided an indispensable basis for any successful attempt to measure the distance of the earth from the sun or the moon, as attempted by his contemporary Aristarchus with inadequate information. Aristarchus is memorable for advancing, also with inadequate data, the heliocentric view of the motion of the earth and other planets of the sun. He also anticipated our Gregorian calendar by adding a fraction of $(1623)^{-1}$ of a day to the current estimate of the year, i.e. $365\frac{1}{4}$ as adopted by Julius Caesar on the advice of a later Alexandrian astronomer Sosigenes.

We can date an estimate of the distance of the earth from the moon, and one as good as any current before the time of Columbus, to a century later than that of Aristarchus and Eratosthenes, i.e. about 150 BC, when Alexandrian astronomy and geography received an enormous impetus from the work of three contemporaries: Hypsicles, Hipparchus, and Marinus. The first seems to have been responsible for introducing his colleagues and pupils to the Babylonian system of angular measurement (360° to a complete circle) and to sexagesimal fractions (p. 58). Hipparchus introduced a new system of mapping the position of stars by guide lines comparable to our familiar system of terrestrial longitude and latitude. He catalogued the positions of some 850 fixed stars in this way. For reasons which we shall see later, he needed for this purpose a table of trigonometrical ratios; and seems to have been the first to have constructed one, probably by the half-angle formulae (pp. 146–7). Marinus is notable as the first to introduce the use of circles of latitude and longitude to map the habitable globe as then known.

We know of the work of Hipparchus and Marinus from that of Claudius Ptolemaios (about AD 150), usually called Ptolemy but with no relation to the dynasty which ruled Alexandria from the death of Alexander to its Roman Annexation at the death of Cleopatra. Ptolemy's work called the *Almagest* in its Arabic translations was the bible of the Moslem geographers who fathered the cartography of the Great Navigations in which Columbus served his apprenticeship. Though Hindu and Moslem mathematicians of the intervening period greatly improved on it, his work remained the corner-stone of practical mathematics till the time of Columbus. He extended the star catalogue of Hipparchus by addition of over two hundred items, not necessarily by his own exertions. We may credit him with being a pioneer of what is called *projective* geometry inasmuch as he gives three recipes for representing contours on a spherical surface by contours on a flat one (Figs. 82–3). Above all, we owe to him a very comprehensive table of trigonometrical values. Indeed, the development of trigonometry was the supreme bequest of the Alexandrian culture to posterity.

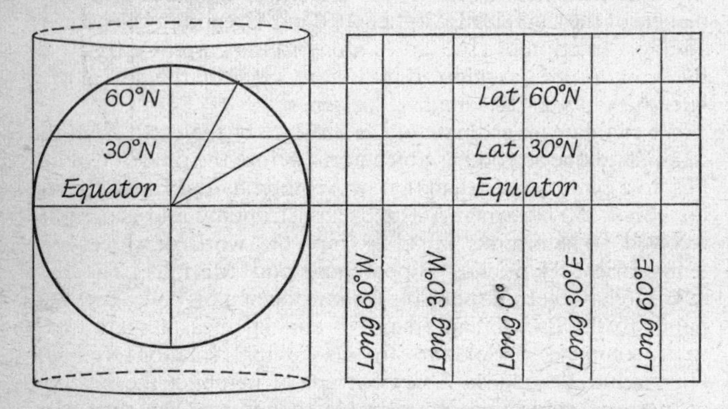

Fig. 83. Cylindrical Projection of the Earth's Land Masses

The two simplest ways of projecting the earth's land contours on a flat surface are first to extend the planes of latitude and longitude on to either a cylinder of paper tangential to the equator or to a paper cone tangential to a latitude midway between the extremes, then to cut the paper in the plane of some great circle of longitude and flatten it out. On the cylindrical projection circles of longitude come out as equally spaced parallel vertical lines, and circles of latitude as parallel horizontal lines becoming nearer as one approaches the poles. This projection preserves the area relations of the land and sea masses but distorts their shapes especially at the poles. No flat projection can preserve both shape and area relations realistically.

Though the intention of its Founding Fathers had little to do with the more mundane tasks of the surveyor, the uses of trigonometry for field work on land are not essentially different from those invoked to measure our distance from the nearer celestial objects. Here we shall steal a march on history by looking first at the construction of a table of sines in modern guise (see Fig. 84), and then at its uses in the hands of the tunnel engineers who took their cue from Heron.

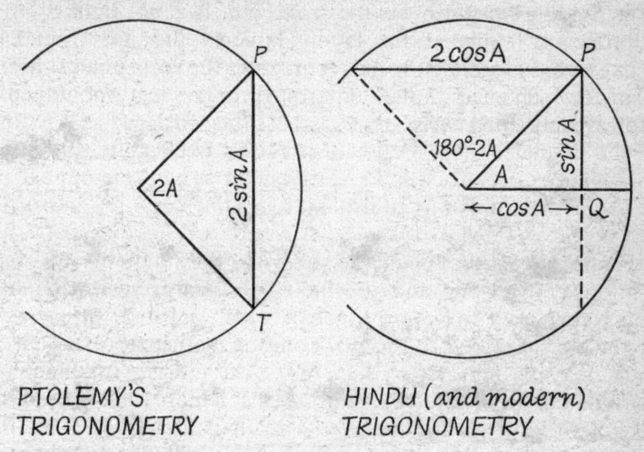

Fig. 84. Alexandrian and Modern Trigonometry
The Circle is of unit radius ($r = 1$).

When one speaks of trigonometry, in contradistinction to drawing-board geometry, one implies at an elementary level, a technique of measurement based on formulae for which it is possible to construct tables of numerical values of length ratios referable to any angle. In learning how to construct such tables, we have seen that a simple relation ($cos^2 A = 1 - sin^2 A$) suffices for the derivation of the cosine from the sine of an angle; and the same is true of the tangent, i.e. $tan A = sin A \div cos A$. In principle, Ptolemy's trigonometry, based on the much earlier work of Hipparchus, and one may guess that of Archimedes before him, subsumes the entire progress of trigonometry until the death of Newton. It undoubtedly gave an enormous impetus to the study of trigonometry by Moslem mathematicians during the period AD 800–1100; but trigonometry transmitted by them to the later European geographers of the Middle Ages followed a pattern traceable to their Hindu predecessors during the four centuries after the collapse of Alexandrian culture.

In turning to Euclid (Chapter 3) as a springboard from which we may dive into trigonometry, we have taken our clues from the properties of the right-angled triangle. In a sense this is inescapable, but we can define our trigonometrical ratios alternatively by reference to a circle of unit radius. When we do so (Fig. 58), the *sine* becomes the *semichord.* It is not unlikely that Hipparchus tabulated the latter. Ptolemy did not (Fig. 84). What he did is merely for the record, and the reader can skip the next few sentences unless interested in ancient monuments. Ptolemy tabulated *chords* by the angles they enclosed as fractions of the diameter of a circle whose radius is 60 units, the latter chosen because he used the sexagesimal system mentioned on p. 58. In effect, this was equivalent to putting $\sin A = \frac{1}{2}$ *chord* $2A$ and $\cos A = \frac{1}{2}$ *chord* $(180° - 2A)$.

From the standpoint of the surveyor, the tabulation of *sines* is more convenient; and the fact that Hindu civilization had undertaken vast irrigation schemes which enlisted the prowess of the surveyor before the flowering of mathematical talent in India may be relevant to their use. Alexandrian trigonometry germinated in a different soil. As the handmaid of scientific geography and of astronomy, its main concern was with the sphere. We might leave it at that, were it not for the fact that the recognition of the *sine* as the *semichord* of a circle of unit radius is the clue to our last but one step, already taken by Ptolemy in his own way, to the construction of a table of trigonometrical ratios, easily translatable into the sort we now use.

In Chapter 3, we have seen (p. 148) how Hipparchus might have made a table of semichords by using the half-angle formula of

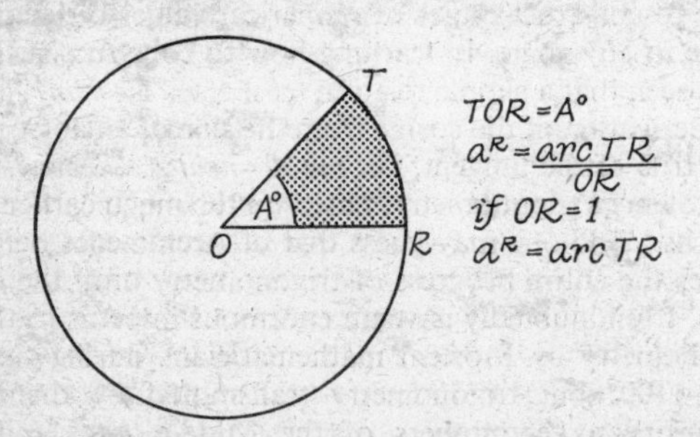

Fig. 85. Radian Measure

$OR = 1$ and $A° \equiv a^R$ (*A degrees = a radians*) i.e. $a = arc\ TR$. If $OR = R$ is not equal to unity, the length of the arc TR is $R \,.\, a^R$.

Fig. 64; but we have not seen how to make a table of *equally spaced intervals*, e.g. in degrees or corresponding fractions of a degree. To complete the story, we need to know two things:

(*a*) what is the value of *sin A* if $A = 1°$;
(*b*) what is the value of $\sin(A + B)$ if, for example, $A = 1\frac{15}{16}$ and $B = 3\frac{3}{4}$.

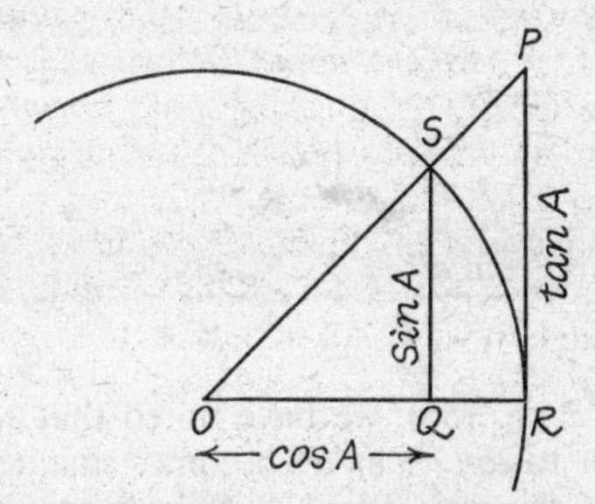

Fig. 86. Trigonometrical Ratios in a Circle of Unit Radius

Here $OR = 1 = OS$; $OQS = 90° = ORP$ and $SOQ = A°$, so that

$$\frac{SQ}{OS} = \sin A = SQ; \quad \frac{OQ}{OS} = \cos A = OQ; \quad \frac{PR}{OR} = \tan A = PR$$

If we join S to R by a straight line (SR):

$$\begin{aligned} SR^2 &= QR^2 + SQ^2 = (OR - OQ)^2 + \sin^2 A \\ &= (1 - \cos A)^2 + \sin^2 A \\ &= 1 - 2\cos A + \cos^2 A + \sin^2 A \\ &= 2 - 2\cos A \end{aligned}$$

By using a method which compels agreement, to find limits within which π lies, Archimedes solved the first problem and opened a vista to the solution of others at a much later date. With the advantage of hindsight, we can interpret his achievement by saying that it introduces us to a new unit of angular measurement. In the circle of Fig. 85, the length of the radius ($OT = 1 = OR$) is 1 unit. We then say that the angle TOR is $A°$ or *a radians* (written as a^R), if a is the length of the arc TR. We trace the entire perimeter of length $2\pi\,(OR) = 2\pi$ by rotating OR through 360°, so that 360° is equivalent to 2π radians. If we take π as approximately $3\frac{1}{7}$:

$$1 \text{ degree} = \frac{\pi}{180} \text{ radians} \simeq \frac{11}{630} \text{ radians.}$$

$$1 \text{ radian} \simeq \frac{90 \times 7}{11} \text{ degrees} \simeq 57\tfrac{3}{11}{}^{\circ}$$

The angle *SOQ* of the circle of unit radius shown in Fig. 86 is in radian measure *a* if *a* is the length of the *arc SR*, and it is clear that:

$$SQ < a^R < PR$$

$$\therefore \quad \frac{SQ}{a^R} < 1 < \frac{PR}{a^R}$$

$$\frac{\sin a^R}{a^R} < 1 < \frac{\tan a^R}{a^R}$$

In evaluating π (p. 155), we have noted that $\sin a^R$ becomes closer and closer to $\tan a^R$ as a^R becomes smaller. Hence as a^R becomes smaller, the ratio $\sin a^R \div a^R$ becomes nearer to unity, and we may write

$$\underset{a \to 0}{Lt} \frac{\sin a}{a} = 1$$

How quickly this ratio approaches unity, i.e. $\sin a \simeq a$ in radian measure, is a matter we can investigate by recourse to the half-angle formula shown in Fig. 64. By definition (as above) 15° in radian measure approximately:

$$\frac{15 \times 11}{630} \simeq 0{\cdot}2618$$

By recourse to the half-angle formula sin 15° $\simeq$ 0·2588. By using the half-angle formulae (pp. 146–8), and a somewhat better approximation for π, we get the values shown in the accompanying table. From these results it is clear that there is no error in the fifth decimal place, if we write:

$$\sin 1^{\circ} = 1^{\circ} \text{ in radian measure}$$

$$\therefore \quad \sin 1^{\circ} = \frac{\pi}{180} = 0{\cdot}01745$$

Degree	Sine	Radian
$15°$	0·2588190	0·2617994
$11\frac{1}{4}°$	0·1950903	0·1963495
$7\frac{1}{2}°$	0·1305202	0·1308997
$5\frac{5}{8}°$	0·0980171	0·0981747
$3\frac{3}{4}°$	0·0654031	0·0654498
$2\frac{13}{16}°$	0·0490676	0·0490873
$1\frac{7}{8}°$	0·0327190	0·0327249
$1\frac{13}{32}°$	0·0245412	0·0245436
$\frac{15}{16}°$	0·0163617	0·0163624
$\frac{45}{64}°$	0·0122715	0·0122718

Evidently, we can therefore take the sines of $\frac{1}{8}$ and $\frac{3}{16}$ degrees to be the radian measure of these angles. Knowing the sines of $1\frac{7}{8}$ and $2\frac{13}{16}$ as in the accompanying table, we can therefore find $sine\ 2° = sine\ (1\frac{7}{8} + \frac{1}{8})°$ or $sine\ 3° = sine\ (2\frac{13}{16} + \frac{3}{16})$, if we are able to devise a formula for $sin\ (A + B)$. Before taking this step, however, let us examine what we may take as a sufficiently accurate value of π, i.e. one which lies between limits sufficiently small for our purpose, e.g. making a table of *sines*, etc., correct to five decimal places.

Apart from its relevance to our theme, i.e. construction of a table of sines, etc., the accompanying table of sines and radians is instructive for another reason. In textbooks of physics, the reader will often find formulae (e.g. in the theory of the pendulum or of the focal length of thin lenses) based on the assumption that we can put x for $sin\ x$ if x is small. This is true only if x is in radian measure, and in radian measure what we think of as sizeable if accustomed to think in degrees may indeed be small. For instance 5° is less than 0·1 radians. It is also worth while to notice an elementary consequence of Dem. 7 in the light of our definition of *circular measure*, i.e. of the radian. If $OT = R$ in Fig. 85 is not our unit of length, the length of the arc TR is $a^R . R$.

Finding a Value for π. In the Old Testament (2 Chron. iv. 2) we read: 'Also he made a molten sea of ten cubits from brim to brim, round in compass, and five cubits the height thereof, and a line of thirty cubits did compass it about.' The circumference was thus six times the radius, or three times the diameter. That is to say, the ancient Hebrews, like the Babylonians, were content with taking π as 3. The Ahmes papyrus shows that about 1500 BC the Egyptians used $\sqrt{10}$ or 3·16. You can get as good a value as this by experiment if you measure the diameter and boundary of all the tins, plates, and saucepans in your house with a measuring tape. The same value was fashionable among the Chinese calendar-makers and engineers. About AD 480 we find that an irrigation engineer called Tsu Ch'ung Chih, who constructed a sort of motor-boat and reintroduced the compass or 'south pointing' instrument, arrived at an estimate of astonishing accuracy for the time. In our notation it is equivalent to saying that π lies between 3·1415926 and 3·1415927. We do not know how he got it. It is not easy to believe that he did so by drawing a large-scale diagram. A possible clue is provided by the fact that the Japanese were using about AD 1700 a method similar

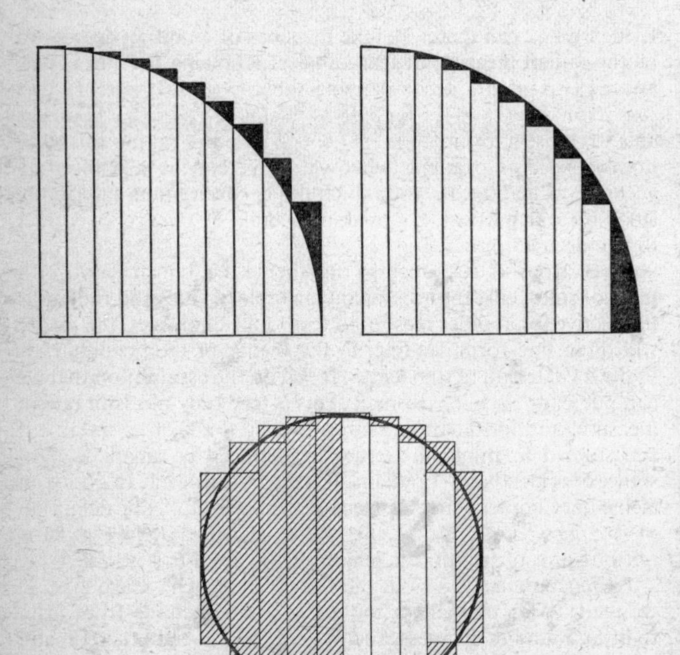

Fig. 87. The Japanese Method for Computing π

to that which was just being applied in Europe at the same time. The Japanese way involves splitting the circle into tiny rectangles, and relies on the fact that the area of a circle is πr^2 (p. 154), where r is the radius, so that if the radius is 1 unit of length, the area of the circle is π units of area.

If you trace a circle on graph paper, you will see that the area of the circle lies between that of two sets of superimposed oblong strips, white and shaded in Fig. 87 at the bottom. In each half of the circle there will be one more of the outer or longer strips than the inner or shorter, drawn as in the two quarter-circles at the top of Fig. 87. The quarter-circle of Fig. 88 is enclosed by five rectangular strips of the same width, and itself encloses four superimposed rectangular strips of the same width. From the way the rectangles are drawn you will see why the fifth inner rectangle vanishes. If the radius of the circle is 1 unit of length, the width of each strip is $\frac{1}{5}$. The area of all the rectangles of the outer series is

$$\begin{aligned} &\tfrac{1}{5}y_0 + \tfrac{1}{5}y_1 + \tfrac{1}{5}y_2 + \tfrac{1}{5}y_3 + \tfrac{1}{5}y_4 \\ &= \tfrac{1}{5}(y_0 + y_1 + y_2 + y_3 + y_4) \end{aligned}$$

In a complete circle the corresponding rectangular strips would amount to four times this, i.e.

$$A_c = \tfrac{4}{5}(y_0 + y_1 + y_2 + y_3 + y_4)$$

Similarly, for the area of all the inner strips in the full circle we may write:

$$A_i = \tfrac{4}{5}(y_1 + y_2 + y_3 + y_4)$$

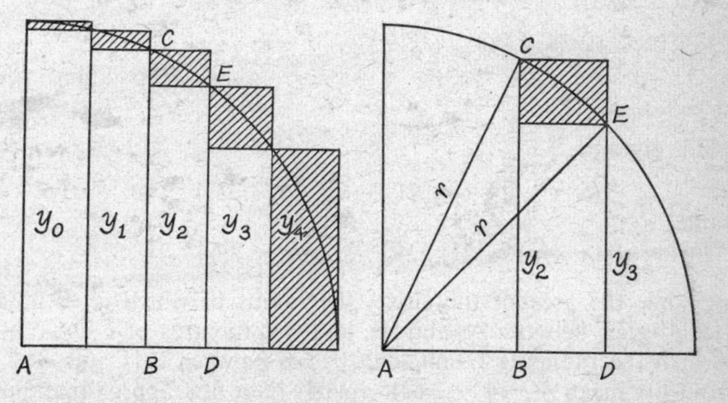

Fig. 88. The Japanese Method for Computing π (*contd.*)

The values of y_1, y_2, etc., depend on using Dem. 4, the Chinese right-angle theorem. In the triangle ABC:

$$r^2 = y_2^2 + (AB)^2$$

Since r, the radius, is 1 and AB is two steps of $\frac{1}{5}$, we can put:

$$1 = y_2^2 + (\tfrac{2}{5})^2$$

$$\therefore \quad 1 - (\tfrac{2}{5})^2 = y_2^2$$

i.e. $$y_2 = \sqrt{1 - \left(\frac{2}{5}\right)^2} = \sqrt{\frac{5^2 - 2^2}{5^2}} = \tfrac{1}{5}\sqrt{5^2 - 2^2}$$

Similarly, in the triangle AED

$$1^2 = y_3^2 + (\tfrac{3}{5})^2$$

$$\therefore \quad y_3 = \tfrac{1}{5}\sqrt{5^2 - 3^2}$$

In the same way $y_1 = \frac{1}{5}\sqrt{5^2 - 1^2}$

$$y_4 = \tfrac{1}{5}\sqrt{5^2 - 4^2}$$

And $y_0 = 1 = \frac{5}{5}$.

So we may write

$$A_c = \tfrac{4}{5}[\tfrac{5}{5} + \tfrac{1}{5}\sqrt{5^2 - 1^2} + \tfrac{1}{5}\sqrt{5^2 - 2^2} + \tfrac{1}{5}\sqrt{5^2 - 3^2} + \tfrac{1}{5}\sqrt{5^2 - 4^2}]$$

$$\therefore A_c = \frac{4}{5^2}[5 + \sqrt{5^2 - 1^2} + \sqrt{5^2 - 2^2} + \sqrt{5^2 - 3^2} + \sqrt{5^2 - 4^2}] \quad \text{(i)}$$

Likewise

$$A_i = \frac{4}{5^2}[\sqrt{5^2 - 1^2} + \sqrt{5^2 - 2^2} + \sqrt{5^2 - 3^2} + \sqrt{5^2 - 4^2}] \quad \text{(ii)}$$

This gives us

$$A_c = \tfrac{4}{25}(5 + \sqrt{24} + \sqrt{21} + 4 + 3) = 3{\cdot}44$$

and

$$A_i = \tfrac{4}{25}(\sqrt{24} + \sqrt{21} + 4 + 3) \qquad = 2{\cdot}64$$

Since the area of the circle, the radius of which is of unit length, lies between A_c and A_i, and π is the area of a circle of which the radius is 1 unit long, π lies between 3·44 and 2·64; and the mean $\frac{1}{2}(3{\cdot}44 + 2{\cdot}64) = 3{\cdot}04$. As a first approximation, we may take: $\pi = 3{\cdot}04 \pm 0{\cdot}40$.

With a similar figure in which the radius is divided into ten equal parts, so that there are ten outer and nine inner rectangles, you should now be able to get for yourself

$$A_c = \frac{4}{10^2}(10 + \sqrt{10^2 - 1^2} + \sqrt{10^2 - 2^2} + \sqrt{10^2 - 3^2}$$
$$+ \sqrt{10^2 - 4^2} + \sqrt{10^2 - 5^2} + \sqrt{10^2 - 6^2}$$
$$+ \sqrt{10^2 - 7^2} + \sqrt{10^2 - 8^2} + \sqrt{10^2 - 9^2})$$

$$A_i = \frac{4}{10^2}(\sqrt{10^2 - 1^2} + \sqrt{10^2 - 2^2} + \sqrt{10^2 - 3^2}$$
$$+ \sqrt{10^2 - 4^2} + \sqrt{10^2 - 5^2} + \sqrt{10^2 - 6^2}$$
$$+ \sqrt{10^2 - 7^2} + \sqrt{10^2 - 8^2} + \sqrt{10^2 - 9^2})$$

In this way, with the aid of a table of square roots, and calling π_n the value of π we get when the radius is divided into n equivalent strips, we can build up a table like this:

$$\pi_5 = 3{\cdot}04 \pm 0{\cdot}40$$
$$\pi_{10} = 3{\cdot}10 \pm 0{\cdot}20$$
$$\pi_{15} = 3{\cdot}12 \pm 0{\cdot}14$$
$$\pi_{20} = 3{\cdot}13 \pm 0{\cdot}10$$

If you work these out, you will have discovered for yourself that there is no need to make a new figure each time. The rule for getting π can be expressed as a pair of series, the sum of the outer rectangles being:

$$\frac{4}{n^2}(n + \sqrt{n^2 - 1^2} + \sqrt{n^2 - 2^2} + \sqrt{n^2 - 3^2} + \sqrt{n^2 - 4^2} \ldots)$$

The sum of the inner rectangles will be the same with the first term within the brackets left out. The rule can be written for the outer rectangle strips in the symbolism of p. 219 (Chapter 4):

$$\frac{4}{n^2}\sum_{r=0}^{r=n} \sqrt{n^2 - r^2}$$

The corresponding rule for the inner rectangle strips is then:

$$\frac{4}{n^2}\sum_{r=1}^{r=n} \sqrt{n^2 - r^2}$$

If you recall earlier discussion on the use of series (Chapter 4), you will ask whether it is possible to get this series into a form so that, when n is very big, you can make it choke off at any convenient point, as you can neglect the later terms in the series representing a recurring decimal fraction. The Japanese actually succeeded in doing this at the end of the seventeenth century. Matsunaga gave an estimate of π which in our notation is a decimal fraction correct to fifty figures. We shall leave the solution of the problem till we can approach it in its proper historical setting. What we have done already is enough to suggest the possibility of finding a series for π which can be stretched out to any order of precision which measurement demands.

Archimedes was content to cite π as lying between the limits $3\frac{10}{71}$ and $3\frac{1}{7}$ whose mean is 3·14185 . . . At the present time π is known to seven hundred correct decimal places. The accompanying tables show values cited at different times by authors in different countries. Vieta obtained his value by working out the limits from a polygon of 393,216 sides. The later values are based on series. As a matter of fact ten decimal places are enough to give the circumference of the earth within a fraction of an inch, and thirty decimals would give the boundary of the entire visible universe within a fraction too small to be measured by the most powerful modern microscope. So you have no need to be disappointed with the result we have obtained for most practical purposes. For dcsigning the best aeroplane engines, it is only necessary to know four decimal places (3·1416).

TABLE OF VALUES OF π

Babylonians and Hebrews and earliest Chinese	3·0
Egyptians (*c.* 1500 BC)	3·16
Archimedes (240 BC) (interested in wheels)	between 3·140 and 3·142
Chinese calendar-makers and engineers:	
Liu Hsing (*c* AD 25)	3·16
Wang Fun (*c.* AD 250)	3·15
Tsu Ch'ung Chih (*c.* AD 480) (interested in machinery)	between 3·1415926 and 3·1415927
Hindus and Arabs:	
Aryabhata (*c.* AD 450)	3·1416
Al Kashi (*c.* AD 1430)	3·1415926535897932

European:

Vieta (*c.* AD 1593)	between $\begin{cases} 3{\cdot}1415926537 \\ 3{\cdot}1415926355 \end{cases}$
Ceulen (*c.* AD 1610)	correct to thirty-five decimal places
Wallis (*c.* AD 1650) and Gregory (*c.* AD 1668)	unlimited series

Japanese:

Takebe (*c.* AD 1690)	unlimited series
Matsunaga (*c.* AD 1720)	correct to fifty decimal places in our notation

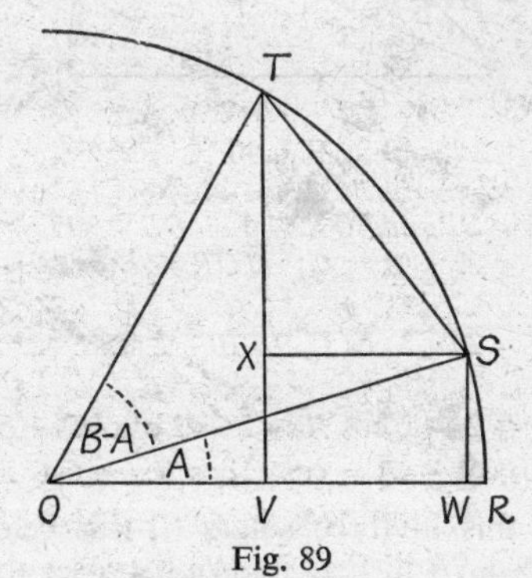

Fig. 89

As in Fig. 86:

$$OR = 1, \text{ so that } OT = 1 = OS$$
$$SOR = A^\circ; \quad TOR = B^\circ$$
$$\cos A = OW; \quad \cos B = OV$$
$$TS^2 = SX^2 + TX^2$$
$$SX = VW = OW - OV = \cos A - \cos B$$
$$TX = TV - XV = TV - SW = \sin B - \sin A$$
$$\therefore \; TS^2 = (\cos A - \cos B)^2 + (\sin B - \sin A)^2$$

Spacing Our Trigonometrical Tables. Being now clear about how to find *sin A* if $A = 1^\circ$ or any fraction of unity, we may now take the final step towards a table of sines (etc.) in equal intervals. All that remains is to find a formula for *sin* $(A + B)$ when $A = 1^\circ$ and B is some value obtainable as on p. 147 by the half-angle formula. We need not do this the hard way by what his successors called Ptolemy's Theorem, or by the equally laborious method of most textbooks still current in the forties of our own century.

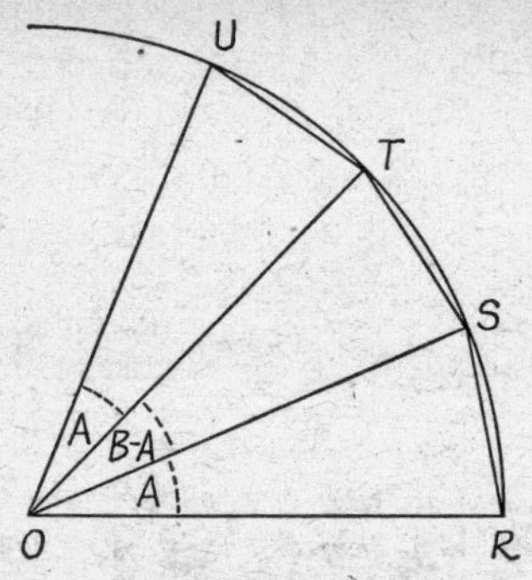

Fig. 90.

In this figure:

$$UT = SR;\ and\ OR = 1 = OS = OT = OU$$
$$UOR = A + B = C; \quad TOR = B; \quad A = C - B$$

From Fig. 89:

$$UT^2 = 2 - 2 \cos C \cos B - 2 \sin C \sin B$$

From Fig. 86:

$$SR^2 = 2 - 2 \cos A = 2 - 2 \cos (C - B)$$
$$\therefore \cos (C - B) = \cos C \cos B + \sin C \sin B$$

We shall do this in three stages (i)–(iii) below. First look again at the chord SR in Fig. 86. We there see that:

$$SR^2 = 2 - 2 \cos A \quad . \quad . \quad . \quad 1 \text{ (i)}$$

Now turn to Fig. 89. We there see that we can express the chord TS in another way:

$$\begin{aligned} TS^2 &= SX^2 + TX^2 \\ &= (\cos A - \cos B)^2 + (\sin B - \sin A)^2 \\ &= \cos^2 A - 2 \cos A \cos B + \cos^2 B + \sin^2 B \\ &\qquad - 2 \sin A \sin B + \sin^2 A \\ &= (\cos^2 A + \sin^2 A) + (\cos^2 B + \sin^2 B) - 2 \cos A \cos \\ &\qquad B - 2 \sin A \sin B \end{aligned}$$

$$\therefore \quad TS^2 = 2 - 2 \cos A \cos B - 2 \sin A \sin B \quad . \quad . \quad \text{(ii)}$$

All that remains is to look at Fig. 90 in which there are three angles: A, B, C, such that $C = A + B$:

From Fig. 89 and (ii)

$$UT^2 = 2 - 2 \cos C \cos B - 2 \sin C \sin B$$

From Fig. 86 and (i)

$$SR^2 = 2 - 2 \cos A = 2 - 2 \cos (C-B)$$

But $UT = SR$, so that

$$2 - 2 \cos C \cos B - 2 \sin C \sin B = 2 - 2 \cos (C-B)$$

$$\therefore \quad \cos (C - B) = \cos C \cos B + \sin C \sin B \quad . \quad . \quad \text{(iii)}$$

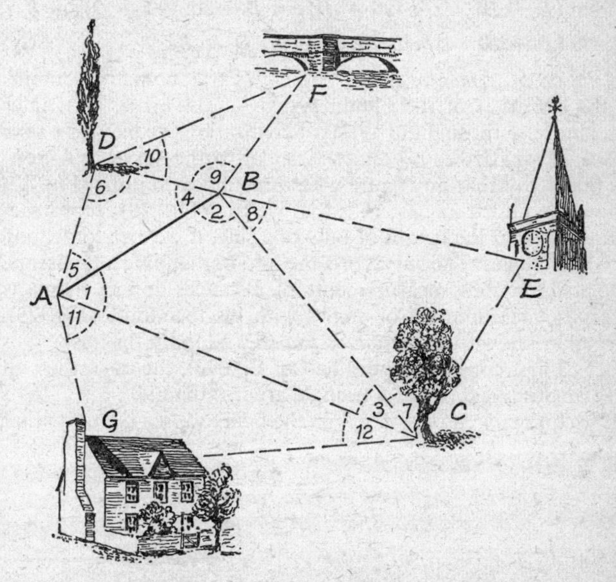

Fig. 91. Triangulation in District Map-Making

When surveying a district, the Surveyor first measures accurately the length of a base line *AB*. From *A* he sights *C*, a conspicuous object (here a tree) at an angle *BAC* (= 1), and from *B* at an angle *ABC* (= 2). He now knows 2 angles and one side of the triangle *ABC*, whence also the angle *BCA* and can get the lengths *BC* and *AC* by use of the sine formulae. He may sight the church *E* from *B* and *C* to determine the angles *CBE* (= 8) and *BEC* (= 7), whence also *CEB*. In the triangle *BEC*, he has already found the distance *BC*, hence he can derive (as before) *BE* and *CE*. He may next sight the Farm *G* from *A* and *C*. He now knows the angles *GAC* (= 11); *ACG* (= 12) whence *AGC*, having also found *AC* as above. By the sine formulae, he can now derive the distances *AG*, *CG*. In the same way, he can locate the tree *D* and the bridge *F*.

As it stands, the last formula can meet all our needs, if the end in view is to space our trigonometrical table in equal intervals. Since we have stated as our goal to find a formula for *sin* $(A + B)$, we shall now derive it from the above. To do so, we put $C = (90 - D)$, so that

$$\cos (90 - D - B) = \cos [90 - (D - B)] = \sin (D + B)$$

$$\therefore \quad \sin (D + B) = \cos (90 - D) \cos B + \sin (90 - D) \sin B$$

$$\therefore \quad \sin (D + B) = \sin D \cos B + \cos D \sin B \quad . \quad . \quad . \quad \text{(iv)}$$

Solution of Triangles. Since Ptolemy's trigonometry contains all the essentials of the Hindu system of surveying, as it later came into use throughout Western civilization, we may here take a look at what rules are most essential to the surveyor's craft. We first recall that any figure with a rectilinear outline (Fig. 12) is divisible into triangles; next that we can (Fig. 28) reconstruct a triangle from the length of only one side, if we also know any two of its angles. The surveyor's method of mapping a landscape is to make as few measurements of distances and as many as necessary measurement of angles with his theodolite (Fig. 25). Formulae known as *solution of triangles* simplify this task.

Let us first consider (upper half of Fig. 92), the case when all three angles (A, B, C) of a triangle are less than 90°.

If we know *three sides* a, b, c in the figure we see (Dem. 4) that

$$\begin{aligned} p^2 &= a^2 - d^2 \\ &= c^2 - (b - d)^2 \end{aligned}$$

$$\therefore \quad a^2 - d^2 = c^2 - b^2 + 2bd - d^2$$

$$\therefore \quad c^2 = a^2 + b^2 - 2bd$$

Since $\cos C = d \div a$ so that $d = a \cos C$

$$\therefore \quad c^2 = a^2 + b^2 - 2ab \,.\, \cos C$$

$$\therefore \quad \cos C = \frac{a^2 + b^2 - c^2}{2ab}$$

In the same way we can show that

$$\cos B = \frac{a^2 + c^2 - b^2}{2ac}$$

$$\cos A = \frac{b^2 + c^2 - a^2}{2cb}$$

So if we know a, b, c we can get $\cos A$, $\cos B$, or $\cos C$, and the tables of cosines give us what A, B, and C are.

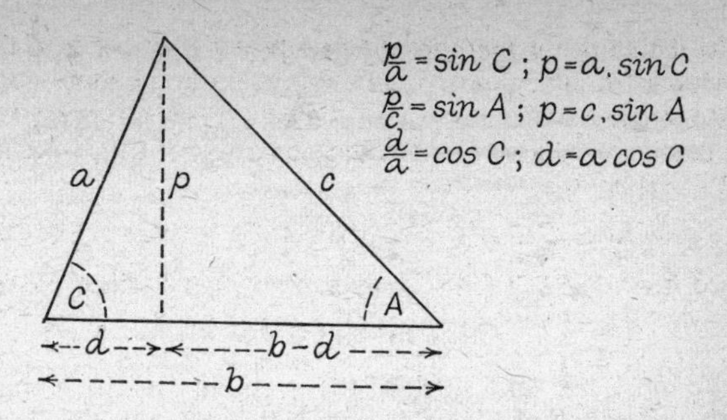

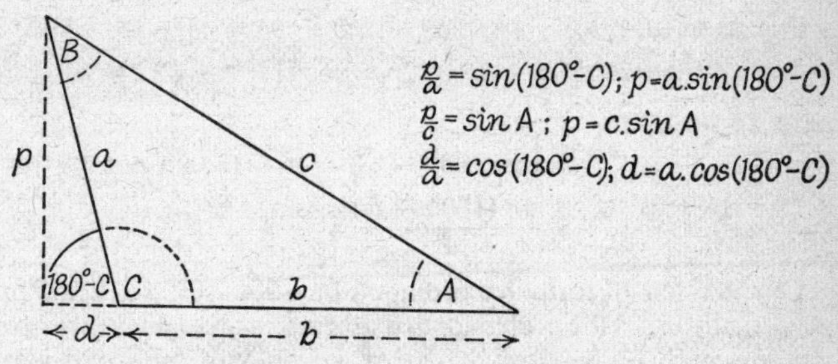

Fig. 92. Solution of Triangles

If we know the lengths of two sides a and b together with the included angle C, we can find the length of the third side c by looking up the tables for cos C, since

$$c^2 = a^2 + b^2 - 2ab \cos C$$

We can then use the preceding formula or a simpler one to get the remaining angles. From Fig. 92 you will see that

$$\sin A = p \div c$$

$$\therefore \quad p = c \sin A$$

$$\sin C = p \div a$$

$$\therefore \quad p = a \sin C$$

$$\therefore \quad c \sin A = a \sin C$$

$$\therefore \quad \frac{\sin A}{a} = \frac{\sin C}{c}$$

$$\therefore \quad \sin \mathrm{A} = \frac{a \sin C}{c}$$

Thus we can find the angle A and hence B, since $B = 180° - (A + C)$.

Similarly, if we know two angles (A, C) and a side (a), we can get the side c, since

$$c = \frac{a \sin C}{\sin A}$$

Since $B = 180° - (A + C)$, we can now get b, since

$$b^2 = a^2 + c^2 - 2ac \cos B$$

We must also take stock of the possibility that one angle (C) in the lower half of Fig. 92 is greater than 90°. We then see that:

$$c^2 = p^2 + (d + b)^2 = p^2 + d^2 + 2db + b^2 = a^2 + b^2 + 2bd$$

$$\therefore \quad c^2 = a^2 + b^2 + 2ab \,.\, \cos (180° - C)$$

Also we see that:

$$a \,.\, \sin (180° - C) = p = C \,.\, \sin A$$

$$\therefore \quad \frac{\mathit{sin}\ (180° - C)}{c} = \frac{\sin A}{a}$$

Our two sets of formulae prompt a question of the sort which perpetually urges the professional mathematician to push forward in the search for *greater generality*, i.e. finding rules of wider scope which are *consistent with and include* rules already known. Our formulae for solution of triangles which do or do not have an angle greater than 90° would be consistent if we were to write:

$$\cos (180° - A) = - \cos A \quad \textit{and} \quad \sin (180° - A) = \sin A$$

So far, any such statement as the last is without meaning. Within the strait-jacket of Euclid's geometry, we have defined *sin A*, *cos A*, *tan A* in terms of $A \leqslant 90°$. The new geometry (Chapter 8) of the Newtonian era offers us an escape from this restriction. With its help we can define *sin* 450° in as meaningful a way as *sin* 45°, and our redefinition is consistent with all that we have learned so far.

One formula of relevance to the surveyor's task, when he has charted a landscape in triangles of specifiable sides, was a discovery by an Alexandrian of uncertain date, most likely about AD 50. This tells us the area of any triangle if we know the lengths of its sides. As mentioned already, the author, Heron, was an outstanding inventor. His discovery of a formula for calculating the area of a triangle from the length of its sides signalizes what is perhaps the biggest break of the time from the shackles of Euclid's way of thinking. It expresses an area as the square root of a 4-dimensional product.

In our notation, the argument proceeds as follows, on the assumption that the sides of a triangle are of lengths a, b, c and that the angles are A (enclosed by b and c), B (enclosed by a and c), C (enclosed by a and b):

$$2\,bc\,.\cos A = b^2 + c^2 - a^2 \quad \text{(Fig. 92)}$$

$$2\left(\sin\frac{A}{2}\right)^2 = 1 - \cos A \quad \text{(Fig. 64)}$$

$$2\left(\cos\frac{A}{2}\right)^2 = 1 + \cos A \quad \text{(Fig. 64)}$$

$$\sin A = 2\sin\frac{A}{2}\,.\cos\frac{A}{2} \quad \text{(Fig. 64)}$$

If we now write s for *half* the sum of the sides of a triangle, so that $2s = a + b + c$:

$$2bc + b^2 + c^2 - a^2 = (b + c + a)\,(b + c - a) = 4s(s - a)$$

$$2bc - b^2 - c^2 + a^2 = (a + b - c)\,(a - b + c) = 4(s - b)\,(s - c)$$

From this we derive:

$$2\left(\sin\frac{A}{2}\right)^2 = \frac{2(s-b)\,(s-c)}{bc}$$

$$2\left(\cos\frac{B}{2}\right)^2 = \frac{2\,s(s-a)}{bc}$$

$$2\sin\frac{A}{2}\,.\cos\frac{A}{2} = 2\frac{\sqrt{s(s-a)\,(s-b)\,(s-c)}}{bc} = \sin A$$

We have seen that we can express the area S of a triangle in the form $S = \frac{1}{2}\,bc.\ \sin A$, so that

$$S = \sqrt{s(s-a)\,(s-b)\,(s-c)}$$

Celestial Surveying. The most direct way of measuring the distance of the earth from the moon is to record simultaneously at two stations on the same longitude the angle that the latter makes at transit with the plumb line, i.e. its zenith distance. The plumb line, the earth's centre, the moon, and their meridian of longitude are each then (Fig. 59) in the same flat section of space. For simplicity (Fig. 93), we shall assume that it is at transit

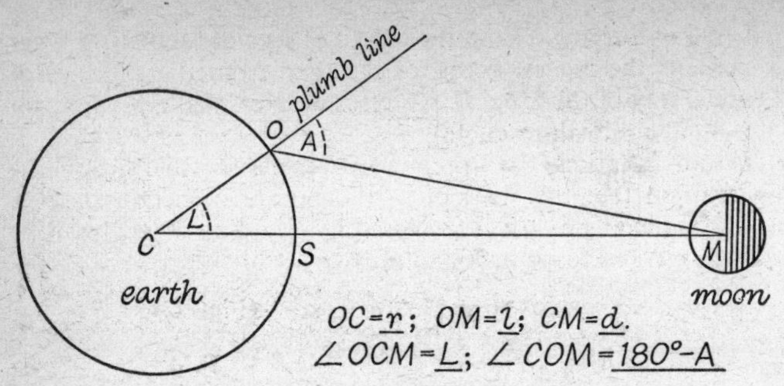

Fig. 93. Moon's So-called Geocentric Parallax

If the moon were as far away as the nearest visible fixed star, the difference between the angle A (its zenith distance at transit) and L (their difference of latitude) would be the same at any two stations on the same great circle of longitude. Actually, such a difference, though very small, was accurately measurable with instruments (like the home-made theodolite of Fig. 25) at the disposal of the Alexandrian astronomers more than a century before the Christian era. In this figure, one station (S) is such that the moon at transit is directly overhead (*z.d.* zero).

directly overhead at one station (S) and that its transit *z.d* at another station (O) is A. In this figure C is the earth's centre, and L is the known difference of latitude between the two stations; $OC = r = CS$ is the known radius of the earth. We know $A(= 180° - COM)$ by direct observation, and $L + COM + OMC = 180°$, so that $OMC = A–L$ and by our sine formula of p. 225:

$$\frac{\sin(A–L)}{r} = \frac{\sin L}{OM}, \textit{ so that } \quad OM = \frac{r \,.\, \sin L}{\sin(A–L)}$$

We can thus calculate OM. What we actually need to find is $CM = d$ (or $SM = d–r$). Our cosine formula tells us that:

$$d^2 = r^2 + OM^2 + 2r\,(OM) \,.\, \cos A$$

As Hipparchus found, $d \simeq 60\ r$. If we take r as 4,000 miles, $d \simeq$ 240,000 miles a now familiar figure in the domain of spacemanship. To get the size of the moon (radius R), all we now need to do is to use our astrolabe (Fig. 25) to measure the angle A which separates the observer from two opposite edges of its disc. We then see from Fig. 94 that

$$\sin(\tfrac{1}{2}A) = \frac{R}{d}$$

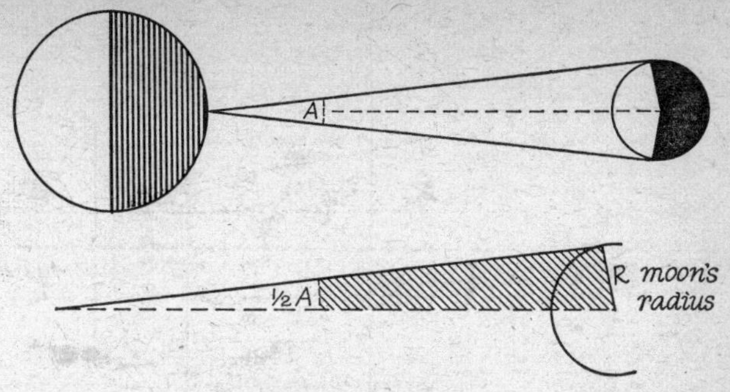

Fig. 94. Size of the Moon

Crisis in Computation. Three-quarters of a century before Hypsicles initiated his pupils into the mysteries of a system of fractions borrowed from the temple sites of Iraq, the new impetus to measurement had provoked a new awareness of the inadequacy of the clumsy numeral signs bequeathed by Athens (Fig. 9) to Alexandria. With what aids (other than the abacus) Archimedes could compute a value for π with an error of roughly 1 in 10,000 we do not know. We do know that he had in the fire several irons other than the precocious preoccupation of Aristarchus with the distance of the earth from the sun and moon. For one thing, he returned to an earlier tradition than exclusive reliance on rule and compass by investigating the construction of one of the class of curves called a *spiral*. This may have some connection with the invention of the screw pump named after him. He was also the first person on record to recognize that an infinite series can choke off. This is the series:

$$S = 1 + \tfrac{1}{4} + \tfrac{1}{16} + \tfrac{1}{64} + \tfrac{1}{256} \cdots$$

By writing this out below, the sum is easy to recognize:

$$4S = 4 + 1 + \tfrac{1}{4} + \tfrac{1}{16} + \tfrac{1}{64} + \tfrac{1}{256} \cdots$$

$$S = \quad 1 + \tfrac{1}{4} + \tfrac{1}{16} + \tfrac{1}{64} + \tfrac{1}{256} \cdots$$

$$\therefore \quad 3S = 4 \quad \textit{and} \quad S = 1\tfrac{1}{3}$$

Oddly enough, his predecessors might have arrived at a recognition of the same general principle on the testimony of their sacrosanct rule and compass drill by considering the following for which $S = 1$:

$$S = \tfrac{1}{2} + \tfrac{1}{4} + \tfrac{1}{8} + \tfrac{1}{16} + \tfrac{1}{32} \cdots$$

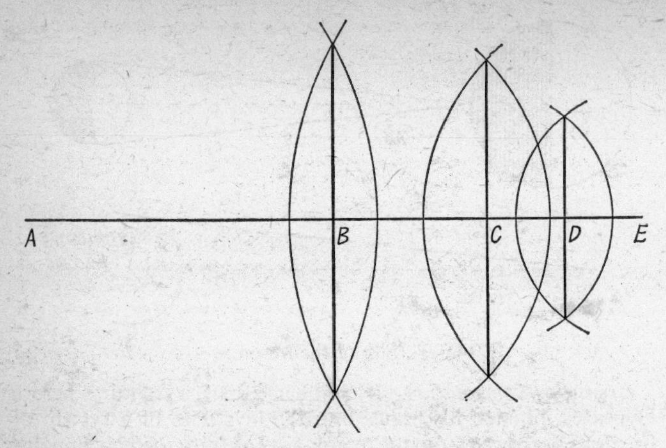

Fig. 95. Geometrical Representation of an Infinite Series Whose Sum has a Finite Value

$$AE = 1 \textit{ unit. } AB = \tfrac{1}{2} = BE; \quad BC = \tfrac{1}{4} = CE; \quad CD = \tfrac{1}{8} = DE$$

$$AE = AB + BC + CD + DE$$

$$1 = \tfrac{1}{2} + \tfrac{1}{4} + \tfrac{1}{8} + \tfrac{1}{16}$$

In the same way, by successive bisection:

$$DE = \tfrac{1}{16} + \tfrac{1}{32} + \tfrac{1}{64} + \tfrac{1}{128}, \text{ and so on}$$

$$1 = \tfrac{1}{2} + \tfrac{1}{4} + \tfrac{1}{8} + \tfrac{1}{16} + \tfrac{1}{32} + \tfrac{1}{64}, \text{ and so on } \textit{ad infinitum.}$$

One can represent this geometrically as follows. Consider (Fig. 95) a line of unit length. We bisect it. We then bisect one of the two segments. We then bisect one of the remaining segments, and so on. All these segments can merely add up to unity. We may set out the proof algebraically thus:

$$S = \tfrac{1}{2} + \tfrac{1}{4} + \tfrac{1}{8} + \tfrac{1}{16} + \tfrac{1}{32} + \tfrac{1}{64} \ldots$$

$$\tfrac{1}{2}S = \quad \ldots \tfrac{1}{4} + \tfrac{1}{8} + \tfrac{1}{16} + \tfrac{1}{32} + \tfrac{1}{64} \ldots$$

Whence $S - \frac{1}{2}S = \frac{1}{2}$ or $\frac{1}{2}S = \frac{1}{2}$, i.e. $S = 1$. However long we go on, the sum of the series thus chokes off at 1. The advantage of series like this for representing measurements to any required order of precision can be seen by comparing the sum of the first five and the first ten terms:

0·5	0·5
0·25	0·25
0·125	0·125
0·0625	0·0625
0·03125	0·03125
0·96875	0·015625
	0·0078125
	0·00390625
	0·001953125
	0·0009765625
	0·9990234375

By taking the first five terms we get 0·97 correct to two decimals. This is roughly 3 per cent less than 1. By taking ten terms we get 0·9990, which is only one in a thousand out; but the rapidity with which the series chokes off was much more difficult for the Alexandrians to recognize with the numbers which they actually used than it is for ourselves using decimal fractions. Their Attic Greek numeral system (Fig. 9) used the first nine letters of the Greek alphabet for 1 to 9, the next nine letters for 10 to 90, the next nine for 100 to 900. To do this they added three archaic letters (digamma, san, koppa) to the ordinary alphabet to make the letters up to 27. If translated into our own more familiar alphabet, this is the same as putting:

a	b	c	d	e	f	g	h	i
1	2	3	4	5	6	7	8	9
j	k	l	m	n	o	p	q	r
10	20	30	40	50	60	70	80	90
s	t	u	v,	etc.				
100	200	300	400,	etc.				

In this notation the number 17 would be *jg*, the number 68 would be *oh*, and 259 *tni*. You may now find it easier to see one possible reason why the Platonists mixed up their notions about God and the number 3. To get beyond 999 they had to start all over again, using the same letters with ticks to signify higher decimal orders.

Archimedes wrote a tract in which he made an estimate of the number of grains of sand in the world. This was by no means a useless performance in an age when people's ideas of how big things could be were confined by the number of letters which they had at their disposal. In the *Sand Reckoner* Archimedes hit on two of the most powerful peculiarities which reside in the modern number script. He proposed that all high numbers should be represented by multiples of simple powers of ten. He also hit upon the law which underlies the modern calculating device called *logarithms*. The rule can be seen by putting side by side any simple geometric series and its parent series, e.g.:

1	2	3	4	5	6	7	8	9	10
2	4	8	16	32	64	128	256	512	1,024

If you want to multiply any two numbers in the bottom series, all you have to do is to look up the corresponding numbers in the parent series and add them. You then get your result by taking the number in the bottom series corresponding with the number represented by their sum in the parent series. Thus to multiply 16 by 32, add the corresponding parent numbers (logarithms, as we call them today), $4 + 5 = 9$. The number (*antilogarithm*, as we call it today) in the bottom series corresponding with 9 in the top series is 512, which is the answer required. Test this rule by building up other series, e.g. 3, 9, 27, 81, 243, 729, etc. In the shorthand of modern algebra, the rule is:

$$a^m \times a^n = a^{m+n}$$

Archimedes did not succeed in reforming the number script of his contemporaries, nor in making tables of logarithms by which any multiplication can be carried through rapidly. Such a change would have meant uprooting the social culture of his time. People were still using the old notation for *low* numbers. His brilliant failure shows that we cannot afford to let the mass of mankind be uneducated. An advance like that proposed by Archimedes must arise from a sense of common need. It is not enough that a few isolated men of genius should recognize what is wanted. The mathematician needs the co-operation of the plain man, just as much as the plain man needs the mathematician if he is to enjoy a punctual system of wheel-driven transport.

The Attic alphabet was a millstone about the necks of the Alexandrians. The first stage of their culture was signalized by tremendous achievements in the art of measurement as applied to astronomy and mechanics. It introduced calculations of appalling magnitude to people who used a number script which introduces an entirely new set of symbols at each decimal order. The second stage of the Alexandrian culture is characterized

	1	2	3	4	5	6	7	8	9	10	20	30	40	50	60	70	80	90
	a	b	c	d	e	f	g	h	i	j	k	l	m	n	o	p	q	r
2 = b	b	d	f	h	j	jb	jd	jf	jh	k	m	o	q	s	sk	sm	so	sq
3 = c	c	f	i	jb	je	jh	ka	kd	kg	l	o	r	sk	sn	sq	tj	tm	tp
4 = d	d	h	jb	jf	k	kd	kh	lb	lf	m	q	sk	so	t	tm	tq	uk	uo
5 = e	e	j	je	k	ke	l	le	m	me	n	s	sn	t	tn	u	un	v	vn
6 = f	f	jb	jh	kd	l	lf	mb	mh	nd	o	sk	sq	tm	u	uo	vk	vq	wm
7 = g	g	jd	ka	kh	le	mb	mi	nf	oc	p	sm	tj	tq	un	vk	vr	wo	xl
8 = h	h	jf	kd	lb	m	mh	nf	od	pb	q	so	tm	uk	v	vq	wo	xm	yk
9 = i	i	jh	kg	lf	me	nd	oc	pb	qa	r	sq	tp	uo	vn	wm	xl	yk	zj
10 = j	j	k	l	m	n	o	p	q	r	s	t	u	v	w	x	y	z	—

Fig. 96. Part of the Alexandrian Multiplication Table
(Roman alphabet substituted for Greek letters)

by a serious attempt to tackle the problem of devising simple and rapid means of calculation.

Theon (about AD 380), the last of the Alexandrian mathematicians, multiplied numbers without the use of the abacus, or at least only using it for the final step, by means of a multiplication table. As there are three decimal orders in the alphabetic script, the complete table involved the multiplication of three sets of nine rows and columns instead of one set of ten rows and columns, like our own. A part of this is given in Fig. 96, and it is sufficient to be able to follow the ensuing example. To multiply 13 by 18, the steps would be:

$$\begin{aligned} 13 \times 18 &= (10 + 3)(10 + 8) & jc \times jh &= (j + c)(j + h) \\ &= 10^2 + 8(10) + 3(10) + 3(8) & &= j(j) + j(h) + c(j) + c(h) \\ &= 100 + 80 + 30 + 24 & &= s + q + l + kd \\ &= 234 & &= tld \end{aligned}$$

Alternatively:

$$\begin{aligned} 13 \times 18 &= (10 + 3)(20 - 2) & jc \times jh &= (j + c)(k - b) \\ &= 10(20) - 2(10) + 3(20) - 3(2) & &= j(k) - j(b) + c(k) - c(b) \\ &= 200 - 20 + 60 - 6 & &= t - k + o - f \\ &= 234 & &= tld \end{aligned}$$

You can develop a sympathetic attitude towards the Alexandrians by making up sums like these and working them out by a multiplication table like that in Fig. 96. Of course, if you want to multiply fairly large numbers you will have to increase the size of your table.

Theon also dealt with a practical problem which has arisen in our attempt to make a table of angle ratios. We then had to use a table of square roots. The method of getting a square root of numbers like $\sqrt{3}$ and $\sqrt{2}$ given on p. 104 was in Attic numerals extremely laborious. Fig. 98 shows the method which Theon used. On the right, as in Fig. 33:

$$(x + a)^2 = x^2 + 2ax + a^2$$

The figure on the left is essentially the same, but instead of a we have written dx, which does not mean d multiplied by x but a quantity very small compared with x (*dwarf* x). As before

$$(x + dx)^2 = x^2 + 2x(dx) + (dx)^2$$

or $$(x + dx)^2 - x^2 = 2x(dx) + (dx)^2$$

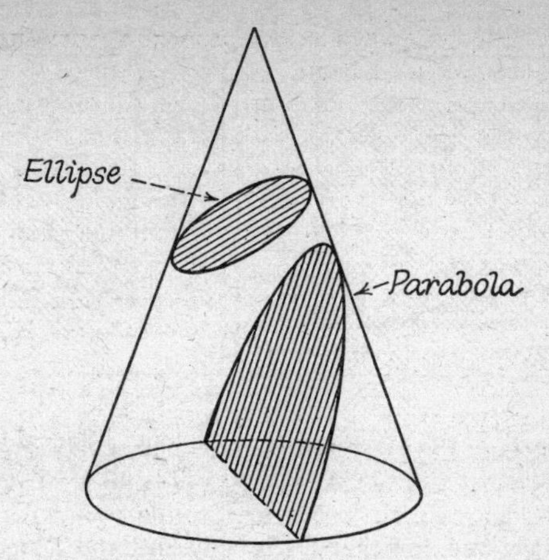

Fig. 97. Conic Sections

Like his contemporary Archimedes, Apollonius, who lived about 230 BC, broke away from Plato's rule and studied curves which cannot be drawn with a compass and rule. In particular, he studied three curves which correspond with the boundary of a slice of a cone. Two are shown here. The ellipse is the figure which represents the orbits of the planets. The *parabola* approximately represents the path of a cannon-ball. A third conic section (the *hyperbola*) will be used, when we describe the expansion of the gas in an internal-combustion engine.

The figures show you that $(dx)^2$ is very small compared with the two rectangles $x(dx)$, so we should not be much out if we put

$$(x + dx)^2 - x^2 = 2x(dx)$$

or

$$dx = \frac{(x + dx)^2 - x^2}{2x}$$

You can see that this is very little out by the following arithmetical example. The quantity 1·01 may be written (1 + 0·01), in which 0·01 may stand for dx and 1 for x, since 0·01 is very small compared with 1. We then get approximately

$$dx = \frac{(1 \cdot 01)^2 - 1^2}{2} = \frac{1 \cdot 0201 - 1}{2} = 0 \cdot 01005$$

The value obtained (0·01005) differs from the original one (dx = 0·01) by only 0·00005.

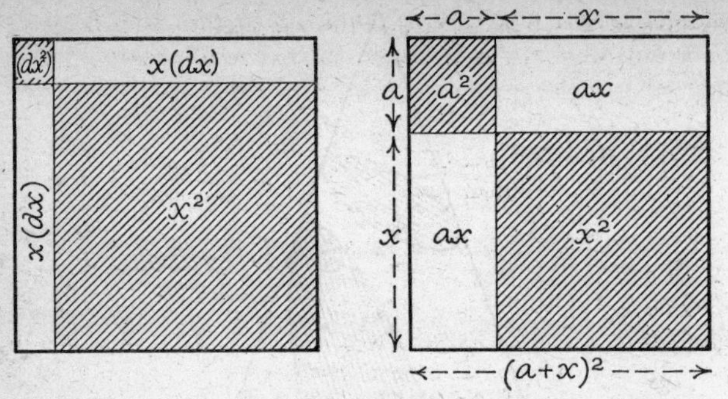

Fig. 98. Theon's Method for Computing a Square Root

To get a square root by using this formula we first make a guess. For instance, we know that $\sqrt{2}$ is between 1 and 2, since 1^2 (= 1) is less than 2 and 2^2 (= 4) is greater than 2. Since $14^2 = 196$, a good guess is 1·4. This is a little too small, so we may put $\sqrt{2} = 1{\cdot}4 + dx$.

$$(1{\cdot}4 + dx)^2 = 2$$

Then from the formula,

$$dx = \frac{(1{\cdot}4 + dx)^2 - (1{\cdot}4)^2}{2(1{\cdot}4)}$$

$$= \frac{2 - (1{\cdot}4)^2}{2(1{\cdot}4)} = \frac{2 - 1{\cdot}96}{2{\cdot}8}$$

$$= 0{\cdot}014 \text{ (approx.)}$$

So we have approximately

$$(1{\cdot}4 + 0{\cdot}014)^2 = 2$$

or

$$1{\cdot}414 = \sqrt{2}$$

This, of course, will be a little out. So we take it as a second approximation, and proceed to a third, putting d^2x for the new dwarf x, i.e.* $(1{\cdot}414 + d^2x)^2 = 2$.

$$\therefore \quad d^2x = \frac{2 - (1{\cdot}414)^2}{2(1{\cdot}414)}$$

* Note carefully d^2x does *not* mean d squared multiplied by x. It is shorthand for second dwarf x. The ^{2} is an adjective, not a verb.

This gives $d^2x = 0{\cdot}0002$. So as a third approximation we can give $\sqrt{2} = 1{\cdot}4142$. Comparing these successive approximations, we find:

$(1{\cdot}4)^2 = 1{\cdot}96$	error 2 per cent	
$(1{\cdot}414)^2 = 1{\cdot}999396$	„	0·03 per cent
$(1{\cdot}4142)^2 = 1{\cdot}99996164$	„	0·002 per cent

We can continue like this as long as we need to do so.

Theon was the last of the Alexandrian mathematicians of importance. His daughter, Hypatia, edited the works of Diophantus and taught mathematics in Alexandria. She was murdered by the monks of St Cyril. Gibbon describes how they scraped her naked body with oyster shells. 'Je me contente de remarquer,' observes Voltaire, 'que St Cyrille était homme et homme de parti, qu'il a pu se laisser trop comporter à son zèle; que, quand on met des belles dames toutes nues, ce n'est pas pour les massacrer; que St Cyrille a sans doute demandé pardon à Dieu de cette action abominable, et que je prie le père des miséricordes d'avoir pitie de son âme.'

Theon's method of arriving at a square root introduces us to a conception which plays a very important part in the modern branch of mathematics called the *differential calculus*. The method used by Archimedes for getting the value of π illustrates the principle which lies at the root of the *integral calculus*. The invention of latitude and longitude by Hipparchus and the curves of Apollonius (Fig. 97), another brilliant Alexandrian, embody the basic conception of the new geometry of the Newtonian century. Diophantus, mentioned in connection with figurate numbers, p. 175, made an attempt to use alphabetic letters, as we use them in algebra. Inevitably, it was unsuccessful. The Attic numeral had already pre-empted the whole Greek alphabet with a different end in view.

Thus the germs of almost every important advance of the sixteenth and seventeenth centuries of our own era can be found in the achievements of the Alexandrians. That they progressed so far and yet failed to advance farther is not explained sufficiently by saying that Alexandrian civilization participated in the downfall of the Roman Empire. It had reached the limits of further growth within the social culture which it had inherited. The next great advance came because a less sophisticated people were equipped with a number script which could meet the requirements of Alexandrian mathematics. The essentially novel feature of the

Hindu culture was that men who were not advanced mathematicians had invented what the most brilliant mathematicians of Alexandria had failed to invent, a symbol (0) for *nothing*. There is no more fitting epitaph for the decay of Alexandrian science and mathematics than two lines from the poem of Omar Khayyám, himself foremost among the Moslem mathematicians who brought together the fruits of the Hindu and Alexandrian contributions to human knowledge:

> . . . the stars are setting, and the caravan
> Starts for the DAWN OF NOTHING, oh, make haste . . .

Discoveries and Tests on Chapter 5

1. Use the formula $\cos^2 A + \sin^2 A = 1$ to find:

   (*a*) cos 40° and sin 50°, given that sin 40° = 0·6428;
   (*b*) cos 75° and sin 15°, given that cos 15° = 0·9659.

2. Using the half-angle formula,

If sin 20° = 0·3420, cos 20° = 0·9397, find sin 10°, cos 10°, tan 20°, and tan 10°.

3. If sin 40° = 0·6428, and cos 40° = 0·7660, find sin 50°, cos 50°, tan 50°, sin 20°, cos 20°, tan 20°.

4. If sin 50° = 0·7660, sin 43° = 0·6820, sin $23\frac{1}{2}$° = 0·3987, find the following:

cos 50°	tan 50°	cos $21\frac{1}{2}$°	tan $21\frac{1}{2}$°
cos 25°	tan 25°	cos 43°	tan 43°
cos 47°	tan 47°	cos 40°	tan 40°
cos $23\frac{1}{2}$°	tan $23\frac{1}{2}$°	cos $66\frac{1}{2}$°	tan $66\frac{1}{2}$°

5. If cos 40° = 0·7660, sin 40° = 0·6428, cos 15° = 0·9659, sin 15° = 0·2588, cos $26\frac{1}{2}$° = 0·8949, and sin $26\frac{1}{2}$° = 0·4462, use the formulae for $\sin (A + B)$ and $\cos (A + B)$ to find:

cos 55°	sin 55°	cos $66\frac{1}{2}$°	sin $66\frac{1}{2}$°
cos $41\frac{1}{2}$°	sin $41\frac{1}{2}$°	cos $56\frac{1}{2}$°	sin $56\frac{1}{2}$°

6. Using the values in the previous examples, see whether you can find what the correct formula is for $\sin (A - B)$ and $\cos (A + B)$. Verify your result by reference to the figure in Chapter 8, p. 367.

7. In books on trigonometry $\frac{1}{\sin A}$ is called cosec A, $\frac{1}{\cos A}$ is called sec A, and $\frac{1}{\tan A}$ is called cot A. These are abbreviations for secant, cosecant, and cotangent. Use the method for demonstrating that $\cos^2 A + \sin^2 A = 1$ to build up the following formulae which are sometimes used in higher mathematics:

$$1 + \cot^2 A = \operatorname{cosec}^2 A$$
$$1 + \tan^2 A = \sec^2 A$$

8. Solve the cliff problem in Fig. 53 by using the sine formula for the solution of triangles, first to get the distance from the nearest point of observation to the top of the cliff, and then the height.

9. If in Fig. 87 you make A greater than 90° and drop the perpendicular p on to the side b extended beyond the corner A, show that the cosine formula becomes

$$a^2 = b^2 + c^2 + 2bc \cos (180° - A)$$

and the sine formula becomes

$$\frac{\sin (180° - A)}{a} = \frac{\sin B}{b} = \frac{\sin C}{c}$$

10. If the formulae for the solution of triangles when A is less than 90° are:

$$a^2 = b^2 + c^2 - 2bc \cos A$$

and $$\sin A = \frac{a \sin B}{b} = \frac{a \sin C}{c}$$

and when A is greater than 90° are:

$$a^2 = b^2 + c^2 + 2bc \cos (180° - A)$$

and $$\sin (180° - A) = \frac{a \sin B}{b}, \text{ etc.,}$$

what is the connection between:

(*a*) $\cos A$ and $\cos (180° - A)$

and (*b*) $\sin A$ and $\sin (180° - A)$

If you could find a geometrical meaning for the sines and cosines of angles greater than 90°, what would you conclude to be the numerical values of the sine, cosine, and tangent of 150°, 135°, 120°? Verify your result, using the addition formulae and the formula $\cos^2 A + \sin^2 A = 1$.

11. Two men start walking at the same time from a crossroads, both walking at 3 miles per hour. Their roads diverge at an angle of 15°. How far apart will they be at the end of two hours?

12. From a base line AB, 500 yards long, the bearings of a flagstaff from A and B are 112° and 63° respectively. Find the distance of the flagstaff from A.

13. From a boat out at sea the elevation of the top of a cliff is found to be 24°. The boatman rows 80 feet straight towards the cliff, and the elevation of the top of the cliff is now 47°. What is the height of the cliff?

14. Three villages, A, B, C are connected by straight, level roads. $AB = 6$ miles, $BC = 9$ miles, and the angle between AB and BC is 130°. What is the distance between A and C?

15. A boat sails 8 miles due South. It then changes its direction and sails for 11 miles in a line bearing 54° East of North. How far will it then be from its starting-point?

16. From the formulae for $\sin(A + B)$ and $\cos(A + B)$ find the formula for $\sin 2A$, $\sin 3A$, $\cos 2A$, and $\cos 3A$.

17. From the formulae for $\sin(A + B)$ and $\sin(A - B)$, $\cos(A + B)$ and $\cos(A - B)$, show that

$$\sin C + \sin D = 2 \sin \frac{C + D}{2} \cos \frac{C - D}{2}$$

$$\cos C + \cos D = 2 \cos \frac{C + D}{2} \cos \frac{C - D}{2}$$

$$\sin C - \sin D = 2 \cos \frac{C + D}{2} \sin \frac{C - D}{2}$$

$$\cos C - \cos D = -2 \sin \frac{C + D}{2} \sin \frac{C - D}{2}.$$

Hence, show how the half-angle formulae can be built up from the formulae for the addition of angles. Hint: $C + D = 2A$, $C - D = 2B$.

18. Using trigonometrical tables, find the limits between which π lies by considering the boundaries and areas of the inscribed and circumscribed figures with seventy-two sides.

19. Use the fact that $\sin x$ is very nearly equal to x, when x is a small angle measured in radians, to obtain values for $\sin \frac{1}{2}°$, $\sin 1°$, and $\sin 1\frac{1}{2}°$, taking π as 3·1416.

20. Since the moon's rim almost exactly coincides with that of the sun in a total eclipse, the angular diameter of the sun (cf. Fig. 89) may be taken as roughly half a degree. Taking the sun's

distance as 93 million miles, find the diameter of the sun, using the fact that for small angles measured in radians $\sin x = x$ and $\cos x = 1$.

21. Tabulate in steps of one degree, the angles 1° to 10° in radians, and tabulate in degrees taking steps of quarter-radians the angles from 0 to 2 radians.

22. In the next few examples take the earth's radius as 3,960 miles.

The ancient Inca capital Quito, Kisumu on Lake Victoria in Kenya, and Pontianak in Borneo are all within half a degree from the equator. The longitude of Quito is 78° W of Greenwich, Kisumu 35° E, and Pontianak 109° E. Find the shortest distance between each two of these places, using the value $3\frac{1}{7}$ for π.

23. Archangel, Zanzibar, and Mecca are all within about a degree of longitude 40° E. The latitude of Archangel is $64\frac{2}{3}$° N. The latitude of Mecca is $21\frac{1}{3}$° N, and of Zanzibar 6° S. What are the distances between them?

24. With the aid of a figure show that the length of a degree measured along latitude L is $x \cos L$, if x is the length of a degree at the equator.

25. If Winnipeg and Plymouth are within a third of a degree of latitude 50° N, find the distance between them if Plymouth is 4° W of Greenwich and Winnipeg is 97° W.

26. Reading and Greenwich have the same latitude, 51° 28′ N, and the longitude of Reading is 59′ W. How far is Reading from Greenwich?

27. Two places are on the same meridian of longitude. The latitude of A is 31° N. B is 200 miles from A. What is the latitude of B?

28. Using the method given on p. 78 for multiplying 19 by 28, illustrate the law of signs by multiplying:

(*a*) 13 by 27 (*c*) 15 by 39
(*b*) 17 by 42 (*d*) 21 by 48
(*e*) 28 by 53

29. Do the multiplications given in Example 28 with the aid of the Alexandrian multiplication table.

30. Sum the following geometrical progressions to n terms:

(*a*) $3 - 9 + 27 \ldots$ (*c*) $2\frac{1}{4} - 1\frac{1}{2} + 1 \ldots$
(*b*) $\frac{1}{4} - \frac{1}{8} + \frac{1}{16} \ldots$ (*d*) $1 - \frac{2}{3} + \frac{4}{9} \ldots$

Check your results arithmetically by finding the sum to five terms.

31. Find (*a*) the $2n$th term and (*b*) the $(2n + 1)$th term of the series, a, $-ar$, ar^2, $-ar^3$, etc.

Things to Memorize

1. $\cos (A - B) = \cos A \,.\, \cos B + \sin A \,.\, \sin B$
2. $\sin (A + B) = \sin A \,.\, \cos B + \cos A \,.\, \sin B$
3. $a^2 = b^2 + c^2 - 2bc \,.\, \cos A \quad \text{if } A \leqslant 90°$

   $a^2 = b^2 + c^2 + 2bc \,.\, \cos (180° - A) \quad \text{if } A > 90°$
4. If radii of length R joining to its centre the extremities of an arc TR enclose an angle A^R in circular measure, its length is $R \,.\, A^R$.
5. When we measure x in radians *sin* x and *tan* x become nearer and nearer to x as x becomes smaller, i.e.:

$$\mathop{\mathfrak{L}t}_{x \to 0} \frac{\sin x}{x} = 1 = \mathop{\mathfrak{L}t}_{x \to 0} \frac{\tan x}{x}$$

6. The following are equivalents in degrees and radians

Degrees	Radians	Degrees	Radians
360	2π	90	$\pi/2$
270	$3\pi/2$	60	$\pi/3$
180	π	45	$\pi/4$
120	$2\pi/3$	30	$\pi/6$

CHAPTER 6

THE DAWN OF NOTHING

How Algebra Began

Among many advantages of adopting a symbol for zero, we have already touched on one in Chapter 2. Without it, we can set no upper limit to the number of signs we require. In terms of the art of calculation as practised in the ancient world, one may say that provision for larger and larger numbers demands the introduction of at least one new symbol for each bar added to the counting frame. Why this is necessarily true, is easy to see if we recall the Roman representation of 32, i.e. XXXII. If the Romans had written it in the form III II, there would have been no way of distinguishing it from 302, 320, 3,020, 3,200, etc. The one simple way of escape from this dilemma is to introduce a sign such as the Maya lozenge (Fig. 8), a dot or a circle for the *empty* column of the abacus. We can then write 32, 302, 320, 3,020 as below:

III II; III$_0$II; III II$_0$; III$_0$II$_0$

Having such a symbol at our disposal, we can abandon the *repetitive* principle of the earliest number scripts illustrated by this example without loss of clarity. If our base is b, we need only $(b - 1)$ other signs e.g. if $b = 10$ the other signs we need are 9 in all. We can then express any namable number however great without enlarging our stock in trade. Once mankind had the zero symbol at its disposal, another advantage was manifest. Its invention liberated the human intellect from the prison bars of the counting frame. The new script was a complete model of the mechanical process one performs with it. With a sign for the empty column, 'carrying over' on slate, paper or parchment is just as easy as carrying over on the abacus. In other words, it was possible thereafter, and for the first time in history, to formulate the simple rules of calculation we now learn as arithmetic in childhood. In mediaeval Europe, the name used for such rules was *algorithms*, a corruption of the name of a thirteenth century Moslem mathematician, spelt Al Khwarismi or Alkarismi.

Dr Needham, a contemporary scholar of eminence, contends that the introduction of the zero symbol (0) happened in China; and it is admissible that the earliest known inscription on which it survives to date comes from the Indo-Chinese borderland. While the debt of the Western World of today to the Ancient civilization of China is vast and too little recognized, there are very good reasons for adhering to the more generally accepted view that it happened in India, whence it spread both East and West. In the East, the zero symbol, first a dot then a circle, was certainly in use before AD 700, most likely before AD 400. Seemingly, the intention was essentially a practical one. The Hindu word for (0) is *sunya*, which means *empty*. The identification of 0 with 'nothing' or zero was an afterthought.

Our knowledge of Hindu mathematics begins with the *Lilavati* of Aryabhata about AD 470. This author discusses the rules of arithmetic, uses the law of signs (p. 92), gives a table of sines in intervals of $3\frac{3}{4}°$, and evaluates π as 3·1416. In short, Hindu mathematics starts where Alexandrian mathematics left off. Just a little later (about AD 630), comes Brahmagupta, who follows the same themes as Aryabhata: calculation, series, equations. These early Hindu mathematicians had already stated the laws of 'ciphers' or *sunya* on which all our arithmetic depends, namely

$$a \times 0 = 0$$
$$a + 0 = a$$
$$a - 0 = a$$

They used fractions freely without the aid of metaphorical units, writing them as we do, except that they did not use a bar. Thus seven-eighths was written $\begin{smallmatrix}7\\8\end{smallmatrix}$. About AD 800 Baghdad became a centre of learning under a Moslem Caliphate. Exiles from the Alexandrian schools which had been closed after the rise of Christianity had brought pagan science into Persia at an earlier date. Greek philosophical works had also been brought thither by banished Nestorian heretics. Jewish scholars were set by the Caliph to translate Syriac and Greek texts into Arabic. The works of Ptolemy, Euclid, Aristotle, and a host of other authors of classical scientific treatises were circulated from Baghdad to the Moslem universities set up in several countries, notably Spain, during the ninth and tenth centuries.

The Arab nomads who conquered and overran the ruins of the Roman Empire had no priesthood. Within the world of Islam timekeeping thus broke asunder from any association with a pre-existing priestly caste. Jewish and Arabic scholars, set to the task of making calendars, improved on the astronomical tables of the Alexandrians and Hindus, and they brought to their task the advantage of the simple number script which the Hindus had invented. Among famous Moslem mathematicians first and foremost comes Alkarismi (Al Khwarismi), who lived in the ninth century AD. Another great mathematician, Omar Khayyám, lived in the twelfth century AD. You may recall the close connection between this new awakening of interest in mathematics and the secular task of keeping time, in the words of the *Rubaiyat*:

> 'Ah but my computations, people say,
> Have squared the Year to human compass, eh?
> If so, by striking from the Calendar
> Unborn tomorrow and dead yesterday . . . .'

The two chief foci for the introduction of Arabic and Hindu mathematics to the backward Nordic peoples of Europe were the Moorish universities in Spain and Sicilian trade in the Mediterranean. A Sicilian coin with the date 1134 Anno Domini is the first extant example in Christendom of the official use of the so-called Gobar numerals, the Hindu numbers as modified by the western Arabs. In Britain, the earliest case is said to be the rent roll of the St Andrew's chapter in 1490. Italian merchants were using them for their obvious advantages in commercial calculations in the thirteenth century. The change did not come about without obstruction from the representatives of custom thought. An edict of AD 1259 forbade the bankers of Florence to use the infidel symbols, and the ecclesiastical authorities of the University of Padua in AD 1348 ordered that the price list of books should be prepared not in 'ciphers', but in 'plain' letters.

Several social circumstances contributed to the diffusion of the Moorish culture. Among others, the Christian religion, in replacing the Roman Pantheon, had taken over the social function of the priests as calendar-makers, and as custodians of the calendar, some monks were interested in mathematics. Thus Adelard of Bath disguised himself as a Moslem (about AD 1120), studied at Cordova, and translated the works of Euclid and Alkarismi, together with Arabic astronomical tables. Gerard of Cremona studied about the same time at Toledo. He translated about ninety Arabic texts, including the Arabic edition of Ptolemy's *Almagest*. The heretical ecclesiastic Paciulo, who had the good fortune not to be burned at the stake, translated the arithmetic of Bhaskara, a twelfth-century Hindu mathematician, and introduced Theon's method of getting square roots.

Outside the confines of the monasteries, the new aids to calculation gained the ear of the rising merchant-class. A foremost figure among the mercantile mathematicians is Leonardo Fibonacci, whose *Liber Abaci* (AD 1228) was the first commercial arithmetic. His name is remembered for a quaint series of numbers, entitled the Fibonacci series. It is:

$$0;\ 1;\ 1;\ 2;\ 3;\ 5;\ 8;\ 13;\ 21;\ \text{etc.}$$

The rule of formation in the symbolism of Chapter 2 is:

$$t_r = t_{r-1} + t_{r-2};\quad t_0 = 0;\quad t_1 = 1$$

As far as its author was concerned, this seems to have been a *jeu d'esprit*. Oddly enough, a use has been found for it quite recently in applying Mendel's laws of heredity to the effects of brother–sister inbreeding. Fibonacci, who was the despair of his teachers as a boy, found mathematics interesting by applying it to the social needs of his class. He came to make up series for fun, when he had learned to use equations to solve practical problems in interest and debt. He was patronized by the atheistical Frederick II under whose encouragement the University of Salerno became a centre from which Jewish physicians carried the Moorish learning into the ecclesiastical centres of learning in Northern Europe. 'Physician and algebraist' was a term still used in Spain until quite recently, just as surgeon and barber went together in the Middle Ages.

Before proceeding to an account of the new arithmetic or Algorithms, it will help us at a later stage if we pause to recognize what latent possibilities of the new numerals immediately impressed themselves on the imagination of the unsophisticated pupils of the Moorish culture. Stifel, whom we have already met as a commentator of the Apocalyptic 'mysterium', was not referring to the number 666 when he declared (AD 1525), 'I might write a whole book concerning the marvellous things relating to numbers'. We have noticed one of these marvellous things in Chapter 2. A new use for the zero symbol suggests itself, if we label our abacus columns as below.

7th column	6th column	5th column	4th column	3rd column	2nd column
1,000,000	100,000	10,000	1,000	100	10
10^6	10^5	10^4	10^3	10^2	10^1

Here n goes down one step when the value of the bead is reduced through dividing by ten. So the index number n of the first column should be one less than 1, i.e. 0. Indeed, we may go farther. One less than 0 is -1, so 1 divided by 10 is 10^{-1}. Thus we can stretch the horizon of the index numbers backwards to any degree of smallness thus:

$$\begin{array}{cccccccc} .\ 10{,}000 & 1{,}000 & 100 & 10 & 1 & \frac{1}{10} & \frac{1}{100} & \frac{1}{1000}\cdots \\ 10^4 & 10^3 & 10^2 & 10^1 & 10^0 & 10^{-1} & 10^{-2} & 10^{-3} \end{array}$$

A second marvellous thing about the new numbers is not so easy to see at a glance. If n and m are whole numbers, Archimedes and Apollonius had long ago recognized the rule we now write in the form $10^n \times 10^m = 10^{n+m}$, or for any base ($b$) in the form $b^n \times b^m = b^{n+m}$. Perhaps because the new number signs so lately introduced into Northern Europe made new rules so much more easy than previously to explore, the first outstanding mathematician of Christendom to come to terms with them fully grasped that we may give a meaning to n and m consistent with the one cited above when n or m or both are rational *fractions*. For his own time an unusually enlightened ecclesiastic, Oresme (about AD 1360) anticipated the elements of several developments of a much later time. Three centuries before Wallis suggested, and Newton adopted, the fractional notation we now use, he recognized the meaning we now attach to $3^{\frac{1}{2}}$ and $5^{\frac{1}{3}}$. He wrote these in his own shorthand as $\frac{1}{2} . 3^p$ and $\frac{1}{3} . 5^p$ respectively. If the rule of Archimedes holds good:

$$3^{\frac{1}{2}} \times 3^{\frac{1}{2}} = 3^{\frac{1}{2}+\frac{1}{2}} = 3 \quad \textit{and} \quad 5^{\frac{1}{3}} \times 5^{\frac{1}{3}} \times 5^{\frac{1}{3}} = 5^{\frac{1}{3}+\frac{1}{3}+\frac{1}{3}} = 5$$

In the so-called *surd* notation of the Moslem mathematicians therefore:

$$3^{\frac{1}{2}} = \sqrt{3} \quad \textit{and} \quad 5^{\frac{1}{3}} = \sqrt[3]{5} \quad \textit{so that} \quad b^{\frac{1}{n}} = \sqrt[n]{b}$$

Oresmus (the Latin form of the name of the scholar-priest) went farther. For what we now write as $4^{\frac{3}{2}} = \sqrt{4^3} = \sqrt{64} = 8$, and in his own shorthand $\boxed{1^p . \frac{1}{2}}\ 4 = 8$, he recognized the more extensive rule which the new numerals make it so easy to see, e.g.:

$$(10^3)^2 = (1000)^2 = 1{,}000{,}000 = 10^6 = 10^{3\times 2}, \text{ i.e. } (b^n)^m = b^{nm}$$

This rule holds good when m in nm is rational but not a whole number, as when we write:

$$6^{5/2} = (6^5)^{\frac{1}{2}} = \sqrt{6^5}, \textit{ since } \sqrt{6^5} \times \sqrt{6^5} = 6^5 \textit{ and } b^{\frac{p}{q}} = (b^p)^{\frac{1}{q}}$$

In translating this discovery of Oresmus into our own contemporary shorthand, we have taken for granted a third marvellous thing about numbers. They can equally well give us the meaning of a^n when a is any number besides 10. This is another way of saying that the advantages of the Hindu numerals have nothing to do with the mysterious properties of 10. On the contrary, the mysterious properties of 10 are merely due to the fact that Hindu numerals were introduced to fit a counting-frame on which the index numbers went up one when the value of

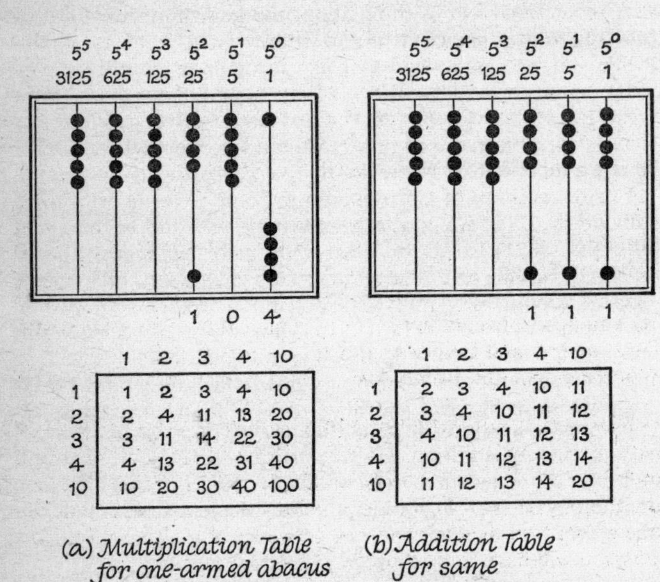

	1	2	3	4	10
1	1	2	3	4	10
2	2	4	11	13	20
3	3	11	14	22	30
4	4	13	22	31	40
10	10	20	30	40	100

	1	2	3	4	10
1	2	3	4	10	11
2	3	4	10	11	12
3	4	10	11	12	13
4	10	11	12	13	14
10	11	12	13	14	20

(a) Multiplication Table for one-armed abacus

(b) Addition Table for same

Fig. 99. The Abacus of the One-armed Man

The principle of the abacus is that the value of a bead in successive columns from left to right is given by:

$$\ldots x^5\ x^4\ x^3\ x^2\ x^1\ x^0$$

In our notation $x = 10$. For the abacus drawn here $x = 5$

104 in the notation of the one-armed abacus would be 1(25) + 0(5) + 4(1), i.e. 29 in the notation of the ten-finger abacus which we use.

111 in the notation of the one-armed abacus would be 1(25) + 1(5) + 1(1), i.e. 31 in the notation of the ten-finger abacus which we use.

the beads went up ten; and 10 itself was originally chosen in the first place simply because man used his ten fingers to count.

Suppose we were all one-armed. We should then count in fives as the military Romans did to some extent by using the intervals V, L, D, for 5, 50, 500. The first column of the type of counting-frame shown in Fig. 99 would carry five beads, each worth one, the second column five beads each worth five, the third column five beads each worth five times five, the fourth five beads each worth five times five times five. When we had counted all the beads on the first column we should switch them back and put one bead in the second column, leaving the first empty. When we had counted five in the second we should switch back all of them and put one out in the third. So if we used 0 for the empty column we should write 1, 2, 3, 4, as we do in the decimal system, but five would be 10, and twenty-five would be 100. Using the words themselves for our own notation, we should have the symbols:

One to five	1	2	3	4	10
Six to ten	11	12	13	14	20
Eleven to fifteen	21	22	23	24	30
. . . . . .					
Twenty-one to twenty-five	41	42	43	44	100
One hundred and twenty-one to one hundred and twenty-five	441	442	443	444	1,000

The multiplication table for this system is given in Fig. 99 with a table for addition. Having worried it out, you may then multiply the number corresponding with 29 ('104') in our notation by 31 (i.e. '111'), using exactly the same method as the one we use ourselves, except that the tables of multiplication and addition will have to be those in Fig. 102. Thus:

```
  104                                 104
  111                                 111
-----     or if you are more used    ----
104          to the other way         104
 104                                 104
  104                               104
------                             ------
12,044                             12,044
```

The number '12,044' here means 1(625) 2(125), 0(25), 4(5), 4(1), i.e. 899, which is what you get by putting

$$\begin{array}{r} 29 \\ 31 \\ \hline 87 \\ 29 \\ \hline 899 \end{array} \quad \text{or} \quad \begin{array}{r} 29 \\ 31 \\ \hline 29 \\ 87 \\ \hline 899 \end{array}$$

The Arithmetic of the Base 2. Laplace, the renowned French astronomer–mathematician who told Napoleon that God is an unnecessary hypothesis, recognized forty years before Babbage, an Englishman, designed the first computing machine, that the number 2 (in our Hindu–Arabic notation) has an immense advantage in terms of the number of *different* operations we need to perform to carry out a computation such as some our parents had to learn (e.g. finding $\sqrt{4235}$) the hard way. In the time of Laplace, a sufficient objection was man – paper – mileage involved. As will be sufficiently clear later, 524,288 in our own (decimal) *notation* would be unity followed by 19 zeros in a binary (i.e. base 2) notation. However, it is not a difficult task to gear the cogs of an abacus on wheels to translate one notation into another; and if one can set it to rotate at sufficient speed, space ceases to be very relevant to the most convenient way of grinding out the result.

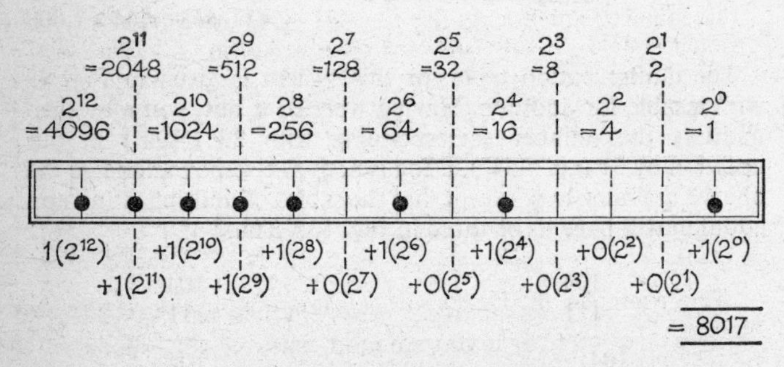

Fig. 100. Abacus for the Base 2

The simplest type of mechanical computer is simply a rotary abacus, as is indeed the distance meter of an automobile; and if we can take advantage of electromagnetic polarity, the simplest base from which to work is the base 2. Just as we require only 9 (= 10 − 1) number signs other than zero, if our base is 10, we need only 1 (= 2 − 1) number sign other than zero if our base is 2. Since this is the base with which an electronic brain works, it is worth while to familiarize ourselves with it thoroughly. Let us first recall the powers of 2 in our own (*decimal*) notation:

$2^0 = 1$; $2^1 = 2$; $2^2 = 4$; $2^3 = 8$; $2^4 = 16$; $2^5 = 32$; $2^6 = 64$; $2^7 = 128$; $2^8 = 256$; $2^9 = 512$; $2^{10} = 1{,}024$, etc.

In the notation of the base 2, our number 32 being 2^5 is unity followed by five zeros, and we may set out our numbers 1–32 in *binary* notation as below:

1 = 1	9 = 1001	17 = 10001	25 = 11001
2 = 10	10 = 1010	18 = 10010	26 = 11010
3 = 11	11 = 1011	19 = 10011	27 = 11011
4 = 100	12 = 1100	20 = 10100	28 = 11100
5 = 101	13 = 1101	21 = 10101	29 = 11101
6 = 110	14 = 1110	22 = 10110	30 = 11110
7 = 111	15 = 1111	23 = 10111	31 = 11111
8 = 1000	16 = 10000	24 = 11000	32 = 100000

Our tables of binary addition and multiplication are as simple as is conceivable, being as below:

Addition

Hindu–Arabic

	0	1	2
0	0	1	2
1	1	2	3
2	2	3	4

Binary

	0	1	10
0	0	1	10
1	1	10	11
10	10	11	100

Multiplication

Hindu–Arabic

	0	1	2
0	0	0	0
1	0	1	2
2	0	2	4

Binary

	0	1	10
0	0	0	0
1	0	1	10
10	0	10	100

Two examples of the schoolbook lay-out for calculation will suffice.

Addition

$$\begin{array}{r} 27 \\ 21 \\ \hline 48 \end{array} = 32 + 16 = 2^5 + 2^4$$

$$\begin{array}{r} 11011 \\ 10101 \\ \hline 110000 \end{array} =$$

$$= 1(2^5) + 1(2^4) + 0(2^3) + 0(2^2) + 0(2^1) + 0(2^0)$$

$$\begin{array}{r} 27 \\ 21 \\ \hline 27 \\ 54 \\ \hline 567 \end{array} \qquad \begin{array}{r} 11011 \\ 10101 \\ \hline 11011 \\ 00000 \\ 11011 \\ 00000 \\ 11011 \\ \hline 1000110111 \end{array}$$

$$567 = 512 + 32 + 16 + 4 + 2 + 1$$
$$= 2^9 + 2^5 + 2^4 + 2^2 + 2^1 + 2^0$$

$$1000110111 = 1\,.\,(2^9) + 0(2^8) + 0(2^7) + 0(2^6) + 1(2^5) + 1(2^4) + 0(2^3) + 1(2^2) + 1(2^1) + 1(2^0)$$

Since we need only two signs for the arithmetic of base 2, we might use (as for the electric charge) + or – instead of 1 and 0. Thus 567 in our notation would then be:

$$+ - - - + + - + + +$$

Algorithms. The scope of the new arithmetic is set forth in the introduction to one of the earliest books on the new *Craft of Nombrynge* (AD 1300) to be written in the English language. 'Here tells that ther ben 7 spices or partes of the craft. The first is called addicion, the secunde is called subtraccion. The thyrd is called duplacion. The 4 is called dimydicion. The 5 is called multiplicacion. The 6 is called diuision. The 7 is called extraccion of the Rote.'

You will already have grasped the fact that the number symbols of the Hindus were different from all other number symbols which preceded them in the Old World. Those of their predecessors had been labels with which you recorded a calculation you were going to carry out, or had already performed, on the counting-frame. The Hindu numerals swept away the necessity for this clumsy instrument. Addition and subtraction could be done as easily 'in one's head' as on the abacus itself. In the language of modern physiology carrying over 'in one's head' means that the brain receives from small changes of tone in the muscles of the eye-socket and fingers with which we count precisely the same sequence of nerve messages as those which accompany carrying over on the counting-frame.

The algorithm of multiplication is reducible to an elementary property of rectangles:

$$a(b + c + d) = ab + ac + ad$$

So multiplying 532 by 7, i.e. 7 times (500 + 30 + 2), is the same thing as

$$7(500) + 7(30) + 7(2)$$

To begin with, this might be written:

$$\begin{array}{r} 532 \\ 7 \\ \hline 14 \\ 210 \\ 3{,}500 \\ \hline 3{,}724 \end{array} \qquad \text{or} \qquad \begin{array}{r} 532 \\ 7 \\ \hline 3{,}500 \\ 210 \\ 14 \\ \hline 3{,}724 \end{array}$$

This was soon shortened to:

$$\begin{array}{r} 532 \\ 7 \\ \hline 3{,}724 \end{array} \qquad \text{or} \qquad \begin{array}{r} 532 \\ 7 \\ \hline 3{,}724 \end{array}$$

Once this step was made by applying the rule of 'carrying over', a simple method for multiplying numbers of any size followed naturally. Thus

$$532\ (732) = (532)\ 700 + (532)\ 30 + (532)\ 2$$

This can be written in a form suitable for addition in either of two ways, the one on the right being better because it is suitable for approximation, especially when decimal fractions are used:

$$\begin{array}{rr} 532 & \\ 732 & \\ \hline 1{,}064 & 2(532) \\ 15{,}960 & 30(532) \\ 372{,}400 & 700(532) \\ \hline 389{,}424 & \end{array} \qquad\qquad \begin{array}{rr} 532 & \\ 732 & \\ \hline 327{,}400 & 700(532) \\ 15{,}960 & 30(532) \\ 1{,}064 & 2(532) \\ \hline 389{,}424 & \end{array}$$

The earliest commercial arithmetics which used the Arabic–Hindu algorithms put down the numbers which are carried over, thus:

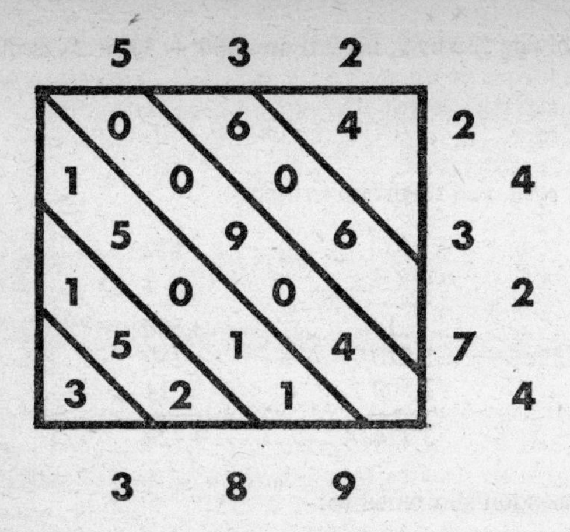

The answer was read off by adding the columns diagonally.

Multiplication, as we now perform it, presupposes that we have access to a table for multiplication. The identification of number symbols with the beads instead of with the columns reduces the size of the table necessary. We need to be able to multiply only up to ten times ten, and this is an undertaking vastly less formidable to the human memory than learning the Alexandrian multiplication tables used by Theon. One can indeed reap the full benefits of the innovation only when the table has been committed to memory, so that continual reference can be dispensed with. This could not happen in Europe till there were new schools to meet the requirements of the mercantile class; and Germany took the lead in this development. The *craft of nombrynge* was so important in Germany during the fourteenth century that it could boast a guild of *Rechenmeister*.

You must not imagine that there is even at the present time any absolute uniformity in the various algorithms in use. In Europe the two alternative ways of carrying out multiplication (left–right and right–left) are both practised today very much as they were used by the Arabs. There is less uniformity in the methods of division, and the one taught in English schools is not exactly like any of the Arabic ones. It is a comparatively late device first used, as far as we know, by Calandri in 1491. If you do not know why we divide in the way we do, the best way to understand it is to use abstract numbers for the value of each bead on the counting-frame, giving the value 1 for the first column, x for the second, x^2 for the third, and x^3 for the fourth, etc. We can then rewrite the previous multiplication:

$$
\begin{array}{l}
5x^2 + 3x + 2 \\
7x^2 + 3x + 2 \\
\hline
35x^4 + 21x^3 + 14x^2 \\
\qquad 15x^3 + 9x^2 + 6x \\
\qquad\qquad + 10x^2 + 6x + 4 \\
\hline
35x^4 + 36x^3 + 33x^2 + 12x + 4
\end{array}
$$

(because $7x^2 \times 5x^2 = 7 \times x \times x \times 5 \times x \times x = 35x^4$, etc.)

(If $x = 10$, this is 389,424)

Division on the abacus is repeated subtraction. Dividing 389,424 by 732 means, how many times can 732 be taken away from 389,424, leaving nothing over? How the method of division which we use corresponds to working with the counting-frame is seen by putting it in the form $(35x^4 + 36x^3 + 33x^2 + 12x + 4) \div (7x^2 + 3x + 2)$, thus:

$$
\begin{array}{rl}
7x^2 + 3x + 2) & 35x^4 + 36x^3 + 33x^2 + 12x + 4 \; (5x^2 + 3x + 2 \\
& 35x^4 + 15x^3 + 10x^2 \\
\hline
& \qquad 21x^3 + 23x^2 + 12x + 4 \\
& \qquad 21x^3 + 9x^2 + 6x \\
\hline
& \qquad\qquad 14x^2 + 6x + 4 \\
& \qquad\qquad 14x^2 + 6x + 4 \\
\hline
& \qquad\qquad 0 \quad 0 \quad 0
\end{array}
$$

Taking away $5x^2$ times $7x^2 + 3x + 2$ empties the fifth column of the frame.

Taking away $3x$ times $7x^2 + 3x + 2$ empties the fourth column.

Taking away 2 times $7x^2 + 3x + 3$ empties the remaining columns.

The fact that we still use the commercial idiom 'borrow one' in subtraction and division reminds us that the laws of calculation were developed in response to the cultural needs of the mercantile class. The use of abstract numbers for the value of the beads shows you why the laws of arithmetic are just the same for the one-armed abacus as for the ten-finger abacus. The use of number symbols which could represent a counting-frame with as many columns as we need is a parable of the growing volume of trade which necessitated operations with large numbers. The popularity of the new methods grew, as Europe absorbed from the East two social inventions which made it convenient to dispense with the counting-frame altogether. Paper and printing, like *sunya*, came from the East. The 'dawn of nothing' was also the dawn of cheap material for writing.

How the search for general rules about the behaviour of numbers was prompted by the flood of light which the new symbols shed is well illustrated by the fact that all the algorithms for fractions now used were invented by the Hindus. Before the introduction of the sexagesimal system (p. 58), treatment of fractions never advanced beyond the level of the Egyptian Rhind papyrus. Fractions were dealt with metaphorically by imagining smaller units, just as we divide tons into hundredweights, and these into pounds, and pounds into ounces. This inability to treat a fraction as a number on its own merits is the explanation of a practice which continued for several millennia. The mathematicians of antiquity went to extraordinary pains to split up fractions like $\frac{2}{43}$ into a sum of unit fractions, e.g. $\frac{2}{43} = \frac{1}{30} + \frac{1}{86} + \frac{1}{645}$ or $\frac{1}{43} + \frac{1}{86} + \frac{1}{129} + \frac{1}{258}$. As this example shows, the procedure was as useless as it was ambiguous. A simple explanation of such apparent perversity is that the first calculators were trying to put side by side two fractions and see whether one was larger than the other in the same way as we might compare two weights like 1 stone 1 pound 1 ounce with 1 hundredweight, 1 stone and 1 pound.

Equipped with their simple and eloquent number symbols, the Hindus broke away completely from this metaphorical way of dealing with fractions. Having rules for rapid calculation without mechanical aids, they experimented with them as with whole numbers. Thus Mahavira (AD 850) gave our rule for dividing one fraction by another in the same words which a school teacher might use today: 'Make the denominator the numerator and then multiply.'

Thus we may write:

$$\frac{a}{b} \div \frac{c}{d} = \frac{a}{b} \times \frac{d}{c} = \frac{ad}{bc}$$

$$e.g. \quad \frac{3}{5} \div \frac{4}{7} = \frac{3}{5} \times \frac{7}{4} = \frac{21}{20}$$

In so far as algebra developed in connection with the practical need for rules to make calculations quickly and easily, it progressed rapidly as soon as people began to use and to write numbers in such a way that the rules were easy to recognize, and, if recognized, easy to apply. The cogency of this consideration is illustrated by the last 'spices' of the craft of 'nombrynge', the 'extraccion of the Rote', or, as we now say, the extraction of square roots. This was the parent of one of the marvellous things relating to numbers discovered by Stifel and his contemporaries. We saw how Stifel stretched the paper counting-frame of the Hindu numbers backwards to represent tenths, hundredths, thousandths, etc., by imaginary columns with index numbers −1, −2, −3, etc., to the right of the units column. If we make a gap or put a dot at the right of it to show which is the units column, the numbers 125 in 1·125 mean 1 tenth, 2 hundredths, 5 thousandths, just as the same numbers read to the left of the dot in 5210·1 mean 5 thousands, 2 hundreds, and 1 ten. The way this practice grew up was directly associated with the rule for using prime numbers to simplify extraction of square roots.

The Hindus and Arabs greatly improved on the simple trigonometrical tables of the Alexandrians in connection with their own studies in astronomy. This, as we have seen, requires tables of square roots. The advantage of stretching the counting-frame to the right of the units column to represent fractions in descending order diminishing by one-tenth at each step, was grasped in principle by the Arabian mathematicians. If they wanted $\sqrt{2}$ they would put it in the form $\sqrt{\frac{200}{100}} = \frac{1}{10}\sqrt{200}$ for a first approximation, $\sqrt{\frac{20000}{10000}}$ or $\frac{1}{100}\sqrt{20000}$ for a second approximation, $\sqrt{\frac{2000000}{1000000}}$ or $\frac{1}{1000}\sqrt{2000000}$ for a third approximation, and so on. Trial at once shows that $\sqrt{200}$ is roughly 14, which, divided by 10 came to be written 1 4. Similarly, the nearest whole number for $\sqrt{2000000}$ is 1,414 and this, divided by 1,000, was written with a gap 1 414 to indicate that the 414 is what we call the decimal fraction part of the square root. Tables of square roots printed in this form are given by the *Rechenmeister* Adam Riese in 1522. Independently as a natural offspring of the same process, Al Kashi of Samarkand gives his value for π as 3 14159 . . . correct to nine decimals about AD 1400. The decimal point to mark the gap was introduced by

Pelazzi of Nice about 1492. In England it is written above the line, in America on the line. In western Europe, a comma on the line marks the gap. The introduction of the typewriter makes it fairly certain that the English custom will be superseded by the American or Continental one.

By the time of Adam Riese, people had grasped the essential truth that the rules which govern the arithmetic of a numeral system based on the principle of position are exactly the same, whatever value we give to the beads of a particular column. In other words, addition, multiplication, division, subtraction, etc., are exactly the same for decimal fractions as for whole numbers, the only precaution necessary being either to arrange the numbers suitably to show where the point should come (e.g. the left-right method of multiplication), or to put in the decimal point at the right place by common sense.

The decimal way of representing fractions greatly simplifies commercial arithmetic, if one's weights and measures conform to a decimal plan. Stevinus, who was a warehouse clerk, put in charge of the provisioning of his army by William of Orange, advocated the legal adoption of such a system as early as 1585. The idea was revived by Benjamin Franklin and others at the time of the American Revolution, and thoroughly carried out in the end by the initiative of the National Assembly of France. England still clings to an antiquated jumble of weights and measures, and has not yet completely succeeded in introducing the Gobar numerals into the House of Commons.

Equations. Alexandrian mathematicians had been forced to pay attention to the art of calculation by the problems they encountered in astronomy and mechanics. The early Hindu mathematicians devoted a great deal of attention to problems involving numbers such as arise in trade. When we speak of this early Hindu work as *algebra*, we must remember that the words algebra and arithmetic are used in school books in a somewhat different sense from that in which they are used in histories of mathematics. What is called arithmetic nowadays does not correspond with the *arithmetike* of the Greeks, dealt with in Chapter 6. The arithmetic of our schools is made up partly of rules for calculation based on the Hindu and Arab algorithms and partly of the solution of numerical problems without using the abstract number symbols of what is ordinarily called algebra. The simple and consistent rules for using abstract numbers and the shorthand symbols for mathematical relations and operations have been evolved very slowly.

As already mentioned, Diophantus was the first person to attempt anything of the kind, and for many centuries mathematicians dealt with problems involving numbers on entirely individualistic lines. Each writer would use a shorthand which he understood himself without attempting to introduce a universal convention. So he was forced to fall back on the language of everyday life when he attempted to explain his methods to other people. Mathematicians use the term 'algebra' to mean rules for solving problems about numbers of one kind or another, whether the rules are written out in full (*rhetorical* algebra), or more or less simplified by abbreviations (*syncopated* algebra), or expressed with the aid of letters and operative signs exclusively (*symbolic* algebra). Some problems in commercial arithmetic which we learn to solve at school correspond with what the mathematician calls rhetorical algebra.

The Arabs used syncopated expressions corresponding with what we would call equations. Individual authors among the first converts to the Moorish learning, like the Dominican friar Jordanus (about AD 1220), replaced words altogether by symbols. His contemporary, Leonardo of Pisa (Fibonacci) did the same. The following examples, which show the transition from pure rhetorical algebra to modern algebraic shorthand, are not given to exhibit a continuous historical sequence so much as to bring into clear historical perspective the fact that size language grew by imperceptible stages out of the language of everyday life.

Regiomontanus, AD 1464:

3 Census et 6 demptis 5 rebus aequatur zero

Pacioli, AD 1494:

3 Census p 6 de 5 rebus ae 0

Vieta, AD 1591:

3 in *A* quad – 5 in *A* plano + aequatur 0

Stevinus, AD 1585:

3 ② – 5 ① + 6 ⊙ = 0

Descartes, AD 1637:

$3x^2 - 5x + 6 = 0$

It goes without saying that the advance from 'rhetorical' discussion of rules for solving problems to symbolism of the modern sort was well-nigh impossible for the Greeks, who had already exhausted the letters of the alphabet for proper numbers. Although the Hindu numerals removed this obstacle to progress, there was at first no social machinery to impose the universal use of devices for representing operators. The only operative symbol which was transmitted to us by the Arabs from Hindu sources is the square-root sign ($\sqrt{}$). In medieval Europe the social machinery which paved the way for this tremendous economy in the language of size emerged in a somewhat surprising way. Our word 'plus' is short for 'surplus'. In the medieval warehouses the marks '+' and '−' were chalked on sacks, crates, or barrels to signify whether they exceeded or fell short of the weight assigned. These signs were introduced into general use by one of the first products of the printing press – Widman's *Commercial Arithmetic*, published in 1489 at Leipzig – and one of the first to use them for solving equations was Stevinus, already mentioned. An English commercial arithmetic by Record, published a century later, introduced '×' and '='. From this point onwards the shorthand first used by Descartes was generally adopted, and mathematics was liberated from the clumsy limitations of everyday speech. Again you will notice how a turning-point in the history of mathematics arose from the common social heritage rather than through an invention of isolated genius.

The transition from rhetorical to symbolical algebra is one of the most important things to understand in mathematics. What is called solving an equation is putting it in a form in which its meaning is obvious. The rules of algebra tell us how to do this. The really difficult step consists in translating our problem from the language of everyday life into the language of algebra. At this point the mathematician himself may go astray, if the mathematician has less opportunity for understanding the problem as stated in the language of everyday life than the plain man who is not a mathematician. Once a problem has been translated into a mathematical sentence (or equation), we can safely trust the mathematician to do the rest. What remains is a problem in the peculiar language of mathematics. A danger lies in trusting the mathematician to do the translation.

The latter-day Alexandrian who gave his name to what we still call *Diophantine equations* realized a difference between the two realms of communication illustrated at the most elementary level by the following example:

> A farmer's livestock consists of pigs and cattle, the number of pigs being three times the number of cattle and the total of the two being T. How many of each does he have?

Formally, one may say that the number of cows is $\frac{1}{4}T$ and the numbers of pigs is $\frac{3}{4}T$; but this is a meaningful problem only if T is a multiple of 4, e.g. 20 (15 pigs and 5 cows) or 36 (27 pigs and 9 cows), etc. Otherwise, the formal solution is irrelevant. The verbal statement is then internally inconsistent with the *implicit* assumption that both answers must be expressible by *whole numbers*.

To understand the art of translating from the language of everyday life to the language of size we have to face a very common difficulty in learning any foreign language. We cannot easily make sense of a sentence in a foreign language merely by looking up the words in the dictionary. Every language has its own particular idiosyncrasies of word order or idiom. So we may here usefully supplement what we have said about language of size, order, and shape in an earlier chapter by the following three rules and two cautions.

Rules. (i) Translate separately each item of information (stated or implied) into the form 'By or with something do something to get something.'

(ii) Combine the statements so as to get rid of any quantities which you do not want to know about. To do this you may have to add to statements explicitly made others which are only implied.

(iii) Get the final statement in the form 'The number I want to know (x, n, r, etc.) can be got (=) by putting down an ordinary number.'

Cautions. (iv) See that all numbers representing the same kind of quantities are expressed in the same units, e.g. if money all pounds or all shillings, etc., if distance all yards or all miles, etc., if time all seconds or all hours, etc.

(v) Check the result.

To illustrate translation of verbal statements about numbers into the idiom of algebraical symbolism we shall now cite some problems which can be put in the form of the simplest sort of equations which the Hindu mathematicians gave rhetorical rules for solving. Before doing so a further word of explanation may not be amiss. When you are fluent in the use of a foreign language, you translate into the correct idiom, one way or the other, straight away. When you are beginning, you have to go step by step. To show you that *solving problems is not a special gift but merely the art of applying fixed rules of grammar*, we shall also go step by step with the problems which follow. Of course, you will not need to construe sentence by sentence when you have got used to the trick of translation. You will then put down the equation which represents the verbal statement in one or two steps.

Example I. The current account of a local trade-union committee is four times the deposit account, the total of the two being £35. How much is there in each account?

First statement: The current account is four times the deposit account, i.e. 'By 4 multiply the number of pounds in the deposit account to get the number of pounds in the current account.'

$$4d = c \quad . \quad . \quad . \quad . \quad . \quad . \quad . \quad . \quad \text{(i)}$$

Second statement: The two together make £35, i.e. 'The number of pounds in the current account must be added to the number of pounds in the deposit account to get £35.'

$$c + d = 35 \quad . \quad . \quad . \quad . \quad . \quad . \quad . \quad . \quad \text{(ii)}$$

Combining both statements, we get

$$4d + d = 35$$

$$\therefore \quad 5d = 35 \quad \text{and} \quad d = \tfrac{35}{5} = 7$$

The deposit account is £7, and the current account is £(35−7) = £28. Check: 4 × £7 = £28.

Example II. A train leaves London for Edinburgh at one o'clock, going at 50 miles per hour. Another train leaves Edinburgh for London at four o'clock, going at 25 miles per hour. If Edinburgh is 400 miles from London, when do they meet?

What we are told allows us to get how far the trains have gone in any time. What we want to get is the time which elapses before they are both the same distance from Edinburgh, or the same distance from London. As this time is assumed to be after the second train has started we will reckon it from four o'clock, i.e. the time (t) required is so many hours after four o'clock.

First statement: Train A leaves London at one o'clock, i.e. 'Add 3 (time between one o'clock to four o'clock) to the number of hours after four o'clock when the trains meet to get the number of hours (T) train A has been travelling.'

$$3 + t = T \quad . \quad . \quad . \quad . \quad . \quad . \quad . \quad (i)$$

Second statement: Train A travels at 50 miles per hour away from London, i.e. 'By 50 multiply the time (T) the train has been running when they meet to get the distance (D) from London when they meet.'

$$50T = D \quad . \quad . \quad . \quad . \quad . \quad . \quad (ii)$$

Third statement: The second train leaves Edinburgh at four o'clock travelling at 25 miles per hour, i.e. 'By 25 multiply the time (t) after four o'clock when the trains meet to get the distance from Edinburgh where they meet.'

$$25t = d \quad . \quad . \quad . \quad . \quad . \quad . \quad . \quad (iii)$$

Fourth statement: The distance from London to Edinburgh is 400 miles, i.e. 'From 400 subtract the distance (d) from Edinburgh where they meet to get the distance (D) from London where they meet.'

$$400 - d = D \quad . \quad . \quad . \quad . \quad . \quad . \quad (iv)$$

Combine (i) and (ii): $50(3 + t) = D$ . . . . . . (v)

Combine (iv) and (iii): $400 - 25t = D$ . . . . . . (vi)

Combine (v) and (vi): $50(3 + t) = 400 - 25t$

Divide both sides by 25 to reduce arithmetic.

$$2(3 + t) = 16 - t \quad \textit{or}$$

$$6 + 2t = 16 - t$$

$$\therefore \quad 2t + t = 16 - 6 \quad \textit{and} \quad 3t = 10$$

$$\therefore \quad t = \tfrac{10}{3} = 3\tfrac{1}{3} \text{ (hours after four o'clock)}$$

$$= 3 \text{ hours 20 minutes after four clock} = 7.20 \text{ pm}$$

Check: $50(3 + 3\frac{1}{3}) + 25 \times 3\frac{1}{3} = 400$

Example III. The race of Achilles and the tortoise.

First statement: The speed of the hero is ten times the speed of the tortoise, i.e. 'By 10 multiply the speed (s) of the tortoise to get the speed (S) of Achilles.'

$$10s = S \quad . \quad . \quad . \quad . \quad . \quad . \quad . \quad (i)$$

Second statement: Achilles starts 100 yards behind the tortoise, i.e. 'To 100 add the distance (d) which the tortoise has gone when he is overtaken to get the distance (D) Achilles has run (in yards) when he catches up.'

$$100 + d = D \quad . \quad . \quad . \quad . \quad . \quad . \quad \text{(ii)}$$

To connect these statements we have to remember that the speed is the distance divided by the time (t), which is obviously the same in both cases (i.e. the time when the tortoise is overtaken is the time when Achilles catches him up). So we may add two more implied statements.

Third statement: The distance which the tortoise goes till he is overtaken must be divided by the time he runs to get his speed.

$$\frac{d}{t} = s \quad . \quad . \quad . \quad . \quad . \quad . \quad . \quad \text{(iii)}$$

Fourth statement: The distance that Achilles goes before he catches up must be divided by the time he runs to get the speed of Achilles.

$$\frac{D}{t} = S \quad . \quad . \quad . \quad . \quad . \quad . \quad \text{(iv)}$$

Combine (i) and (iii) $\qquad \dfrac{10d}{t} = S \quad . \quad . \quad . \quad . \quad . \quad . \quad \text{(v)}$

Combine (ii) and (iv) $\qquad \dfrac{100 + d}{t} = S \quad . \quad . \quad . \quad . \quad . \quad . \quad \text{(vi)}$

Combine (v) and (vi) $\qquad \dfrac{10d}{t} = \dfrac{100 + d}{t}$

Multiply both sides by t

$$\begin{aligned} 10d &= 100 + d \\ 10d - d &= 100 \\ \therefore \quad 9d &= 100 \\ d &= \tfrac{100}{9} \\ &= 11\tfrac{1}{9} \text{ (yards)} \quad \text{Check this.} \end{aligned}$$

Example IV. When I am as old as my father is now, I shall be five times as old as my son is now. By then my son will be eight years older than I am now. The combined ages of my father and myself are 100 years. How old is my son? (*Week-End Book*.)

First statement: When I am as old as my father I shall be five times as old as my son is now, i.e. my father is now five times as old as my son is now. This means, 'By 5 multiply my son's age (s) to get my father's age (f).'

$$5s = f \quad . \quad . \quad . \quad . \quad . \quad . \quad . \quad . \quad . \quad \text{(i)}$$

Second statement: When I am as old as my father my son will be eight years older than I am now. Split this up thus: (A) From my father's age (f) take mine (m) to get how long it will be (l) before I am as old as he is.

$$f - m = l \quad . \quad . \quad . \quad . \quad . \quad . \quad . \quad \text{(A)}$$

(B) These l years must be added to my son's age (s) to get how old he will be (S years) when I am as old as my father is now.

$$l + s = S \quad . \quad . \quad . \quad . \quad . \quad . \quad . \quad . \quad \text{(B)}$$

(C) Eight must be added to my present age to get my son's age l years hence.

$$m + 8 = S \quad . \quad . \quad . \quad . \quad . \quad . \quad \text{(C)}$$

Combine (B) and (C) $\quad m + 8 = l + s \quad . \quad . \quad . \quad . \quad . \quad . \quad \text{(D)}$

Combine (A) and (D) $\quad m + 8 = f - m + s$

or $\quad 2m + 8 = f + s \quad . \quad . \quad . \quad . \quad . \quad . \quad \text{(ii)}$

Third statement: The combined ages of my father and myself are 100 years, i.e. 'My father's age must be added to mine to get 100 (years).'

$$m + f = 100$$

or $\quad m = 100 - f$

Combine (i) and (ii) $\quad 5s + s = 2m + 8$

$$6s = 2m + 8 \quad . \quad . \quad . \quad . \quad . \quad \text{(iv)}$$

Combine (i) and (iii) $\quad 100 - 5s = m \quad . \quad . \quad . \quad . \quad . \quad . \quad \text{(v)}$

Combine (iv) and (v) $\quad 6s = 2(100 - 5s) + 8$

$$6s = 200 - 10s + 8$$

$$6s + 10s = 200 + 8$$

$$16s = 208$$

$$s = \tfrac{208}{16}$$

13 (years) Check this.

i.e. my son is 13 years old now.

Example V. (An early Hindu problem from the *Lilavati* of Aryabhata, *c.* AD 450.) 'A merchant pays duty on certain goods at three different places. At the first he gives $\frac{1}{3}$ of his goods, at the second $\frac{1}{4}$ of what he has left, and $\frac{1}{5}$ of the remainder at the third. The total duty is twenty-four coins. What had he at first?'

First statement: At the first place he gives a third of his goods away, i.e. 'From what he had (x) take $\frac{1}{3}$ of its worth to get what he had left (y) when he got to the second place.'

$$x - \tfrac{1}{3}x = y$$

$$\text{or} \qquad \tfrac{2}{3}x = y \qquad \ldots\ldots \qquad \text{(i)}$$

Second statement: At the second place he pays a quarter of its worth, i.e. 'From what he had when he arrived take $\frac{1}{4}$ of its worth to get what he had (z) when he went on.'

$$y - \tfrac{1}{4}y = z$$

$$\text{or} \qquad \tfrac{3}{4}y = z \qquad \ldots\ldots \qquad \text{(ii)}$$

Third statement: He paid one-fifth of the residue at the third place, and this made the total duties up to twenty-four coins, i.e. 'To $\frac{1}{5}$ of what he had when he got there add the duty he paid at the second ($\frac{1}{4}y$) and the duty he paid at the first ($\frac{1}{3}x$) to get 24.'

$$\tfrac{1}{5}z + \tfrac{1}{4}y + \tfrac{1}{3}x = 24 \qquad \ldots\ldots \qquad \text{(iii)}$$

Combine (ii) and (iii):

$$(\tfrac{3}{4} \times \tfrac{1}{5})y + \tfrac{1}{4}y + \tfrac{1}{3}x = 24$$

$$\text{or} \qquad \tfrac{2}{5}y + \tfrac{1}{3}x = 24 \qquad \ldots\ldots \qquad \text{(iv)}$$

Combine (i) and (iv):

$$(\tfrac{2}{5} \times \tfrac{2}{3})x + \tfrac{1}{3}x = 24$$

$$\text{or} \qquad \tfrac{3}{5}x = 24$$

$$x = \frac{5 \times 24}{3}$$

$$= 40$$

i.e. he had forty coins' worth. Check this.

. . . .

These examples have been worked out step by step with great detail to show you that solving problems by algebra is simply translation according to fixed grammatical rules. As stated, you need not go through all these steps when you have become fluent in the use of number language. Once you are at home in it, you will find it much quicker to put down first of all an abstract number for the one which you want to find and then write down everything you are told about it till you have a sentence which stands by itself. For instance, Example V may be worked out more snappily like this:

Let son's age be y years:
Then father's age is $5y$ years:

My age is $(100 - 5y)$ years; and

$$5y - (100 - 5y) + y = 100 - 5y + 8$$
$$\therefore \quad 16y = 208$$
$$\therefore \quad y = 13$$

All the problems which we have translated so far can finally be boiled down to a mathematical sentence which only contains one abstract number standing for the unknown quantity which we are looking for. This can often be done even when the problem is about two unknown quantities, provided the connection between them is simple and obvious. For instance, here is a problem about three unknown quantities which presents no difficulties.

Example VI. In a tool-box there are three times as many tacks as nails, and three times as many nails as screws. The total number of tacks, nails, and screws in the box is 1,872. How many of each are there? You can translate this thus. The number of nails is one-third the number of tacks ($n = \frac{1}{3}t$). The number of screws is one-third the number of nails ($s = \frac{1}{3}n$). The number of all three ($t + n + s$) is 1,872, i.e.:

$$t + \tfrac{1}{3}t + \tfrac{1}{3}(\tfrac{1}{3}t) = 1{,}872$$
$$t(1 + \tfrac{1}{3} + \tfrac{1}{9}) = 1{,}872$$
$$\tfrac{13}{9}t = 1{,}872$$
$$t = \frac{9 \times 1{,}872}{13} = 1{,}296$$

Thus the number of tacks is 1,296, the number of nails one-third of this number, i.e. 432, and the number of screws one-third of 432, i.e. 144. (Check: 1,296 + 432 + 144 = 1,872.)

When more than one unknown quantity occurs in a problem we can boil down the verbal statement to a mathematical sentence containing a single abstract number, only provided that one of the unknown quantities is so many times another, or differs from the other by a known amount. When we cannot do this we can still solve the problem, provided we can make as many distinct equations as there are unknown quantities. For instance, here is a simple problem of this sort.

Example VII. Two pounds of butter and three pounds of sugar cost two shillings and sevenpence, i.e. 31 pence. Three pounds of butter and two pounds of sugar cost three shillings and three-pence. What is the price of each? This problem means:

(i) That twice the cost of a pound (b) of butter (in pence) added to three times the cost of a pound (s) of sugar (in pence) amounts to 31 pence, i.e.:

$$2b + 3s = 31$$

(ii) That three times the cost of a pound of butter added to twice the cost of a pound of sugar amounts to 39 pence, i.e.:

$$3b + 2s = 39$$

We have now two equations and two abstract numbers, and can get rid of either of them by a simple trick which is called solving 'simultaneous equations'. We can do anything we like to one side of an equation so long as we do exactly the same to the other side, and if we multiply both sides of the first equation by three and both sides of the second by two we now have two equations in which one term containing an abstract number is identical, viz.:

$$6b + 9s = 93$$
$$6b + 4s = 78$$

Subtracting $6b + 4s$ from $6b + 9s$ is the same thing as subtracting 78 from 93, and the results of the two subtractions are therefore the same, i.e. $5s = 15$. Hence $s = 3$ (pence). We can get b by putting the value of s into either of the original equations; thus $2b + 9 = 31$, i.e. $2b = 22$ and $b = 11$ (pence). The price of sugar is therefore threepence, and of butter elevenpence a pound.

The general rule for solving two simultaneous equations where a, b, c, d, e, f stand for known numbers and x and y for unknown numbers in the final statement of the problem may be put in this way. If:

$$ax + by = c$$
$$dx + ey = f$$

Then to get rid of x multiply the first by d and the second by a:

$$dax + dby = dc$$
$$dax + eay = fa$$

Subtract the second from the first:

$$(db - ea)y = dc - fa$$

We have now a simple equation with only one abstract number y, the others being stated in the problem itself. Alternatively, of course, we might, if it involved multiplying by smaller numbers, prefer to multiply the first equation by e and the second by b, getting rid of y and leaving x

$$(ea - db)x = ce - bf$$

* * * *

Although the Hindus and Arabs made little use of operative symbols, which stand in modern algebra for mathematical verbs, when we translate a problem from everyday speech into the language of size and order as we have done in the foregoing examples, they gave grammatical rules which are substantially those which have been given in Chapter 2. Alkarismi distinguished between two general rules. The first he called *al-muqabalah*, or, as our textbooks say, collecting like terms. In modern shorthand this is the rule for avoiding redundancy, illustrated by:

$$q + 2q = x + 6x - 3x$$
$$\therefore \quad 3q = 4x$$

The other rule, the name for which has been assimilated into our own language, was *al-gebra*, i.e. transferring quantities from one side of an equation to another, e.g. in our shorthand:

$$bx + q = p$$
$$\therefore \quad bx = p - q$$

Alkarismi gives the rule which we now use for solving equations containing the square of an unknown number which we are trying to find. The method he gives is essentially the same as one first used by Diophantus. An actual example which is given by Alkarismi is:

$$x^2 + 10x = 39$$

The rule which Alkarismi gives is based upon a simple Euclidean figure (Fig. 101). Suppose you draw a square on a line x units long and continue two adjoining sides of the square 5 units farther, completing the two rectangles with adjacent sides 5 and

x units. You now have an L-shaped figure, the area of which is:

$$x^2 + 5x + 5x = x^2 + 10x$$

If we complete the square with sides 5 units in the figure on the left of Fig. 101, as has been done on the right, the area is now:

$$x^2 + 10x + 25$$
$$= (x + 5)^2$$

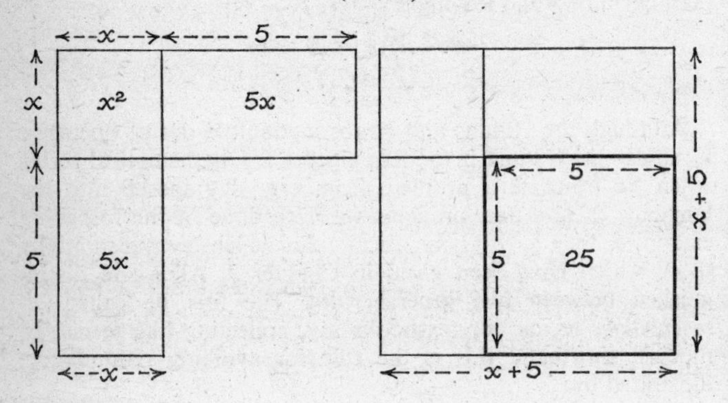

Fig. 101. Solution of a Quadratic Equation by the Completed-square Method of Alkarismi

(i) $x^2 + 10x = 39$

(ii) $x^2 + 10x + 25 = 25 + 39$
$(x + 5)^2 = 64 = 8^2$
$x + 5 = 8$

as we know from Dem. 4. The original equation tells us that:

$$x^2 + 10x = 39$$
$$x^2 + 10x + 25 = 39 + 25$$
$$= 64$$
$$\therefore \quad (x + 5)^2 = 8^2$$
$$\therefore \quad x + 5 = 8$$
$$\therefore \quad x = 8 - 5$$
$$\therefore \quad x = 3$$

Alkarismi thus gives the rule for solving such equations. Calling the number by which x is multiplied (10 in this illustration) the 'roots', 'You halve the number of the "roots" which in the present instance gives 5. This you multiply by itself. The product is 25. Add this to 39. The sum is 64. Now take the square root of this, which is 8, and subtract from it half the number of "roots", which is 5. This is the root of the square you sought for.'

Nowadays we call the number corresponding with 10 in this equation the *coefficient* of x. Replacing it by an abstract number at the beginning of the alphabet to signify that it is a number we already know, while replacing 39 in the same way, we say that, if:

$$x^2 + bx = c$$

$$x = \sqrt{\frac{b^2}{4} + c} - \frac{b}{2}$$

You will recognize $\frac{b^2}{4}$ as the square of half the coefficient of x, or, as Alkarismi would say, the result of multiplying half the number of the 'roots' by itself.

We still call this rule for finding the value of x in an equation containing x^2 *completing the square* to remind us of the fact that the algebra of equations developed from the hieroglyphic way of solving a problem by scale diagrams like those of Fig. 101, and we call equations like the one which we have just solved '*quadratic equations*', from the Latin word *quadratum*, for a four-sided figure, though modern books on elementary algebra no longer use the figure to suggest the rule. Here is a problem which you can solve by the rule we have just given. There are easier ones given at the end of the chapter for practice.

Example VIII. Two hikers go out for the day, one walking a quarter of a mile an hour faster than the other. The faster one reaches the end of this journey half an hour earlier than the slower of the two. Both walk 34 miles. At what rate does each walk?

First statement: The one who gets there first walks a quarter of a mile per hour faster, i.e. 'To the speed of the slower (m miles per hour) add $\frac{1}{4}$ mile per hour to get the speed of the faster (n miles per hour),' or:

$$\tfrac{1}{4} + m = n$$

$$\therefore \quad n = \frac{4m + 1}{4} \quad \ldots \ldots \text{(i)}$$

Second statement: The faster walker took half an hour less, i.e. 'From the time (h hours) taken by the slower walker take $\frac{1}{2}$ to get the time (H hours) taken by the faster one,' or:

$$h - \tfrac{1}{2} = H \quad \ldots \ldots \text{(ii)}$$

Third statement: The faster one walking H hours at n miles per hour covers 34 miles, and the slower one walking h hours at m miles per hour covers 34 miles. This means 'Divide 34 miles by the time taken by each hiker to get the speed at which he travels,' i.e.:

$$n = 34 \div H$$

$$\therefore \quad H = \frac{34}{n} \quad \ldots\ldots\ldots \quad \text{(iii}a\text{)}$$

$$m = 34 \div h$$

$$\therefore \quad h = \frac{34}{m} \quad \ldots\ldots\ldots \quad \text{(iii}b\text{)}$$

Combine (iii) with (ii):

$$\frac{34}{m} - \frac{1}{2} = \frac{34}{n} \quad \ldots\ldots\ldots \quad \text{(iv)}$$

Combine (iv) with (i):

$$\frac{34}{m} - \frac{1}{2} = \frac{34 \times 4}{4m + 1}$$

Apply the diagonal rule:

$$68 + 271m - 4m^2 = 272m$$

$$\therefore \quad -m - 4m^2 = -68$$

$$\therefore \quad m^2 + \tfrac{1}{4}m = 17$$

Apply Alkarismi's rule:

$$m = \sqrt{\tfrac{1}{4}(\tfrac{1}{4})^2 + 17} - \tfrac{1}{2}(\tfrac{1}{4})$$

$$= \sqrt{\frac{1{,}089}{64}} - \frac{1}{8}$$

$$= \tfrac{33}{8} - \tfrac{1}{8}$$

$$= 4$$

Hence the speed (m) of the slower of the two is 4 miles per hour, and that of the faster is $4\frac{1}{4}$ miles per hour. (Check: The faster goes at $4\frac{1}{4} = \frac{17}{4}$ or $\frac{34}{8}$ miles per hour, i.e. takes 8 hours. The slower takes $8\frac{1}{2}$ hours, in which time he does $8\frac{1}{2} \times 4 = 34$ miles.)

The solution of equations by this rule brought the Moslem mathematicians face to face with a limitation inherent in Greek geometry. According to the law of signs:

$$-a \times -a = a^2 \quad \text{also} \quad +a \times +a = a^2$$

$$\therefore \quad \sqrt{a^2} = +a \quad \text{or} \quad -a$$

or, as it is frequently written, $\pm a$. So every square-root sign represents an operation which gives two results, e.g.:

$$100 = (\pm 10)^2$$
$$49 = (\pm 7)^2$$

If you go back to the equation which Alkarismi used to illustrate the rule, you will see that

$$x = 8 - 5$$

or $$x = -8 - 5$$

i.e. $$x = 3$$

or $$x = -13$$

Alkarismi recognized that both these answers check up as follows:

$$3^2 + 10(3) = 9 + 30 = 39$$
$$(-13)^2 + 10(-13) = 169 - 130 = 39$$

We have simply neglected the second answer because we have not yet found a physical meaning for a negative answer, which is the alternative to the one illustrated in Fig. 101.

In the social context of a time when Moslem culture was a beacon in the darkness of Christendom, the bewilderment caused by the difficulty of finding any conceivable meaning for two answers to a problem of this sort is easy to understand. Had bank overdrafts been a familiar fact to people not as yet accustomed to live on hire purchase, or had the devout Keepers of the Calendar had the foresight to name a zero year before or after the *Hegira* (that of the Prophet's flight from Mecca), it would have been easier to tailor the practical problem to the dictates of the mathematical dilemma. Had there been a date (BC 0 = AD 0) between BC 1 and AD 1, it might have been possible to sell the idea more easily to devout Christians. Actually, it was very difficult to devise a practical problem which yields two equally meaningful answers of opposite sign before the invention – in Newton's old age – of thermometers with a zero mark on the scale. Here is an example to show how easily one could do so thereafter.

Example IX. In a storage room, the temperature twenty minutes after a previous reading was 2° higher; and the record showed that the product of the two readings was 15°. What was the initial reading? If we write t for the initial reading:

$$t(t + 2) = 15, \textit{ so that } t^2 + 2t - 15 = 0$$

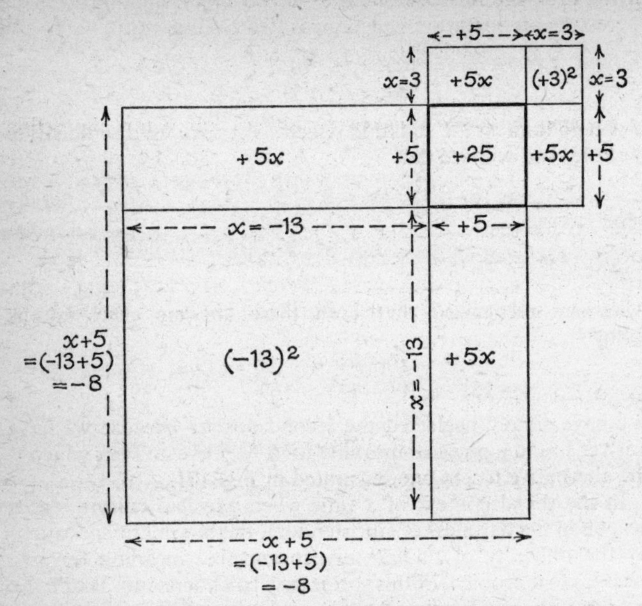

Fig. 102. Alkarismi's Problem in Co-ordinate Geometry

In the previous figure we do not know what x is. The figure is merely used to suggest the numerical procedure. In the Co-ordinate Geometry of Chapter 8 you will see that if we label quantities drawn upwards or rightwards as positive we must label quantities measured downwards or leftwards with the minus sign. The equation ($x^2 + 10x = 39$) tells us that:

$$x + 5 = 8 \text{ or } -8$$

This means that the area of the whole square whose side is $x + 5$ is 64 square units. The lower large square is made up of — (*a*) two rectangles whose area is:
together:

$$2(5)\,(-13) = -130 \text{ square units}$$

(*b*) two squares whose area together is:

$$(-13)\,(-13) + (+5)\,(+5)$$

or $+194$ square units. The total area is:

$$194 - 130 = 64$$

If we write the formula of p. 271 as on p. 277:

$$t = \frac{-2 \pm \sqrt{4 + 60}}{2} = \frac{-2 \pm 8}{2} = +3^\circ \text{ or } -5^\circ$$

Example X. While it was extremely difficult, if not impossible, in the time of Alkarismi to cite examples of practical problems which lead to a negative number as a meaningful answer, it was not impossible to show that the use of both square roots can lead to a correct conclusion, as illustrated by the following problem: a number is multiplied by itself. The result is added to six. On taking away five times the number, nothing is left. What was the number? In algebraic shorthand the riddle can be translated

$$x^2 - 5x + 6 = 0$$

We may write this

$$x^2 - 5x = -6$$

By applying Alkarismi's rule and the law of signs, the solution is

$$x = \sqrt{-6 + (-\tfrac{5}{2})^2} - (-\tfrac{5}{2})$$

$$x = \sqrt{-6 + \tfrac{25}{4}} + \tfrac{5}{2}$$

$$= \pm \tfrac{1}{2} + \tfrac{5}{2}$$

$$= +2 \quad \text{or} \quad +3$$

You will see that both these answers are consistent with the riddle proposed:

$$2^2 - 5(2) + 6 = 4 - 10 + 6 = 0$$
$$3^2 - 5(3) + 6 = 9 - 15 + 6 = 0$$

A quadratic equation may have solutions which are neither positive nor negative numbers. This complication worried the Italian Cardan towards the end of the period of transition which intervened between the Alexandrian era and the invention of coordinate geometry. Playing with riddles like the one we have just set, Cardan stumbled upon a new sort of answer, such as we get when we change the numbers in the last riddle thus: 'A number is multiplied by itself. The result is added to five. On taking away twice the number, nothing is left. What was the number?' The translation in this case is

$$x^2 - 2x + 5 = 0$$
$$x^2 - 2x = -5$$
$$x = \sqrt{-5+1} + 1$$
$$= 1 \pm \sqrt{-4}$$
$$x = 1 \pm 2\sqrt{-1}$$

This raises the question: What on earth is the square root of -1? As a purely grammatical convention it is clearly the number which when multiplied by itself gives -1. A number which did this would give the correct answer as you can see by the two following checks, one when $\pm$ means $+$ throughout, and the other in which it means $-$,

$$\therefore \quad 1 \pm 2\sqrt{-1}$$
$$1 \pm 2\sqrt{-1}$$
$$\overline{}$$
$$1 \pm 2\sqrt{-1}$$
$$\pm 2\sqrt{-1} + 4(-1)$$
$$\overline{}$$
$$x^2 = 1 \pm 4\sqrt{-1} - 4$$
$$\therefore x^2 - 2x + 5 = 1 \pm 4\sqrt{-1} - 4 - 2(1 \pm 2\sqrt{-1}) + 5 = 0$$

Although this shows us that our answer was perfectly grammatical, it does not get us any nearer to saying what the square root of -1 is. All we can say about it at present is that it is a part of speech which can be used grammatically in the kind of sentences which we call quadratic equations. The first mathematicians who encountered quantities such as $\sqrt{-5}$ called them imaginary numbers, which left them in the clouds. To get clearer insight into what they may mean we need a new geometry which is the theme of Chapter 8.

Alkarismi's rule for the solution of a quadratic equation is sometimes given in a more general form for the solution of equations in which the coefficient of x^2 is not *one*, as in the examples given. The equation

$$ax^2 + bx + c = 0$$

can be written

$$x^2 + \frac{bx}{a} = -\frac{c}{a}$$

Applying Alkarismi's rule:

$$x = \sqrt{-\frac{c}{a} + \left(\frac{b}{2a}\right)^2} - \frac{b}{2a}$$

$$\therefore \quad x = \frac{-b \pm \sqrt{b^2 - 4ac}}{2a}$$

As a numerical example which you can check, the solution of:

$$3x^2 - 7x = 6$$

is

$$x = \frac{7 \pm \sqrt{49 + 72}}{6} = \frac{7 \pm 11}{6}$$

$$\therefore \quad x = 3 \text{ or } -\tfrac{2}{3}$$

Series. The new number language brought to light new things about the Natural History of the natural numbers. It is not at all surprising that the Hindus and Arabs revived interest in the ancient Chinese number lore, and made some interesting discoveries of their own. Thus Aryabhata gives the rules for finding the sums of various series of numbers such as:

$$\begin{array}{llll} 1 & 2 & 3 & 4\ldots \\ 1^2 & 2^2 & 3^2 & 4^2\ldots \\ 1^3 & 2^3 & 3^3 & 4^3\ldots \end{array}$$

We have already seen in Chapter 4 how we may use the sum of the series of the top row (i.e. the triangular numbers) as a clue to the build-up of the sum of the series in the next two; and there can be little doubt that the pioneers of Hindu mathematics followed the same trail. In the same context (p. 196) we have also seen how the study of the triangular numbers leads to the family of series commonly called *Pascal's triangle* after the founder of the modern theory of choice and chance in the first half of the seventeenth century of our own era. Actually it was known to, if not first discovered by, the Moslem astronomer, poet, and mathematician Omar Khayyám about five hundred and fifty years earlier. It is figured in the Precious Mirror of the Four Elements written about AD 1300 by the Chinese mathematician Chu Shi Kei, who lived at the time when the Mogul Empire was sprawling into Eastern Europe. Below, we recall the first eight rows:

$$
\begin{array}{ccccccccccccccc}
 & & & & & & & 1 & & & & & & & \\
 & & & & & & 1 & & 1 & & & & & & \\
 & & & & & 1 & & 2 & & 1 & & & & & \\
 & & & & 1 & & 3 & & 3 & & 1 & & & & \\
 & & & 1 & & 4 & & 6 & & 4 & & 1 & & & \\
 & & 1 & & 5 & & 10 & & 10 & & 5 & & 1 & & \\
 & 1 & & 6 & & 15 & & 20 & & 15 & & 6 & & 1 & \\
1 & & 7 & & 21 & & 35 & & 35 & & 21 & & 7 & & 1
\end{array}
$$

Reading diagonally downwards from right to left we have the series of 'unity, the source of all', the natural numbers, the simple triangular numbers, and successive higher orders of triangular numbers, as you will see by referring back to p. 196, Chapter 4. Reading horizontally we have:

$$
\begin{array}{lllllll}
1 & & & & & & \\
1 & 1 & & & & & \\
1 & 2 & 1 & & & & \\
1 & 3 & 3 & 1 & & & \\
1 & 4 & 6 & 4 & 1 & & \\
1 & 5 & 10 & 10 & 5 & 1 & \\
1 & 6 & 15 & 20 & 15 & 6 & 1, \text{etc.}
\end{array}
$$

There are, as Michael Stifel might have said, many marvellous things relating to these numbers. The first is that they show us how to write out fully the expression $(x + a)^n$ without multiplying out. If n is an integer successive multiplication of $(x + a)$ leads to a simple rule, as shown below:

$$
\begin{array}{ll}
x + a & = (x + a)^1 \\
x + a & \\
\hline
x^2 + ax & \\
\qquad ax + a^2 & \\
\hline
x^2 + 2ax + a^2 & = (x + a)^2 \\
\qquad x + a & \\
\hline
x^3 + 2ax^2 + a^2x & \\
\qquad ax^2 + 2a^2x + a^3 & \\
\hline
x^3 + 3ax^2 + 3a^2x + a^3 & = (x + a)^3 \\
\qquad\qquad\qquad x + a & \\
\hline
\end{array}
$$

$$\begin{array}{l} x^4 + 3ax^3 + 3a^2x^2 + a^3x \\ \phantom{x^4 + {}} ax^3 + 3a^2x^2 + 3a^3x + a^4 \\ \hline x^4 + 4ax^3 + 6a^2x^2 + 4a^3x + a^4 \qquad = (x + a)^4 \\ \phantom{x^4 + 4ax^3 + 6a^2x^2 + 4a^3x + {}} x + a \\ \hline x^5 + 4ax^4 + 6a^2x^3 + 4a^3x^2 + a^4x \\ \phantom{x^5 + {}} ax^4 + 4a^2x^3 + 6a^3x^2 + 4a^4x + a^5 \\ \hline x^5 + 5ax^4 + 10a^2x^3 + 10a^3x^2 + 5a^4x + a^5 \qquad = (x + a)^5 \end{array}$$

Tabulating these results, we get:

$$\begin{aligned} (x + a)^1 &= x + a \\ (x + a)^2 &= x^2 + 2ax + a^2 \\ (x + a)^3 &= x^3 + 3ax^2 + 3a^2x + a^3 \\ (x + a)^4 &= x^4 + 4ax^3 + 6a^2x^2 + 4a^3x + a^4 \\ (x + a)^5 &= x^5 + 5ax^4 + 10a^2x^3 + 10a^3x^2 + 5a^4x + a^5 \end{aligned}$$

The numbers in front of each term, or 'coefficients', in these expressions are series in Omar Khayyám's triangle. Thus we should expect $(x + a)^6$ to be:

$$x^6 + 6ax^5 + 15a^2x^4 + 20a^3x^3 + 15a^4x^2 + 6a^5x + a^6$$

You will find that it is so. There is thus a simple rule for writing down $(x + a)^n$. It is called the Binomial Theorem. If you go back to Chapter 4, p. 196, you will recall that the series:

$$1,\quad 4,\quad 6,\quad 4,\quad 1,$$

is the same as

$${}^4C_0 \quad {}^4C_1 \quad {}^4C_2 \quad {}^4C_3 \quad {}^4C_4$$

i.e.

$$1,\ \frac{4}{1},\ \frac{4\,.\,3}{2\,.\,1},\ \frac{4\,.\,3\,.\,2}{3\,.\,2\,.\,1},\ 1.$$

Similarly, the coefficients of $(x + a)^6$ may be written:

$$1,\quad 6,\quad 15,\quad 20,\quad 15,\quad 6,\quad 1$$

$${}^6C_0 \quad {}^6C_1 \quad {}^6C_2 \quad {}^6C_3 \quad {}^6C_4 \quad {}^6C_5 \quad {}^6C_6$$

$$1,\ \frac{6}{1},\ \frac{6\,.\,5}{2\,.\,1},\ \frac{6\,.\,5\,.\,4}{3\,.\,2\,.\,1},\ \frac{6\,.\,5\,.\,4\,.\,3}{4\,.\,3\,.\,2\,.\,1},\ \frac{6\,.\,5\,.\,4\,.\,3\,.\,2}{5\,.\,4\,.\,3\,.\,2\,.\,1},\ 1$$

So we can write down the result of multiplying out $(x + a)^n$ as

$$x^n + n\,.\,ax^{n-1} + \frac{n(n-1)}{2\,.\,1}a^2x^{n-2} + \frac{n(n-1)(n-2)}{3\,.\,2\,.\,1}a^3x^{n-3}$$

$$+ \frac{n(n-1)(n-2)(n-3)}{1\,.\,2\,.\,3\,.\,4}a^4x^{n-4} \ldots + a^n$$

This expression is very long-winded. As in Chapter 4, we can shorten it by recourse to the following symbols:

$$n^{(r)} = 1 \, . \, n \, . \, (n-1)(n-2)(n-3) \ldots (n-r+1), \textit{ so that } n^{(0)} = 1 = n^{(1)}$$

$$r! = 1 \, . \, r \, . \, (r-1)(r-2) \ldots 3 \, . \, 2 \, . \, 1 = r^{(r)}$$

$$n_{(r)} = n^{(r)} \div r!$$

We may then write the theorem of Omar Khayyám as below:

$$(x+a)^n = \sum_{r=0}^{r=n} n_{(r)} \, . \, x^{n-r} \, . \, a^r$$

$$(x-a)^n = \sum_{r=0}^{r=n} (-1)^r \, . \, n_{(r)} \, . \, x^{n-r} \, . \, a^r$$

In the mathematical advances of Newton's time, the Binomial theorem turned out to be an exceedingly valuable instrument for a great variety of uses. One use which you can test out for yourself is an arithmetical device. To find $(4{\cdot}84)^8$ it is not necessary to go through a long series of multiplication sums. We can put:

$$\begin{aligned}(4{\cdot}84)^8 &= (4 \times 1{\cdot}21)^8 \\ &\ 4^8 \times (1{\cdot}21)^8 \\ &\ 4^8 \times (1 + \tfrac{21}{100})^8\end{aligned}$$

Applying the binomial theorem, we get:

$$\left(1 + \frac{21}{100}\right)^8 = 1 + 8\left(\frac{21}{100}\right) + \frac{8 \, . \, 7}{2 \, . \, 1}\left(\frac{21}{100}\right)^2 + \frac{8 \, . \, 7 \, . \, 6}{3 \, . \, 2 \, . \, 1}\left(\frac{21}{100}\right)^3 + \frac{8 \, . \, 7 \, . \, 6 \, . \, 5}{4 \, . \, 3 \, . \, 2 \, . \, 1}\left(\frac{21}{100}\right)^4, \text{ and so on}$$

$$= 1 + 8(0{\cdot}21) + 28(0{\cdot}0441) + 56(0{\cdot}009261), \text{ and so on}$$

The advantage of this is that we have a series of numbers getting smaller and smaller, and we can stop adding more terms at any convenient place. For instance:

$$(1{\cdot}01)^{10} = 1 + 10(0{\cdot}01) + \frac{10 \, . \, 9}{2 \, . \, 1}(0{\cdot}0001) + \frac{10 \, . \, 9 \, . \, 8}{3 \, . \, 2 \, . \, 1}(0{\cdot}000001), \text{ and so on}$$

$$= 1 + 0{\cdot}1 + 0{\cdot}0045 + 0{\cdot}000120 + 0{\cdot}0000021, \text{ and so on}$$

The answer correct to 7 places is:

$$1{\cdot}1046221$$

We shall now use the numbers of Pascal's triangle to get a different rule discovered by a Scots mathematician (Gregory) who was a contemporary of Newton.

Gregory's Formula. The last example leads us to a very suggestive revaluation of the *vanishing triangle* (p. 184) by which we can always represent a number series of a family expressible in figurate form. Let us first set out such a family as below:

rank (r)	0	1	2	3	4	5	6	7 . . .
$s_{r.0}$	2	2	2	2	2	2	2	2 . . .
$s_{r.1}$	0	2	4	6	8	10	12	14 . . .
$s_{r.2}$	0	2	6	12	20	30	42	56 . . .
$s_{r.3}$	0	2	8	20	40	70	112	168 . . .
$s_{r.4}$	0	2	10	30	70	140	252	420 . . .

We now work backwards to get our vanishing triangle thus:

0 2 10 30 70 140 252 420
2 8 20 40 70 112 168
6 12 20 30 42 56
6 8 10 12 14
2 2 2 2
0 0 0
0 0
0

Let us now label this again in a schema, which we can recognize after the event:

s_0 s_1 s_2 s_3 s_4 s_5 s_6
$\Delta^1 s_0$ $\Delta^1 s_1$ $\Delta^1 s_2$ $\Delta^1 s_3$ $\Delta^1 s_4$ $\Delta^1 s_5$
$\Delta^2 s_0$ $\Delta^2 s_1$ $\Delta^2 s_2$ $\Delta^2 s_3$ $\Delta^2 s_4$
$\Delta^3 s_0$ $\Delta^3 s_1$ $\Delta^3 s_2$ $\Delta^3 s_3$
$\Delta^4 s_0$ $\Delta^4 s_1$ $\Delta^4 s_2$
. . . .
. . .

From the way in which we build up our vanishing triangle by subtraction it is clear that:

$$s_{r+1} = s_r + \Delta^1 s_r; \quad \Delta^1 s_{r+1} = \Delta^1 s_r + \Delta^2 s_r; \quad \Delta^2 s_{r+1} = \Delta^2 s_r + \Delta^3 s_r, \text{ etc.}$$

Hence, we may write:

$$s_1 = s_0 + \Delta^1 s_0$$

$$s_2 = s_1 + \Delta^1 s_1 = (s_0 + \Delta^1 s_0) + (\Delta^1 s_0 + \Delta^2 s_0)$$
$$= s_0 + 2\Delta^1 s_0 + \Delta^2 s_0$$

$$s_3 = s_2 + \Delta^1 s_2 = (s_0 + 2\Delta^1 s_0 + \Delta^2 s_0) + (\Delta^1 s_1 + \Delta^2 s_1)$$
$$= (s_0 + 2\Delta^1 s_0 + \Delta^2 s_0) + (\Delta^1 s_0 + \Delta^2 s_0) + (\Delta^2 s_0 + \Delta^3 s_0)$$
$$= s_0 + 3\Delta^1 s_0 + 3\Delta^2 s_0 + \Delta^3 s_0$$

Proceeding thus, we find:

$$s_4 = s_0 + 4\Delta^1 s_0 + 6\Delta^2 s_0 + 4\Delta^3 s_0 + \Delta^4 s_0$$
$$s_5 = s_0 + 5\Delta^1 s_0 + 10\Delta^2 s_0 + 10\Delta^3 s_0 + 5\Delta^4 s_0 + \Delta^5 s_0$$
$$s_6 = s_0 + 6\Delta^1 s_0 + 15\Delta^2 s_0 + 20\Delta^3 s_0 + 15\Delta^4 s_0 + 6\Delta^5 s_0 + \Delta^6 s_0$$

In short, we may write in the compact symbolism of p. 198:

$$s_r = s_0 + r_{(1)}\,\Delta^1 s_0 + r_{(2)}\,\Delta^2 s_0 + r_{(3)}\,\Delta^3 s_0, \text{ etc.}$$

In the foregoing vanishing triangle:

$s_0 = 0;\ \Delta^1 s_0 = 2;\ \Delta^2 s_0 = 6;\ \Delta^3 s_0 = 6;\ \Delta^4 s_0 = 2;\ \Delta^5 s_0 = 0$ and $\Delta^n s_0 = 0$ if $n \geqslant 4$. According to our formula:

$$s_r = 0 + r_{(1)}\,.\,2 + r_{(2)}\,.\,6 + r_{(3)}\,.\,6 + r_{(4)}\,.\,2 + 0$$
$$= 0 + 2r + \frac{6\,.\,r(r-1)}{2\,.\,1} + \frac{6\,.\,r(r-1)(r-2)}{3\,.\,2\,.\,1} + \frac{2r(r-1)(r-2)(r-3)}{4\,.\,3\,.\,2\,.\,1}$$
$$= 0 + 2r + 3r(r-1) + r(r-1)(r-2) + \frac{r(r-1)(r-2)(r-3)}{12}$$
$$= \frac{r^4 + 6r^3 + 11r^2 + 6r}{12}$$

Check $s_6 = \dfrac{6^4 + 6^4 + 11\,.\,6^2 + 6^2}{12} = \dfrac{2\,.\,6^4 + 2\,.\,6^3}{12}$

$$= \frac{2\,.\,6^3\,.\,7}{12} = 252$$

$$s_7 = \frac{7^4 + 6\,.\,7^3 + 11\,.\,7^2 + 42}{12} = 420$$

Here is another example, the series whose terms are the sums of the cubes of the natural numbers:

rank (r)	0	1	2	3	4	5	6
series (S_r)	0	1	9	36	100	225	441

$\Delta^1 s_r$	1	8	27	64	125	216
$\Delta^2 s_r$	7	19	37	61	91	
$\Delta^3 s_r$	12	18	24	30		
$\Delta^4 s_r$	6	6	6			
$\Delta^5 s_r$	0	0				
$\Delta^6 s_r$	0					

In this example our formula gives:

$$s_r = 0 + r_{(1)} . 1 + r_{(2)} . 7 + r_{(3)} . 12 + r_{(4)} . 6$$

$$= r + \frac{7r(r-1)}{2.1} + \frac{12\,r(r-1)(r-2)}{3.2.1} + \frac{6\,r(r-1)(r-2)(r-3)}{4.3.2.1}$$

$$= \frac{r(7r-5)}{2} + 2r(r-1)(r-2) + \frac{r(r-1)(r-2)(r-3)}{4}$$

$$= \frac{2r(7r-5) + 8r(r-1)(r-2) + r(r-1)(r-2)(r-3)}{4}$$

$$= \frac{14r^2 - 10r + 8r^3 - 24r^2 + 16r - r^4 - 6r^3 + 11r^2 - 6r}{4}$$

$$= \frac{r^4 + 2r^3 + r^2}{4} = \frac{r^2(r+1)^2}{4}$$

This as we already know (p. 108) is the square of the rth triangular number; and the reader can use the formula for finding expression for the polygonal figurates of Fig. 78 and for the corresponding 3-dimensional series. In the form given, the Gregory recipe has no pay-off in the world's work, but it suggested to a fellow countryman of Gregory in the first half of the eighteenth century a formula, *Maclaurin's theorem*, for generating a large family of useful series.

Chapter 6

Discoveries and Tests

1. In Mr Wells's story, the two-toed sloth was the dominant species of Rampole Island. Suppose the two-toed sloth had evolved a brain as good as Mr Blettsworthy's, it would have a number system with a base 2, 4, or 8. Make multiplication tables for the systems of numeration it might use. Multiply 24 by 48 in all three systems, and check your results. You will find it easier if you assume that it first learnt to calculate with a counting-frame and make diagrams of the three different kinds of counting-frames.

2. You already know and can check by multiplication that:

$$(x + a)(x + b) = x^2 + (a + b)x + ab$$

The expression $x^2 + 5x - 6$ is built up in the same way, a being -1 and b being 6. So it is the product of the two factors $(x - 1)(x + 6)$. You can check this by dividing $x^2 + 5x - 6$ by $(x - 1)$ or by $(x + 6)$, and by multiplying $(x - 1)$ and $(x + 6)$ together. In the same way, write down the factors of the following by noticing how they are built up, checking each answer first by division and then by multiplication:

(*a*) $a^2 + 10a + 24$
(*b*) $p^2 + 5p + 6$
(*c*) $x^2 - 3x + 2$
(*d*) $m^2 + 4m + 3$
(*e*) $x^2 - 10x + 16$
(*f*) $f^2 + f - 20$
(*g*) $t^2 - 3t - 40$
(*h*) $q^2 - 10q + 21$
(*i*) $c^2 - 12c + 32$
(*j*) $n^2 + 8n - 20$
(*k*) $h^2 + 12h + 20$
(*l*) $z^2 + z - 42$
(*m*) $y^2 - y - 42$
(*n*) $b^2 - b - 20$

3. By direct multiplication you can show that:

$$(ax + b)(ax - b) = a^2x^2 - b^2$$

and $$(ax + by)(ax - by) = a^2x^2 - b^2y^2$$

Notice that the expression $4x^2 - 25$ is built up in this way from $(2x - 5)(2x + 5)$, etc., and write down the factors of:

(*a*) $x^2 - 36$
(*b*) $9x^2 - 25$
(*c*) $4x^2 - 100$
(*d*) $100y^2 - 25$
(*e*) $64x^2 - 49$
(*f*) $81x^2 - 64$

(*g*) $25x^2 - 16$
(*h*) $49p^2 - 16q^2$
(*i*) $256t^2 - 169s^2$
(*j*) $4p^2 - 9q^2$
(*k*) $p^2 - 81q^2$
(*l*) $25n^2 - 9$
(*m*) $36t^2 - 16s^2$
(*n*) $9a^2 - 49b^2$

Using the 'surd' sign ($\sqrt{\ }$), write down the factors of:

(*o*) $3 - x^2$
(*p*) $2 - 3x^2$
(*q*) $5x^2 - 3$
(*r*) $x^2 - 2$
(*s*) $2a^2 - 3$
(*t*) $7a^2 - 3b^2$

4. By direct multiplication show that:

$$(ax + b)(cx + d) = acx^2 + (ad + bc)x + bd$$

Notice that the expression $6x^2 - 7x - 20$ is built up in the same way from the two factors $(3x + 4)(2x - 5)$, in which $a = 3$, $b = 4$, $c = 2$, and $d = -5$. In this way find the factors of the following and check by numerical substitution:

(*a*) $3x^2 + 10x + 3$
(*b*) $6x^2 + 19x + 10$
(*c*) $6p^2 + 5p + 1$
(*d*) $3t^2 + 22t + 35$
(*e*) $6n^2 + 11n + 3$
(*f*) $6q^2 - 7q + 2$
(*g*) $11p^2 - 54p + 63$
(*h*) $20x^4 + x^2 - 1$
(*i*) $15 + 4x - 4x^2$
(*j*) $6n^2 - n - 12$
(*k*) $15x^2 + 7x - 2$
(*l*) $7x - 6 - 2x^2$
(*m*) $15 - 4x - 4x^2$
(*n*) $7x - 6x^2 + 20$

5. By direct multiplication you can show that:

$$(ax + by)(cx + dy) = acx^2 + (ad + bc)xy + bdy^2$$

The expression $6x^2 + 7xy - 20y^2$ is built up in the same way by multiplying the two factors $(3x - 4y)(2x + 5y)$, in which $a = 3$, $b = -4$, $c = 2$, and $d = 5$. In this way write down the factors of the following, and check your answers by division and multiplication:

(*a*) $6a^2 + 7ax - 3x^2$
(*b*) $15a^2 - 16abc - 15b^2c^2$
(*c*) $6a^2 - 37ab - 35b^2$
(*d*) $2a^2 - 7ab - 9b^2$
(*e*) $6f^2 - 23fg - 18g^2$
(*f*) $21m^2 + 13ml - 20l^2$
(*g*) $12n^2 - 7mn - 12m^2$
(*h*) $36p^2 + 3pqr - 5q^2r^2$
(*i*) $14d^2 + 11de - 15e^2$
(*j*) $3t^2 - 13ts - 16s^2$
(*k*) $9m^2 + 9mn - 4n^2$
(*l*) $6q^2 - pq - 12p^2$
(*m*) $4l^2 - 25lm + 25m^2$

6. In arithmetic you are used to reducing a fraction like $\frac{14}{21}$ to

the simpler form $\frac{2}{3}$ by noticing that it is built up as $\frac{2 \times 7}{3 \times 7}$. In the same way use your knowledge of how to find factors to simplify the following fractions, checking by numerical substitution:

(a) $\dfrac{x+y}{x^2+2xy+y^2}$

(b) $\dfrac{(x-y)}{x^2-y^2}$

(c) $\dfrac{(x+y)}{x^2-y^2}$

(d) $\dfrac{x-y}{x^2-2xy+y^2}$

(e) $\dfrac{ax+ay}{ax^2-ay^2}$

(f) $\dfrac{42x^2yz}{56xyz^2}$

(g) $\dfrac{x^2+3x+2}{x^2+5x+6}$

(h) $\dfrac{x^2+2x+1}{x^2+3x+2}$

(i) $\dfrac{x^2-1}{2x^2+3x-5}$

(j) $\dfrac{9x^2-49}{3x^2+14x-49}$

(k) $\dfrac{a^4b+ab^4}{a^4-a^3b+a^2b^2}$

(l) $\dfrac{8a^3-1}{4a^2-4a+1}$

(m) $\dfrac{2x^3-3x^2+4x-6}{x^3-2x^2+2x-4}$

7. Express the following in the simplest form:

(a) $\dfrac{a}{b}+\dfrac{a}{c}$

(b) $\dfrac{a^2b+ab^2}{a+b}$

(c) $x+y-\dfrac{9x^2-4y^2}{3x+2y}$

(d) $a+2b+\dfrac{4b^2}{a-2b}$

(e) $\dfrac{a}{x+1}+\dfrac{3a}{2x+2}-\dfrac{5a}{4x+4}$

(f) $\dfrac{a+2b}{3}-\dfrac{a-3b}{4}$

(g) $\dfrac{7a}{4x+8y}-\dfrac{3a}{2x+4y}$

(h) $x^2+2xy+y^2-\dfrac{x(x^2+3xy+4y^2)}{x+y}$

8. Express as a single fraction in its simplest form:

(a) $\dfrac{1}{x+1}+\dfrac{1}{x-1}$

(b) $\dfrac{1}{x+1}-\dfrac{1}{x-1}$

(e) $\dfrac{x}{x-y}-\dfrac{y}{x+y}$

(f) $\dfrac{x-y}{x+y}+\dfrac{xy}{x^2-y^2}$

(c) $\frac{1}{a-b} - \frac{1}{a+b}$ (g) $\frac{x}{2y} - \frac{x-y}{2(x+y)}$

(d) $\frac{a}{a-b} + \frac{b}{a+b}$ (h) $\frac{y-5}{y-6} - \frac{y-3}{y-4}$

(i) $\frac{x^2+y^2}{x^2-y^2} - \frac{y}{x-y} + \frac{x}{x+y}$

(j) $\frac{x+p}{x+q} + \frac{x+q}{x+p} - \frac{2(x-p)(x-q)}{(x+p)(x+q)}$

(k) $\frac{1}{t^2-6t+5} - \frac{2}{t^2+2t-3} + \frac{1}{t^2-2t-15}$

(l) $\frac{1}{n} + \frac{1}{n-1} - \frac{1}{n+1} + \frac{2n}{n^2-1}$

(m) $\frac{t}{t^2-1} + \frac{1}{t-1} - \frac{1}{t+1}$

9. If, following the suggestion of Stevinus, we call $\frac{1}{100}$, 10^{-2}, illustrate by using concrete numbers for a and b the following rules:

$$10^a \times 10^b = 10^{a+b}$$
$$10^a \div 10^b = 10^{a-b} \text{ (}a\text{ greater than }b\text{ or less than }b\text{)}$$
$$(10^a)^b = 10^{ab} = (10^b)^a$$

Test the general rules:

$$n^a \times n^b = n^{a+b}, \text{ etc.}$$

using numbers other than 10.

10. Applying the diagonal rule of p. 97 and the methods of the preceding examples, solve the following equations:

(a) $\frac{x+2}{3} - \frac{x+1}{5} = \frac{x-3}{4} - 1$

(b) $\frac{x+a}{a+b} + \frac{x-3b}{a-b} = 3$

(c) $\frac{1}{2x-3} + \frac{x}{3x-2} = \frac{1}{3}$

(d) $\frac{3}{x-1} - \frac{2}{x-2} = \frac{1}{x-3}$

(e) $\frac{2x-1}{x-2} - \frac{x+2}{2x+1} = \frac{3}{2}$

$$(f)\ \frac{x}{x-2} - \frac{x}{x+2} = \frac{1}{x-2} - \frac{4}{x+2}$$

11. (*a*) A man travelled 8 miles in $1\frac{3}{4}$ hours. If he rode part of the way at 12 miles per hour and walked the rest at 3 miles per hour, how far did he have to walk? (*b*) A train usually travels at 40 miles an hour. On a journey of 80 miles a stoppage of 15 minutes takes place. By travelling the rest of the way at 50 miles an hour the train arrives on time. How far from the start did the stoppage take place?

(*c*) A manufacturer cuts the price of his goods by $2\frac{1}{2}$ per cent. By what percentage must the sales increase after the cut to produce an increase of 1 per cent in the gross receipts?

12. Solve the following equations:

$$(a)\ x^2 + 11x - 210 = 0$$
$$(b)\ x^2 - 3x = 88$$
$$(c)\ 12x^2 + x = 20$$
$$(d)\ (3x + 1)(8x - 5) = 1$$
$$(e)\ 3x^2 - 7x - 136 = 0$$
$$(f)\ 2(x - 1) = \frac{x-1}{x+1}$$
$$(g)\ x(x - b) = a(a - b)$$
$$(h)\ \frac{1}{x+1} - \frac{1}{2+x} = \frac{1}{x+10}$$

13. (*a*) Find three consecutive whole numbers the sum of whose squares is 110.

(*b*) A square lawn with one side lying North and South has a border 6 feet wide taken off its south edge. It is then lengthened by adding a strip 3 feet wide on the West side. If the present area of the lawn is 500 square feet, what was the length of the side of the original lawn?

(*c*) The circumference of the hind wheel of a wagon is 1 foot more than the circumference of the front wheel. If the front wheel makes twenty-two more revolutions than the hind wheel in travelling a mile, find the radius of each wheel.

(*d*) In a right-angled triangle *ABC* the hypotenuse is 9 inches longer than one of the other sides *AC*. The remaining side is 2 inches less than half *AC*. Find the lengths of the sides.

14. The working of Example IV on p. 264 can be done in another way. The three statements are:

(i) $5s = f$ (ii) $2m + 8 = f + s$ (iii) $m = 100 - f$

From (i) substitute $5s$ for f in (ii) and (iii). Then:

$$2m + 8 = 5s + s$$

and

$$m = 100 - 5s$$

Rearranging, we get: $2m - 6s = -8$ . . . . . . (iv)

$$m + 5s = 100. \quad \ldots \quad \text{(v)}$$

We have here two unknown quantities and two equations. We can get rid of one unknown by combining the two equations. We combine the equations by subtracting the left-hand and right-hand sides separately. Before we do this we must decide which unknown we will try to get rid of or eliminate. The easiest one to eliminate will be m. To get m to vanish when we add the left-hand sides we must multiply (v) by 2. We then obtain:

$$2m - 6s = -8 \quad \ldots \quad \text{(vi)}$$

$$2m + 10s = 200 \quad \ldots \quad \text{(vii)}$$

Subtracting both sides, we get:

$$-16s = -208$$

or

$$16s = 208$$

$$s = 13$$

In the present instance this is all we need to know, but if we wanted to know m, we put $s = 13$ in either (iv) or (v), and we have then a simple equation which we can solve for m.

Usually it is necessary to multiply both equations by a different factor, as in the following example:

$$3x + 4y = 15 \quad \ldots \quad \text{(i)}$$

$$2x + 5y = 17 \quad \ldots \quad \text{(ii)}$$

To eliminate x multiply (i) by 2 and (ii) by 3:

$$6x + 8y = 30$$

$$6x + 15y = 51$$

Subtract both sides: $-7y = -21$

$$y = 3$$

Put the value obtained for y in (i):

$$3x + 12 = 15$$
$$3x = 3$$
$$x = 1$$

Check by putting the values obtained for x and y in (ii):

$$2 + 15 = 17$$

In order to find two unknown quantities we must have two equations which make two different statements about them.

The method for solving simultaneous equations can be summarized thus:

Step I. Arrange your equations so that like terms (e.g. x's) come under one another.

Step II. Decide which unknown to eliminate.

Step III. Multiply each term of the first equation by the coefficient of the selected unknown in the second equation and vice versa.

Step IV. Subtract the left-hand sides and the right-hand sides of your equations.

Step V. Solve the resulting simple equation.

Step VI. Substitute the value obtained in one of your original equations and thus find the second unknown.

Step VII. Check by substituting for both unknowns in the other original equation.

Equations involving three unknowns can be solved in a similar manner. Three equations are required. By taking them in two pairs we can eliminate one unknown from each pair, and thus obtain two equations with two unknowns.

$$2x + 3y = 4z$$
$$3x + 4y = 5z + 4$$
$$5x - 3z = y - 2$$

Rearrange:

$$2x + 3y - 4z = 0 \quad . \quad . \quad . \quad . \quad . \quad \text{(i)}$$
$$3x + 4y - 5z = 4 \quad . \quad . \quad . \quad . \quad . \quad \text{(ii)}$$
$$5x - y - 3z = -2 \quad . \quad . \quad . \quad . \quad \text{(iii)}$$

Eliminate y from (i) and (iii) by multiplying (iii) by -3:

$$2x + 3y - 4z = 0$$
$$-15x + 3y + 9z = 6$$
$$17x - 13z = -6 \quad . \quad . \quad . \quad . \quad \text{(iv)}$$

Eliminate y from (ii) and (iii) by multiplying (iii) by -4:

$$3x + 4y - 5z = 4$$
$$-20x + 4y + 12z = 8$$
$$23x - 17z = -4 \quad . \quad . \quad . \quad . \quad . \quad (v)$$

(iv) and (v) can now be solved for x and z as before. By substituting the values obtained in (i) y can be found and all three values checked from (ii) and (iii).

Solve the following equations:

(*a*) $x = 5y$
$x - y = 8$

(*b*) $3y = 4x$
$8x - 5y = 4$

(*c*) $x = 5y - 4$
$10y - 3x = 2$

(*d*) $60x - 17y = 285$
$75x - 19y = 390$

(*e*) $x + y = 23$
$y + z = 25$
$z + x = 24$

(*f*) $2x + 7y = 48$
$5y - 2x = 24$
$x + y + z = 10$

15. The following problems lead to simultaneous equations:

(*a*) The third term of an AP is 8, and the tenth term is 30. Find the seventh term.

(*b*) The fourth term of an AP is $-\frac{1}{8}$ and the seventh term is $\frac{1}{64}$. Find the first term.

(*c*) In a room, twice the length is equal to three times the breadth. If the room were 3 feet wider and 3 feet shorter it would be square. Find the measurements of the room.

(*d*) A hall seats 600 people, with chairs in rows across the hall. Five chairs are taken out of each row to provide a passage down the middle. In order to seat the same number of people, six more rows have to be added. Find the original number of chairs in a row.

(*e*) Two towns P and Q are 100 miles apart on a railway line. There are two stations R and S between them. The distance between R and S is 10 miles more than the distance between P and R, and the distance between S and Q is 20 miles more than the distance between R and S. Find the distance between R and S in miles.

16. Give the nth term of the following series:

(*a*) by using triangular numbers, and (*b*) by using vanishing triangles:

(i) 1, 6, 15, 28, 45
(ii) 1, 6, 18, 40, 75
(iii) 1, 20, 75, 184, 365, 636
(iv) 1, 7, 19, 37, 61, 91
(v) 1, 4, 10, 19, 31, 46
(vi) 1, 5, 13, 25, 41, 61

17. Expand the following:

(*a*) by the binomial theorem, and (*b*) by direct multiplication:

(i) $(x + 2)^5$
(ii) $(a + b)^3$
(iii) $(x + y)^4$
(iv) $(2x + 1)^6$
(v) $(3a - 2b)^4$
(vi) $(x - 1)^7$

Check your results by repeated division.

18. Using the binomial theorem, calculate to four decimal places:

(i) $(1{\cdot}04)^3$
(ii) $(0{\cdot}98)^5$
(iii) $(1{\cdot}12)^4$
(iv) $(5{\cdot}05)^3$

Things to Memorize

1. If $x^2 + ax + b = 0$:

$$x = \frac{a \pm \sqrt{a^2 - 4b}}{2a}$$

2. $a^0 = 1$, whatever a is, apart from 0:

$$a^{-n} = \frac{1}{a^n}$$

3. $$(a + b)^n = a^n + na^{n-1}b + \frac{n(n-1)}{1.2}a^{n-2}b^2 + \frac{n(n-1)(n-2)}{1.2.3}a^{n-3}b^3 \ldots + b^n$$

Number Games to Illustrate Algebraic Symbolism

1. Think of a number. Multiply the number above it by the number below it. Add 1. Give me the answer. The number you thought of is (the square root of the answer).

The explanation of this puzzle is as follows. Let the number thought of be a. The number above is $a + 1$, the number below is $a - 1$, and $(a - 1)(a + 1) = a^2 - 1$. Add 1 and you get a^2. By taking the square root of the number given you get the number thought of.

This sort of game can go on indefinitely, and will provide good practice in the use of symbols and factorization. Here are some more.

2. Think of a number less than 10. Multiply it by 2. Add 3. Multiply the answer by 5. Add another number less than 10. Give the answer. To tell what the numbers thought of are, subtract 15 from the answer. Then the digit in the tens place is the first number thought of, and the digit in the units place is the second number thought of.

In algebraic language:

$$(2a + 3)5 + b = 10a + b + 15$$

3. Think of a number. Square it. Subtract 9. Divide the answer by the number which is three more than the number you first thought of. Give the answer. What is the number thought of? Explain this in algebraic language.

4. Think of a number. Add 2. Square the answer. Take away four times the first number. Give the answer. What is the number thought of? Explain this, and make up some more for yourself.

5. Express in symbolic form the statement: if each of two different numbers is exactly divisible by a whole number x, their sum and their difference are each also divisible by x. Hence explore the following rules for the decimal notation:

(*a*) A number is divisible by 5 if its last digit is 5 or 0.

(*b*) A number is exactly divisible by 3 if the sum of its digits is exactly divisible by 3, and by 9 if the sum of its digits is exactly divisible by 9.

(*c*) A number is divisible by 4, if the last two; by 8, if the last three; by 16 if the last four digits are divisible by the same number (i.e. 4, 8, 16 as the case may be). N.B. Use (*b*) and (*c*) in the last example to find a rule for factors of 6 and 12.

(*d*) From the fact that 1,001 is exactly divisible by 7, 11, and 13, justify the rule that a number of six digits is divisible by one of these numbers if the difference between the number represented by the first three and last three digits (in their appropriate order) is so divisible. Extend the rule to any numbers of more than three digits.

CHAPTER 7

MATHEMATICS FOR THE MARINER

OUR LAST chapter dealt with the unique contribution of the Hindus and of their Moslem pupils to the mathematical lore of the Western World, partly during the period when Sicily and the greater part of what is now Portugal and Spain were under Moorish rule, and more especially during the three centuries during which Jewish scholarship kept alive the bequest of Moslem culture throughout Christendom as a whole. The topic of this chapter is not a unique contribution of the East, but the offspring of a fertile marriage between the mathematical astronomy of Alexandria and the new number lore. It is not essential to the understanding of any subsequent chapter; but it should be welcome to any reader with an interest in navigation by sea or by air, to any reader with an interest in the night sky and with opportunities for observing it in the open country, and to any reader with a taste for making simple instruments like those of Figs. 25 and 103. Before persevering, the reader who does not intend to skip it, should recall Exercise 10 at the end of Chapter 5, and Figs. 59 and 62 in Chapter 3.

Since our topic is the mathematical background of navigation and scientific geography, let us recall some of the facts familiar to master mariners long before the Christian era. Most important of such is that the difference between the *zenith distance* (Fig. 24) at transit of any fixed star at a particular place A and at a second place B is the same; and this difference is what we now call the difference of terrestrial latitude between A and that of B. For instance, the following figures give the transit z.d. of two bright stars at Memphis in Egypt and at London (UK):

	Sirius	Aldebaran	Latitude
Memphis	$46\frac{1}{2}°$ (S)	$14°$ (S)	$30°$ (N)
London	$68°$ (S)	$35\frac{1}{2}°$ (S)	$51\frac{1}{2}°$ (N)
Difference	$21\frac{1}{2}°$	$21\frac{1}{2}°$	$21\frac{1}{2}°$

This fact, equally true of the noon sun on *one and the same day of the year*, summarizes the essential link between astronomy and geography, between star map-making and earth map-making. It explains why we need to be able to locate a celestial body before we can locate our position on the earth's surface, and why, as we learned in Chapter 5, the invention of star maps preceded the introduction of maps of the habitable globe laid out gridwise by latitude and longitude. There we have seen that Hipparchus and his contemporaries laid the firm foundations; but the Moslem World made many improvements on what it had learned from the *Almagest* of Ptolemy. In the domain of practice, they greatly increased the number of localities for which it was possible to cite a reliable figure for longitude by observations (Fig. 111) of eclipses and conjunctions of planets – the only methods which Columbus, Amerigo Vespucci, and Magellan had at their disposal. From a theoretical viewpoint, they were able to simplify the Ptolemaic recipes in two ways. They adopted the Hindu (Fig. 84) trigonometry and they had at their disposal for construction of tables the immensely powerful resources of the Hindu numeral system.

When one speaks of a star map, one does not necessarily mean a picture. Just as the gazetteer citing latitude and longitude of localities at the end of the *Encyclopaedia Britannica* or *Chambers*' contains all the data for locating any town on a pictorial map, a corresponding catalogue of the co-ordinates of stars meets all the needs of the mariner. Those which Moslem science transmitted to the West were the necessary technical foundation for the Great Navigations, and we shall see later that map-making played a great part in stimulating the discovery of new mathematical tools in the period which followed them. In AD 1420 Henry, then Crown Prince of Portugal, built an observatory on the headland of Sagres, one of the promontories which terminate at Cape St Vincent, the extreme southwest point of Europe. There he set up a school of seamanship under one Master Jacome from Majorca, and for forty years devoted himself to cosmographical studies while equipping and organizing expeditions which won for him the title of Henry the Navigator. For the preparation of maps, nautical tables, and instruments he enlisted Arab cartographers and Jewish astronomers, employing them to instruct his captains and assist in piloting his vessels. Peter Nunes declares that the prince's master mariners were well equipped with instruments and those rules of astronomy and geometry 'which all map-makers should know'.

By 1483 Portuguese ships had occupied Madeira, rounded the

West Coast of Africa and built a formidable still intact fort at El Mina on the coast of what is now Ghana. In the expedition responsible for this, Columbus was a junior officer. Several circumstances thereafter conspired to exploit more fully than had been heretofore possible in the Moslem World the principles of scientific navigation, as transmitted by them. Too few of us realize that printing has a special relevance to the way in which spherical trigonometry was made adaptable to the mariner's needs. Printing from moveable type developed rapidly in the half century between the death of Henry the Navigator and the Columbian voyages, and made available to master mariners up to date forecasts of eclipses as well as star catalogues. Half a century later, portable clocks, sufficiently reliable over an interval of twenty-four hours, simplified the determination of longitude. Fertilized by the spread of the printed word, the new spectacle industry was flourishing and gave birth to the telescope within a century from the death of Columbus himself.

In what follows we shall see that one can make a star map of the dome of heaven in terms of:

(*a*) circles of *declination* in the same plane as circles of terrestrial latitude and as such referable to a plane (*the celestial equator*) in which the earth's equator lies (Fig. 105);

(*b*) circles of *right ascension* intersecting at the imaginary celestial poles in line with the earth's axis, and therefore comparable to circles of terrestrial longitude which intersect at the earth's poles (Fig. 109).

We shall also see that it is possible to construct such a map and to use it to locate stars, or to locate our position on the earth's surface thereby, without recourse to any mathematics other than very elementary geometry. In fact, such maps, the flat projection of which we call a *planisphere*, subserve all the requirements of modern navigation at sea or in the air. However, a map (see Fig. 110) constructed in this way eventually becomes unreliable. Indeed, we may credibly assume that the challenge which impelled Hipparchus to compute a table of trigonometrical ratios was the need to make one that is *durable*. The use of the word durable in this context calls for explanation of what we mean when we speak of the *fixed stars*. Though their rising and setting positions do not change appreciably in the course of a lifetime, their relations to the horizon change in the course of centuries owing to the phenomenon called *Precession of the Equinoxes*.

In modern terms, this implies that the earth's axis wobbles like that of a spinning-top when slowing down. In doing so, it traces a complete circle in about 25,000 years.

In these days of mechanized printing and computing, periodic revisions are of trivial inconvenience compared with the use of a system (Fig. 125) which Hipparchus introduced and Ptolemy elaborated. From their point of view, that of the earth-bound observer, precession signified a rotation of the plane of the ecliptic, i.e. the sun's annual track, at a fixed angle (about $23\frac{1}{2}°$) to the plane of the celestial equator. If mapped in planes of so-called *celestial latitude* parallel to the ecliptic plane, and in circles of so-called *celestial longitude* intersecting at the poles of the ecliptic on a line passing through the earth's centre at right angles to the ecliptic plane, our star map or catalogue has two advantages from the Alexandrian point of view. Celestial latitudes of fixed stars remain the same, and celestial longitudes change at a regular

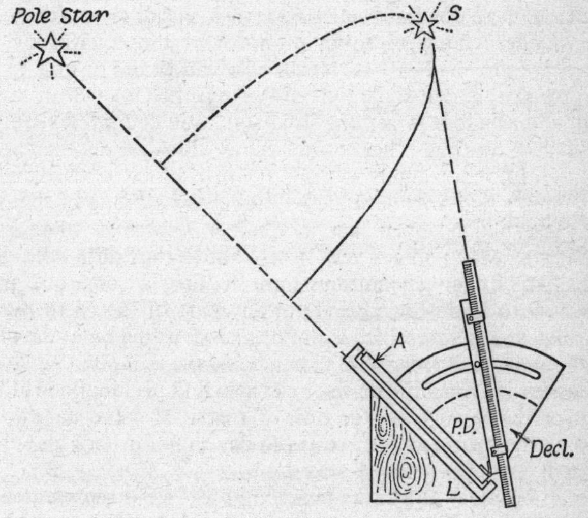

Fig. 103

A simple 'equatorial' made with a piece of iron pipe and wood. The pipe which serves for telescope rotates around the axis *A* fixed at an angle *L* (latitude of the place) due North. When it is clamped at an angle *PD* (the 'polar distance' of the star or 90° – declin) you can rotate it about *A* as the star (*S*) revolves, keeping *S* always in view.

speed of approximate 1° in seventy years. Thus no revision is necessary, if we note the date of compilation.

When hand copying was the only means of issuing tables and computation relied on the Attic Greek number signs, this was a comprehensible advantage. It is probable that Hipparchus used the simpler system to make a basic compilation, and converted the values into the alternative by a method which involves the formulae which we now call *solution of spherical triangles*. Ptolemy (Fig. 84) worked with *chords*. Moslem astronomers and geographers used *semichords* or as we now say *sines*; and their practice became the basis of celestial surveying in the Columbian era.

The Construction of a Star Map

On a country walk we recognize our whereabouts by familiar landmarks. The mariner's landmarks are the stars. If you were stranded on an uninhabited island after weeks of fever, you could tell whether you were south or north of the equator by the appearance of the heavens. North of it you would be able to see the pole star, and at some time of the year the Great Bear. South of it you would not see the pole star, but you would see the Southern Cross and other constellations which are not seen in the latitude of London and New York. If you live south of the equator, you cannot find your latitude by taking the altitude of the pole star, as explained on p. 141 in Chapter 3, because there is no bright star which shines very nearly above the South Pole. Star maps, which the Alexandrians were the first to construct, show you how to find your latitude in any part of the earth by the bearings of any star which is not obscured, as the pole star often is obscured, by clouds. With their aid and the accurate knowledge of Greenwich time, the mariner can also find his longitude at any hour of the night from the time of transit of some star. So he does not have to wait for noon each day to find where his ship is located.

In an ordinary globe, the position of a place is represented by where two sets of circles cross. One set of *great* circles, which all intersect at the poles, have the same radius as the globe itself. These are the meridians of longitude which are numbered according to the angles which they make with each other at the poles when the globe is viewed from above or below, and equally, of course, by the fraction of the circumference of the equator (or of any circle bounding a flat slab parallel to the slab bounded by the equator) cut off by them. Thus the angle between Long. 15° W and 45° W at the pole is also the angle formed by joining to its centre the ends of the arc which is $\frac{45° - 15°}{360°}$ or $\frac{1}{12}$ of the circumference of any circle parallel to the equator. The other set of circles are the small* circles of latitude. These are numbered by the angle between any point on the boundary, the earth's centre, and a point on the equator in the same meridian of longitude as the point of the boundary chosen. Thus the angle between Lat. 15° N and Lat. 45° N is the angle at the earth's centre made by an arc $\frac{1}{12}$ of the circumference of a circle of longitude (or of the equator itself). Circles of latitude form the boundaries of flat slabs ('planes') drawn at right angles to the polar axis round which the earth rotates, i.e. the axis round which the sun and stars appear to turn. Circles of longitude bound the rim of flat slabs ('planes') which intersect at the poles.

This way of mapping out the world arose from the discovery that the fixed stars all appear to rotate at the same rate in circular arcs lying in parallel layers at *right angles* to the line which joins the eye to the pole in the heavens (i.e. approximately speaking, to the pole star in the northern hemisphere). The elevation of this line above the horizon (which, as we saw on p. 141, is the latitude of the observer) is different at different places. It becomes greater as we sail due North, i.e. towards the pole star, and less as we go due South, i.e. away from the pole star. At any particular place at transit, one and the same star always has the same elevation above the pole, and hence makes the same angle with the zenith as it crosses the meridian. As illustrated above, the fact that the stars appear to rotate in circular arcs lying in parallel layers at right angles to the polar axis is shown by the fact that the difference between the meridian zenith distances of any star at two places is the same as the difference between the elevation of the pole as measured at the same two places, and it can be demonstrated directly by fixing a shaft pointing straight at the celestial pole and fixing a telescope (or a piece of steel tube) so that it can rotate at any required angle about the shaft itself as axis (Fig. 103). If the telescope is now clamped at such an angle as to point to a particular star, we can follow the course of the star throughout

* One circle of latitude (0°, the equator) is a great circle, i.e. a circle with the same radius as the sphere on which it is traced.

the night by simply rotating it on its free axis without lowering it or raising it. If it is set by very accurate modern clockwork so that it can turn through 360° in a sidereal day (i.e. the time between two meridian transits of any star whatever), it will always point to the same star. The fact that any particular star crosses the meridian exactly the same number of minutes after or before any other particular star suggested to the astronomers of

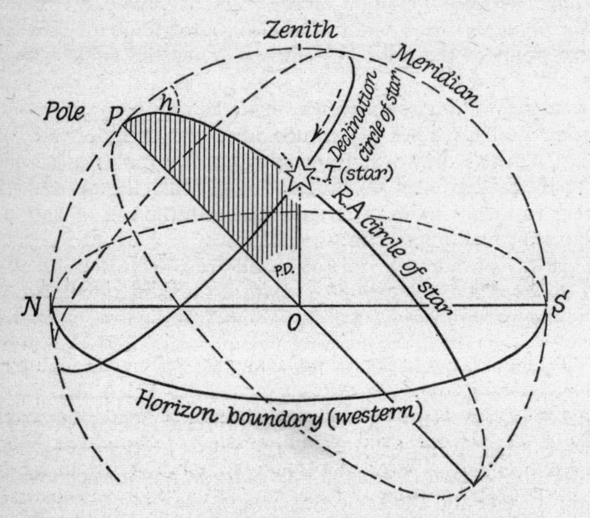

Fig. 104. Apparent Rotation of the Celestial Sphere

The position of a star (T) in the celestial sphere may be represented by a point where a small circle of declination which measures its elevation above the celestial equator intersects a large circle of Right Ascension (RA). All stars on the same declination circle must cross the meridian at the same angular divergence from the zenith and are above the observer's horizon for the same length of time in each twenty-four hours. The arc PT or flat angle POT measures the angular divergence of the star from the pole (polar distance) and is hence 90° – declin. All stars on the same great circle of RA cross the meridian at the same instance. The angle between two RA circles measures the difference between their times of transit. The angle h between the plane of the meridian and the RA circle of the star is the angle through which it has rotated since it crossed the meridian. If h is 15° it crossed the meridian 1 hour ago. So h is called the hour angle of the star. If the hour angle is h degrees, the star made its transit $h \div 15$ hours previously.

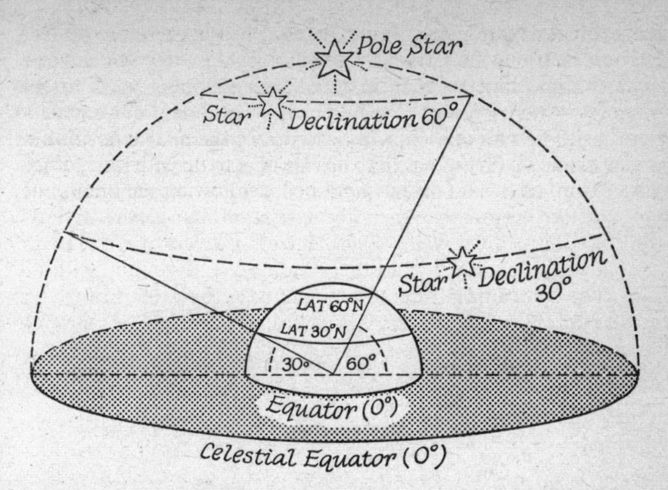

Fig. 105. Small Circles of Latitude on the Terrestrial and of Declination on the Celestial Sphere

antiquity that they were spaced out on great fixed circles all intersecting at the celestial poles.

Thus each star can be given a position in a great imaginary sphere of the heavens, fixed by the crossing of two circles (Fig. 104), a great circle of *Right Ascension* (comparable to our meridian of longitude) cutting all other similar circles at the celestial poles, and a small circle of *Declination* (comparable to our parallels of latitude), all lying on planes at right angles to the polar axis. A circle of declination is numbered in the same way as circles of latitude by the angle which the two points where it and the celestial equator are cut by its RA meridian make at the centre of the earth (Fig. 105). What we call the earth's polar axis is only the axis about which the stars appear to rotate, and what we call the plane of the earth's equator is therefore only the slab of the earth where it is cut by the plane of the heavenly equator. The line which joins an observer to the centre of the earth (see Fig. 59) goes through his zenith, cutting his parallel of latitude and a corresponding declination circle in the heavens. Any star on such a declination circle will pass directly above an observer anywhere on the corresponding circle of latitude once in twenty-four hours. Once the stars had been mapped out in this way to act as landmarks in navigation, it was therefore an easy step to map out the globe in a similar way.

It is important to bear in mind that mapping stars in this manner is merely a way of telling us in what direction we have to look or to point a telescope in order to see them. The position of a star as shown on the star map has nothing to do with how far it is away from us. If you dug straight down following the plumbline, you would eventually reach the centre of the earth; and the bottom of a straight well, as viewed from the centre of the earth, if that were possible, would therefore be exactly in line with the top. It has the same latitude and longitude as the latter, though it is not so far away from the earth's centre. The latitude or longitude

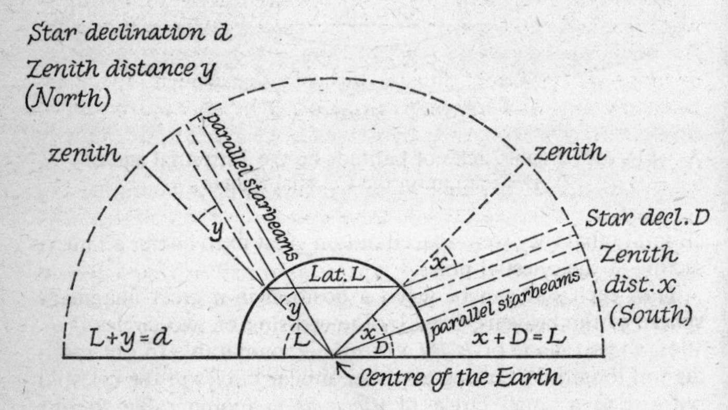

Fig. 106

Two stars on the northern half of the celestial sphere, one crossing the meridian north, the other south of the zenith. If the star crosses north: declin = lat. of observer + meridian zenith distance. If it crosses south: lat. of observer = declination + meridian zenith distance, i.e. declin = latitude of observer − meridian zenith distance.

of the bottom of a mine is the latitude and longitude of the spot where the line joining it with the centre of the earth continued upwards cuts the earth's surface. So the declination and right ascension of a star measure the place where the line joining the earth's centre and the star cuts an imaginary globe whose radius extends to the furthermost stars. In a total eclipse, the sun and the moon have the same declination and RA just as the top and bottom of a mine have the same latitude and longitude. This means that the sun and moon are directly in line with the centre of the earth like the top and the bottom of a mine.

At noon, the sun's shadow lies on the line joining the North and South points of the horizon. This is also the observer's meridian of longitude which joins the North and South Poles. The sun itself lies at noon on the imaginary circular arc or celestial meridian which passes through the celestial pole and the zenith. The celestial pole is in line with the earth's pole and the earth's centre. The zenith (Fig. 59), is in line with the observer and the earth's centre. So the celestial meridian, the earth's centre, and the observer's meridian of longitude all lie in the same flat slab of space. When, at the moment of its meridian 'transit', the sun or a star is at its highest point in the heavens, we can therefore apply the rules of flat geometry which show that the observer's latitude is connected in a very simple way with the declination of any heavenly body and its zenith distance. By observing the zenith distance of a star at meridian transit, we can at once find its declination, if we know our latitude, and conversely, if we have once determined its declination we can always determine our latitude. As there is always some star near the meridian this means that the mariner can determine his latitude at any time of the night with the aid of the star map or tables giving the declination of stars. For a star which transits north of the zenith (Fig. 107) in a northerly latitude the formula is simply

$$\text{Declination} = \text{Observer's latitude} + \text{meridian z.d.}$$

For a star which transits south of the zenith the formula is

$$\text{Declination} = \text{Observer's latitude} - \text{meridian z.d.}$$

The first formula holds for all situations, if we reckon zenith distances measured south of the zenith as negative and latitudes or declinations south of the terrestrial or celestial equators as negative.

We are now familiar with the fact that the observer's latitude is the same thing as the elevation of the pole above the horizon. When there was no bright pole star, as in Alexandrian times, this was done by taking the average altitude of any star near the pole when it was at its highest point above and its lowest point below the meridian and this could be done before people had begun to interpret what they were doing as we do.

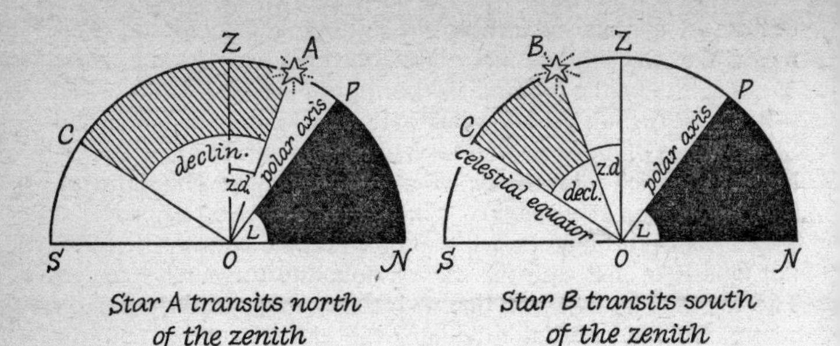

Fig. 107. Latitude, Declination, and Zenith Distance at Meridian Transit

As explained in Chapter 3 (Fig. 59), the observer's (*O*) latitude is the angle between the horizon and the celestial pole (*P*), i.e. $<PON$. Hence, the angle *ZOP* is 90° − Lat. (*L*).

For a star (*A*) which transits north of the zenith:

$$ZOP = AOP + \text{zenith distance (z.d.)}$$

Since the star's declination is the angle it makes with the celestial equator which is at right angles to the polar axis:

$$\begin{aligned} \text{declin.} &= 90° - AOP \\ &= 90° - (ZOP - \text{z.d.}) \\ &= 90° - (90° - \text{lat.}) + \text{z.d.} \end{aligned}$$

$$\therefore \quad \text{declin.} = \text{lat.} + \text{z.d.}$$

For a star (*B*) which transits south of the zenith:

$$\text{declin.} + \text{z.d.} = 90° - ZOP = 90° - (90° - \text{lat.})$$

$$\therefore \quad \text{declin.} = \text{lat.} - \text{z.d.}$$

Just as we can find the latitude of a place by observing the zenith distance of any star at meridian transit if we know its declination from the star map, we can also get our longitude by observing the times of meridian transit of any star, if we know its right ascension (Figs. 108–10) on the star map and have a chronometer set by standard time. Circles of terrestrial longitude are nowadays numbered in degrees from 0° to 180° east and west of the Greenwich prime meridian (0°). Right ascension is always reckoned *east* of the celestial meridian on which the 'First Point of Aries' lies. This is the Celestial Greenwich, and is denoted by the astrological symbol ♈. It is the position the sun occupies at the vernal equinox (March 21st). The celestial sphere appears to turn through 360° in 24 hours. So it is more convenient to number the circles of RA in hours and minutes, from 0 to 24 hours. Since it appears to rotate from east to west, a star which has the RA 13 hours 21 minutes (e.g. Spica in the constellation of Virgo) crosses the meridian 13 hours and 21 minutes after the sun crosses the meridian at the same place on the day of the vernal equinox. That is to say, it crosses the meridian at 1.21 am local time. If at the moment of transit, the Radio announcer gave Greenwich

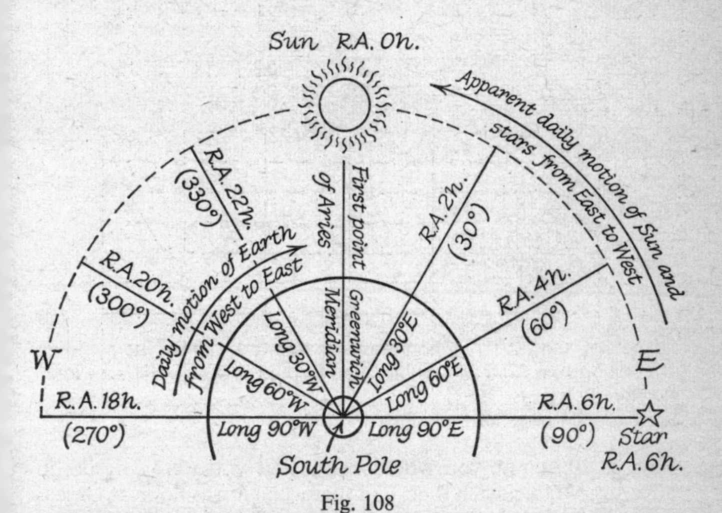

Fig. 108

Noon at Greenwich on March 21st. Showing relation of RA, longitude and time. At noon the RA meridian of the sun in the celestial sphere is in the same plane as the longitude meridian of the observer. If you are 30° W of Greenwich the earth must rotate through 30° or $\frac{1}{12}$ of a revolution, taking 2 hours before your meridian is in the plane of the sun's, or the sun must appear to travel through 30° before its meridian is in the same plane as yours. Hence, your noon will be 2 hours behind Greenwich.

A clock set by Greenwich time will record 2 pm when the sun crosses your meridian, i.e. at noon local time. If the date is March 21st when the sun's RA is O, a star of right ascension 6 hours will cross the meridian at 6 pm local time. If it crossed at 8 pm Greenwich time your clock would be 2 hours slow by Greenwich, so your longitude would be 30° W. The figure shows the anticlockwise rotation of the stars looking northwards, so the South Pole is nearest to you.

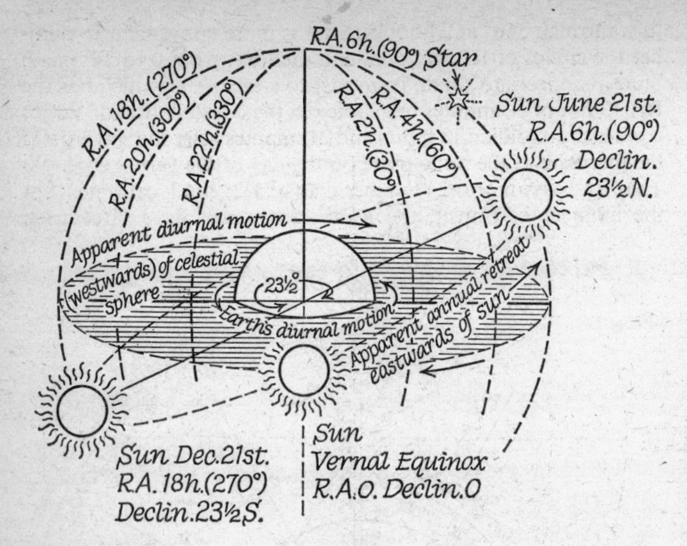

Fig. 109

The star shown (RA 6 hours) makes its transit above the meridian at noon on June 21st, and midnight on December 21st, i.e. it is a winter star like Betelgeuse.

time as 10.21 pm, you would know that you were 3 hours in advance of Greenwich time, and that when it was noon at Greenwich your local time was 3.0 pm. So your longitude (see Fig. 62, p. 143) would be $3 \times 15° = 45°$ East of Greenwich.

At other times of the year you would have to make allowance for the fact that the sun's position relative to the earth and fixed stars changes through 360° or 24 hours of RA in $365\frac{1}{4}$ days. The exact values of the sun's RA on each day of the year are given in the nautical almanacs, for the preparation of which modern Governments maintain their public observatories. Without tables, you can calculate the time reckoned from local noon roughly as follows (Fig. 110). Since the stars reach the meridian a little earlier each night, the sun appears to retreat farther east, and its RA increases approximately by $\frac{360}{365}$ or 1° or $\frac{1}{15}$ hour (4 minutes) in time units per day. Suppose, then, that Betelgeuse transits at a certain time on March 1st. The sun then has 20 days to re-

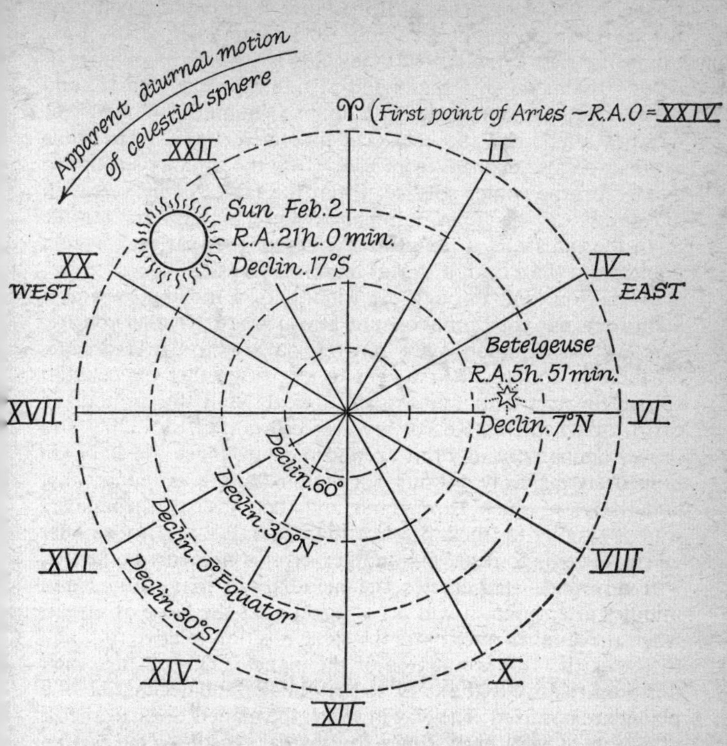

Fig. 110. Star Map (or Planisphere) to Illustrate Relation of Right Ascension to Local Time of Transit

If the sun's RA is x it transits x hours after the zero RA circle or First Point of Aries. If the star's RA is y, it transits y hours after ♈. The star therefore transits $y - x$ hours after the sun, i.e. its local time of transit is $(y - x)$. Hence:

Star's RA − sun's RA = Local time of transit

It may happen that the difference is negative, as in the example in the figure, the local time of transit being −15.9, i.e. 15 hours 9 minutes before noon, which is the same as 8 hours 51 minutes after noon (8.51 pm). The figure shows that the sun transits 3 hours before ♈, and the star 5 hours 51 minutes after, making the time of transit as stated. The orientation is the same as in Fig. 114.

treat east before it reaches the First Point of Aries, i.e. it will cross the meridian 80 minutes (1 hour 20 minutes) before the First Point of Aries. If the RA of Betelgeuse is 5 hours 51 minutes, it crosses the meridian 5 hours and 51 minutes later than ♈, and hence 1 hour + 20 minutes + 5 hours 51 minutes = 7 hours 11 minutes after noon. So the local time is 7.11 pm.* The sun's declination also changes from $+23\frac{1}{2}°$ on the summer solstice to $-23\frac{1}{2}°$ on the winter solstice. From the tables of its values in *Whitaker's* or the *Nautical Almanac* you can get your latitude from the sun's z.d. at noon on any day of the year, just as you could from the z.d. of a star at meridian transit.

When you look through the window of a moving train at a stationary one, you cannot be sure, as you pass it, whether you are, it is, or both you and it are moving relatively to the landscape. So there is nothing at first sight to tell us whether the celestial sphere revolves daily and the sun retreats annually around the earth, or whether the earth revolves on its own axis daily and moves annually in its orbit around the sun. Since the stars are immensely far away, all our calculations hold good either way, and the view which Hipparchus and the Moslem astronomers took is simpler for most practical purposes. But it is not simpler when we come to deal with another class of heavenly bodies.

If any fixed star crosses the meridian so many hours and minutes after noon, it will do so again after the lapse of a year, when the sun has once more the same position relative to it and to the earth. This is not true of the planets. They change their positions relative to the stars, so that the RA and declination of a planet are not fixed. The way in which the planets do so attracted attention in very early times for several reasons. One is that two of them are exceedingly conspicuous – much brighter than the most brilliant stars. Another is that they all move near the belt through which the sun and moon appear to revolve. When

* For simplicity no reference is made to the correction called '*the equation of time*' explained in the Almanacs. In Britain, radio time is *Greenwich mean time*, plus one hour in 'summer time', which differs from Greenwich local time by a few minutes varying in the course of the year. Mean time is used because the solar day (noon to noon) varies in length throughout the year, and a clock cannot be made to keep in step with it. So an average solar day is taken as the basis of time and the time of noon is shifted backwards and forwards by a certain number of minutes at different seasons. The difference is tabulated.

as at *conjunction*, they have the same RA as the moon, and if they also have the same declination within $\frac{1}{4}°$ they are eclipsed, *i.e.* 'occulted' by it. Such events were watched for in ancient times. When there were no portable timepieces, the sky, with the moon or a planet as hour-hand, was the only clock available for identifying simultaneous moments at places far apart on the earth's surface. Eclipses and conjunctions were first used, originally those involving the moon. In this way it could be seen that noon was not simultaneous at all places. When tables of the moon's position had been worked out, the angular distance of the moon from any bright star could also be used.

Before chronometers were invented, there was no practicable way of finding longitude at sea except by using such celestial signals. Most of the methods are complicated, but the one based on an eclipse of the moon is readily explained. This was how the pilots of Columbus, Amerigo Vespucci, and Magellan, trained in Moslem astronomy, were able to define the position of America on the world map (see Fig. 111). So knowing exactly where the planets are located was a matter of some practical importance in the period of the Great Navigations, when Copernicus and Kepler showed that their positions can be calculated more accurately and far more simply if we reject the common-sense view of the priestly astronomers.

To calculate the courses of the planets reliably the flat geometry which we have learned so far is quite useless, for a simple reason which you will see easily enough if you consider the behaviour of the planet *Venus*. On one night of the year the difference between the RA of the sun and that of any particular fixed star will be 12 hours. So it will cross the meridian at midnight. On such occasions the earth and the celestial poles will lie in the same plane as the sun and the star. Venus never crosses the meridian after dark. It is always seen setting just after sunset, or rising just before sunrise (Fig. 112). According to the modern Copernican view, this is merely because it revolves *between* the sun and the earth. As the earth turns away from the sun it becomes visible only after our meridian has passed it by. As the earth turns towards the sun it ceases to be visible with the naked eye, because the sun rises, before our meridian reaches it. Thus we cannot find the RA or declination of Venus and trace the way it changes by finding its meridian zenith distance and time of transit. We can calculate them by finding its zenith distance when its position in the celestial sphere has apparently revolved through

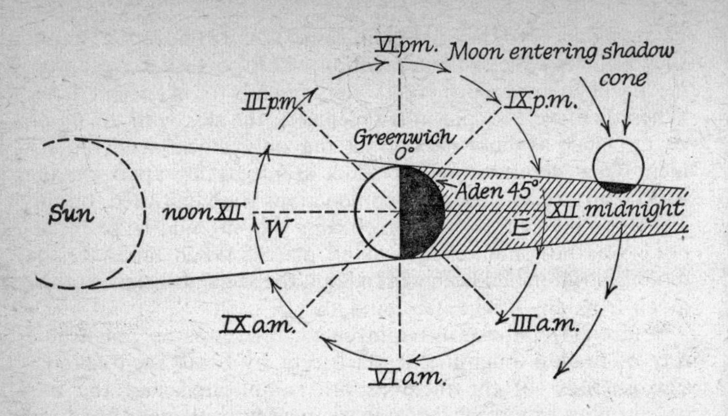

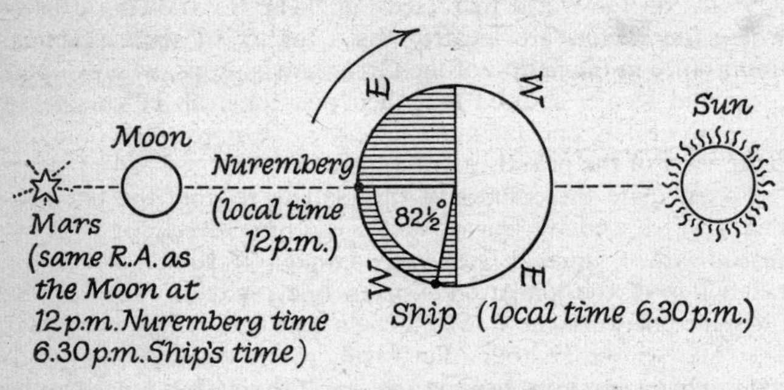

Fig. 111. Finding Longitude by Lunar Eclipse or Planetary Conjunction

Before the invention of the chronometer knowledge of local time and estimates of longitude depended upon observation of the interval between noon and some celestial signal such as an eclipse of the moon, or the occultation of a planet or star by the moon's disc. The interval was measured with hour glasses or crude clocks which would not keep time accurately over a long voyage.

Above: Moon entering eclipse at IX pm *local* time at Aden and nearly midnight local time at Calcutta. South Pole nearest reader.

Below: How Amerigo Vespucci found his longitude by a conjunction of the moon and Mars.

When the RA of the moon is the same as that of the planet they are

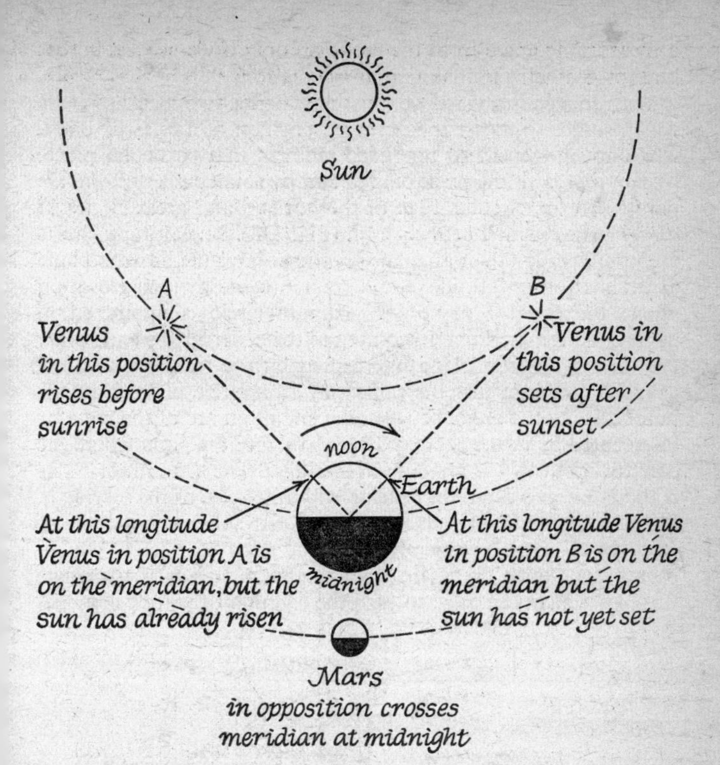

Fig. 112

The inferior planets, Mercury and Venus, have always passed the meridian when they become visible or have not yet reached it when they cease to be visible. Earth seen with South Pole towards the reader.

[*Caption continued*]

said to be in conjunction. If their declination is also the same, the planet will be occulted by the moon's disc. If the declination differs by a small angle it is still possible to gauge when the RA of the two is the same by the naked eye. The exact moment of the conjunction can be determined by successive observations of their local co-ordinates (azimuth and zenith distance). From these the RA can be calculated by the spherical triangle formula given below.

a measurable angle from the meridian only if we use a different kind of geometry in making our calculations.

With this end in view, we must now define two measurements which suffice to locate the observed position of a celestial body, when not in transit. At any fixed moment in a particular place, we can represent the position of a star by small circles of *altitude* parallel to the circular edge of the horizon and great circles of *azimuth* intersecting at the zenith (Fig. 113). The altitude circles are numbered by their angular elevation above the horizon plane as declination or latitude circles are numbered by their elevation above the equator plane. An azimuth circle is numbered in degrees off the meridian by joining to the observer the ends of an arc on the horizon plane intercepted by the meridian and the azimuth circle, in just the same way as a circle of longitude is numbered by the angle between the end of an arc of the equator intercepted by it, the centre of the earth, and the point where the equator is cut by the Greenwich meridian. The azimuth of a star is therefore its East/West bearing with reference to the meridian. If you have mounted your home-made astrolabe or theodolite of Fig. 103 to revolve vertically on a graduated base set so that 0° points due south or north, the azimuth of a star is the angle through which you have to turn the sighting tube (or telescope)

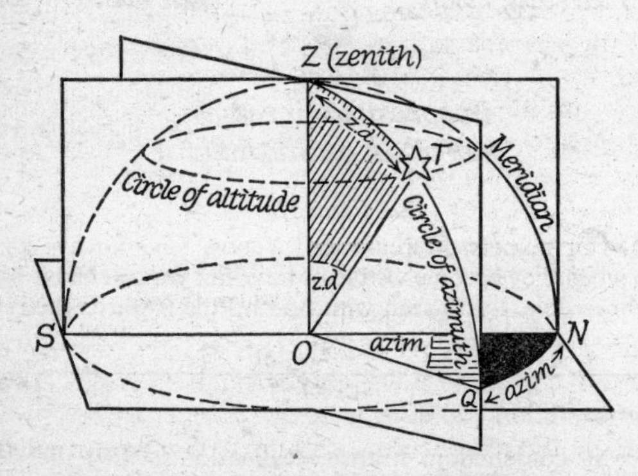

Fig. 113. Local Co-ordinates of a Star

The horizon bearing or altitude is 90° – z.d. The zenith distance z.d. is measured by the arc *TZ* or the flat angle *ZOT* in the azimuth plane. The meridian bearing or azimuth is the arc *NQ* which in degrees is the flat angle *NOQ*, or the angle between the meridian plane *NZS* and the azimuth plane *ZOQ*.

on its base, and the altitude is obtained by subtracting from 90° the zenith distance. If the protractor is numbered reversibly from 0° to 90° and 90° to 0°, you can, of course, read off the altitude at once.

If you point your telescope towards a star and turn it freely a little later to the place where the star is then seen, you sweep out an arc on the celestial sphere like the course of a ship on the high seas. If you think about the way in which ships actually sail, you will see that they never sail in the straight lines of Euclid's geometry. They have to sail in arcs of circles on the curved surface of the globe. The shortest arc between two points is the one which is least bent, hence the one with the greatest radius, which is the radius of the globe itself. So the shortest distance between two points for purposes of navigation is not the Euclidean straight line, but the 'great' circle which passes through them.

If two ships travel from *A* to *B*, one in the most direct course, the other without changing its latitude till it gets to the longitude meridian of *B*, and then along the latter until it reaches *B*, the two courses form a three-sided figure, of which the sides are circular arcs. Similarly, when a star appears to move along its declination circle in the heavens we have to rotate our eye horizontally along an arc of altitude and vertically along an arc of azimuth to follow its course. The apparent movement of the star and the movement of our eye or our telescope trace out a triangle with curved sides in the celestial dome. In reaching any point in the heavens a star has to rotate about the polar axis through a certain angle away from the meridian, while our telescope has turned through a definite angle about the zenith axis, and the star has turned through a certain angle along its declination circle, while our telescope has been tilted through a definite angle to the horizon plane. Since the polar axis makes a fixed angle to the horizon plane (Fig. 59), we should expect that all these quantities are connected, just as a ship's shortest course is connected with the latitude of its port of departure and destination. For both problems we need to

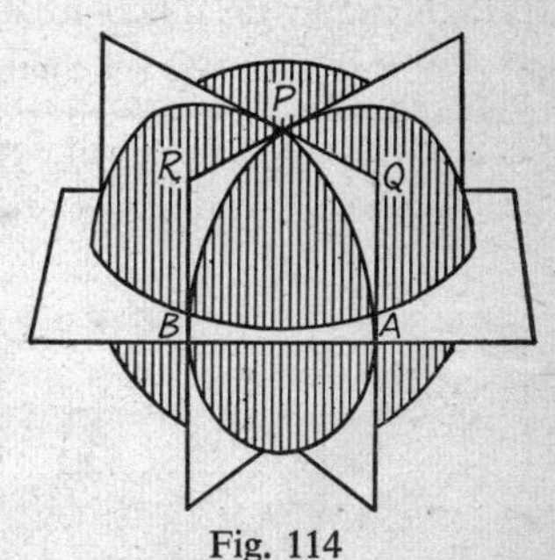

Fig. 114

Intersecting Flat Planes on which three Great Circles of a Sphere lie.

know what sort of connections to expect between the parts of figures which have curved sides.

Spherical Triangles

Fig. 114 shows a globe in which three flat planes have been sliced through two meridians of longitude (along *PA* and *PB*) and through the equator (*AB*). Each of these planes cuts the surface of the terrestrial sphere in a complete circle, the centre of which is the centre of the sphere. Where they intersect on the surface, they make the corners of a three-sided figure of which the sides are all arcs of *great circles*, i.e. circles with the same centre and the same radius as the sphere itself. Such a figure is called a spherical triangle. It has three sides, *PA*, *PB*, and *AB*, which we shall call *b* (opposite *B*), *a* (opposite *A*) and *p*. It has also three angles *B*, *A*, and *P* (*PBA*, *PAB*, and *APB*). What you already know about a map will tell you how these angles are measured. The angle *APB* is simply the difference of longitude between the two points *A* and *B* marked on the equator, and it is measured by the inclination of the two planes which cut from pole to pole along the axis of the globe. You will notice therefore that, since the earth's axis is at right angles to the equator plane, the plane of *AB* is at right angles to the plane of *PA* and of *PB*; and since we measure angles where two great circles traced on a sphere cut one another by the angle between the planes on which the great circles themselves lie, the spherical angle *PAB* is a right angle, and so is *PBA*. Thus the three angles of the spherical triangle are together greater than two right angles, an important difference between spherical triangles and Euclid's triangles. In practice, of course, it is a lot of trouble to draw a figure like Fig. 114 showing the planes of the circles traced on the surface of a sphere, and the angle between two intersecting *great* circles (i.e. like the equator or meridians of longitude with centre at the sphere centre). So we

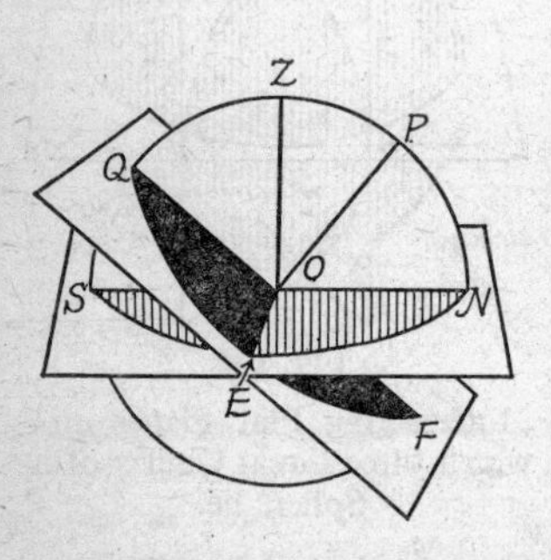

Fig. 115. The angles between two arcs (*QE* and *SE*) of great circles on sphere is the angle (*ZOP*) between the poles of great circles *Z* and *P*.

measure them in one of three other ways which involve only flat geometry, which we have already learnt. These are:

(*a*) The geometry method: The angle BPA between the spherical sides PB and PA is the same as the flat angle RPQ *between the tangents* RP and QP which touch PB and PA at their common point, i.e. the 'pole' P of the equatorial circle.

(*b*) The geography method: Remembering that BPA is simply the number of degrees of longitude between A and B, you will see that it is simply the number of degrees in the arc cut off where the great circles on which PB and PA lie intersect any circle of latitude, i.e. any circle of which the plane is at right angles to the line joining the two poles where the great circles intersect above and below.

(*c*) The astronomy method: This is illustrated in the next figure (Fig. 115). It shows the intersection of the plane of the celestial equator $FEQO$ with the plane of the horizon $NESO$ on the celestial sphere, on the surface of which lie the pole of the celestial equator at P and the pole of the horizon circle, i.e. the zenith at Z. The angle QES between the arcs QE and SE is the angle QOS between their intersecting planes, and

$$QOS = 90° - QOZ$$
$$POZ = 90° - QOZ$$

So the angle between two spherical arcs is the angle between the poles of the great circles on which they lie.

The next thing to learn about spherical triangles is how to measure their sides. A spherical triangle is not primarily a figure which represents the distances between three objects at its corners but merely the difference in direction of three objects as seen from one centre. The side of a spherical triangle is the angle through which the eye or the telescope has to be turned to get directly from one corner to the other. The sides of spherical triangles, like their angles, are always measured in degrees or radians. Thus in Fig. 114 the side a (PB) and the side b (PA) are each equivalent to the latitude of the north pole $\left(90° \text{ or } \frac{\pi}{2} \text{ radians}\right)$, and the side p (AB) is the difference of longitude between A and B, and just happens in this special case to be the same as the angle P. In Fig. 115 the points Q, E, S form the apices of a spherical triangle on the celestial sphere. Two of the sides s (QE) and q (SE) are right angles. The third side e (SQ) is also equivalent to the angle E, which is equal to the flat angle POZ. The spherical angles Q and S are right angles, and you see again that the three angles of a spherical triangle make up more than two right angles.

A moment ago, we saw that the latitude and longitude at the top and bottom of a mine are the same, because latitude and longitude measure the direction of an object as it would be seen from the centre of the earth, and, as seen from the centre of the earth, the top and bottom of the mine lie in exactly the same direction. Thus any three points in space can be represented by the corners of a spherical triangle traced on the surface of a sphere which encloses all three. If all three points happen to be at the same distance from the centre of observation, we can express the sides of such a triangle in actual lengths instead of degrees or radians.

The way in which this is done can be illustrated by a simple calculation which does not involve a spherical triangle, since parallels of latitude (other than the equator) are not great circles. If we know the number of degrees or radians in the angle joining the ends of an arc to the centre of the circle on which it lies, we also know its length. A difference of longitude of one degree anywhere along the equator is one three hundred and sixtieth of the earth's circumference, i.e. if the earth's radius (R) is taken as 3,960 miles, it is

$$2\pi \times 3{,}960 \div 360$$
$$\simeq \tfrac{44}{7} \times 11 = 69 \text{ miles (approximately)}$$

Without any allowance for the slight flattening of the earth at the Poles, this is equivalent to a degree of latitude measured along a meridian of longitude or a degree measured along any great circle at the earth's surface.

A difference of longitude equivalent to one degree measured along any other parallel of latitude is easily determined as illustrated in Fig. 116, in which AB is n degrees of longitude measured along the latitude circle L and DC is n degrees along the equator. The circumference of the equatorial circle of which DC is an arc is $2\pi \,.\, OC$. One degree corresponds to $2\pi \,.\, OC \div 360$ and n degrees to $2\pi n \,.\, OC \div 360$. Hence:

$$DC = 2\pi n \,.\, OC \div 360 \text{ and } OC = 360 \,.\, DC \div 2\pi n$$

Likewise:

$$AB = 2\pi n \,.\, QB \div 360 \text{ and } QB = 360 \,.\, AB \div 2\pi n$$

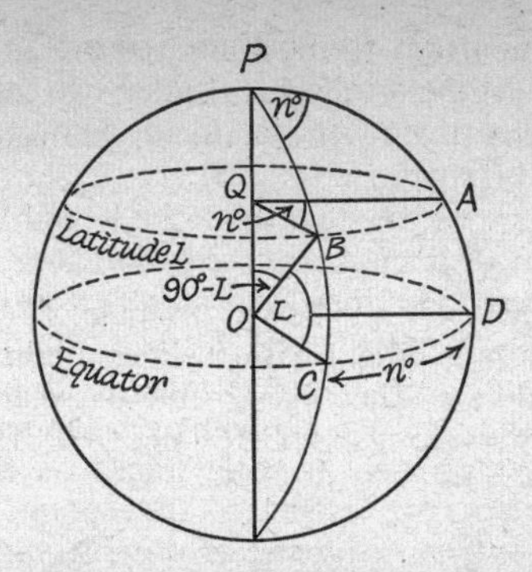

Fig. 116. How to Calculate the Length of n Degrees of Longitude Measured along any Parallel of Latitude

CD are two points on the equator, *AB* two points on a parallel of latitude, *PO* is half the earth's axis, *COD* the plane of the equator, *QAB* the plane of the latitude of *A* and *B*, and *OB* is the earth's radius (*R*).

The angle $QOB = 90° - L$, and since the plane of *QAB* is at right angles to the polar axis, *QOB* is a right-angled triangle in which $OB = R = OC$ and

$$\sin QOB = QB \div OB$$
$$\therefore \quad \sin (90° - L) = QB \div OC$$
$$\therefore \quad \cos L = QB \div OC$$
$$\therefore \quad QB = OC \cos L$$
$$360 \,.\, AB \div 2\pi n = 360 \; DC \,.\, \cos L \div 2\pi n$$
$$\therefore \quad \text{arc } AB = \text{arc } DC \cos L$$
$$= n \times 69 \times \cos \mathrm{L} \text{ miles (approximately)}$$

Solution of Spherical Triangles

General expressions for 'solution' of spherical triangles analogous to those given in Chapter 5, pp. 224–6, can be found, and are used constantly in astronomy and mathematical geography for the reasons already given. They depend on the corresponding formulae for flat triangles, and the only difficulty in understanding them is due to the fact that it is not easy to make clear diagrams of solid figures on a flat page.

The most important formula for solving spherical triangles tells us how to get the third side (a) when two sides (b and c) and the angle between them (A) are already known. The analogous formula for flat triangles is (p. 224):

$$a^2 = b^2 + c^2 - 2bc \cos A$$

Fig. 117 shows you a spherical triangle ABC formed by the intersection of three circles with their common centre O at the centre of the sphere. The edges of the flat planes in which the sides a, b, c lie meet along OA, OBQ, and OCP. The edges AQ and

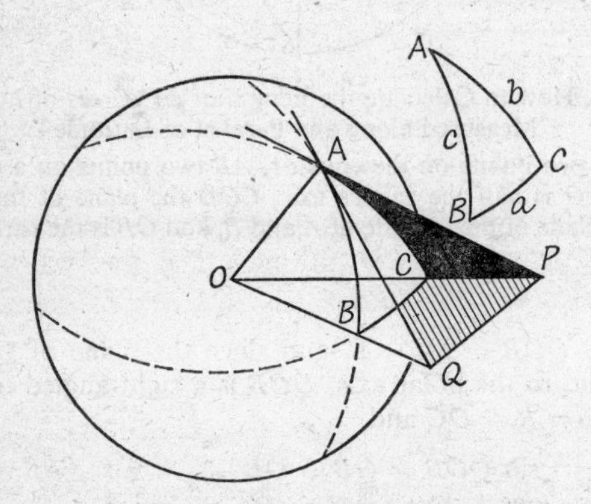

Fig. 117. Solution of Spherical Triangles

AP just graze the great circles of the arcs c and b at A, i.e. AQ is the tangent to c and AP to b. So OAQ and OAP are really right angles, though it is impossible to draw them as such in the flat. The edges of three planes in which the arcs a, b, c lie forms a flat triangle PAQ, of which the apical angle PAQ is equivalent to the angle A of the spherical triangle according to the definition of how we measure the angles of a spherical triangle.

To get a clear picture of the way in which the parts of the spherical triangle are related to the parts of the four flat triangles which form the sides of a little pyramid in which it lies, cut out a paper model drawn as in Fig. 118 and fold the parts together along the three edges where you are told to do so. All the facts required are indicated in the figures. When you have made the model, take it to pieces as in Fig. 119 so that you can see each

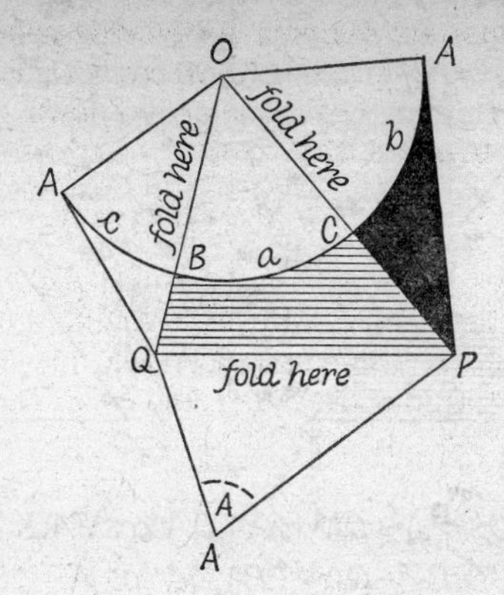

Fig. 118. Solution of Spherical Triangles (*continued*)

triangle in the position which is most familiar. You then see, as indicated in Fig. 119, by the rule for flat triangles:

$$PQ^2 = PO^2 + QO^2 - 2PO \,.\, QO \cos a$$
$$PQ^2 = PA^2 + QA^2 - 2PA \,.\, QA \cos A$$

$$\therefore (PO^2 - PA^2) + (QO^2 - QA^2) - 2PO \,.\, QO \cos a + 2PA \,.\, QA \cos A = 0$$

$$\therefore 2PO \,.\, QO \cos a = 2AO^2 + 2PA \,.\, QA \cos A$$

Divide through by $2PO \,.\, QO$, then

$$\cos a = \frac{AO}{PO} \cdot \frac{AO}{QO} + \frac{PA}{PO} \cdot \frac{QA}{QO} \cos A$$
$$= \cos POA \cos QOA + \sin POA \sin QOA \cos A$$
$$= \cos b \cos c + \sin b \sin c \cos A$$

The formula for getting the third side (a), when you know the other two (b and c) and the included angle A, is, therefore:

$$\underline{\cos a = \cos b \cos c + \sin b \sin c \cos A}$$

Of course, once you know the way this is pieced together you just learn it by heart, or refer to a book when you want to use it.

We can adapt Fig. 118 to demonstrate another formula directly; but should the reader find it difficult to memorize, there there is no need for alarm. We can also do so by recourse to

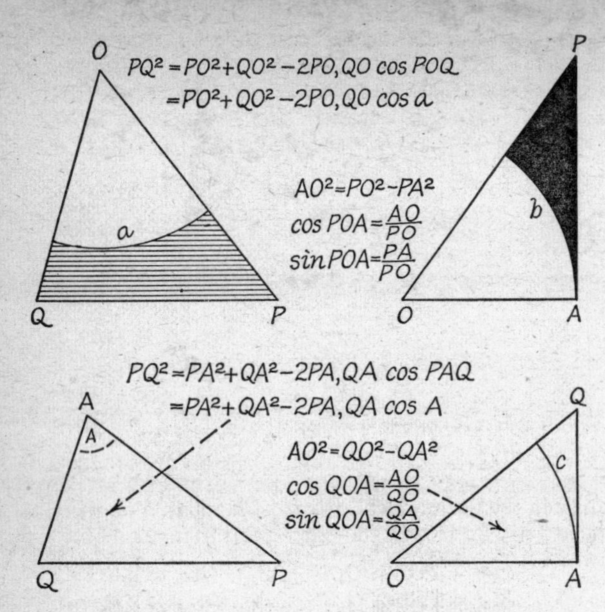

Fig. 119. Solution of Spherical Triangles (*continued*)
The Cosine Formulae

the last one cited, as shown below (p. 327). As we unfold the model in Fig. 117 the moving point A_1 revolves about the line OO describing a circular arc in a plane at right angles to this line OQ and to the plane OQP till it comes to the position A (Fig. 120). Similarly it describes a circular arc about the line OP till it comes to the position A_2. So

$$A_1UO = 90° = A_2VO$$

When the figure is folded back into its original position A is directly above a point W in the plane OQP and W is the point where the generating radii A_1U and A_2V meet. From the first figure we see that:

(i) $$\frac{A_1U}{A_2O} = \sin c; \quad \text{and} \quad \frac{A_2V}{A_1O} = \sin b$$

$$A_1U = AO \sin c; \quad \text{and } A_2V = AO \sin b$$

The angle AUW is the angle between the planes which include the arcs AB and BC respectively. Hence AUW is equivalent to $\angle B$ of the spherical triangle. From the second figure we see that:

$$\text{(ii)} \qquad \frac{AW}{A_1U} = \sin B; \quad \text{and similarly } \frac{AW}{A_2V} = \sin C$$

$$\therefore \quad AW = A_1U \sin B; \quad \text{and } AW = A_2V \sin C$$

Combining (i) and (ii), we get:

$$AO \sin c \sin B = AW = AO \sin b \sin C$$

$$\therefore \quad \frac{\sin c}{\sin C} = \frac{\sin b}{\sin B}$$

In the same way each is equal to sin a/sin A.

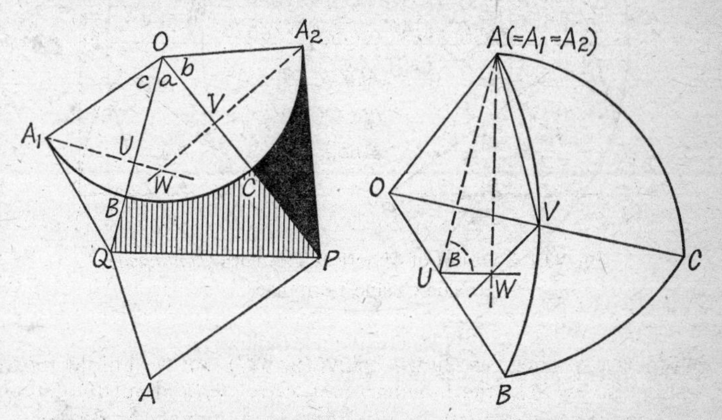

Fig. 120. Solution of Spherical Triangles – The Sine Formulae

Calculating a Ship's Direct Course

The reader may not want to go through any more of this sort of thing until able to see what earthly good comes out of it. The first example is of a kind which you can amuse yourself with for hours, if you have any map which gives the distances of sea routes on the dotted line indicating their courses. Remember, if you do so, that they are usually given in sea miles (60 sea miles = 1° of a great circle), and take into account the fact that a ship has to steer round a few corners before it gets away from, or when it is approaching, land. Thus the map gives for the course from Bristol to Kingston in Jamaica 4,003 sea miles, about 25 miles in excess of the calculation which follows. This makes no allowance for getting out of the Bristol Channel and into port (see Fig. 121).

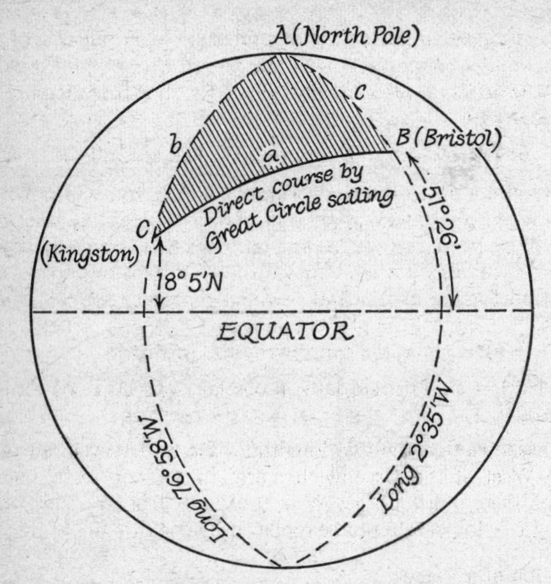

Fig. 121. Great-circle Sailing

The details are explained in the text.

The latitude of Bristol is 51° 26′ N of the equator, and therefore 38° 34′ from the pole, along the great circle of longitude 2° 35′ W. The latitude of Kingston is 18° 5′, i.e. it is 71° 55′ from the pole along the great circle of longitude 76° 58′ W. The arc joining the pole to Bristol (*c*), the arc joining the pole to Kingston (*b*), and the arc (*a*) of the great circle representing the course from Bristol to Kingston form a spherical triangle, of which we know two sides (*b* and *c*), and the included angle *A*, which is the difference of longitude 76° 58′ − 2° 35′ = 74° 23′ between the two places. So we can find *a* by putting:

$\cos a = \cos 71°\ 55' \cos 38°\ 34' + \sin 38°\ 34' \sin 71°\ 55' \cos 74°\ 23'$

From the tables:

$$\cos a = 0{\cdot}3104 \times 0{\cdot}7819 + 0{\cdot}6234 \times 0{\cdot}9506 \times 0{\cdot}2692$$
$$= 0{\cdot}4022$$

Thus a is approximately $66\frac{1}{3}°$ of a great circle, i.e. a circle of the earth's complete circumference. The length of 1° of the earth's circumference is approximately 69 miles. So the distance is approximately

$$66\tfrac{1}{3} \times 69 = 4{,}577 \text{ land miles } (3{,}980 \text{ sea miles})$$

Before going on to the next example, which you will not find difficult when you have worked out a few examples like this with the aid of the list of longitudes and latitudes of ports given at the end of most atlases, notice that you need not trouble to do the subtraction for the arcs b and c which represent the *polar distances* (90° – lat.) of the places. Since $\sin (90° - x) = \cos x$ and $\cos (90° - x) = \sin x$, we can rewrite the formula:

$$\cos (\text{dist.}) = \sin \text{lat.}_1 \sin \text{lat.}_2 + \cos \text{lat.}_1 \cos \text{lat.}_2 \cos (\text{long.}_1 - \text{long.}_2)$$

This assumes that the two longitudes are both measured East or both West of the Greenwich meridian. Of course, if one is measured East and the other West the angle A is their sum, and $\cos (\text{long.}_1 - \text{long.}_2)$ should be replaced by $\cos (\text{long.}_1 + \text{long.}_2)$.

Declination of a Planet

As we have seen, the declination (or RA) of a planet like Venus or Mercury cannot be obtained by meridian observations, and the declination (or RA) of an outer planet like Mars or Jupiter can be determined by meridian observations only at periods when it transits during the hours of darkness. So it is impossible to trace the positions of any planet relative to the fixed stars in all parts of its course unless we can determine their declinations and right ascensions by some other method. It is easy to obtain the former in precisely the same way as we get a ship's course in great-circle sailing.

The local position of every star at any instant can be placed at the corner of a spherical triangle (Fig. 122) like the Bristol–Kingston triangle of the last example. One side (b), like the polar distance of Kingston, is the arc between the celestial pole and the zenith along the prime meridian. The elevation of the pole is the latitude (L) of the observer. So $b = 90° - L$. One side (c) is the

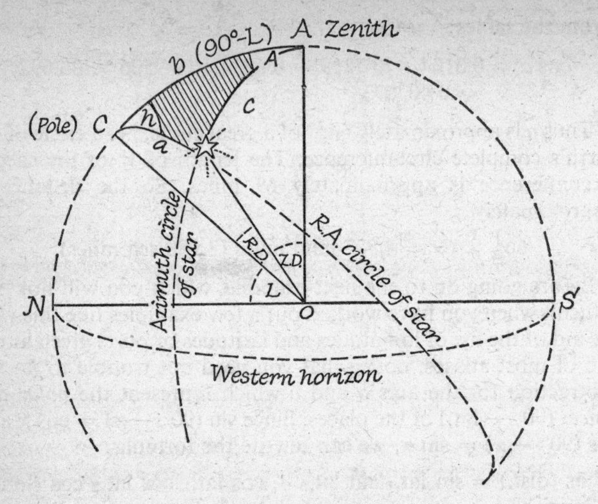

Fig. 122. The Star Triangle

Compare with Figs. 104 and 113.

arc between the star and the zenith on its own great circle of azimuth. This arc is its zenith distance (c = z.d.). The angle A between its azimuth circle and the prime meridian which cuts it at the zenith is its azimuth (A = azim.). Between the ends of these two arcs passes the great circle of right ascension which joins the star to the celestial pole, and the length of the arc between the star and the pole on its RA circle is its polar distance. Since the celestial pole is 90° from the celestial equator, the star's polar distance, which is the third side of our spherical triangle, is the difference between one right angle and its declination ($a = 90° -$ declin.). Applying the formula we have

$$\cos(90° - \text{declin.}) = \cos(90° - \text{lat.}) \cos \text{z.d.} + \sin(90° - \text{lat.}) \sin \text{z.d.} \cos \text{azim.}$$

This may be rewritten

$$\sin(\text{declin.}) = \sin \text{lat.} \cos \text{z.d.} + \cos \text{lat.} \sin \text{z.d.} \cos \text{azim.}$$

In applying this you have to remember that we have reckoned azimuth from the north point. If the star transits south, the azimuth may be greater than 90°, and we reckon it from the south point. Since $c^2 = a^2 + b^2 - 2ab \cos C$ becomes

$$c^2 = a^2 + b^2 + 2ab \cos (180° - C)$$

when C is greater than 90° (*See* pp. 226 and 239), the formula for the spherical triangle becomes

$$\text{sin declin.} = \text{sin lat. cos z.d.} - \text{cos lat. sin z.d. cos azim.}$$

This means that if you know the azimuth and zenith distance of a heavenly body at one and the same time, you can calculate its declination without waiting for it to reach the meridian. You cannot as easily get the latitude of the place from an observation on a star of known declination directly by using the same formula; but if you have the z.d. and azimuth of any two stars taken at one and the same place, you can calculate its latitude provided you have an almanac or star map to give the declination of the stars. In general, the arithmetic takes less time than waiting about for one bright and easily recognized star to cross the meridian.

Another application of the formula just derived gives the direction of a heavenly body when rising or setting at a known latitude, or conversely the latitude from the rising or setting of a star. At the instant when a heavenly body is rising or setting its zenith distance is 90°. Since $\cos 90° = 0$ and $\sin 90° = 1$, the formula then becomes

$$\text{sin declin.} = \text{cos lat. cos azim.}$$

On the equinoxes when the sun's declination is 0°,

$$\text{cos lat. cos azim.} = 0$$
$$\therefore \quad \text{cos azim.} = 0$$
$$\therefore \quad \text{azim.} = 90°$$

That is to say, the sun rises due East and sets due West in all parts of the world that day. To find the direction of the rising or setting sun at latitude $51\frac{1}{2}°$ N (London) on June 21st, when the sun's declination is $23\frac{1}{2}°$ N, we have only to put

$$\sin 23\tfrac{1}{2}° = \cos 51\tfrac{1}{2}° \text{ cos azim.}$$

From the tables, therefore:

$$0{\cdot}3987 = 0{\cdot}6225 \text{ cos azim.}$$
$$\therefore \quad \text{cos azim.} = 0{\cdot}6405$$
$$\text{azim.} = 50\tfrac{1}{6}°$$

Thus the sun rises and sets $50\frac{1}{6}°$ from the meridian on the north side, or $90° - 50\frac{1}{6}° = 39\frac{5}{6}°$ north of the East or West point. Conversely, of course, you can use the observed direction of rising and setting to get your latitude.

Another Thing About Spherical Triangles

How the declination of a planet changes is much less interesting than how its RA changes, because the way in which its RA changes is so much easier to explain if we assume that the earth and the planets revolve around the sun, as Aristarchus believed and Copernicus taught. This is not a book on astronomy, so we cannot spend any time on how Copernicus was led to this conclusion, but you should be able to worry it out from a book on astronomy if you understand clearly how the position of a planet is determined.

If you look at the star triangle shown in Fig. 122, you will see that the angle C between the arc which represents the star's polar distance and the arc b, which is the angle between the observer and the earth's pole (90° – lat.), is the angle through which the star has rotated since it was last on the meridian. Since the celestial sphere appears to rotate through 360° in 24 hours, i.e. 15° an hour, this angle C is sometimes called the hour angle of the star, because you can get the time (in hours) which has elapsed since the star made its transit by dividing the number of degrees by 15. If you know when the star crossed the meridian by local time, you also know how long has elapsed since the sun crossed the meridian because time is reckoned that way, and if you know the sun's RA on the same day, you know how long has elapsed since ♈ crossed the meridian. Thus all you have to do to get the star's RA is to add to its time of transit the sun's RA. So to get its RA from the star's altitude and azimuth at any observed time, we need to determine one of the other angles of a spherical triangle of which we already know two sides and the angle between them.

For flat triangles the formula we should use (Chapter 5, p. 225) is:

$$\sin C = c\frac{\sin A}{a}\left(\text{or } \sin C = \sin c\frac{\sin B}{b}, \text{ if we knew } B\right)$$

For spherical triangles (Fig. 120) the corresponding sine formulae are

$$\sin C = \frac{\sin c \sin A}{\sin a}\left(\text{or } \sin C = \frac{\sin c \sin B}{\sin b}\right)$$

Since some readers may find Fig. 120 difficult to visualize, we may derive the sine formulae from the cosine formula of Fig. 119 by recalling the rule:

$$\cos^2 x = 1 - \sin^2 x$$

First, however, you should notice that if we can get a when b, c, and A are known from:

$$\cos a = \cos b \cos c + \sin b \sin c \cos A$$

we can also get c when a, b, and C are known from

$$\cos c = \cos a \cos b + \sin a \sin b \cos C$$

The demonstration of the sine formula is as follows. Rearrange the first of these, and then square both sides:

$$-\cos A \sin b \sin c = \cos b \cos c - \cos a$$
$$\cos^2 A \sin^2 b \sin^2 c = \cos^2 b \cos^2 c - 2 \cos a \cos b \cos c + \cos^2 a$$

Now make the substitution:

$$(1 - \sin^2 A) \sin^2 b \sin^2 c = (1 - \sin^2 b)(1 - \sin^2 c) - \cos a \cos b \cos c + (1 - \sin^2 a)$$

$$\therefore \sin^2 b \sin^2 c - \sin^2 A \sin^2 b \sin^2 c = 1 - \sin^2 b - \sin^2 c + \sin^2 b \sin^2 c - 2 \cos a \cos b \cos c + 1 - \sin^2 a$$

After taking away $\sin^2 b \sin^2 c$ from both sides this becomes:

$$-\sin^2 A \sin^2 b \sin^2 c = 2 - \sin^2 a - \sin^2 b - \sin^2 c - 2 \cos a \cos b \cos c$$

Just by looking at this you can see that the right-hand side would be the same if we had started with:

$$\cos c = \cos a \cos b + \sin a \sin b \sin C$$

in which case we should have found

$$-\sin^2 C \sin^2 a \sin^2 b = 2 - \sin^2 a - \sin^2 b - \sin^2 c - 2 \cos a \cos b \cos c$$

Hence we can put

$$-\sin^2 C \sin^2 a \sin^2 b = -\sin^2 A \sin^2 b \sin^2 c$$

Dividing by $-\sin^2 b$, we get:

$$\sin^2 C \sin^2 a = \sin^2 A \sin^2 c$$

$$\therefore \qquad \sin C \sin a = \pm \sin A \sin c$$

$$\text{or} \qquad \sin C = \pm \frac{\sin A \sin c}{\sin a}$$

This is a little long-winded, but it is not difficult to apply, when you have regained your breath.

In our original triangle of Fig. 122 A is the azimuth, c is the zenith distance, and a the polar distance (90° – declin.) of the star, i.e. sin a = cos declin. Hence:

$$\sin \text{hour angle} = \frac{\sin \text{azim.} \sin \text{z.d.}}{\cos \text{declin.}}$$

Suppose one of the stars in Orion is found to have the hour angle 10° when it is west of the meridian at 8.40 pm local time. It crossed the meridian $\frac{10}{15}$ hour = 40 minutes before, i.e. at exactly eight o'clock, and its RA is greater than that of the sun by 8 hours. If the sun's RA on that day were 21 hours 50 minutes, the sun would transit 2 hours 10 minutes before ♈, i.e. ♈ would transit at 2.10 pm, and the star 8 hours 0 minutes – 2 hours 10 minutes = 5 hours 50 minutes after ♈. So its RA would be 5 hours 50 minutes.

The same formula also tells you how to calculate the times of rising and setting of stars in any particular latitude. At rising or setting the z.d. of a heavenly body is 90°, and sin 90° = 1. So the formula becomes:

$$\sin \text{hour angle} = \frac{\sin \text{azim.}}{\cos \text{declin.}}$$

The azimuth of a rising or setting star can be found from the formula already given, i.e.:

$$\cos \text{azim.} = \frac{\sin \text{declin.}}{\cos \text{lat.}}$$

As an example we may take the time of sunrise on the winter solstice in London (Lat. $51\frac{1}{2}$°). By the last formula the azimuth of the rising and setting sun is $50\frac{1}{6}$° from the south point on the winter solstice. So at sunrise and sunset:

$$\sin \text{hour angle} = \frac{\sin 50\frac{1}{6}^\circ}{\cos(-23\frac{1}{2}^\circ)}$$
$$= \frac{0{\cdot}7679}{0{\cdot}9171}$$
$$= 0{\cdot}8373$$

Since 0·8373 is the sine of 56° 51′, the time which elapses between setting and rising and meridian transit (i.e. noon, since it is the sun with which we are dealing) is $56\frac{5}{6} \div 15$ hours, i.e. 3 hours 47 minutes. Thus sunrise would occur at 8.13 am, and sunset at 3.47 pm. Daylight lasts roughly $7\frac{1}{2}$ hours. This calcula-

The earliest geometrical problems arose from the need for a calendar to regulate the seasonal pursuits of settled agriculture. The recurrence of the seasons was recognized by erecting monuments in line with the rising, setting, or transit of celestial bodies. This photograph, taken at Stonehenge, shows how the position of a stone marked the day of the summer solstice when the sun rises farthest north along the eastern boundary of the horizon

MATHEMATICS IN MILITARY AFFAIRS

The great medieval mathematician Tartaglia applies mathematics to artillery in a book published in 1546 at Venice (*above*). The picture opposite is the frontispiece of the *Arithmeticall Militare Treatise* of the brothers Digges, published in London in 1572

The following four prints taken from old books show one way in which solving the problem of motion had become a technical necessity in the period of Stevinus and Galileo

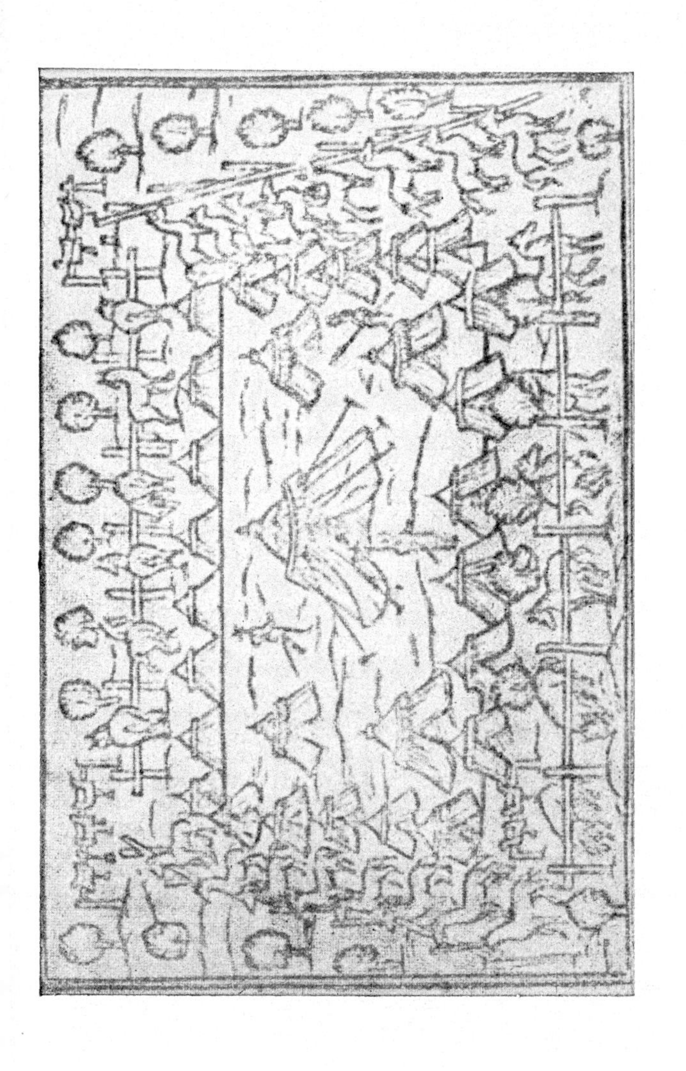

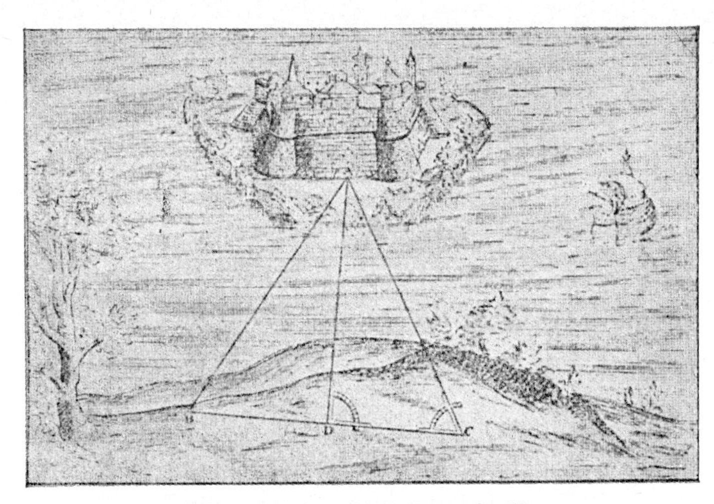

MILITARY MATHEMATICS:

From Bettino's *Apiaria* (Bologna 1645)

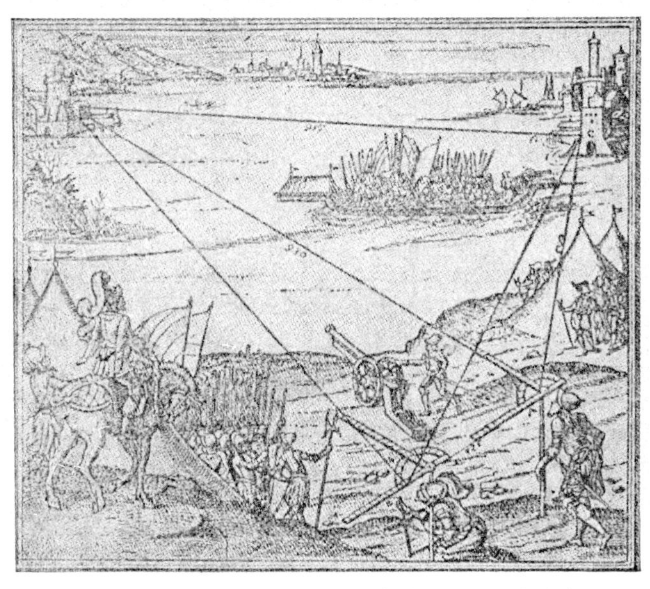

From Zubler's work on geometric instruments (1607)

tion differs by about 6 minutes from the value given in Whitaker's Almanac. This is partly due to approximations made in the arithmetic and partly due to other things about which you need not worry, because you will not find it difficult to put in the refinements when you understand the basic principles.

The same formula would apply to calculating the times of sunset or sunrise on June 21, when the lengths of day and night are reversed. As hinted in Exercise 10 on p. 239 and explained more fully at the end of this chapter on p. 335,

$$\sin A = \sin (180° - A)$$

Since $\sin A = \sin (180° - A)$, 0·8373 may be either sin 56° 51′ or sin (180° − 56° 51′), i.e. sin (123° 9′). A figure shows you at once which value to take. An equatorial star rising due east passes through 90° in reaching the meridian. A star south of the equator passes through a smaller and a star north of the equator through a larger arc. So if *the declination of a heavenly body is north* (*like the sun on June* 21), *we take the solution as sin* (180° − *A*), *and if south as sin A*. Thus the hour angle of sunrise and sunset on June 21 would be ($123\frac{1}{6} \div 15$) hours = 8 h. 13 m., i.e. sunrise would be at 3.47 am and sunset at 8.13 pm.

Theory of the Sun-dial

One last example of the use of spherical triangles is provided by what has now become a garden ornament. The sun-dial which you see in gardens or on the walls of old churches in European countries is an invention based on the same principles as great-circle sailing.

Such a sun-dial was invented by Moslems who made considerable progress in the study of spherical trigonometry, and it is quite different from the shadow clocks. The angle which the shadow of a vertical pole makes with the meridian is the sun's azimuth, and the azimuth angle, through which the shadow rotates when the sun has turned through a given angle, is not the same at all seasons. It depends on the sun's declination. So the length of a working hour as recorded by the shadow clock was not the same fraction of a day at different times of the year.

The Moorish astronomers saw that this could be put right by setting the pole of the shadow clock along the earth's polar axis. When the style is set (Fig. 123) so that you would see the pole star if your eye followed the line along its upper edge, the scale can be graduated in divisions corresponding with equivalent intervals

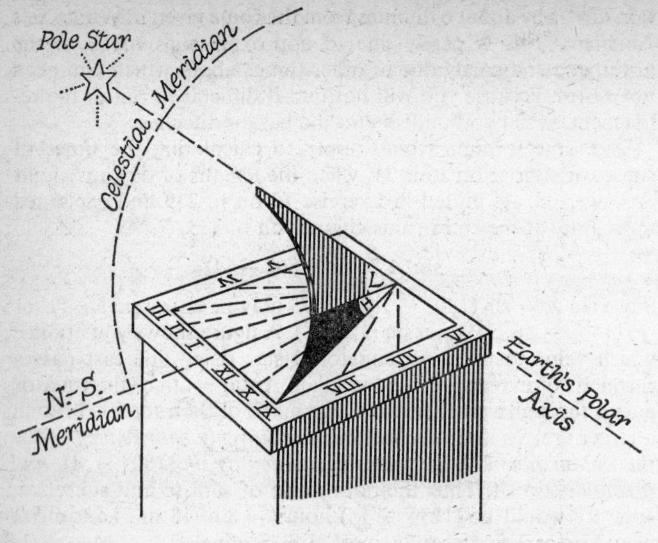

Fig. 123. The Moorish Sun-dial

at all seasons of the year. That is to say, the 'style' or pointer of the sun-dial is set along the meridian, and its edge is elevated at an angle equivalent to the latitude of the place for which it is made. So a sun-dial which keeps time in Seville, where a great Moorish university flourished in the tenth century of our era, will not keep time correctly in London.

The reason for this is explained by Fig. 124. Whatever the declination or RA of the sun happens to be on any day, it appears to revolve around the polar axis, where the planes of all the great circles of Right Ascension intersect. Suppose it has rotated through an hour angle C; any ray which strikes the edge of the style lies completely on the same plane as the great circle of RA on which the sun happens to lie. This plane slices the horizon along the line which joins the observer O to the point where the sun's RA circle cuts the great circle which bounds the horizon. The arc (c) of the horizon circle between this point and the meridian is measured by the plane angle H through which the shadow turns while the sun revolves through an hour angle C. If C is the same, $c = H$ is the same. In other words, when the sun has turned through an angle equivalent to x hours of solar time ($15x°$), the shadow has turned through an angle H, which is the same in summer or winter.

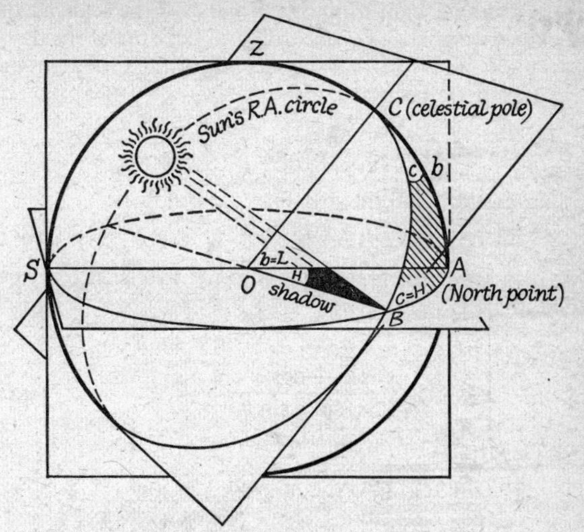

Fig. 124. The Sun-dial Triangle

COB is the plane of the sun's RA circle, intersecting the meridian plane *AZS* along *OC*, so that the style of the sun-dial lies in this line. The edge of the shadow is where the plane of the sun's RA circle cuts the horizon plane *ABS* along *OB*. The angle *C* is the angle through which the sun has revolved around *CO* after crossing the meridian plane.

To make a sun-dial we have first to cut a style so that its upper edge is inclined to the base at an angle $L = b$, the latitude of the place where it will be used. Then the base must be put in line with the meridian, so that the upper edge points due North. All that remains is to graduate the scale in hours. On a summer holiday you can get some amusing practice in trigonometry by doing this.

The theory of graduating the scale depends on a third formula for the solution of spherical triangles. In Fig. 124 you will see that the spherical triangle *ABC* is right-angled. Since the meridian plane is at right angles to the horizon plane, the angle *A*, which is the angle between these two planes, is 90°. In this triangle the parts which concern us are C, the hour angle of the sun, $c = H$, the angle of the shadow, and $b = L$, the latitude of the place. The last we know, and the first we can supply. The second is what we want to know to mark off the scale according to the values we give C, i.e. according to the time of the day.

If $A = 90°$, $\sin A = 1$, so the second formula for right-angled spherical triangles becomes:

$$\sin C = \frac{\sin c}{\sin a} \quad \text{(i)}$$

And since $\cos 90° = 0$ the first formula becomes:

$$\cos a = \cos b \cos c \quad \text{(ii)}$$

If we also apply the first formula to the two sides a, b and the included angle C

$$\cos c = \cos a \cos b + \sin a \sin b \cos C$$

$$\therefore \quad \cos C = \frac{\cos c - \cos a \cos b}{\sin a \sin b} \quad \text{(iii)}$$

Combining (i) and (iii) we get:

$$\frac{\sin C}{\cos C} = \frac{\sin c}{\sin a} \times \frac{\sin a \sin b}{\cos c - \cos a \cos b}$$

$$\therefore \quad \tan C = \frac{\sin c \sin b}{\cos c - \cos a \cos b}$$

Multiply both sides by $\sin b$, then:

$$\tan C \sin b = \frac{\sin c \sin^2 b}{\cos c - \cos a \cos b}$$

Substitute for $\cos a$ according to (ii):

$$\tan C \sin b = \frac{\sin c \sin^2 b}{\cos c - \cos c \cos^2 b}$$

$$= \frac{\sin c \sin^2 b}{\cos c (1 - \cos^2 b)}$$

$$= \frac{\sin c \sin^2 b}{\cos c \sin^2 b}$$

$$= \frac{\sin c}{\cos c}$$

$$= \tan c$$

Thus we have:

$$\tan\text{(shadow angle)} = \sin\text{(latitude)} \times \tan\text{(hour angle)}$$

To get the angle we have to mark off the edge of the shadow at 2.30 pm when the hour angle is $2\frac{1}{2}$ hours, or $2\frac{1}{2} \times 15° = 37\frac{1}{2}°$ for a sun-dial which will keep time in London, we put:

$$\begin{aligned}\tan\text{ shadow angle} &= \sin 51\tfrac{1}{2}° \times \tan 37\tfrac{1}{2}° \\ &= 0{\cdot}7826 \times 0{\cdot}7673 \\ &= 0{\cdot}6005\end{aligned}$$

According to the tables of tangents $\tan 31° = 0{\cdot}6009$ and $\tan 30{\cdot}9° = 0{\cdot}5985$. So the required angle within a tenth of a degree is 31°.

You will now be able to work out the other graduation lines for the scale yourself according, of course, to the latitude of your home. If it happens to be New York, you must put sin 41° for sin lat.

Trigonometrical Ratios of Large Angles

In using the formulae of this chapter correctly, several points which have not been dealt with arise. One has been mentioned in connection with the times of sunset and sunrise on the solstices (p. 325). The south point on the horizon is 180° from the north point. So if the azimuth of a star measured from the north point is A, it is $(180° - A)$ measured from the south point. In the the trigonometry of flat figures (p. 226) we have seen that when an angle A of a triangle is less than 90°:

$$a^2 = b^2 + c^2 - 2bc\cos A \quad . \quad . \quad . \quad . \quad . \quad . \quad . \quad \text{(i)}$$

When A is more than 90°,

$$a^2 = b^2 + c^2 + 2bc\cos(180° - A) \quad . \quad . \quad . \quad \text{(ii)}$$

Similarly, if two angles A and B are both less than 90°:

$$\frac{\sin A}{a} = \frac{\sin B}{b} \quad . \quad . \quad . \quad . \quad . \quad . \quad . \quad \text{(iii)}$$

If one of these angles of a flat triangle, e.g. A, is greater than 90°:

$$\frac{\sin(180° - A)}{a} = \frac{\sin B}{b} \quad . \quad . \quad . \quad . \quad . \quad . \quad \text{(iv)}$$

If we agree to say that cos (180° − A) means the same thing as − cos A, (or that − cos (180° − A) means the same thing as cos A), the single rule (i) includes (ii). Similarly, if we agree to say that sin (180° − A) means the same thing as sin A, the single rule (iii) includes (iv), whether a triangle has three angles none of which is greater than 90°, or one of which is greater than 90°. If we decide to do this we should say that cos 45° has the value + 0·7071 and cos 135° has the value − 0·7071. So if the answer to a problem is cos A = − 0·7071 and tables tell us that cos 45° = + 0·7071, we conclude that A is 180° − 45° = 135°. Similarly, sin 30° has the same value as sin 150° (+ 0·5); and the value 0·5 given in tables of angles from 0° to 90° as sin 30° should be read sin 30° or sin 150°. Thus we only need the one rule for the declination of a star, that when the azimuth angle is always reckoned from the south point (whether greater or less than 90°):

$$\text{sin declin.} = \text{sin lat. cos z.d.} - \text{sin z.d. cos lat. cos azim.}$$

At first sight, this seems a paradox, because we have been used to thinking of sines, cosines, and tangents as ratios of side lengths in right-angled triangles of which no angle can be greater than 90°, and it therefore seems useless to talk of the sine or cosine of an angle of 150°. If we had started with figures drawn on a sphere, as all figures really are drawn on the earth's spherical surface, it would not necessarily seem so silly. The three angles of a spherical triangle make up more than two right angles, and one of the angles of a right-angled spherical triangle may be greater than 90°. Look at the matter in another way. It is easy to draw a geometrical diagram to show the meaning of a^n when $n = 1$ (a line), $n = 2$ (a flat square), $n = 3$ (a cube). We cannot make the same sort of figure to show the meaning of the operator 4 in the expression a^4. Yet we can do the same thing with a^4 in arithmetic as we can do with a^1, a^2, a^3. So why should we say that sin 150° means nothing, merely because we cannot draw a flat right-angled triangle having one angle of 150°, or that cos 110° means nothing because we cannot draw a flat right-angled triangle with an angle of 110°?

After all, we can look on sine, cosine, and tangent as different ways of measuring an angle, and if so, it is not so odd that angles bigger than 90° should have sines, cosines, and tangents like angles less than 90°. Indeed, the measurement of an angle by its sine, cosine, or tangent does not necessarily imply drawing a triangle at all. An angle is ordinarily defined by the number of units of length in a circular arc of radius 1 unit. To say that the angle between two lines is one degree means that if we draw a circle of 1 inch radius around the point where they cross, the length of the small arc between the points where they cut the circle is $(2\pi \div 360)$ inches long.

The sine of an angle may also be defined in units of length, as we have done in Chapter 6, p. 212. If PT is the chord joining the ends of the arc $2A$, of a circle of unit radius $\frac{1}{2}PT$ (PQ in Fig. 84), the half chord is the sine of the half arc A. The line PT has just as much right to be called the chord of the large arc $360° - 2A$ as of the small arc $2A$, and the arc $\frac{1}{2}(360° - 2A) = 180° - A$ has just as much right to be called the half angle whose sine is the half chord PQ as A itself. Accordingly, this way of measuring sines means that

$$\sin A = \sin (180° - A)$$

The foregoing remarks show that we can accommodate the trigonometry of Ptolemy and the more convenient, though at first sight more restricted, trigonometry of our Hindu–Arabic teachers, only if we can give to *sin A* and *cos A* when $A > 90°$ a meaning *consistent with all the formulae* we have derived on the assumption $A \leqslant 90°$. The issue first raised its ugly head in the context of the surveyor's recipe for solution of drawing-board triangles. There we have seen that two sets of formulae are identical for triangles in which A does and does not exceed a right angle, if we write:

$$\sin (180° - A) = \sin A \textit{ and } \cos (180° - A) = -\cos A\,. \quad \text{(i)}$$

Perhaps, we may best start by noticing that it means the same thing (i.e. rotation through four right angles or two π radians) to speak of an angle of 360° and an angle of 0°. This implies $\cos 360° = 1 = \cos 0°$, $\sin 360° = 0 = \sin 0$. If anything, *sin* or *cos* $(B + A)$ can therefore mean only the same thing as *sin* or *cos* $(\pm A)$ when $B = 360°$. Hence, our two formulae for *cos* $(B + A)$ and *sin* $(B + A)$ on pp. 223–4 become:

$$\cos (360° + B) = \cos 360° \,.\, \cos B - \sin 360° \,.\, \sin B = \cos B$$
$$\sin (360° + B) = \sin 360° \,.\, \cos B + \cos 360° \,.\, \sin B = \sin B$$

This is consistent with our assumptions so far. Let us now turn to the half angle and double angle formulae on Fig. 64, i.e.:

$$\sin 2A = 2 \sin A \,.\, \cos A; \quad \cos 2A = 1 - 2 \sin^2 A$$
$$2 \sin^2 (\tfrac{1}{2}A) = 1 - \cos A; \quad 2 \cos^2 (\tfrac{1}{2}A) = 1 + \cos A$$

Here we may express *cos* $(180° - A)$ in the form *cos* $2(90° - \frac{1}{2}A)$, so that:

$$\cos(180° - A) = 1 - 2\cos^2(\tfrac{1}{2}A) = 1 - (1 + \cos A)$$
$$= -\cos A$$

in agreement with (i) above.

The corresponding *sine* formula is consistent with the conclusion that *sin* $(180° - A) =$ *sin* A, since we may write:

$$\sin(180 - A) = \sin 2\left(90 - \frac{A}{2}\right) = 2\sin\left(90 - \frac{A}{2}\right).\cos\left(90 - \frac{A}{2}\right)$$

$$\therefore \sin(180 - A) = 2\cos\frac{A}{2}.\sin\frac{A}{2}$$

$$\therefore \sin^2(180 - A) = 4\cos^2\frac{A}{2}.\sin^2\frac{A}{2} = (1 + \cos A)(1 - \cos A)$$

$$\therefore \sin^2(180 - A) = 1 - \cos^2 A = \sin^2 A$$

The last test is *not* conclusive, because either $(\sin A)^2$ or $(-\sin A)^2 = \sin^2 A$; but we may use the addition formula with the same end in view as follows, noticing *en passant* that *sin* $(180°) =$ *sin* $(90° + 90°) = 0$ and *cos* $(180°) = \cos(90° + 90°) = -1$; and these equalities are consistent with our earlier condition that $A \leqslant 90°$:

$$\sin(180 - A) = \sin(90 + 90 - A) = \sin 90.(\cos 90 - A) +$$
$$\cos 90.\sin(90 - A)$$
$$= \cos(90 - A) = \sin A$$

If all our formulae are to make sense, we must find a meaning for *cos* or *sin* $(-A)$ consistent with the addition formulae, in which event, and in conformity with (i) above:

$$\sin B = \sin(180 - B) = \sin 180.\cos(-B) + \cos 180.\sin(-B)$$
$$= -\sin(-B)$$
$$\cos B = -\cos(180 - B) = -\cos 180.\cos(-B) +$$
$$\sin 180.\sin(-B) = \cos(-B)$$

We may then write:

$$\left.\begin{aligned}\sin(A - B) &= \sin A\cos B - \sin B\cos A\\ \cos(A + B) &= \cos A\cos B - \sin A\sin B\end{aligned}\right\} \quad . \quad . \quad \text{(ii)}$$

It is easy to test (p. 134) that these formulae are true when $A = 45° = B$ or when $A = 60°$ and $B = 30°$; and we may also test their consistency with the $2A$ formulae above as below:

$$\sin A = \sin(2A - A) = \sin 2A \,.\, \cos A - \cos 2A \,.\, \sin A$$
$$= (2 \sin A \,.\, \cos A) \cos A - (1 - 2\sin^2 A) \,.\, \sin A$$
$$= 2 \sin A \,.\, \cos^2 A - \sin A + 2 \sin^3 A$$
$$= 2 \sin A (1 - \sin^2 A) - \sin A + 2 \sin^3 A$$
$$= 2 \sin A - 2 \sin^3 A - \sin A + 2 \sin^3 A$$
$$= \sin A$$

Similarly, we may test the formula for *cos* $(A - B)$ as follows:

$$\cos A = \cos 2A \cos A + \sin 2A \sin A$$
$$= \cos A (1 - 2 \sin^2 A) + 2 \sin^2 A \,.\, \cos A$$
$$= \cos A (1 - 2 \sin^2 A + 2 \sin^2 A) = \cos A$$

Yet another test of the consistency of the *difference formulae* with what we already know is the following:

$$\cos A = \sin(90 - A) = \sin 90 \,.\, \cos A - \cos 90 \,.\, \sin A = \cos A$$
$$\sin A = \cos(90 - A) = \cos 90 \,.\, \sin A + \sin 90 \,.\, \cos A$$
$$= \sin A$$

or again we may write:

$$\sin A = \sin(90 - 90 + A)$$
$$= \sin 90 \,.\, \cos(90 - A) - \cos 90 \,.\, \sin(90 - A) = \sin A$$
$$\cos A = \cos(90 - 90 + A)$$
$$= \cos 90 \,.\, \cos(90 - A) + \sin 90 \,.\, \sin(90 - A) = \cos A$$

It thus appears that our interpretation of the meaning of *cos A* and *sin A* when $A > 90°$ in terms consistent with (1) above, is likewise consistent with all the formulae we had previously derived on the assumption that $A \leqslant 90°$. In our next chapter, we shall bring the meaning of the operator (-1) as defined in Chapter 2 more explicitly into the picture. The reader should now be able to use the addition formulae *sin* or *cos* $(A + B)$ to show that *sin* $(90 + A)$ = *cos A* and *cos* $(90 + A) = -$*sin A*.

. . . . .

Readers who may wish to know how one can switch from locating a star in declination and RA to locating it in celestial latitude and longitude will find the clues in Fig. 125. It involves solution of two spherical triangles:

(*a*) the pole of the ecliptic (E), the celestial pole (P) with the intersection (♈) of the zero meridians of RA and longitude form one triangle in which: the arc EP is the obliquity of the ecliptic, approximately $23\frac{1}{2}°$ and the arc E♈ $= 90° = P$♈;

(*b*) E, P, and S the position of a star (of known declination and RA) form another triangle in which we know EP ($23\frac{1}{2}°$), $PS = (90° -$ declin. of the star) and EPS, if we also know $E♈P$.

The solution of the first triangle leads to the result $♈PE = 90° = ♈EP$, and $EPS = ♈PE + ♈ES = 90° +$ RA (of the star). The solution of the second triangle yields:

$ES = 90° -$ latitude (of the star) and $PES = 90° -$ its longitude

One use of the result is that it allows us to date an ancient monument of known astronomical orientation. Owing to precession, longitude changes at the steady rate of $50\frac{1}{4}$ seconds of an arc per

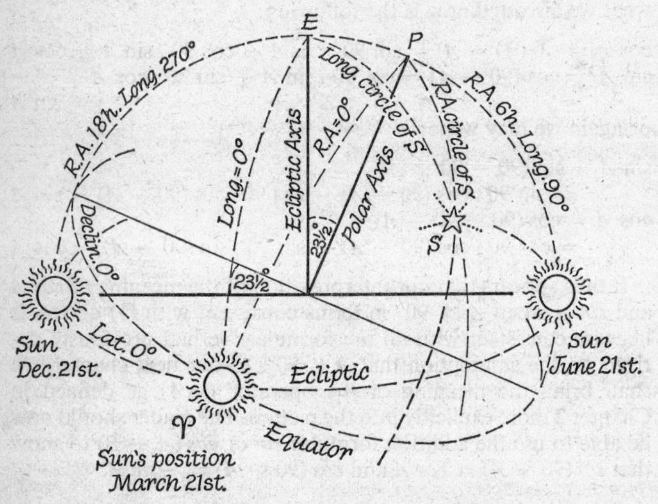

Fig. 125. The Two Ways of Star Mapping

Celestial latitude and longitude give the direction of a star as seen from the earth's centre with reference to the plane of the ecliptic (or earth's orbit about the sun). Declination and RA give its direction as seen from the earth centre with reference to the plane of the celestial equator (at right angles to the earth's polar axis). Since the position of the stars with reference to the ecliptic plane is fixed the latitude of a star is not affected by precession. Longitude is reckoned like RA from the meridian which passes through the node ♈. Since this slowly retreats at $50\frac{1}{4}$ seconds of an arc per year, the longitude of any star increases by $50\frac{1}{4}$ seconds per year.

year. Whence the longitude of a star is easily calculated for a date in the past (e.g. 2500 BC). Its declination at that date can then be found from the following formula in which its latitude is as at the present time:

$$\sin(\text{declin.}) = \cos 23\tfrac{1}{2}^\circ . \sin(\text{lat.}) + \sin 23\tfrac{1}{2}^\circ . \cos(\text{lat.}) . \sin(\text{long.})$$

Chapter 7

Tests

1. What is the sun's declination and right ascension on March 21st, June 21st, September 23rd, and December 21st?

2. What *approximately* is the sun's RA on July 4th, May 1st, January 1st, November 5th? (Work backwards or forwards from the four dates given. Check by *Whitaker*.)

3. With a home-made astrolabe (Fig. 12, Chapter 2) the following observations were made at a certain place on December 25th:

Sun's Zenith Distance (South)	Greenwich Time (pm)
$74\frac{1}{2}^\circ$	12.18
74°	12.19
74°	12.20
$73\frac{1}{2}^\circ$	12.21
$73\frac{1}{2}^\circ$	12.22
74°	12.23
$74\frac{1}{2}^\circ$	12.24

What was the latitude and longitude of the place? (Look it up on the map.)

4. Find the approximate RA of the sun on January 25th, and hence at what local time Aldebaran (RA 4 hours 32 minutes) will cross the meridian on that night. If the ship's chronometer then registers 11.15 pm at Greenwich, what is the longitude of the ship?

5. If the declination of Aldebaran is 16° N (to the nearest degree) and its altitude at meridian transit is 60° from the southern horizon, what is the ship's latitude?

6. The RA of the star *a* in the Great Bear is nearly 11 hours. Its declination is 62° 5′ N. It crosses the meridian 4° 41′ north of the zenith on April 8th at 1.10 am by the ship's chronometer. What is the position of the ship?

7. Suppose that you were deported to an island, but had with you a wrist watch and *Whitaker's Almanack*. Having set your

watch at noon by observing the sun's shadow, you noticed that the star *a* in the Great Bear was at its lower culmination about eleven o'clock at night. If you had lost count of the days, what would you conclude to be the approximate date?

8. On April 1st, 1895, the moon's RA was approximately 23 hours 48 minutes. Give roughly its appearance, time of rising, and transit on that date.

9. If the sun and the Great Bear were each visible throughout one whole 24 hours, and one only, in the year at a given locality on the Greenwich meridian, how could you determine the distance from London, if you also remembered that the earth's diameter is very nearly 8,000 miles, and the latitude of London is very nearly 51°?

10. If you observed that the sun's noon shadow vanished on one day of the year and pointed South on every other, how many miles would you be from the North Pole?

11. On January 1st the sun reached its highest point in the heavens when the BBC programme indicated 12.17 pm. It was then 16° above the southern horizon. In what part of England did this happen?

12. The approximate RA and declination of Betelgeuse are respectively 5 hours 50 minutes and 7½° N. If your bedroom faces East, and you retire at 11.0 pm, at what time of the year will you see Betelgeuse rising when you get into bed?

13. On April 13th, 1937, the shortest shadow of a pole was exactly equal to its height, and pointed North. This happened when the radio programme timed for 12.10 began. In what county were these observations made?

14. With a home-made astrolabe the following observations were made in Penzance (Lat. 50° N, Long. 5½° W) on February 8th:

	Least Zenith Distance	Greenwich Mean Time
Betelgeuse	42½° South	9.9 pm
Rigel	58½° ,,	8.24 ,,
Sirius	67½° ,,	10.0 ,,

Find the declination and RA of each star, and compare your results with the table in *Whitaker's Almanack* (p. 140). Note Greenwich 'Mean' Time is not the true solar time reckoned from noon at Greenwich, but differs by a few minutes, the difference ('equation of time') being tabulated for each day of the year in *Whitaker* or the *Nautical Almanac*. On February 8th true noon at Greenwich occurred 14 minutes later than noon by Greenwich Mean Time.

15. By aid of a figure show that if the hour angle (h) of a star is the angle through which it has turned since it crossed the meridian (or if the sign is minus the angle through which it must turn to reach the meridian):

RA of star (in hours) = sun's RA (in hours)
— (hour angle in degrees ÷ 15) + *local* time (in hours)

16. With the aid of a map on which ocean distances are given, work out the distances by great-circle sailing from ports connected by direct routes. What are the distances between London and New York, London and Moscow, London and Liverpool?

17. On April 26th, the sun's RA being 2 hours 13 minutes, the bearings of three stars were found to be as follows, by a home-made instrument:

	Azimuth	Zenith Distance	Local Time
Pollux	W 80° from S	45°	9.28 pm
Regulus	W 28° " S	41°	9.39 "
Arcturus	W 7° " N	31°	12.50 "

Find the declination and RA of each star and compare these rough estimates made at Lat. $50\frac{3}{4}$° N with the accurate determinations given in *Whitaker's Almanack*.

18. To get the exact position of the meridian a line was drawn between two posts in line with the setting sun on July 4th at a place Lat. 43° N. At what angle to this line did the meridian lie? (*Whitaker* gives the sun's declination on July 4th as 23° N.) What was the approximate time of sunset?

19. If the RA of Sirius is 6 hours 42 minutes and its declination is $16\frac{3}{5}$° S, find its local times of rising and setting on January 1st at:

Gizeh	Lat. 30° N
New York	Lat. 41° N
London	Lat. $51\frac{1}{2}$° N

Check with planisphere.

Things to Memorize

1. If z.d. is measured north of the zenith the sign is +, if measured south of the zenith it is —. If latitude or declination is measured

north of the equator the sign is +, if south −. In all circumstances:

declination = lat. of observer + *meridian* zenith distance

2. Star's RA = time of transit + sun's RA on same day of year.

(sun's RA 0. Declin. 0 March 21st
RA 12. Declin. 0 September 23rd
RA 6. Declin. $+23\frac{1}{2}°$ June 21st
RA 18. Declin. $-23\frac{1}{2}°$ December 21st)

3. In a spherical triangle:

(*a*) $\cos a = \cos b \cos c + \sin b \sin c \cos A$

(*b*) $\dfrac{\sin A}{\sin a} = \dfrac{\sin B}{\sin b} = \dfrac{\sin C}{\sin c}$

4. For a heavenly body, azimuth being measured from the *south* point

(*a*) sin (declin.) = cos (z.d.) sin (lat.) − sin (z.d.) cos (lat.) cos (azim.)

(*b*) $\sin(\text{hour angle}) = \dfrac{\sin(\text{z.d.})\sin(\text{azim.})}{\cos(\text{declin.})}$

(*c*) When setting and rising (z.d. = ± 90°)

$$\cos(\text{azim.}) = 1\,\frac{\sin(\text{declin.})}{\cos(\text{lat.})}$$

$$\sin(\text{hour angle}) = \pm\frac{\sin(\text{azim.})}{\cos(\text{declin.})}$$

CHAPTER 8

THE GEOMETRY OF MOTION

WHEN DEFINING a figure bounded by curved lines, it is possible to do so in either of two ways. We may define a circle, an ellipse, a parabola, and a hyperbola *timelessly* as sections of a cone. This is what the Platonists did, whence their designation as *Conic Sections*. We can, however, define such curves as the track (so-called *locus*) of a *moving* point, e.g. a circle as the locus of a point moving in a plane at a fixed distance from a fixed point (i.e. its centre). This viewpoint was not alien to Alexandrian mathematicians. Archimedes described one sort of spiral (Fig. 128) in terms of a line rotating at fixed angular speed while a point on it moves away from the centre at a fixed linear speed. No doubt several circumstances conspired to prevent him from developing this line of inquiry in a systematic way. Among others was the fact that motion, as a practical issue, impinged on their habits of thought only in the domain of astronomy. In the setting of Columbus and Cabot it emerged as a manifold challenge in circumstances without parallel in earlier civilizations accustomed to abundant slave labour as a substitute for mechanical power and to long-distance voyages only southwards along the African coast or northwards to Britain in quest of tin.

Explorers, such as the early Phoenician and Greek navigators who supplied the raw materials of Alexandrian geography, had indeed followed routes mainly along coast-lines which closely followed a north–south direction. On unfamiliar sea-lanes, they ventured beyond sight of land no farther than the range of familiar birds which nested on its shores. Knowledge of latitude therefore sufficed for the pilot's needs. When Columbus, Cabot, and Magellan set their courses westward across the uncharted expanse of the Atlantic or Pacific, knowledge of longitude became equally essential. At a time when mechanical clocks driven by a falling weight were still seen only on public buildings, clock technology therefore received the impetus of a powerful new demand. Portable ones with a coiled-spring drive were not available much before 1540 about the time when the cartographer mathematician Gemma Frisius formulated the need for a seaworthy clock for determining longitude (p. 143) from local noon.

During the two centuries before Columbus an innovation from the East was changing the pattern of European warfare. Aside from the fact that Archimedes seems to have applied the principle of the lever to design of large-scale catapults for siege defence, short-range marksmanship had set the mathematicians of antiquity no inviting problem; but the introduction of cannon, which was of decisive importance in the campaigns of the Spanish and Portuguese Conquistadors, raised a compelling one. To aim successfully, one had to know the path (so-called *trajectory*) of the missile. Galileo completed the solution of the problem a century after Columbus sighted the Caribbean; but the issue was prominent in military manuals (see the illustrated section in the centre of this book) published throughout the century after printing from movable type first made possible the distribution of nautical almanacs to an increasingly literate profession of Master Mariners.

In short, the measurement of motion forced itself on the attention of thinking people in the century of Columbus and Cabot to an extent outside the experience of earlier civilizations in circumstances which enlisted a reservoir vastly larger than hitherto of mathematical talent. Prominent in the foreground were problems of map-making. Map-making had prompted Ptolemy at least among the later Alexandrians to investigate what we now call projective geometry, a branch of inquiry stimulated by the introduction of perspective art in the century of Columbus; but the notion of tracing the locus of a point like the course of a ship in a grid of latitude and longitude does not seem to have been articulate till the publication of a treatise by Oresme, a French ecclesiastic at a time (about AD 1360) when Moslem cartography was beginning to penetrate the universities of Christendom.

Doubtless because algebraic symbolism other than the surd sign was then non-existent, the possibility was still-born. Accordingly, it was not easy to formulate a concept, that of *function*, so essential to its exploitation. However, the century of Frisius and Mercator had put the spotlight on the grid concept at a time when map-making, marksmanship, and the swing of the pendulum set the pace of mathematical inventiveness. In such a setting, it was inevitable that someone would eventually revive the issue, as did two Frenchmen, Fermat (1636) and Descartes, independently. Descartes published his *new geometry* somewhat later than did Fermat. That he overshadowed the latter is probably due to the fact that his treatise was the first in which algebraic symbols assume an aspect familiar to ourselves.

When we speak of one variable quantity as a function of another, we may distinguish two classes: *discrete* and *continuous*. If T_n denotes a triangular number (p. 82), we speak of it as a *discrete* function of n (a natural number) because the latter increases by separate steps $(n - 1)$, n, $(+ 1)$, etc, and we exclude

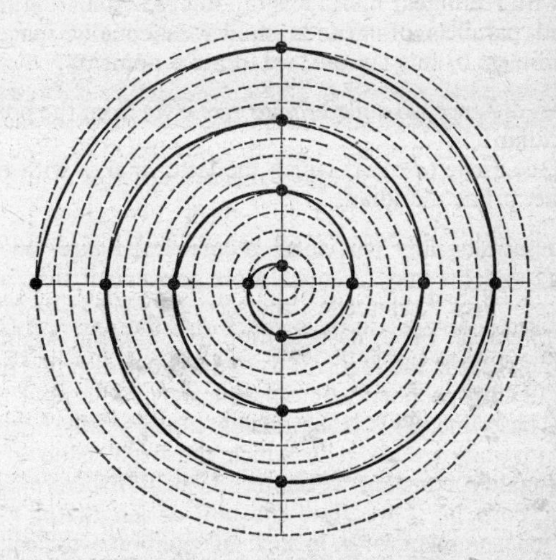

Fig. 128. The Archimedean Spiral

the possibility that T_m has any meaning when m lies between n and $n - 1$ or between n and $(n + 1)$. If we speak of x^2 as a *continuous* function of x we imply (*inter alia*)* that x may have any real value, for instance, $\sqrt{2}$. Co-ordinate geometry, of which Fermat and Descartes laid the foundations, deals only with functions of the second kind.

The term co-ordinates refers to the relevant measurements which locate a point in some fixed *framework*. In a plane two

* We mean also that it is *differentiable* in a sense defined later (Chapter 10).

suffice, in space three. This is one meaning conveyed by saying that a flat figure is 2-dimensional, a solid figure 3-dimensional. For dealing with the former, two frameworks are most convenient. One is like a mercator projection map with equally spaced parallels of latitude and meridians of longitude at right angles. In this lay-out, two distances define the position of a point, one (x) from the prime (Greenwich) meridian (so-called *Y-axis*) along a parallel of latitude, one (y) from the equator (so-called *X-axis*) along a meridian. An alternative framework (Fig. 129) recalls a polar view of the northern or southern hemisphere with rectilinear meridians of longitude radiating from the pole and parallels of latitude shown as equally spaced concentric circles. In this lay-out, we locate a point by:

(i) its distance, so-called *radius vector* (r) from the pole along a meridian;
(ii) the angle (a or A) which the latter makes with the base line, i.e. prime meridian.

For describing any particular geometrical figure, one of the foregoing systems may be much more convenient than another. For the spiral of Archimedes, the second (*polar co-ordinates*) is vastly preferable. From its definition (p. 343), the Archimedian spiral (Fig. 128) is the locus of a point whose distance from the centre is directly proportional to the angle the radius vector makes with the base line. This is expressible algebraically in the form (*polar* equation) $r = K \, . \, a$ in which the fixed value of K, the proportionality constant (Chapter 2, p. 97), determines how close the coils lie if we state the unit of length for r and of angular measurement for a. In such an equation, we speak of a as the *independent variable* and r, a *function* of a, as the *dependent* variable. If we write it in the form $a = K^{-1} \, . \, r$, we speak of a as the dependent one, and with equal propriety a as a function of r.

The polar system of co-ordinates came into use much later than the alternative mentioned above and called *Cartesian* after Descartes. He himself operated in one quadrant only, i.e. the northeast one in a Mercator map; but his successors very speedily recognized what simplifications (Fig. 130) result from labelling:

(i) measurements parallel to the X-axis as positive ($+x$) to one side (arbitrarily chosen as the *right* of the Y axis) and negative ($-x$) to the other (*left* of it);

(ii) measurements parallel with the Y-axis as positive $(+y)$ above the X-axis and negative $(-y)$ below it.

The genesis (Fig. 130) of what one calls the Cartesian equation of the circle will clarify this convenience. If we recall that $(x - a) = -(a - x)$ so that $(a - x)^2 = (x - a)^2$, our representation of the circle, invokes nothing but the Pythagorean rule.

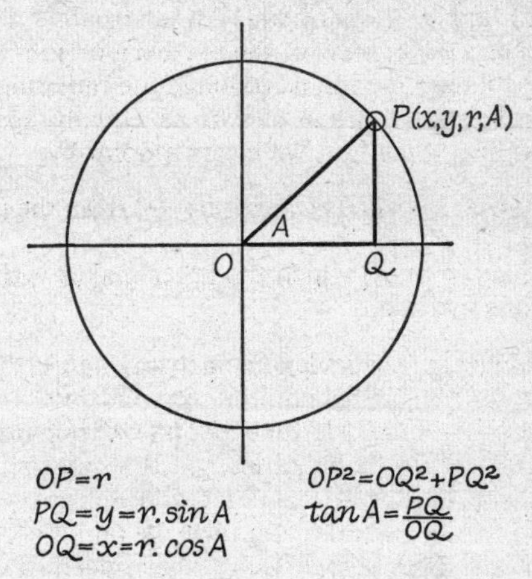

Fig. 129. The Relation between Polar and Cartesian Co-ordinates

$$r^2 = x^2 + y^2 \quad \textit{and} \quad \tan A = y \div x$$

Within the restricted domain of the north-east quadrant of the Cartesian grid (upper half of Fig. 130) we derive the so-called Cartesian equation of the circle, y being a function of x as:

$$(y - b)^2 = R^2 - (x - a)^2$$

$$\therefore \quad y = b \pm \sqrt{R^2 - (x - a)^2}$$

In this expression, a is the x-co-ordinate of the centre of the circle and b its y-co-ordinate. If we take as our circle, centre (lower half of Fig. 130 and Fig. 131), the intersection of the X-axis and Y-axis (so-called *origin*, corresponding to the point on the equator where the Greenwich meridian crosses), and label the

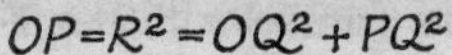

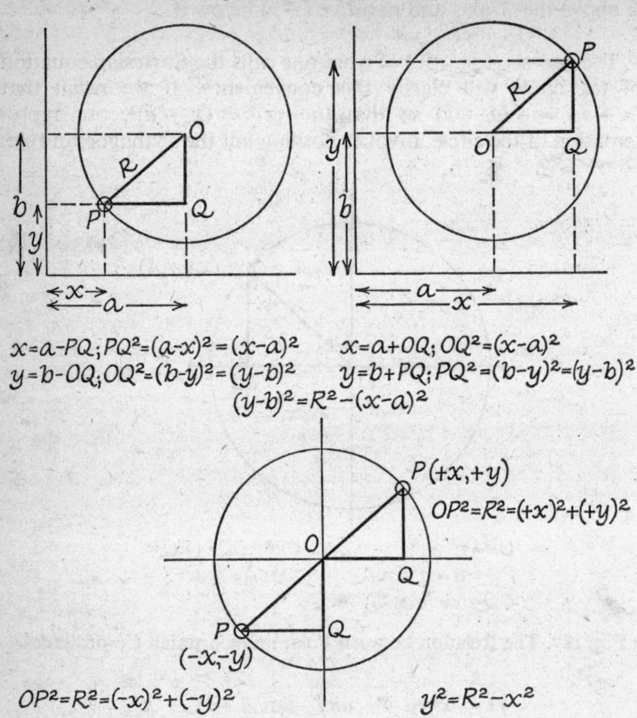

Fig. 130. Cartesian Equation of the Circle

Above: Centre at $x = a$ and $y = b$.

Below: Centre at *origin* ($x = 0 = y$).

separate halves of our axes as positive and negative, the appropriate expression assumes the simpler form:

$$y = \sqrt{R^2 - x^2}$$

We shall see later that labelling our axes in both positive and negative halves has a very considerable pay-off other than to simplify the description of a figure by recourse to an algebraic equation. Before going farther, let us first be clear about what one means by saying that y in the foregoing formulae is a 2-*valued function* of x. For each value of x there are here alternative values of y except when $x = R + a$ or $x = a - R$ in the upper part of Fig. 130 and $x = R$ or $-R$ in the lower. With this behind us, we should now realize that the equation which defines a figure depends both on what *framework* (Cartesian or polar) *of reference* we choose and on *where we place the figure in the chosen framework.*

Let us now reverse our approach. We have assumed the existence of a figure and have deduced its equation from its known geometrical properties. We may likewise assume knowledge of an equation and ask how we disclose what figure it represents. We call this process *plotting a graph*; and the first step is to make a table, of which a few entries will suffice to clarify the recipe. Let our equation be as for a circle of radius 5 with centre at the origin, so that:

$$y^2 = 25 - x^2$$

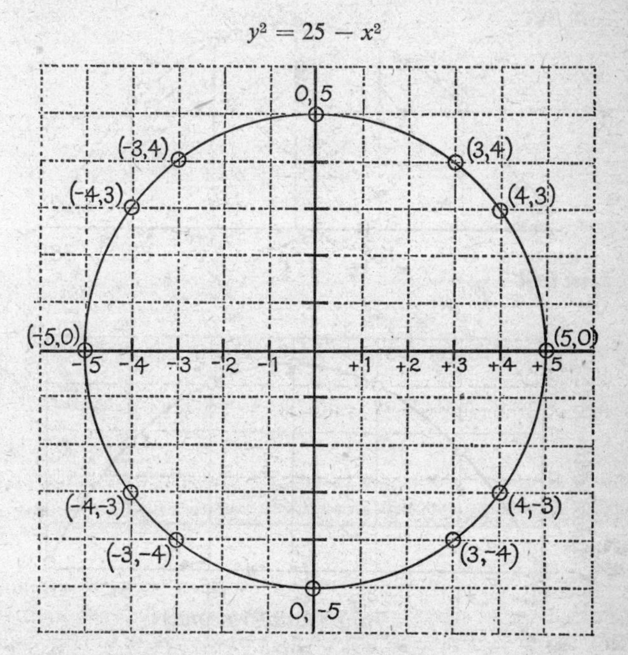

Fig. 131. Plotting the Cartesian Equation of the Circle

We then see that $y = +4$ or -4, if $x = +3$ or -3; $y = +3$ or -3 if $x = +4$ or -4, etc. For simplicity, we need tabulate whole number values of x only if they tally with whole values of y. A miniature table will then be:

$x =$	-5	-4	-3	0	$+3$	$+4$	$+5$
$y =$	0	± 3	± 4	± 5	± 4	± 3	0

We no longer need to labour uselessly, as we should have had to do in the lifetime of Newton, in order to make measurements of our co-ordinates (x, y) at right angles to the axes. Instead, we can purchase, in any store which sells writing matter, graph paper spaced in equal intervals horizontally and vertically. Having made out a double entry table, as above, we have then: (*a*) to line up our *X*- and *Y*-axes at right angles; (*b*) place points cor-

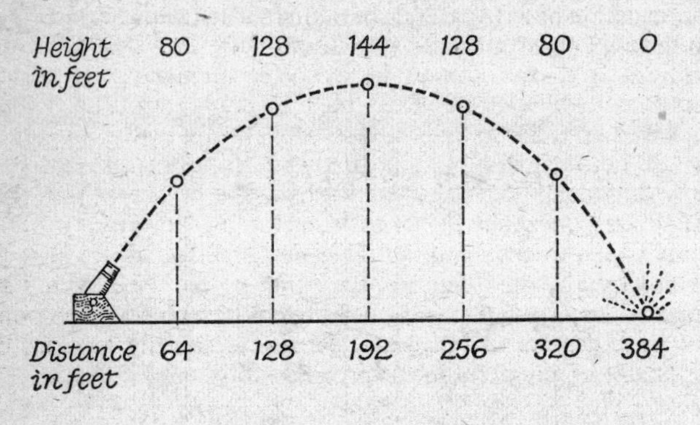

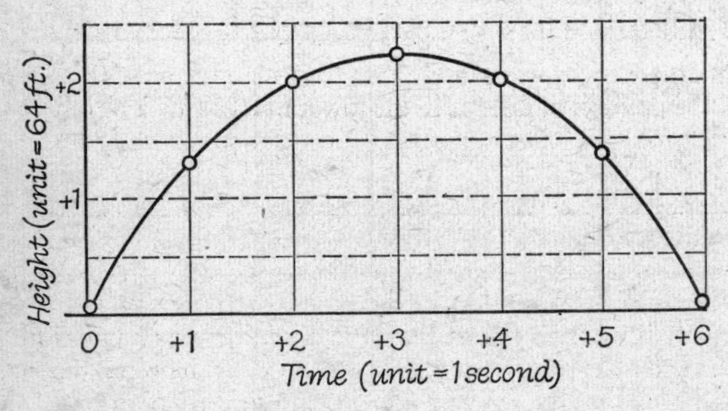

Fig. 132. Path of the Cannon-Ball – putting Time into Geometry

responding to paired numerical values of x and y in the grid accordingly. The final step is to join the points *smoothly*. The last statement is a clue to refining our definition of what we mean by a *continuous* function. Its visual representation as a curve has *no corners*.

To appreciate one way in which the new algebraic geometry brought motion into the picture, let us now look at Galileo's solution of the cannon-ball problem from the viewpoint of consumer (or consumed), i.e. how we can use the equation of its *trajectory* to forecast correctly where it will be after so many seconds from its ejection. Fig. 132 is a scale diagram of the problem. It shows the progress of the cannon-ball, vertically and horizontally. You will see that each position of the moving cannon-ball, like each position of the moving ship, has two co-ordinates, the horizontal distance ($+x$) to the right of the cannon corresponding with its longitude east, and the vertical distance upward from the ground ($+y$) corresponding to its latitude north. The difference between the scale diagram of the cannon-ball and an ordinary map is that we have put into it something we do not include in a map. Maps are made for different ships travelling at different speeds. So they take no account of time. They are concerned only with showing us the position of the ship. In the scale diagram of the cannon-ball we have shown the time which elapses between the firing of the cannon and the instant when the ball passes the horizontal distances corresponding with degrees of longitude. So we can equally give the co-ordinates of any point in the progress of the projectile as:

(*a*) vertical height $= +y$, horizontal distance $= +x$
(*b*) vertical height $= +y$, time since firing $= +t$

In other words, we can use our Cartesian framework to represent the passage of time as in the lower half of Fig. 132, where one unit of measurement along the y axis stands for 64 feet and one unit along the x axis stands for 1 second. The curved line of this figure is called a *parabola*. A projectile moves in a path which corresponds almost perfectly with the parabola as defined by the mathematician, when it moves in a vacuum. In air the correspondence is less perfect, and certain corrections would have to be made if we used the mathematical parabola as a paper model for marksmanship. In the figure drawn we have taken values which illustrate the mathematical definition. Like all calculating devices, it is only an *approximate* description of what happens in the real world. On tabulating values we can reconstruct the equation of the parabola by a little experimentation with the figures. Thus we find:

x	y	$\frac{3x}{2}$	x^2	$\frac{x^2}{4}$	$\left(\frac{3x}{2} - \frac{x^2}{4}\right)$
0	0	0	0	0	0
1	$1\frac{1}{4}$	$1\frac{1}{2}$	1	$\frac{1}{4}$	$1\frac{1}{4}$
2	2	3	4	1	2
3	$2\frac{1}{4}$	$4\frac{1}{2}$	9	$2\frac{1}{4}$	$2\frac{1}{4}$
4	2	6	16	4	2
5	$1\frac{1}{4}$	$7\frac{1}{2}$	25	$6\frac{1}{4}$	$1\frac{1}{4}$
6	0	9	36	9	0

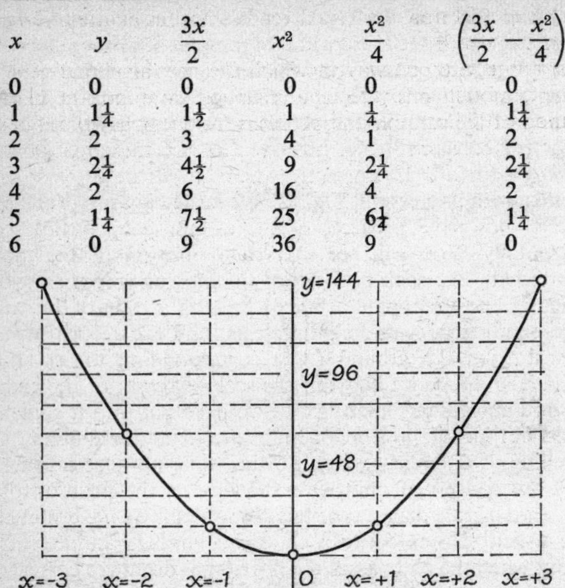

Fig. 133. The Parabola $y = (4x)^2$

The equation of the parabola shown in Fig. 132 thus turns out to be:

$$y = \frac{3x}{2} - \frac{x^2}{4}$$

We may shift the origin to make the maximum value of y (i.e. $2\frac{1}{4}$ in units of 64 feet) lie on the Y axis, if we substitute for x a new co-ordinate $X + 3$, so that $X = 0$ when $x = 3$. The foregoing equation then becomes:

$$y = \frac{3(X+3)}{2} - \frac{(X+3)^2}{4} = \frac{9}{4} - \frac{X^2}{4}$$

$$\therefore \quad \frac{-X^2}{4} = y - \frac{9}{4}$$

If we shift the origin along the Y-axis by the substitution $Y = y - \frac{9}{4}$, so that $Y = 0$ when $y = \frac{9}{4}$:

$$Y = \frac{-X^2}{4}$$

The shape and scale of the curve described in this way is exactly the same as that of Fig. 132. All that has changed is its position within the grid, viz.:

(i) it lies symmetrically about the Y-axis, and wholly below the X-axis except where it cuts the origin;

(ii) the maximum value of Y is zero and corresponds to $X = 0$.

Fig. 133 shows the curve described by the last equation rotated through two right angles (π radians). Its equation is:

$$Y = \frac{X^2}{4}$$

Before examining the effect of rotation of the axes on the form of the equation which describes a curve, let us look at another example of the way in which a shift of the origin along both axes simplifies the algebraic expression. In Fig. 134, the same equation describes two curves called *hyperbolae*, each being the mirror image of the other. We can obtain this twin curve by plotting the following function which contains three *numerical constants* (3, 10, and -2):

$$y = \frac{3x + 10}{x - 2}$$

Below are a few tabulated values for tracing the course of the corresponding curves:

$x =$
$-14, -6, -2, \quad 0, \ +1, +2, \ +3, \ +4, +6, +10, +18 \ldots$
$y =$
$+2, +1, -1, -5, -13, \ \infty, +19, +11, +7, \ +5, \ +4 \ldots$

We shall now shift our Y-axis by putting $(x - 2) = X$. The

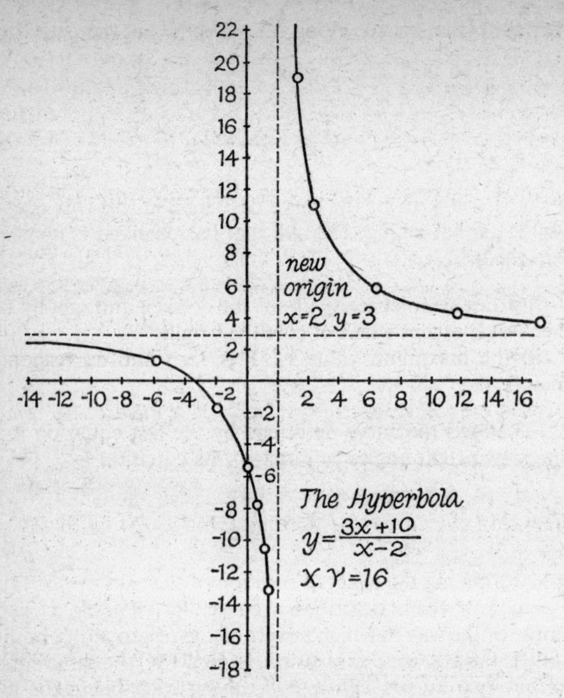

Fig. 134. Two Ways of Describing a Hyperbola

equation then becomes:

$$y = \frac{3(X + 2) + 10}{X} = \frac{3X + 16}{X} = 3 + \frac{16}{X}$$

$$\therefore \quad (y - 3) = \frac{16}{X}$$

To derive this, we have shifted our Y-axis 2 units to the right, so that $X = 0$ when $x = 2$. We can also shift the X-axis by putting $Y = (y - 3)$, so that $Y = 0$ when $y = +3$. Our equation referred to the new grid then becomes:

$$Y = \frac{16}{X} \quad or \quad XY = 16$$

In this form, the hyperbola is the graph of inverse proportion, as defined on p. 97. So described, we see that one limb becomes closer and closer to the X-axis as y approaches 0, and the other limb becomes closer and closer to the Y-axis as x approaches 0. We then say that one limb is *asymptotic* to the Y-, the other *asymptotic* to the X-axis.

Our first numerical example is illustrative of a hyperbola whose equation is:

$$y = \frac{ax + b}{x + C}$$

To reduce it to the form $xy = k$, we first shift the Y-axis by putting $X = x + C$ so that $x = -C$ when $X = 0$, and

$$y = \frac{(aX - aC) + b}{(X + C) - C} = \frac{aX + b - aC}{X} = a + \frac{b - aC}{X},$$

so that
$$y - a = \frac{b - aC}{X}$$

We now shift the X-axis by putting Y for $y - a$ so that $Y = 0$ when $y = +a$. Then

$$Y = \frac{b - aC}{X} \quad \textit{and} \quad XY = (b - aC)$$

Since $X = 0$ when $x = -C$ and $Y = 0$ when $y = +a$, what we have done to tailor the equation of the curve to this pattern is to transfer the origin to the point where $x = -C$ and $y = +a$.

Having seen how the equation of one and the same figure may change as the result of *translation* of the axes, i.e. moving the origin along one or both of them, let us now examine how it may change as the result of *rotating* the axes with or without change of origin. Three cases especially call for comment: (a) rotation through an angle of 90° ($\frac{1}{2}\pi$ radians); (b) rotation through 45° ($\frac{1}{4}\pi$ radians); (c) rotation through an angle of 180° (π radians).

The effect of rotation through 90° is simply to bring the former Y- into the same position as the new X-axis and vice versa. If the rotation is anti-clockwise the positive half of the Y-axis coincides with the negative half of the X-axis. Algebraically, this is equivalent to interchange of the two variables with appropriate change of sign. If the original co-ordinates of a point were x_1, y_1 and the new ones x_2, y_2, the rotation entails: $x_2 = y_1$ and $y_2 = -x_1$. The outcome is that the axis of symmetry of the figure changes accordingly, as will be clear if the reader plots the change for the case $y_1 = x_1^2$, in which event $y_2^2 = x_1$, so that $y_2 = \pm\sqrt{x_2}$. The figure corresponding to the last equation is still a parabola of precisely the same shape as the parabola described by $y_1 = x_1^2$. Only its orientation is different, that is to say:

(i) it is tangent to the *Y*- instead of to the *X*-axis;

(ii) it lies wholly on the right side of the *Y*-axis instead of wholly above the *X*-axis;

(iii) it is symmetrical on either side of the *X*- instead of the *Y*-axis.

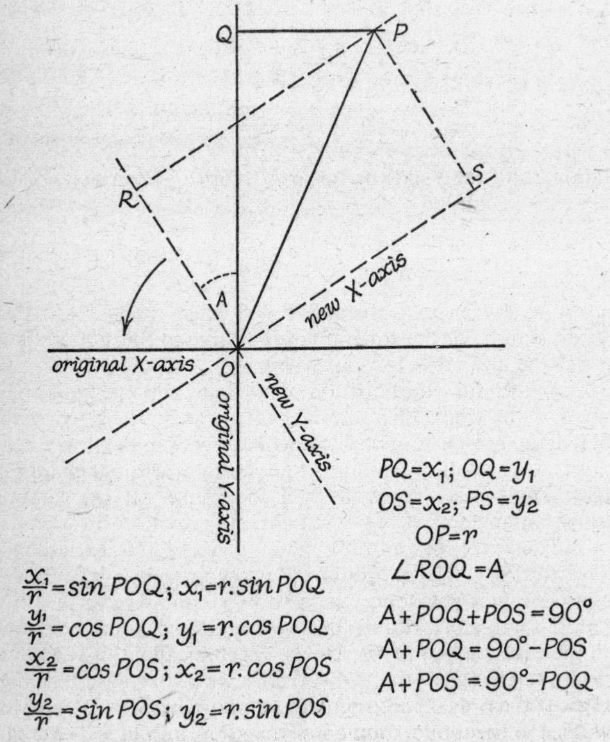

Fig. 135. Rotation of *X*- and *Y*-Axes through A° (anti-clockwise)

The effect of rotation of the axes through 45° is of special interest in connection with the hyperbola whose limbs are asymptotic to the axes. All that is relevant to prescribing a general recipe for such a transformation is in Fig. 135 where the co-ordinates of a point P are x_1, y_1 in the original framework and x_2, y_2 in the other. Since the rotation (*anti-clockwise* in the figure) is about the origin, the radius vector r ($=OP$) of the point P is the same in both frameworks. Our figure shows that:

$$\sin (A + POQ) = \cos (POS) = \sin A \,.\, \cos (POQ) + \cos A \,.\, \sin (POQ)$$

$$\therefore r \cos (POS) = x_2 = \sin A \,.\, r \,.\, \cos (POQ) + \cos A \,.\, r \,.\, \sin (POQ) = \sin \mathrm{A} \,.\, y_1 + \cos A \,.\, x_1$$

$$\cos (A + POQ) = \sin (POS) = \cos A \,.\, \cos (POQ) - \sin A \,.\, \sin (POQ)$$

$$\therefore r \,.\, \sin (POS) = y_2 = \cos A \,.\, r \,.\, \cos (POQ) - \sin A \,.\, r \,.\, \sin (POQ) = \cos A \,.\, y_1 - \sin A \,.\, x_1$$

$$\therefore x_2 = \cos A \,.\, x_1 + \sin A \,.\, y_1$$
$$y_2 = -\sin A \,.\, x_1 + \cos A \,.\, y_1$$

If $A = 90°$, $\sin A = 1$ and $\cos A = 0$, so that $x_2 = y_1$ and $y_2 = -x_1$ as stated above. If $A = 45°$:

$$\sin A = \frac{1}{\sqrt{2}} = \cos A, \text{ so that } x_2 = \frac{1}{\sqrt{2}}(x_1 + y_1) \text{ and } y_2 = \frac{1}{\sqrt{2}}(y_1 - x_1)$$

If we wish to express the result of a 45° rotation of the framework in which the equation of the hyperbola is $x_1 \,.\, y_1 = K$, we need to express x_1 and y_1 in terms of x_2 and y_2. We can derive this as follows:

$$x_2 + y_2 = \frac{2}{\sqrt{2}} \,.\, y_1 = \sqrt{2} \,.\, y_1, \text{ so that } y_1 = \frac{1}{\sqrt{2}}(x_2 + y_2)$$

$$x_2 - y_2 = \frac{2}{\sqrt{2}} \,.\, x_1 = \sqrt{2} \,.\, x_1, \text{ so that } x_1 = \frac{1}{\sqrt{2}}(x_2 - y_2)$$

Hence, if $x_1 y_1 = K$ is the equation of the hyperbola in one framework, the effect of a rotation of the grid through 45° *anti-clockwise* is:

$$x_1 y_1 = K = \frac{1}{\sqrt{2}}(x_2 + y_2) \,.\, \frac{1}{\sqrt{2}}(x_2 - y_2) = \tfrac{1}{2}(x_2^2 - y_2^2)$$

$$\therefore\; x_2^2 - y_2^2 = 2K \;\; and \;\; y_2 = \sqrt{x_2^2 - 2K}$$

If we rotate the axes through two right angles, $cos\ A = -1$ and $sin\ A = 0$, so that:

$$\cos A \,.\, x_1 + \sin A \,.\, y_1 = -x_1 \;\; and \;\; -\sin A \,.\, x_1 + \cos A \,.\, y_1 = -y_1$$

$$\therefore\; x_2 = -x_1 \;\; and \;\; y_2 = -y_1$$

If we rotate the axes through three right angles (270°), $cos\ A = 0$ and $sin\ A = -1$, so that:

$$x_2 = -y_1 \;\; and \;\; y_2 = x_1$$

The last two results are easy to visualize, and the first of the two discloses the transformation of the descending parabola of Fig. 132 into the ascending one of Fig. 133. For Fig. 132 we may write:

$$y_1 = -\frac{x_1^2}{4}$$

If we rotate the axes through 180°, we derive

$$y_2 = \frac{x^2}{4}$$

One other way in which the equation of a figure may assume different disguises draws attention to a concealed assumption in the derivation of all preceding expressions, viz. that the scale, i.e. unit of length, of the two axes of the grid is the same. If we make the scale of one axis different from that of the other, the equation which describes the figure will be different. Consider the parabola:

$$y = 4x^2 - 7$$

If we write $X = 2x$, so that $X = 2$ when $x = 1$ and $X = 1$ when $x = 0{\cdot}5$, our equation becomes $y = X^2 - 7$ and $y - 7 = X^2$. If we shift the origin 7 units upwards, so that $Y = y - 7$, we obtain the simplest possible form of the equation of the parabola, i.e. $Y = X^2$.

Such a transformation involves a *change of scale* on the X-axis and of *origin* on the Y-. Fig. 136 illustrates a change of scale on the Y-axis and a shift of origin along both. The following equation again describes a parabola:

$$y = 3x^2 - 5x - 7$$

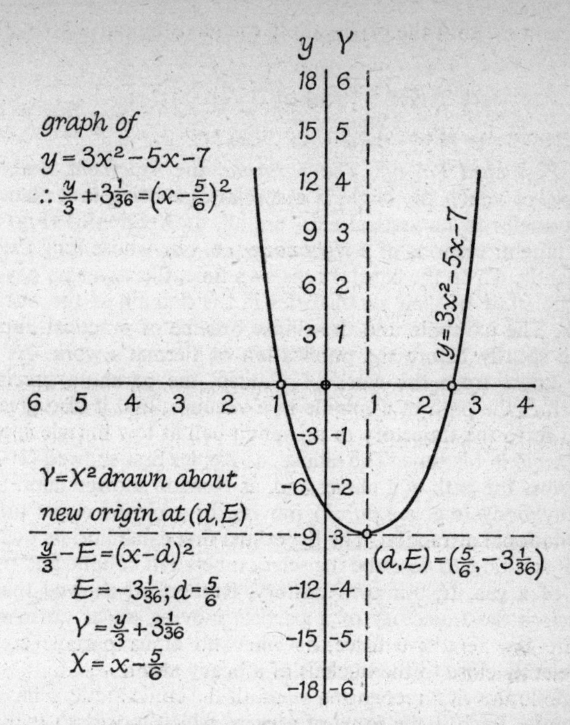

Fig. 136. Effect of Change of Scale and Origin on Equation of Parabola

If we change the scale of the Y-axis so that $3Y_1 = y$, we may write this as:

$$Y_1 = x^2 - \frac{5x}{3} - \frac{7}{3}$$

We can simplify the expression on the right as follows:

$$x^2 - \frac{5x}{3} - \frac{7}{3} = \left(x^2 - \frac{2(5x)}{6} + \frac{5^2}{6^2}\right) - \frac{5^2}{6^2} - \frac{7}{3} = \left(x - \frac{5}{6}\right)^2 - \frac{109}{36}$$

$$\therefore\ Y_1 + \frac{109}{36} = \left(x - \frac{5}{6}\right)^2$$

We can now shift the origin along each axis by putting:

$$Y_2 = Y_1 + \frac{109}{36} \quad and \quad X = x - \frac{5}{6}$$

$$\therefore \; Y_2 = X^2$$

A Postponed Pay-off. The *parabola*, the *hyperbola*, and the *ellipse*, of which the circle is a special case being in a plane at right angles to its vertical axis, are all, as Apollonius (Fig. 97) first taught, sections of a *right* cone (i.e. one whose long axis is perpendicular to the base). In his own time, there was no pay-off for the study of their peculiarities in the domain of the world's work. The parabola and the ellipse became of practical importance shortly before the publication of Fermat's work. As we now know from the work of Galileo, the parabola precisely describes the path of a missile in a vacuum; and it also gives a good fit to the trajectory of a cannon-ball at low muzzle speeds realizable in his time. The ellipse, as Kepler first showed (1609), describes the path of a planet and, as Newton showed later, that of any body (e.g. a *sputnik*) moving in a closed curve under gravitational attraction. The hyperbola first came into its own in Newton's time to describe the relation between volume and pressure of a gas. In our own century, Rutherford showed that it describes the trajectory of a particle moving under an inverse square law repulsive force, as when an alpha particle comes sufficiently close to the nucleus of a heavy atom.

Apollonius first recognized what all the conic sections have in common. Each is the *locus* of a point which moves so that the ratio of its distance (f) from a fixed point (the *focus* F) to its distance (d) from a straight line (the *directrix*) is constant. It is usual to label the numerical value of this fixed ratio as e, so that $f \div d = e$ and $f = ed$. Given this definition of a curve, we may conveniently choose as our Y-axis the *directrix* itself, and as our X-axis the horizontal line on which lies the focus, as in Fig. 137, so that $d = x$ and

$$y^2 + (K - x)^2 = f^2 = d^2e^2 = e^2x^2$$

$$\therefore \; y^2 = (e^2 - 1)x^2 + 2Kx - K^2$$

One calls this the *general equation of the conic*, and it is usual to speak of the fraction e as the *eccentricity* of the latter. By definition, the numerical constant K is the distance of the focus from the *directrix*. Whether the general equation describes a hyperbola,

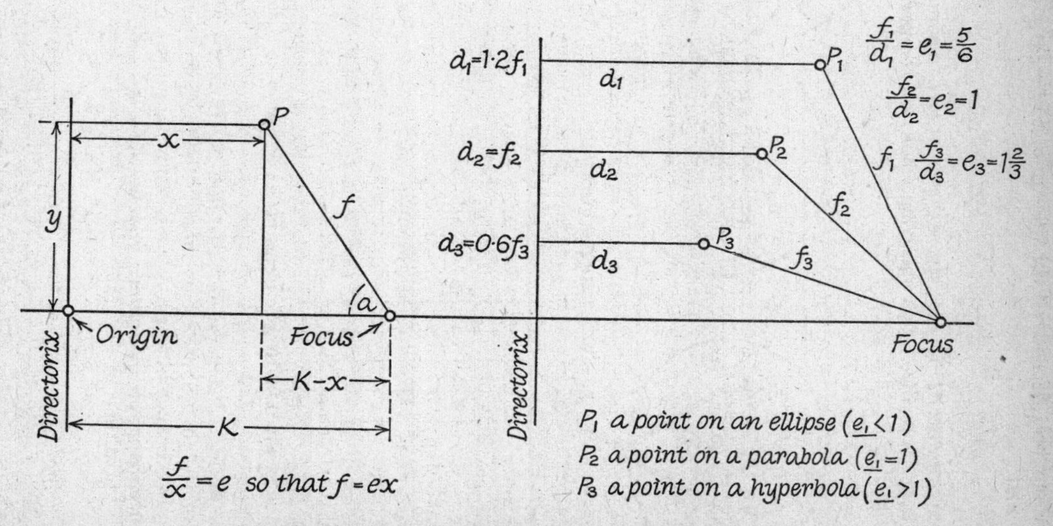

Fig. 137. The Locus of a Point on a Conic Section

ellipse or parabola depends on the value of e. If $e = 1$, $(e^2 - 1) = 0$, so that

$$y^2 = K(2x - K) = K \,.\, X, \textit{ if } X = 2x - K$$

As we have now seen (Fig. 133), this is the equation of a parabola symmetrical about the X-axis and lying wholly on the positive side of the Y-axis.

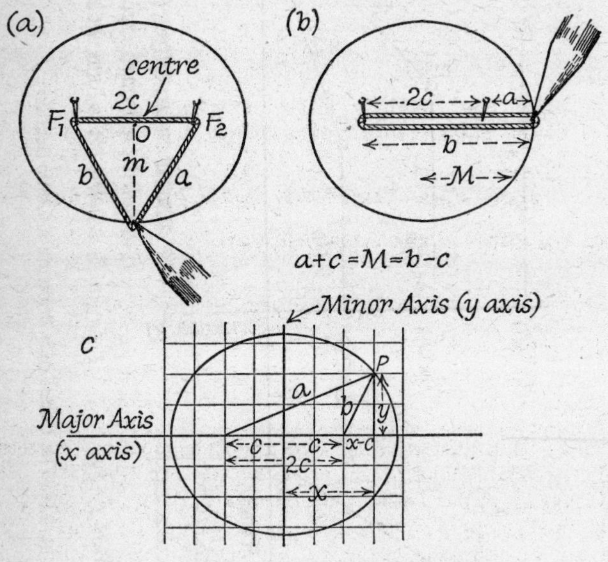

Fig. 138. How to Construct an Ellipse

Before asking what curve the equation describes when e is greater or less than unity, let us simplify the foregoing general equation by a change of origin. We now know that this does not effect the shape of the latter. For brevity, we may write $A = e^2 - 1$, so that:

$$\frac{y^2}{A} = x^2 + \frac{2K}{A}x - \frac{K^2}{A} = x^2 + \frac{2k}{A}x + \frac{K^2}{A^2} - \frac{K^2}{A^2} - \frac{K^2}{A}$$

$$\therefore \quad \frac{y^2}{A} = \left(x + \frac{K}{A}\right)^2 - \frac{K^2(1 + A)}{A^2}$$

$$\therefore \quad \frac{y^2}{A} = X^2 - \frac{K^2(1 + A)}{A^2} \quad \textit{if } X = x + \frac{K}{A}$$

Once more for brevity, we may write:

$$M^2 = \frac{K^2(1 + A)}{A^2} = \frac{K^2 e^2}{(e^2 - 1^2)},$$

$$\textit{so that } M = \frac{Ke}{e^2 - 1} \textit{ or } \frac{Ke}{1 - e^2}$$

$$\therefore y^2 = AX^2 - AM^2$$

When $e > 1$, so that $(e^2 - 1) = A$ is positive, the above is one of the guises (p. 354) in which we have met the *hyperbola*. When $e < 1$, so that A is negative, we may conveniently write it as:

$$\frac{-y^2}{AM^2} = \frac{-X^2}{M^2} + 1, \textit{ so that } \frac{X^2}{M^2} - \frac{y^2}{AM^2} = 1$$

Since $-A = 1 - e^2$, this is equivalent to:

$$\frac{X^2}{M^2} + \frac{y^2}{(1 - e^2)M^2} = 1$$

If $(1 - e^2)M^2 = m^2$:

$$\frac{X^2}{M^2} + \frac{y^2}{m^2} = 1$$

When $e = 0$, so that $m^2 = M^2$, the foregoing equation becomes that of a circle of radius M, i.e. $X^2 + y^2 = M^2$. Since $(-X)^2 = X^2$ and $(-y)^2 = y^2$, there will be on either side of the Y-axis, numerically equal values of X corresponding to one and the same value of y, and on either side of the X-axis numerically equal values of y corresponding to one and the same value of X. Thus the curve (lower half of Fig. 138) described is, like the circle, symmetrical about both axes. That it is also, like the circle, a closed curve is clear from the following considerations. Let $(M + a) > M$. If therefore $X = M + a$,

$$\frac{y^2}{m^2} = 1 - \frac{(M + a)^2}{M^2}$$

Since the quantity on the right is negative, y^2 is *negative*, i.e. no *real* value of y satisfies the equation. Similarly, we may put $y = m + a$, in which event X^2 is negative and no real value of X satisfies the equation.

Now $y^2 = m^2$ and $y = +m$ or $-m$ when $X = 0$. Likewise, $X^2 = M^2$ and $X = +M$ or $-M$ when $y = 0$. In short, the curve lies wholly within the range

$$y = \pm m \text{ when } X = 0; \quad X = \pm M \text{ when } y = 0$$

Since $M > m$, the horizontal range is greater and one speaks of the distance M along the X-axis on either side of the origin as the *major semi-axis*. Similarly, one calls the distance m along the Y-axis on either side of the origin as the *minor semi-axis*.

Needless to say, the figure (lower half of Fig. 138) about which we have been last speaking is the *ellipse*, whose Cartesian equation with origin at the mid-point of the major and minor axes is:

$$\frac{X^2}{M^2} + \frac{y^2}{m^2} = 1$$

We have thus seen that one and the same equation for the conic describes all three main types of conic section, according as the value of one of its numerical constants (e) satisfies one of three conditions:

$$e > 1 \text{ hyperbola}$$
$$e = 1 \text{ parabola}$$
$$e < 1 \text{ ellipse}$$

As is commonly true of *closed* curves (such as the circle or ellipse) and spirals, the polar equation of the ellipse is easy to derive. If we place the origin at F, and take the X-axis as base line, in Fig. 137, let us first recall the meaning of our numerical constants, viz.:

$$Ke = (1 - e^2)M \text{ and } (1 - e^2)M^2 = m^2$$

$$\frac{m^2}{M} = \frac{(1 - e^2) \,.\, M^2}{M} \text{ and } Ke = \frac{m^2}{M}$$

In Fig. 137, $f = ex$, so that

$$K - x = K - \frac{f}{e} = f \,.\, \cos a$$

$$\therefore\ Ke = f + ef \,.\, \cos a = f(1 + e \,.\, \cos a)$$

$$\therefore\ f = \frac{Ke}{1 + e \,.\, \cos a} = \frac{m^2}{M(1 + e \,.\, \cos a)}$$

In the figure f is the *radius vector* of the point P, whence usually written $f = r$, in a polar equation. With origin at the focus, the polar equation is therefore:

$$r = \frac{m^2}{M(1 + e \,.\, \cos a)}$$

For the do-it-yourself enthusiast, there is a very simple method of generating an ellipse. If on paper, one needs two pins, a pencil, and a loop of cotton. If to make a flower plot on a grass lawn, three pegs and a loop of stout cord. The upper half of Fig. 138 shows the lay-out and procedure. The distance between the two pins is $2c$. If a and b are the distances between the pencil point and one or other of the two pins (F_1 and F_2) the total length of the loop is $2c + a + b$. When the pencil point makes the two halves of the loop lie side by side:

$$(a + c) = M \quad \textit{and} \quad (b - c) = M, \quad \textit{so that} \quad 2M = a + b$$

For economy of space, we may introduce a numerical constant $(c \div M) = k$, so that $c = kM$. If we choose as origin a point midway between F_1 and F_2, Fig. 138 shows that:

$$a^2 = y^2 + (x + c)^2 = y^2 + x^2 + 2cx + c^2$$
$$b^2 = y^2 + (x - c)^2 = y^2 + x^2 - 2cx + c^2$$

$$\therefore\ a^2 - b^2 = (a + b)(a - b) = 4cx$$

Since $(a + b) = 2M$:

$$a - b = \frac{2c}{M} \,.\, x = 2kx$$

$$\therefore\quad (a - b)^2 = a^2 - 2ab + b^2 = 4k^2x^2$$

But from above we see that:

$$a^2 + b^2 = 2(y^2 + x^2 + c^2)$$

$$\therefore\ 2ab = 2y^2 + 2x^2 + 2c^2 - 4k^2x^2$$

$$\therefore\ a^2 + 2ab + b^2 = (a + b)^2 = 4y^2 + 4x^2 + 4c^2 - 4k^2x^2$$

Since $(a + b) = 2M$, $(a + b)^2 = 4M^2$, and

$$4M^2 = 4(y^2 + x^2 + k^2M^2 - k^2x^2)$$

$$\therefore\ M^2 = y^2 + (1 - k^2)x^2 + k^2M^2$$

$$\therefore\ (1 - k^2)M^2 = y^2 + (1 - k^2)x^2$$

$$\therefore \frac{x^2}{M^2} + \frac{y^2}{(1 - k^2)M^2} = 1$$

This we have seen to be the Cartesian equation of the ellipse, k being its *eccentricity* and $(1 - k^2)M^2 = m^2$ in the foregoing symbolism.

As a numerical example of the properties of the ellipse, let us consider the moon's orbit. The values (in miles) of the minor and major semi-axes are:

$$m = 238{,}470 \text{ and } M = 238{,}833$$

Since $m^2 = (1 - e^2)M^2$:

$$e = \sqrt{\frac{M^2 - m^2}{M^2}}$$

The Cartesian equation is expressible as:

$$y = \frac{m}{M}\sqrt{M^2 - x^2}$$

$$\therefore y = \frac{238{,}470}{238{,}833}\sqrt{238{,}833^2 - x^2}$$

This formula suffices to plot the orbit completely.

Oscillatory Motion. One of the most rewarding consequences of the division of the Cartesian framework into four quadrants was an immense simplification of the mathematics of oscillatory motion. This depends on the possibility of extending our definition of the trigonometrical ratios to embrace angles of any size. The need to do so has already emerged (p. 325) in connection with the solution of spherical triangles. There we have had occasion to recognize the advantages of identifying the sine and cosine respectively with the length of two lines (Fig. 58) in a circle of unit radius. In the right-hand upper quadrant of Fig. 139 of the Cartesian grid, these two lines ($x = \cos A$, $y = \sin A$) are the co-ordinates of the point Q on a circle of unit radius with centre at the origin; and it is consistent with this definition to define the sines and cosines of POR, POS, POT, and POU respectively as the y and x co-ordinates of the points R, S, T, U located on the boundary of the unit circle by *anti-clockwise* rotation of OP through $(90° + A)$, $(180° - A)$, $(180° + A)$ and $(270° + A)$.

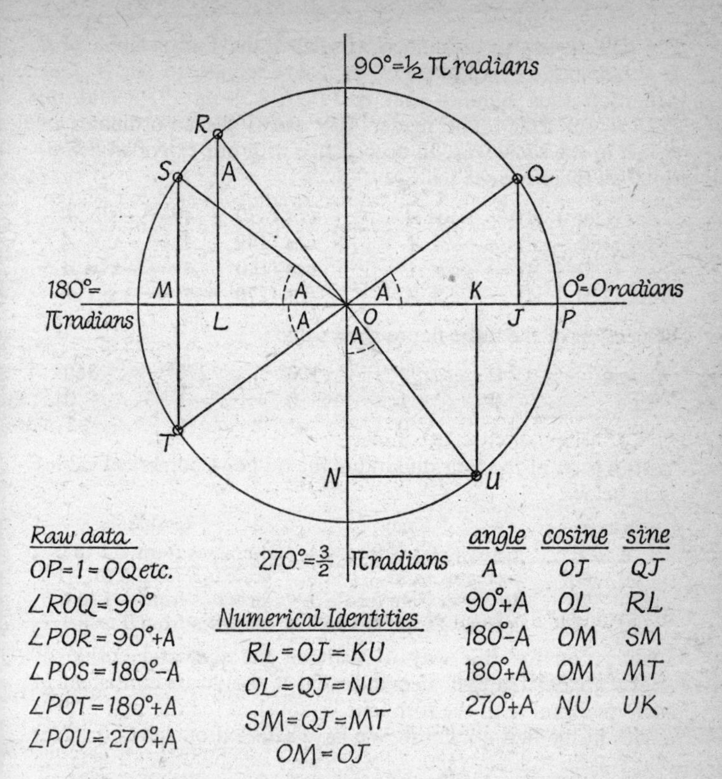

Fig. 139. Trigonometrical Ratios in the Four Quadrants of the Cartesian Grid

The x co-ordinates of any point in the quadrant enclosing angles between 90° and 180° are negative, but the y co-ordinates are positive. Those of any point between 180° and 270° are both negative. In the fourth quadrant x co-ordinates are positive, y negative. We may summarize these statements thus:

Quadrant	Sine	Cosine
First (0°–90°)	+	+
Second (90°–180°)	+	−
Third (180°–270°)	−	−
Fourth (270°–360°)	−	+

Fig. 139 shows that the numerical value of the Y co-ordinate of R is equal to *cos A* and that of its X co-ordinate to sin A. Our definition thus signifies that *cos POR* = *−sin POQ* and *sin POR* = *cos POQ*. The reader may assess the co-ordinates of S, T, U in the same way. In accordance with our extended definition, we therefore see that:

$\sin(90 + A) = +\cos A$	$\cos(90 + A) = -\sin A$
$\sin(180 - A) = +\sin A$	$\cos(180 - A) = -\cos A$
$\sin(180 + A) = -\sin A$	$\cos(180 + A) = -\cos A$
$\sin(270 + A) = -\cos A$	$\cos(270 + A) = +\sin A$

Inspection of the same figure shows that:

angle	0	90°	180°	270°	360°
sine	0	+1	0	−1	0
cosine	+1	0	−1	0	+1

Within each of the four quadrants the range of numerical values is as below:

Range	Sine	Cosine
0°–90°	increases from 0 to +1	decreases from +1 to 0
90°–180°	decreases from +1 to 0	decreases from 0 to −1
180°–270°	decreases from 0 to −1	increases from −1 to 0
270°–360°	increases from −1 to 0	increases from 0 to +1

Justification for this way of defining the sine and cosine of angles greater than 90° demands that it should be consistent in each quadrant with the formulae established for the sum of two angles in the first quadrant. We have satisfied ourselves (p. 226) that this is true for (180° − A). So we need concern ourselves only with (90° + A) in the second and with the third and fourth quadrants. In accordance with the formulae for ($B + A$):

$$\sin(90 + A) = \sin 90 \,.\, \cos A + \cos 90 \,.\, \sin A = +\cos A$$
$$\cos(90 + A) = \cos 90 \,.\, \cos A - \sin 90 \,.\, \sin A = -\sin A$$

When $(A + B) = (180 + A)$, the formulae for *cos* and *sin* $(A + B)$ becomes:

$$\sin(180 + A) = \sin 180 \,.\, \cos A + \cos 180 \,.\, \sin A = -\sin A$$
$$\cos(180 + A) = \cos 180 \,.\, \cos A - \sin 180 \,.\, \sin A = -\cos A$$

Similarly, we derive also in accordance with Fig. 139:

$$\sin(270 + A) = \sin 270 \,.\, \cos A + \cos 270 \,.\, \sin A = -\cos A$$
$$\cos(270 + A) = \cos 270 \,.\, \cos A - \sin 270 \,.\, \sin A = +\sin A$$

Thus the results of applying the formulae for *cos* or *sin* $(B + A)$, are consistent with the signs and numerical values of the co-ordinates $x = cos\ A$ and $y = sin\ A$ in each quadrant of the Cartesian grid. Our extended definition also leads to corresponding formulae for *cos* $(B - A)$ and *sin* $(B - A)$. These follow from the fact that cosines are positive and sines are negative in the fourth quadrant, so that:

$$\cos(-B) = \cos B \quad \textit{and} \quad \sin(-B) = -\sin B$$

We may therefore write:

$$\begin{aligned} \cos(A - B) &= \cos A \,.\, \cos(-B) - \sin A \,.\, \sin(-B) \\ &= \cos A \,.\, \cos B + \sin A \,.\, \sin B \\ \sin(A - B) &= \sin A \,.\, \cos(-B) + \cos A \,.\, \sin(-B) \\ &= \sin A \,.\, \cos B - \cos A \,.\, \sin B \end{aligned}$$

The reader should memorize these formulae, and test them: (*a*) for angles not greater than 90°; (*b*) for $A = 180°$, $A = 270°$, and $A = 360°$.

We have now seen that:

(*a*) a table for sines and cosines of all angles from 0° to 90° inclusive contains all their *numerical* values for the entire range 0°–360°;

(*b*) within the range 0°–360°, each numerical value (other than 0 and 1) turns up four times, twice with a positive and twice with a negative sign.

It should need no further explanation to convey the truism that the ratios for $n \,.\, 360°$ are the same as for 360°, if n is a whole number. In that event *sin* $(n \,.\, 360) = 0$ and $\cos(n \,.\, 360) = +1$, so that:

$$\begin{aligned} \sin(n \,.\, 360 + A) &= \sin(n \,.\, 360) \cos A + \cos(n \,.\, 360) \sin A \\ &= \sin A \\ \cos(n \,.\, 360 + A) &= \cos(n \,.\, 360) \cos A - \sin(n \,.\, 360) \,.\, \sin A \\ &= \cos A \end{aligned}$$

If we use *radian* measure we may write these expressions thus:

$$\sin(2n\pi + a) = \sin a \quad \textit{and} \quad \cos(2n\pi + a) = \cos a$$

In short, as an angle x increases from 0 to any exact multiple of 360° (2π radians) its sine and cosine pass through the same range of *numerical* values, positive and negative values alternating.

If we plot $y = sin\ x$ and $y = cos\ x$, we therefore obtain as in Fig. 140 curves which recall a wave, or any other regular periodic, motion. Such curves have alternative positive peaks (*maxima*) when $y = +1$ and negative peaks (*minima*) when $y = -1$ and the same *period*, i.e. $360° = 2\pi$ radians.

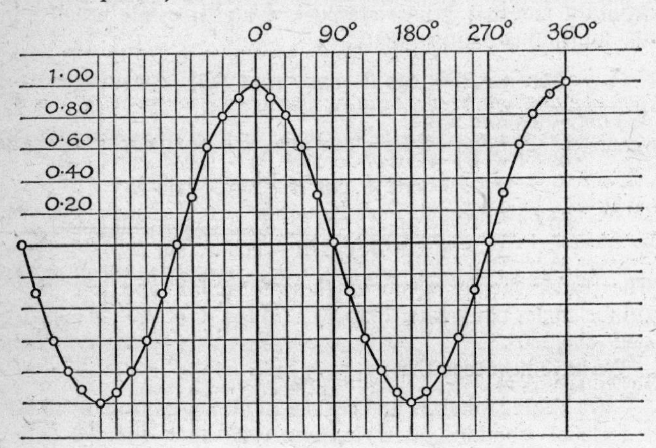

Fig. 140. Graph of $y = cos\ x$

Note: that of $y = sin\ x$ with peaks at 90°, 270°, etc., and zero values at 0°, 180°, etc., is otherwise identical.

The equations $y = cos\ x$ and $y = sin\ x$ define curves which are identical with respect to:

(*a*) *Period* (*P*), i.e. the interval between two peaks on the same side of the *X*-axis, or twice the distance between adjacent points, where $x = 0$;

(*b*) *Amplitude* (*A*), i.e. the numerical value of the distance between a peak and the *X*-axis, i.e. half the distance between the *Y* co-ordinates of peaks of opposite sign.

In each case: (*a*) the period is $x = 360°$ *or* 2π radians; (*b*) $y = \pm 1$ at a peak. The only difference between the two curves concerns the position of the peaks. If $y = cos\ x$, $y = +1$ when $x = 0$, and $y = 0$ when $x = 90°$ or $\frac{\pi}{2}$ radians. If $y = sin\ x$, $y = 0$ when $x = 0$, and $y = +1$ when $x = 90°$ or $\frac{\pi}{2}$ radians.

To exploit the properties of this class of curves to describe wave motion and undamped oscillations (e.g. swing of a pendulum or alternating current), we need to provide for a change of scale by substituting for y, $y \div A$ and for x, Kx, the assumption being that A and K are numerical constants. Our equations then become:

$$y = A \,.\, \cos Kx \quad \textit{and} \quad y = A \,.\, \sin Kx$$

The maximum values of y occur when $Kx = 0$ so that $\cos Kx = 1$ and $y = A$ or when $Kx = \frac{\pi}{2}$ radians and $\sin Kx = 1$ so that $y = A$. Thus A is the numerical value of the amplitude as defined above. Now the period P is the interval (in radians) between Kx and $Kx + 2\pi$, and we may write this as $K(x + P)$ so that:

$$Kx + 2\pi = Kx + KP \quad \textit{and} \quad 2\pi = KP$$

$$K = \frac{2\pi}{P}$$

Our equations last cited thus become:

$$y = A \,.\, \cos\left(\frac{2\pi}{P} \,.\, x\right) \quad \textit{and} \quad y = A \,.\, \sin\left(\frac{2\pi}{P} \,.\, x\right)$$

In these, we may write $x = t$ for time (secs.), $P = T$ and $y = f(t)$ meaning that y is a function of the time, so that:

$$f(t) = A \,.\, \cos\left(\frac{2\pi}{T} \,.\, t\right) \quad \textit{and} \quad f(t) = A \,.\, \sin\left(\frac{2\pi}{T} \,.\, t\right)$$

Given the numerical values of A and T by observation, we have thus all the information we need to fit a curve as in Fig. 141 to data embodying the type of regular periodic motion, called *simple harmonic*.

The Equation of the Straight Line. Since we can now give a meaning to $y = \cos x$ or $y = \sin x$ in each quadrant of the Cartesian grid, we can also describe lines that cross it at any angle A. It follows from what has gone before that $\tan A = \sin A \div \cos A$ will be positive in the first and third quadrants, since $\sin A$ and $\cos A$ have then like signs, and negative in the second and fourth, since they then have opposite signs. If a line which slopes upwards to the right cuts the X-axis at an angle A, $\tan A$ is positive. If it slopes upwards to the left making an angle

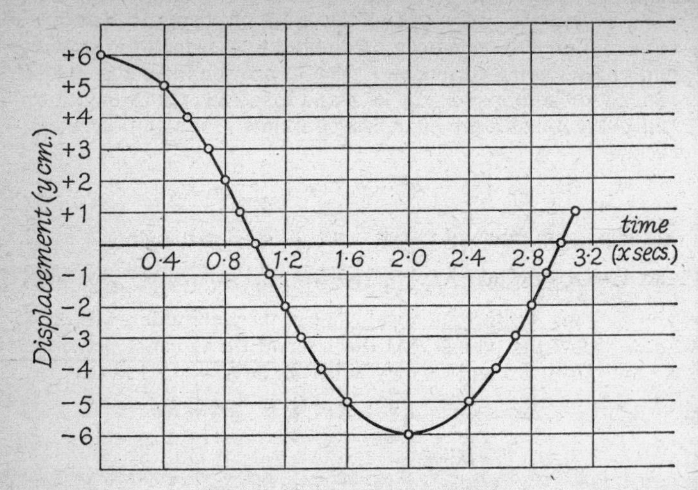

Fig. 141. Simple Harmonic Motion

A with the X-axis, *tan A* is negative. This is consistent with the rule $\tan (180° - A) = -\tan A$.

To deal exhaustively with a line in the Cartesian grid it suffices to visualize four situations:

(i) Cutting the Y-axis at $y = +b$, sloping upwards to the right (Fig. 142):

$$\tan A = \frac{(y - b)}{x}, \quad \textit{so that} \quad (y - b) = (\tan A)x$$

$$\therefore \; y = (\tan A)x + b$$

(ii) Cutting the Y-axis at $y = -b$, sloping upwards to the right (Fig. 142):

$$y \div \left(x - \frac{b}{\tan A}\right) = \tan A, \quad \textit{so that} \quad y = \tan A\left(x - \frac{b}{\tan A}\right)$$

$$\therefore \; y = (\tan A)\,x - b$$

(iii) Cutting the Y-axis at $y = +b$, sloping upwards to the left (Fig. 143):

$$\tan A = \frac{(y - b)}{-x}, \quad \textit{so that} \quad y - b = -(\tan A)x$$

$$\therefore \; y = -(\tan A)x + b$$

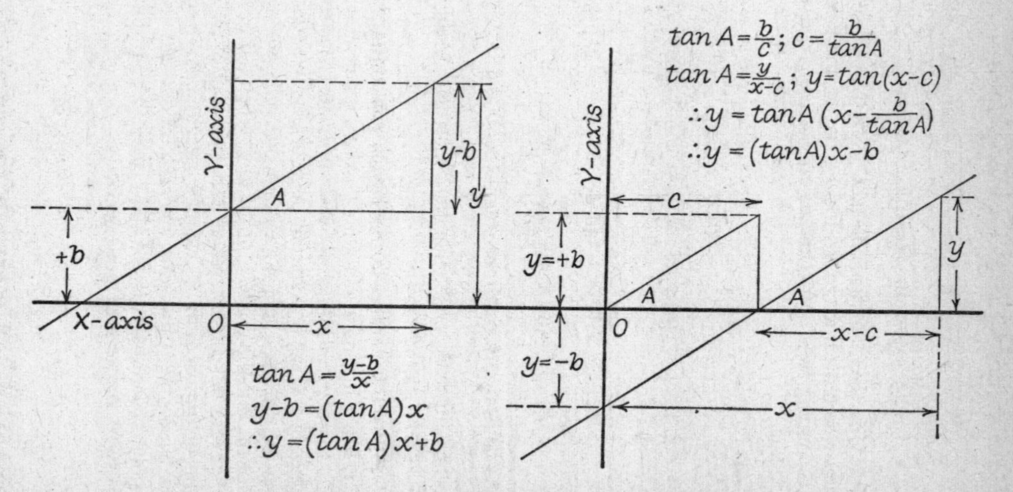

Fig. 142. Equations of a Line in Two Dimensions

(iv) Cutting the Y-axis at $y = -b$, sloping upwards to the left (Fig. 143):

$$y \div -\left(x + \frac{b}{\tan A}\right) = \tan A, \textit{ so that } y = -\tan A\left(x + \frac{b}{\tan A}\right)$$

$$\therefore\ y = -(\tan A)x - b$$

In these formulae the numerical constant $\pm b$ is the y-co-ordinate of the point where the line crosses the Y-axis, i.e. the y-co-ordinate corresponding to $x = 0$. The line crosses the X-axis where $y = 0$, so that:

$$\text{(i) } x = \frac{-b}{\tan A}$$

$$\text{(ii) } x = \frac{+b}{\tan A}$$

$$\text{(iii) } x = \frac{+b}{\tan A}$$

$$\text{(iv) } x = \frac{-b}{\tan A}$$

When the line passes through the origin $b = 0$ and:

$$y = (\tan A)x, \text{ if sloping up to the right}$$

$$y = -(\tan A)x, \text{ if sloping up to the left}$$

For a line which is parallel to the Y-axis, x is constant for all values of y. For a line parallel to the X-axis, y is constant for all values of x. In polar co-ordinates, one co-ordinate (A) is constant for all values of r, if the line goes through origin. In anticipation of Chapter 10, we may here notice that one calls *tan A* in the equation of the line the *gradient* of the line. When one speaks of the gradient of a hill in common speech, one may mean *sin A*.

With this definition in mind, we may usefully take the opportunity to supplement previous remarks about the relation between polar co-ordinates (r, A) and Cartesian co-ordinates in two dimensions. As we see from Fig. 129:

$$r^2 = x^2 + y^2 \quad \textit{and} \quad \frac{y}{x} = \tan . A$$

If *tan* $A = b$, it is customary to use the convention:

$$\tan A = b \equiv \tan^{-1} . b = A$$

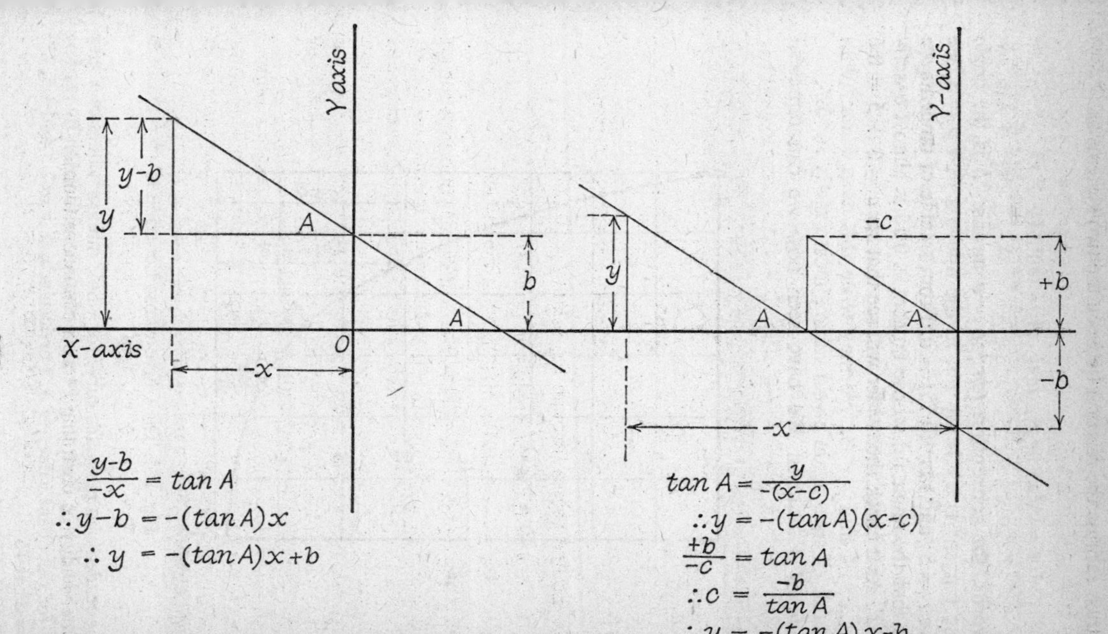

Fig. 143. Equations of a Line in Two Dimensions

We then express the relation between the two systems as:

$$r = \sqrt{x^2 + y^2} \quad and \quad A = \tan^{-1}(y \div x)$$

In words, the convention $tan^{-1}\ b = a$ means: 'b is the angle whose tangent is a'. If we put in the above $x = 4$ $y = 3$, we derive $r = 5$; and $tan^{-1}(0{\cdot}75) = a$. From a table of tangents we learn that the angle (a) whose tangent is 0·75 is almost exactly 36° 52′. As a check the reader will see that $sin\ a = 3 \div 5 = 0{\cdot}6$ when $x = 4$ and $y = 3$. We may write this as $a = sin^{-1}(0{\cdot}6)$. The table of sines again give a value almost exactly 36° 52′.

Damped Vibrations. We have seen how we can represent

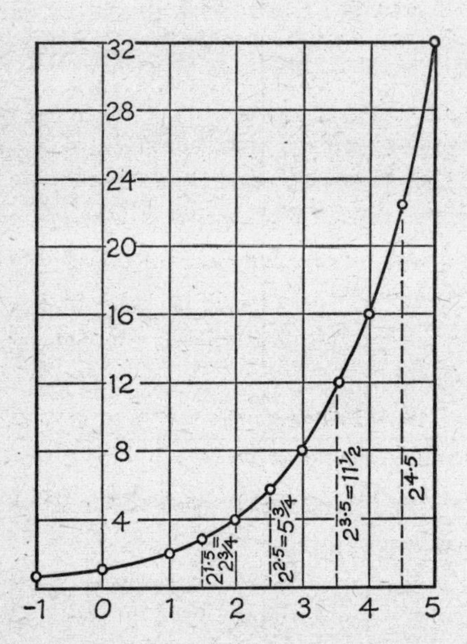

Fig. 144. The Curve whose Equation is

$y = 2^x$

The reader who plots this curve for whole number values of x may test the validity of regarding 2^x as a continuous function by reading off y for fractional values of x. For instance, if $x = 2{\cdot}5$ and $1{\cdot}5$, $2^{2\cdot5}$ $2^{1\cdot5} = 2^4 = 16$.

simple harmonic motion, i.e. oscillatory motion with a fixed period and fixed amplitude, by curves whose equations are:

$$f(t) = A \,.\, \cos\left(\frac{2\pi}{T} \,.\, t\right) \quad \textit{or} \quad f(t) = A \,.\, \sin\left(\frac{2\pi}{T} \,.\, t\right)$$

In these, the numerical constant A signifies that the amplitude remains the same, as does the period T. The displacement of the pendulum and the periodic changes of voltage produced by an A-C *sinusoidal* generator are examples of motion of this sort. More often in everyday life, we meet oscillatory phenomena (e.g. a weight attached to a vertical spring or a condenser discharge) with amplitude decreasing to zero. To provide curves to fit such so-called *damped* oscillations we need to invoke another class of curves, called *exponential.* Simple examples of this kind of curves are those of which the equations are:

$$y = 2^x \text{ and } y = 2^{-x}$$

Fig. 144 shows the curve whose equation is $y = 2^x$. To construct it we need to recall the law of indices first formulated by Archimedes and elaborated by Oresmus and by Stifel, viz.:

$$2^a \times 2^b = 2^{a+b}; \quad \text{e.g.} \quad 8 \times 32 = 2^3 \times 2^5 = 256 = 2^{3+5} = 2^8$$

If this rule is true, when a and b are fractions:

$$2^{\frac{1}{2}} \times 2^{\frac{1}{2}} = 2 = \sqrt{2}\,\sqrt{2}, \quad \textit{so that} \quad 2^{\frac{1}{2}} = \sqrt{2}$$

$$2^{\frac{1}{3}} \times 2^{\frac{1}{3}} \times 2^{\frac{1}{3}} = 2 = \sqrt[3]{2}\,\sqrt[3]{2}\,\sqrt[3]{2}, \quad \textit{so that} \quad 2^{\frac{1}{3}} = \sqrt[3]{2}$$

We also recall that:

$$(2^a)^b = 2^{ab}; \quad \text{e.g.} \quad 8^2 = (2^3)^2 = 2^6 = 64$$

Both rules will be true, if we write:

$$2^{1 \cdot 5} = 2^{\frac{3}{2}} = (2^{\frac{1}{2}})^3 = (\sqrt{2})^3 \quad \textit{and} \quad 2^{2 \cdot 5} = 2^{\frac{5}{2}} = (\sqrt{2})^5$$

For instance, we may put:

$$2^{1 \cdot 5} \times 2^{2 \cdot 5} = 2^{\frac{3}{2}} \times 2^{\frac{5}{2}} = (\sqrt{2})^3 \times (\sqrt{2})^5$$
$$= (\sqrt{2})^8 = 2^4 = 2^{1 \cdot 5 + 2 \cdot 5} = 16$$

Our graph gives us a method of checking this rule. For instance, we can read off as accurately as possible on a graph plotted like Fig. 144 for whole number values of x:

$$2^{1 \cdot 5} \simeq \frac{11}{4}; \quad 2^{2 \cdot 5} \simeq \frac{23}{4}; \quad 2^{3 \cdot 5} \simeq \frac{23}{2}$$

Hence, to the same order of precision we have:

$$2^{1\cdot5} \times 2^{2\cdot5} = 15\cdot8 \text{ (\textit{true value} 16)}$$
$$2^{1\cdot5} \times 2^{3\cdot5} = 31\cdot\dot{6} \text{ (\textit{true value} 32)}$$

Recalling that $2^{-x} = 1 \div 2^x$, the reader will find it instructive to plot a similar graph for $y = 2^{-x}$. Its shape is the mirror image of the curve $y = 2^x$, and is a particular case of a more general type which describes, among other physical phenomena, the rate of cooling (Fig. 145) of a fluid. If we write T for temperature at

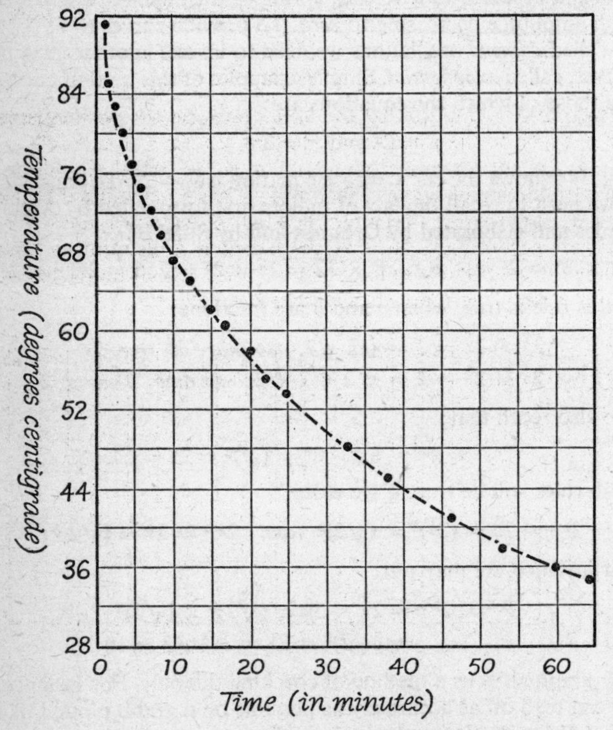

Fig. 145. The Law of Cooling

Graph of cooling of water in a 1 pint saucepan between 92° and 33° C. Room temperature 17° C.

time t and T_0 for initial temperature, i.e. when $t = 0$, the law of cooling is:

$$T = T_0 \, . \, A^{-kt}$$

In this expression $A^{-kt} = A^{\circ} = 1$, when $t = 0$, so that $T = T_0$ when $t = 0$, this being what we mean by calling T_0 the initial temperature.

We now have the clue we need for describing a damped vibration. We have been led to talk about how a curve can represent a cooling-off process by asking how we can tailor an expression invoking *sin x* or *cos x* to represent an oscillatory motion of which the amplitude more or less gradually falls off to zero. We can do so if we write (Fig. 146):

$$y = A \, . \, 2^{-kx} \, . \cos Kx$$

This expression implies that the initial peak on the positive side of the X-axis occurs when $x = 0$, so that $\cos Kx = 1$ and $2^{-kx} = 2^{\circ} = 1$, i.e. $y = A$ when $x = 0$, if A is the initial amplitude. The introduction of 2^{-kx} as a factor of $A \, . \cos Kx$ signifies that the vertical displacement becomes less as x increases and 2^{-kx} decreases accordingly. In 2^{-kx} the constant k determines how rapidly the amplitude damps, since 2^{-3x} will fall off more steeply than 2^{-2x} and 2^{-2x} more steeply than 2^{-x}, etc.

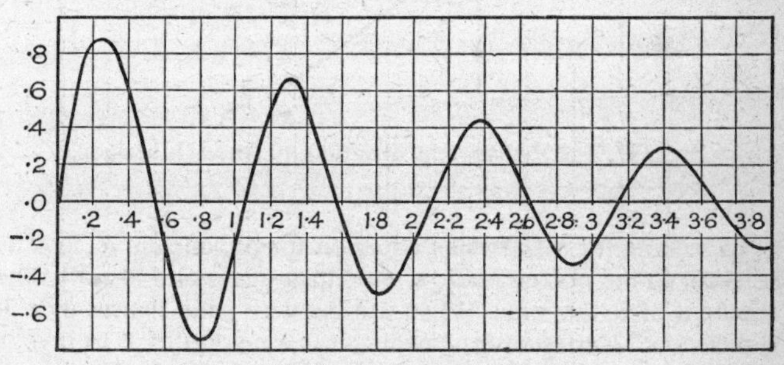

Fig. 146. Graph of a Damped Oscillation

$y = 3^{-\frac{x}{3}} \, . \sin 6x$

Solid Figures. For prescribing solid figures one of three systems of co-ordinates is most useful:

(*a*) *Spherical:* two angles l and L corresponding to longitude and latitude on a globe and a *radius vector* of length R which follows the plumb line joining any point P to its centre 0;

(*b*) *Cylindrical:* polar co-ordinates A and r in the horizontal plane and a distance $\pm z$ from the origin on the vertical Z-axis;

(*c*) *Cartesian:* x and y co-ordinates in the horizontal plane and z on the Z-axis at right angles to it as above.

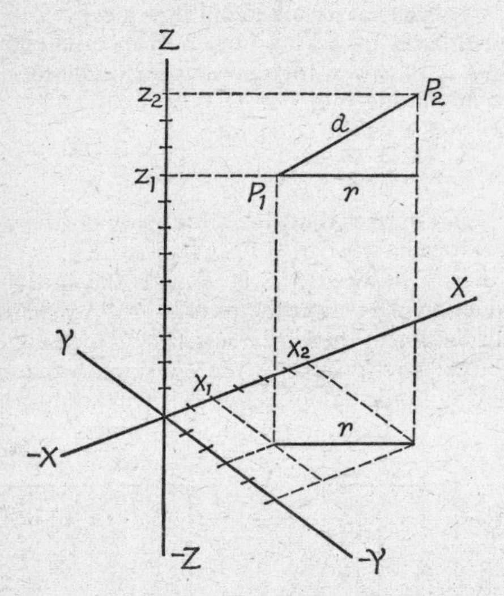

Fig. 147. The Pythagorean Relation in Three Dimensions

In seeking for a Cartesian equation for a plane curve, and its relation to the polar equation, we return again and again to the theorem of Pythagoras. When dealing with solid figures in a 3-dimensional Cartesian framework, it is a helpful trick to invoke what we may call the Pythagorean relation in three dimensions. We imagine a point P (Fig. 149) connected by a radius vector R to the origin (0) and by a line parallel to the Z-axis to the XY plane at a second point Q whose co-ordinates are x and y. Thus OQ is the radius vector (r) of the point Q in the XY plane and $x^2 + y^2 = r^2$ as in Fig. 129. Now OP, PQ, and OQ ($= r$) form a right-angle triangle of which $OP = R$ is the hypotenuse and $PQ = z$ is

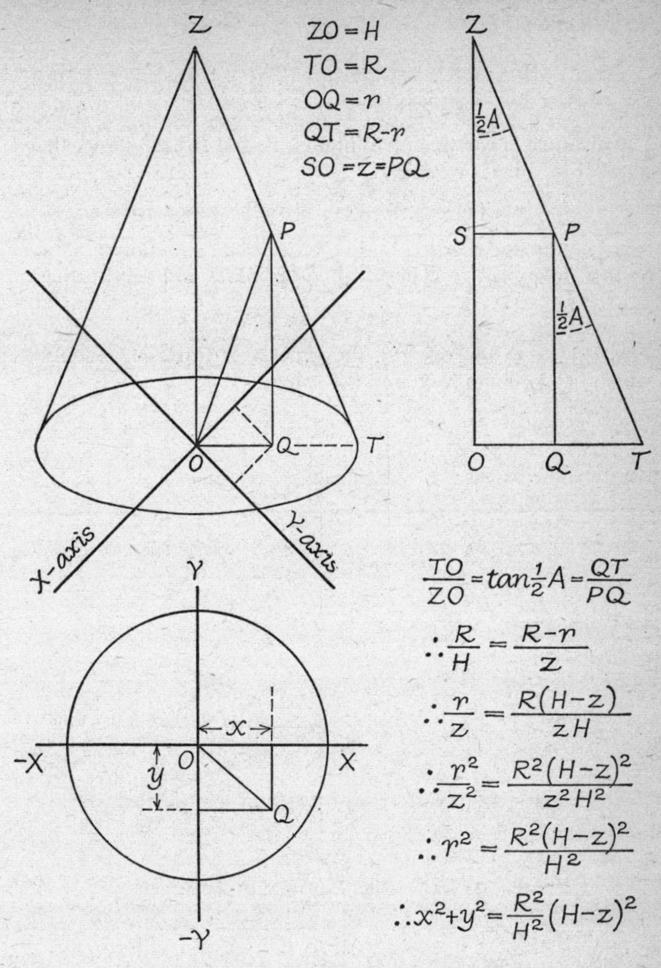

$$\frac{TO}{ZO} = \tan\tfrac{1}{2}A = \frac{QT}{PQ}$$

$$\therefore \frac{R}{H} = \frac{R-r}{z}$$

$$\therefore \frac{r}{z} = \frac{R(H-z)}{zH}$$

$$\therefore \frac{r^2}{z^2} = \frac{R^2(H-z)^2}{z^2H^2}$$

$$\therefore r^2 = \frac{R^2(H-z)^2}{H^2}$$

$$\therefore x^2+y^2 = \frac{R^2}{H^2}(H-z)^2$$

Fig. 148. Cartesian Equation of the Cone
Mid-point of base as Origin.

the z-co-ordinate of P, its other co-ordinates being x and y. In the triangle OPQ:

$$OP^2 = OQ^2 + PQ^2, \text{ so that } R^2 = r^2 + z^2$$

$$\therefore\ R^2 = x^2 + y^2 + z^2$$

The distance (d) between two points (P_1 and P_2) in 3-dimensional space (Fig. 147) is given by:

$$d^2 = (x_2 - x_1)^2 + (y_2 - y_1)^2 + (z_2 - z_1)^2$$

It is sometimes easier to obtain the Cartesian equation of a figure by first defining it in cylindrical co-ordinates and substituting:

$$y = r \,.\, \sin A; \quad x = r \,.\, \cos A$$

Figs. 148 and 149 disclose the genesis of Cartesian equations which describe the cone and the sphere:

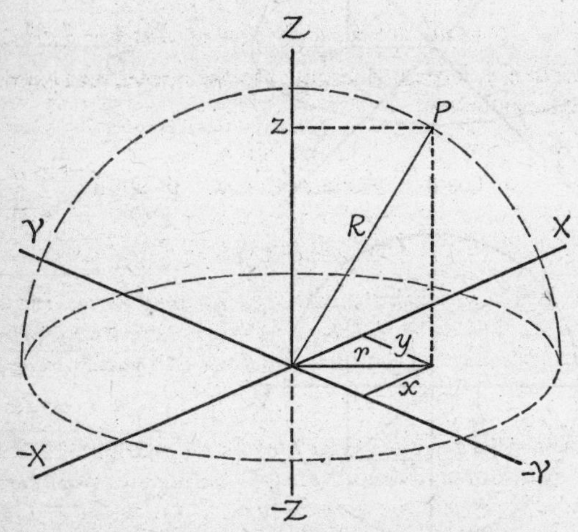

Fig. 149. Cartesian Equation of the Sphere Centre as Origin.

Cone: vertical height H, radius of base R, vertical angle A (*in section*), origin at centre of base:

$$x^2 + y^2 = \frac{R^2}{H^2}(H - z)^2$$

Sphere: radius R, centre at origin:

$$R^2 = x^2 + y^2 + z^2 \quad or \quad z = \sqrt{R^2 - x^2 - y^2}$$

Parametric Equations. To describe a flat geometrical figure, it is sometimes convenient to express the two variables x and y in terms of a third one in the same plane. Though this would entail little if any advantage, we might, for instance, describe the circle of radius r and centre as origin by the two so-called *parametric* equations in which A is the auxiliary variable:

$$x = r \cos A \quad and \quad y = r \sin A$$

One with more relevance to practical affairs is the use of parametric equations of a parabola in connection with the trajectory of the cannon-ball. If t is the time which has elapsed from firing, v the initial velocity at a horizontal angle a and g is a numerical constant (*acceleration under gravity*):

$$x = vt \,.\, \cos a \quad and \quad y = vt \,.\, \sin a - \tfrac{1}{2}gt^2$$

Needless to say, we can eliminate t in the above, and express y in terms of x, as below:

$$\therefore\ t = \frac{x}{v \,.\, \cos a} = \frac{x}{v \,.\, \sin a} \,.\, \frac{\sin a}{\cos a} = \frac{x}{v \,.\, \sin a} \,.\, \tan a$$

$$\therefore\ y = x \,.\, \tan a - \frac{g}{2} \,.\, \frac{x^2}{v^2 \,.\, \cos^2 a}$$

Since a and v are fixed quantities, we may write for brevity $\tan a = K$ and $(g \div 2v^2 \,.\, \cos^2 a) = C$. We then derive the equation of a parabola in the form exhibited earlier (Fig. 132):

$$y = Kx - Cx^2$$

A curve whose equation is most easily expressible in parametric form is of great importance in connection with the design of a clock. A pendulum swinging freely from its pivot, does so in a circular arc. If the amplitude is then considerable, its period varies appreciably. One can ensure complete independence of the two, if one forces it to move in an arc of the curve called a *Cycloid.* The clock-maker does so by placing suitably bent jaws of metal on either side of the point of suspension. The Swiss mathematician James Bernoulli, who first clearly recognized the usefulness of polar co-ordinates and the role of figurate numbers in the realm of choice and chance, discovered (about AD 1700) two unique properties of the cycloid related to its role in clock design. It is the curve along which a particle starting from rest and sliding along it under the influence of gravity to another point not in the same vertical line does so in the shortest time. It is likewise the curve along which a particle starting from rest and sliding down under the influence of gravity alone always takes the same time to reach the same terminal point regardless of the position of the point where motion began.

The two properties (so-called *brachistochrone* and *tautochrone*) last mentioned, refer to a curve which is concave from above. If drawn so that it is a convex from above the X-axis, as in Fig. 150, the cycloid is the curve traced by a pin on the rim of a rolling cart-wheel, and it illustrates one of the several immense advantages of using circular (radian) measure of angles in contradiction to degrees. By definition, an angle *a radians* formed at the centre of a circle of unit radius by two lines which cut off an arc of its circumference is equal to the length of the arc itself. If the length of the radius is R, the length of the arc is $R \, . \, a$. Given this definition, we see (Fig. 150) that:

$$x = R(a - \sin a)$$
$$y = R(1 - \cos a)$$

Graphical Solution of Equations. So far we have concerned ourselves only with the equation as a device for describing a geometrical figure and with the use of a graph to suggest the sort of algebraic expression we speak of as a scientific law. We shall now consider the use of a graph as a means of solving an equation or pair of equations. When we use the graph of a function in this way, we can make the best of a freedom we have not been able to enjoy hitherto. Unless we change the scale of distance on the two axes equally, any such change alters the shape of the figure described by one and the same equation, but we are free to choose the scale on either axis at will, when our only intention is to use a curve or curves to solve an equation or pair of equations. Indeed, it is best to choose the scale so that corresponding points are nearly equally spaced in both directions. This means that:

(*a*) if y is growing much more rapidly than x, as when $y = x^2$, we shall make our unit on the Y-axis smaller than our x unit of length;

(*b*) if x is growing more rapidly than y, as when $y = \sqrt{x}$, we

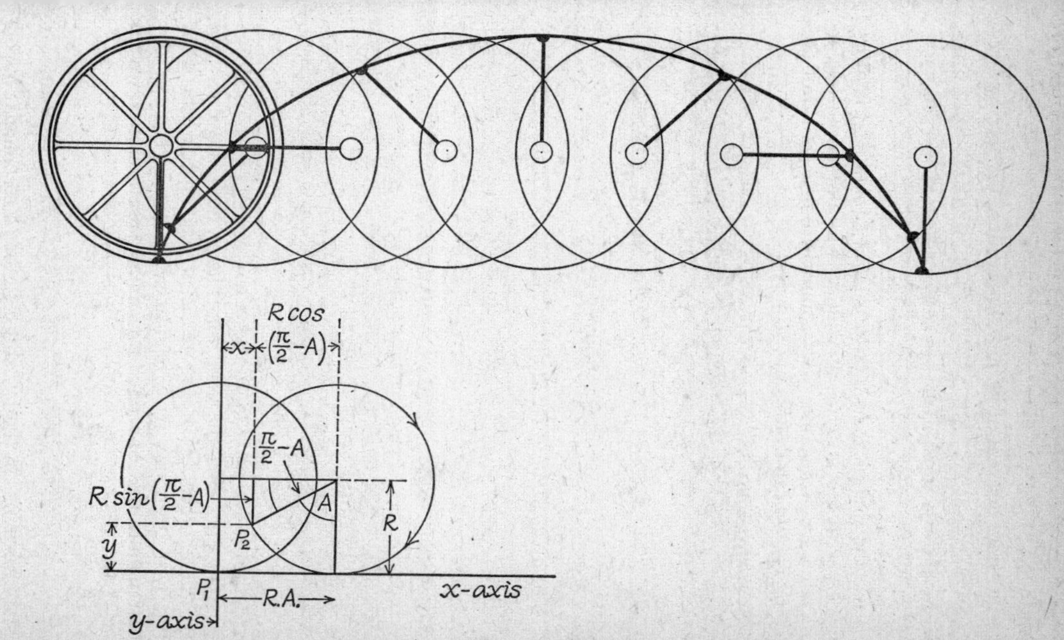

Fig. 150. Genesis of the Cycloid

shall make our unit on the X-axis smaller than our y unit of length.

To illustrate one use of graphical procedure for solving algebraic equations, let us first consider the pair of equations:

$$x^2 + 2xy + y^2 = 49$$
$$x^2 - xy + y^2 = 13$$

To use the graphical method in such a situation, it is more simple to express y in terms of x alone, with or without one or more numerical constants. So we must here first get rid of the terms in xy. We can do this in two ways:

By *subtraction*:

$$\begin{array}{r} x^2 + 2xy + y^2 = 49 \\ x^2 - xy + y^2 = 13 \\ \hline \therefore\ 3xy = 36 \\ \therefore\ xy = 12 \text{ and } y = \dfrac{12}{x} \end{array}$$

Also by *addition*, if we multiply the second of the pair by 2:

$$\begin{array}{r} x^2 + 2xy + y^2 = 49 \\ 2x^2 - 2xy + 2y^2 = 26 \\ \hline 3x^2 + 3y^2 = 75 \end{array}$$

$$\therefore\ x^2 + y^2 = 25 \quad \text{and} \quad y^2 = 25 - x^2$$

We have now two statements about y:

$$y = \frac{12}{x} \quad \text{and} \quad y^2 = 25 - x^2$$

One describes a hyperbola, the other a circle, and both mean the same thing only when the same points lie on both, i.e. where they cross one another; but it is more easy to read off co-ordinates of points with accuracy, if we can convert one or both of our equations into that of a line by a suitable substitution. If we put $Y = y^2$ and $X = x^2$ our two equations now become

$$Y = \frac{144}{X} \quad \text{and} \quad Y = 25 - X$$

The expression on the left still describes a rectangular hyperbola, but that of the second describes a straight line. To plot the

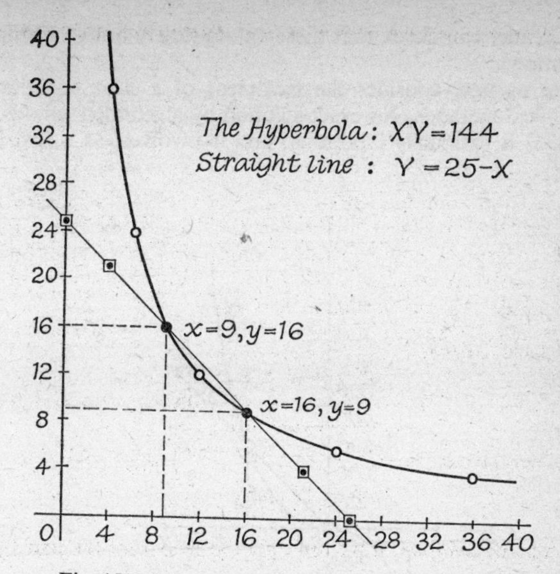

Fig. 151. A Graphical Solution of the Equations

$$x^2 + 2xy + y^2 = 49$$
$$x^2 - xy + y^2 = 39$$

hyperbola, as in Fig. 151, an abbreviated table will take shape as follows:

$X =$	4	9	16	36
$Y =$	36	16	9	4

Specimen points on the equation of the line will be:

$X =$	0	5	15	20	25
$Y =$	25	20	10	5	0

As we see from Fig. 151, the straight line cuts the hyperbola at two points:

$$X = 9,\ Y = 16 \quad \textit{and} \quad X = 16,\ Y = 9$$

Since $x = \sqrt{X}$ and $y = \sqrt{Y}$:

$$x = \pm 3 \text{ if } y = \pm 4 \quad \textit{and} \quad x = \pm 4 \text{ if } y = \pm 3$$

The reader can check that these values satisfy both the original equations.

Let us now consider the treatment of a *cubic* equation by graphical methods. An exact solution by algebraic methods (see p. 433) is extremely laborious; and it involves, as a first step,

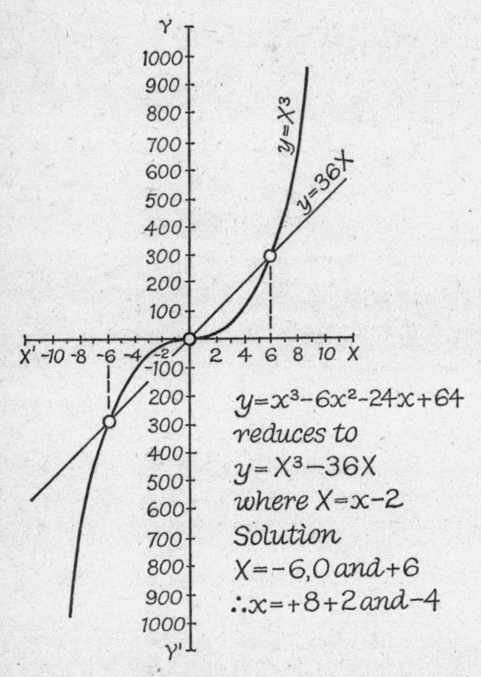

Fig. 152. A Graphical Solution of a Cubic Equation

giving the equation a new look which we can turn to advantage, if we prefer to rely on a graphical procedure. Let us consider the general type of cubic:

$$Ax^3 + Bx^2 + Cx + D = 0$$

We may plot $y = Ax^3 + Bx^2 + Cx + D$, in which event real solutions are values of x when $y = 0$, i.e. when the curve cuts the

X-axis. Alternatively, we might treat the equation in the following terms:

$$Ax^3 + Bx^2 = -(Cx + D) \text{ or } Ax^3 = -(Bx^2 + Cx + D)$$

We may then plot $y = Ax^3 + Bx^2$ and $y = -(Cx + D)$ or $y = Ax^3$ and $y = -(Bx^2 + Cx + D)$. The solutions are then values of x for which the pairs of lines intersect. We can make this dual procedure still more sensitive, if we use a substitution $X = x - a$ to get rid of the term in x^2. To illustrate the procedure (Fig. 152), let us consider the equation:

$$x^3 - 6x^2 - 24x + 64 = 0$$

If $X = x - a$, so that $x = X + a$:

$$x^3 = X^3 + 3aX^2 + 3a^2X + a^3$$
$$-6x^2 = -6X^2 - 12aX - 6a^2$$
$$-24x = -24X - 24a$$

Thus we may write:

$$X^3 + (3a - 6)X^2 + (3a^2 - 12a - 24)X + (a^3 - 6a^2 - 24a + 64) = 0$$

We may get rid of X^2 by writing $(3a - 6) = 0$, so that $a = 2$, in which event:

$$3a^2 - 12a - 24 = 12 - 24 - 24 = -36$$
$$a^3 - 6a^2 - 24a + 64 = 8 - 24 - 48 + 64 = 0$$

Our equation then becomes:

$$X^3 - 36X = 0, \textit{ so that } X^3 = 36X$$

If we plot separately $y = X^3$ and $y = 36X$, the solutions for X are values of x, for which the two graphs intersect. These happen to be:

$$X = -6,\ 0,\ +6$$

Since $x = X + 2$:

$$x = -4,\ +2 + 8$$

Check:

$$8^3 - 6(8^2) - 24(8) + 64 = 8(64) - 6(64) - 3(64) + 64 = 0$$
$$2^3 - 6(2^2) - 24(2) + 64 = 8 - 24 - 48 + 64 = 0$$
$$(-4)^3 - 6(4^2) - 24(-4) + 64 = -64 - 96 + 96 - 64 = 0$$

Solution by Iteration. To solve some equations by exact methods is either impossible or extremely laborious; and it may then be preferable or essential to invoke a procedure called *iteration* which ensures a solution of sufficient precision for all practical purposes. Iteration implies successive approximation, and its practicability presupposes that our first approximation is not far out. Graphical methods can ensure that it need not be. There are many recipes for iterative solution. The simplest, which invokes the binomial theorem, is practicable when an equation contains only whole number powers of the unknown and numerical constants. This depends on the following considerations. First we suppose that $x = a$ is a correct solution, but that our guess is b with an error (e), the sign of which is irrelevant at this stage. Thus we may write $a = b + e$. If our equation is a cubic with terms in x^3, x^2, and x, we may then write:

$$x^3 = b^3 + 3b^2e + 3be^2 + e^3$$
$$x^2 = b^2 + 2be + e^2$$

If our guess is good, in the sense that e is small compared with b, $(3be^2 + e^3)$ will be small compared with $b^3 + 3b^2e$, and e^2 will be small compared with $b^2 + 2be$. For instance, if $b = 2$ and $e = 0{\cdot}1$:

$$b^3 + 3b^2e = 8 + 12(0{\cdot}1) = 9{\cdot}2 \quad \textit{and} \quad 3be^2 + e^3 = 6(0{\cdot}1)^2 + 0{\cdot}001 = {\cdot}061$$

$$b^2 + 2be = 4{\cdot}4 \quad \textit{and} \quad e^2 = 0{\cdot}01$$

On the assumption that our guess is good, we may therefore say:

$$x^3 \simeq b^3 + 3b^2e; \quad x^2 \simeq b^2 + 2be$$

On substituting these values in the original equation, the only unknown will be e, which we can therefore determine. We then have a new estimate $c = b + e$ which we can deal with in the same way – and so on. Alternate values of e (e_1, e_3, etc., or e_2, e_4, etc.) will have opposite signs. The following numerical example must here suffice to illustrate the procedure. Our equation will be the cubic:

$$x^3 + 6x^2 - 6x - 63 = 0$$

We shall assume that a graph has shown us that one solution lies near $x = 2$. So we write first $x = 2 + e_1$, and the equation then becomes:

$$(2 + e_1)^3 + 6(2 + e_1)^2 - 6(2 + e_1) - 63 = 0$$

If we neglect all powers of e_1, higher than e_1 itself, we may write this as:

$$(8 + 12e_1) + 6(4 + 4e_1) - 6(2 + e_1) - 63 = 0 = 30e_1 - 43$$

Hence, $e_1 = 43 \div 30 = 1{\cdot}43$. To simplify further work we may write $2 + e_1 = 2 + 1{\cdot}4 = 3{\cdot}4$. We now have a new estimate and may write $x = 3{\cdot}4 + e_2$, and our original equation becomes:

$$(3{\cdot}4 + e_2)^3 + 6(3{\cdot}4 + e_2)^2 - 6(3{\cdot}4 + e_2) - 63 = 0$$

If we reject higher powers of e_2 as before:

$$(3{\cdot}4)^3 + 3(3{\cdot}4)^2e_2 + 6(3{\cdot}4)^2 + 12(3{\cdot}4)e_2 - 6(3{\cdot}4) - 6e_2 - 63 = 0$$

The solution of this yields $e_2 \simeq -0{\cdot}36$. We now have a third estimate $x \simeq 3{\cdot}4 - 0{\cdot}36 = 3{\cdot}04$. If we continue the process, we get nearer and nearer a true solution which the reader can check, i.e. $x = 3$.

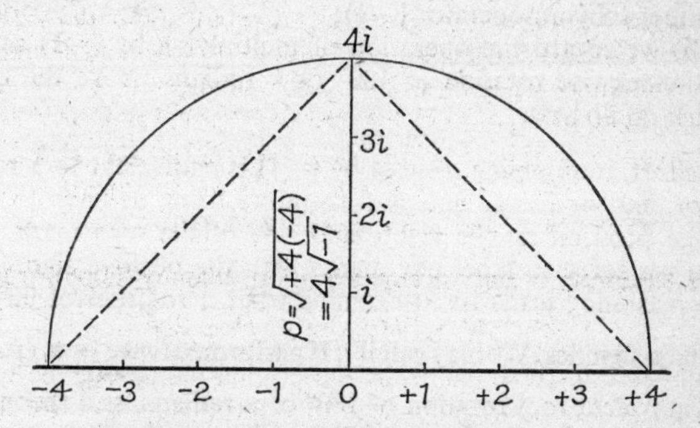

Fig. 153. Cartesian Interpretation of $\sqrt{-1}$.

The Not-so-imaginary i. From the time of Omar Khayyám onwards, familiarity with quadratic equations made mathematicians of the Western World uncomfortably aware of the existence of a class which yield no solution in familiar terms. Let us recall Alkarismi's recipe for the general type:

$$x^2 + bx + c = 0 \;\; if \;\; x = \frac{-b + \sqrt{b^2 - 4c}}{2} \text{ or } \frac{-b - \sqrt{b^2 - 4c}}{2}$$

If c is positive and $b^2 < 4c$ in the above, the sign of $b^2 - 4c$ is negative, e.g. when $x^2 - 8x + 32 = 0$:

$$x = \frac{8 \pm \sqrt{64 - 128}}{2} = \frac{8 \pm \sqrt{-64}}{2} = \frac{8 \pm 8\sqrt{-1}}{2}$$

It is usual to write $\sqrt{-1}$ as i, so that if $x^2 - 8x + 32 = 0$, $x = 4 + 4i$ or $4 - 4i$. Customarily, we speak of such expressions as *complex numbers*, and refer to $4i$ as the *imaginary part*. If we are interested only in the solution of quadratic equations, we can hope to derive no dividend from such anomalies; but we shall later be able to put them to more than one good use in other domains.

We have seen that $(-a)^2 = +a^2$ so that $(-1)^2 = +1 = (+1)^2$. So there is no ordinary number which corresponds with $\sqrt{-1}$; but we have also seen (p. 94) that we can interpret satisfactorily the sign rule $(-) \times (-) \equiv (+)$ only if:

(*a*) we regard $-a$ as the result of multiplying a signless number a by an operator (-1);

(*b*) we regard the operation of multiplying by (-1) as an anti-clockwise rotation of 180° or π radians above the zero mark on an axis.

In short, multiplying a twice by (-1) is equivalent to a rotation of 360° or 2π radians, and therefore bring us back to our original position on the right-hand side of the zero mark, i.e. where we agree to label our marks with the positive sign. The same reasoning leads us to examine what a rotation of 90° or $\frac{\pi}{2}$ radians signifies. We may call it j. If performed twice $j^2 \equiv (j\,.\,j.)$, it is equivalent to a rotation of 180° or π radians, and therefore to so-called multiplication by -1. In other words, $j^2 = -1$ and we may write $j.$ with a dot to signify that it is the *operation* denoted by $\sqrt{-1}$, i.e. $j \equiv i$ as defined above; and Fig. 153 shows that this is entirely consistent with the way in which we label the Cartesian 2-dimensional grid.

To grasp this we first recall that the *angle in a semicircle is a right angle*. So we can form a circle of radius 4 units through the points:

$$x = +4, y = 0; \quad x = 0, y = \pm 4; \quad x = -4, y = 0$$

Now we next recall (Fig. 53 and Dem. 4) that the perpendicular

(p) drawn from the vertex to the hypotenuse of a right-angled triangle cuts it into segments a and b in accordance with the relation:

$$p^2 = ab \quad and \quad p = \sqrt{ab}$$

In Fig. 153 the segments are $a = +4$ and $b = -4$, so that $ab = (+4)(-4) = -4^2$ and $p = \sqrt{-4^2} = \pm 4i$. Since the radius of the circle is 4 units of length, p, $+x$, and $-x$ are each numerically equal to 4, so that the operation denoted by i corresponds to an *anti-clockwise* rotation of the segment of the X-axis labelled positive through $\frac{\pi}{2}$ radians and the operation denoted by $-i$ to a rotation through $\frac{3\pi}{2}$ radians.

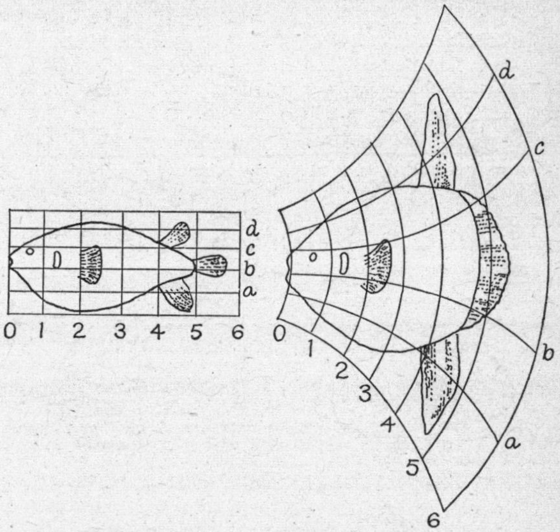

Fig. 154. Curvilinear Transformation of Shape

On the left is typical species of the genus *Diodon*. On the right, says Professor D'Arcy Thompson, 'I have deformed its vertical co-ordinates into a system of concentric circles and its horizontal co-ordinates into a system of curves which approximately and provisionally are made to resemble a system of hyperbolas. The old outline transformed in its integrity to the new network appears as a manifest representation of the closely allied but very different-looking sunfish Orthagoriscus.'

Curvilinear Co-ordinates. One recipe for representing the position of a point is by means of the kind of maps which geographers call Flamsteed or Mollweide projection maps, and mathematicians call curvilinear co-ordinates. The meridians of longitude on the curved surface of the globe are not parallel like the circles of latitude. They converge at the poles. The Mercator projection map in which the meridians of longitude are represented by parallel straight lines distorts the relative sizes of the continents and oceans, making countries like Greenland which lie far north of the equator appear to be much larger than they actually are. Maps like the Flamsteed projection map, in which the meridians are represented by curved lines converging towards the poles, correct this distortion. A very suggestive application of

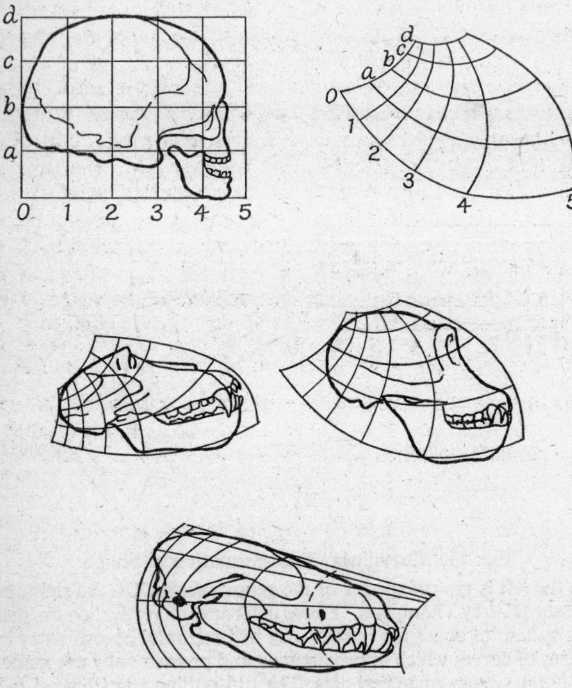

Fig. 155. Curvilinear Transformations of Shape
From *Growth and Form* by Professor D'Arcy Thompson.

curvilinear co-ordinates in Figs. 154–5 is taken from a book *Growth and Form* by Professor D'Arcy Thompson. A skull or body of one species drawn in a framework of curvilinear co-ordinates looks exactly like the skull or body of another species drawn in a Cartesian framework. The use of this method may conceivably provide a clue to the laws of growth in the evolution of species.

Exercises on Chapter 8

In all these exercises on graphs, in case you have not already realized it in reading the text, the following point is very important to remember. If we are using a graph to represent the *shape* of a geometrical figure correctly, we must always use the same unit of measurement for x and y. Thus $r^2 = x^2 + y^2$ is only the equation of a circle if x and y are both measured in the same units, e.g. centimetres or inches. If we are merely using a graph to solve an equation, our units need not be the same length. Thus if y is very large compared with x, it may be convenient to make one unit of x 1 centimetre, and one unit of y 0·01 centimetre, and all that matters is that we remember that a given distance measured along one axis has a different value from a given distance measured along the other axis. Of course, in graphs which represent physical laws the same applies, since we please ourselves what units we use.

Before you attempt any of the succeeding examples, show that the circle is the figure which corresponds with the equation:

$$y = \sqrt{25 - x^2}$$

To do this make a table of all values of y corresponding with values of x between -5 and $+5$ in steps of $\frac{1}{2}$, using a table of square roots. Thus when $x = -4\frac{1}{2}$:

$$\begin{aligned} y &= \sqrt{25 - (\tfrac{9}{2})^2} \\ &= \tfrac{1}{2}\sqrt{19} = \pm\tfrac{1}{2}(4{\cdot}36) \\ &= \pm 2{\cdot}18 \end{aligned}$$

Mark on your graph the two points which are $\pm 2{\cdot}18$ units measured along the y axis and $-4\frac{1}{2}$ along the x axis. If you use 1 centimetre as the unit and graph paper divided by lines 0·1 centimetre apart, y will be two large and practically two small divisions. Do the same for all the other values tabulated, and draw a line running smoothly between the points.

1. An elevator in a sixty-storey building makes the following journeys, starting at the lowest floor: up twenty storeys, down four, up eight, down three, down seventeen, up ten, down one, up five, up eleven, down twenty-four. Where is it at the end of this period?

2. Draw a graph of a circle with radius 4 centimetres: (*a*) with its centre at the origin; (*b*) with its centre at the point $x = 2$, $y = 3$. What is the Cartesian equation in each case?

3. Show that the angle which the tangent to a circle at any point makes with the x axis is $-\frac{x_r}{y_r}$, when x_r and y_r are the co-ordinates of the point and the centre is the origin.

From the figure you have just drawn see whether you can show that when a positive angle a is measured in radians sin a is less than a and tan a is greater than a.

4. If x is any length measured in inches and y is the same length measured in feet, then:

$$12y = x$$

Plot a 5-point graph connecting inches and feet. Read off from the graph how many inches there are in $1\frac{3}{4}$ feet, 3·6 feet, and 4·1 feet.

5. Make similar graphs for converting centimetres into inches, pounds into dollars, and pints into pounds of water.

6. Draw the graph $y = 3x + 4$, then draw at sight (i.e. without making a table):

$$3y = 5x + 6\left(\text{i.e. } y = \frac{5x}{3} + 2\right)$$

$$y = 32x + 40$$

7. On the Fahrenheit scale water freezes at 32° and boils at 212°. On the Centigrade scale it boils at 100° and freezes at 0°. Show that the formula for converting degrees Centigrade into degrees Fahrenheit is:

$$F = \frac{9C}{5} + 32$$

Draw the graph of this equation, and from the graph read off the Centigrade equivalent of normal blood temperature (98·4° F.).

8. If two straight lines are described by the equations:

$$(a)\ y = x + 3$$
$$(b)\ y = \sqrt{3}x + 2$$

at what angles are they inclined to the x axis?

9. If a set of straight lines are described by the equations:

$$y = mx + 1$$
$$y = mx + 2$$
$$y = mx + 3, \text{ etc.}$$

what do you know about these straight lines?

10. What does the quantity C represent graphically in the equation $y = mx + C$?

11. What would be the equation of a straight line drawn parallel to the x axis?

12. Mr Evans sets out for a walk. He walks 3 miles in the first hour, $2\frac{1}{2}$ miles in the next, 2 miles in the next hour. He rests for three-quarters of an hour and then goes on at a uniform rate of $2\frac{1}{2}$ miles an hour for 3 hours. Find from a graph the distance walked by Mr Evans in $1\frac{1}{2}$ hours, $3\frac{1}{2}$ hours, and $5\frac{1}{2}$ hours.

Mr Davies starts from the same point 2 hours later than Mr Evans, and cycles at a uniform rate of 6 miles per hour. Show Mr Davies's journey on the graph you have just drawn and read off the distance from the start when Mr Davies overtakes Mr Evans.

13. Plot the straight lines:

$$2y + 3x = 31$$
$$3y + 2x = 39$$

on the same graph. Read off from your graph what must be the values of x and y so that both equations shall be true at the same time. Turn back to Chapter 6, p. 268 *et seq.*, and you will see that you have found a graphical method for solving problems involving two simultaneous equations.

14. Solve the problems in Example 14, Chapter 6, graphically, and compare your solutions with those you have obtained previously.

15. Draw a careful graph of $y = x^2$, taking values of x between -10 and 10 and making the y unit equal to the x unit. Use your graph:

(*a*) to make a table of the square roots of all the numbers from 1 to 100;

(*b*) to make a table of the squares of all the numbers from 1 to 100.

16. Make graphs connecting:

(*a*) the area of an equilateral triangle with its base;

(*b*) the area of an isosceles triangle with base angles of 45° with its base;

(*c*) the area of a right-angled triangle with one angle of 30° with its base.

17. For the length of a pendulum in centimetres and the number of seconds per complete swing, the following values were found:

seconds	0·7	0·8	0·9	0·5	0·6	0·4
length	49	64	81	25	36	16

Check this with a button, a piece of cotton, and a watch, taking the average of ten complete swings in each case. Draw the graph. What is its equation?

18. Find the formula connecting area (y) and radius (x) of a circle by drawing a graph based on the area obtained by counting squares (Japanese method on p. 217). Having satisfied yourself that the correct formula is:

$$y = cx^2$$

read off from the graph what c is.

19. Solve the following simultaneous equations graphically:

(*a*) $xy = 0$, $3x + 4y = 12$ (*b*) $x^2 + y^2 = 25$, $x + y = 7$ (*c*) $(x - y)^2 = 1$, $(3x - 5y)^2 = 1$

20. Find the curves of which the equations are:

$$y = \pm \tfrac{5}{4}\sqrt{16 - x^2}$$

$$y = \pm \tfrac{4}{7}\sqrt{49 - x^2}$$

To what measurement in the figure do the numbers refer?

21. Graph the Cartesian and polar equations of the ellipse whose major axis is 3 and minor axis 2 units of length.

22. Using the formulae for $\sin (A + B)$, $\cos (A + B)$, pp. 242 and 336, put $\sin 180° = \sin (90° + 90°)$, $\cos 180° = \cos (90° + 90°)$, $\sin 270° = \sin (180° + 90°)$, etc. Hence, show:

(*a*) $\sin 180° = 0$, $\sin 270° = -1$, $\sin 360° = 0$

(*b*) $\cos 180° = -1$, $\cos 270° = 0$, $\cos 360° = +1$

Substituting these values show that:

$$\sin (90^\circ + A) = +\cos A, \cos (90^\circ + A)$$
$$= -\sin A, \tan (90^\circ + A) = -\cot A$$

$$\sin (180^\circ + A) = -\sin A, \cos (180^\circ + A) =$$
$$= -\cos A, \tan (180^\circ + A) = +\tan A$$

$$\sin (270^\circ + A) = -\cos A, \cos (270^\circ + A)$$
$$= +\sin A, \tan (270^\circ + A) = -\cot A$$

$$\sin (360^\circ + A) = +\sin A, \cos (360^\circ + A)$$
$$= +\cos A, \tan (360^\circ + A) = +\tan A$$

23. Using the formulae for $\sin (A - B)$, $\cos (A - B)$ in Chapter 8, pp. 368–9, show that:

$$\sin (90^\circ - A) = +\cos A, \cos (90^\circ - A)$$
$$= +\sin A, \tan (90^\circ - A) = +\cot A$$

$$\sin (180^\circ - A) = +\sin A, \cos (180^\circ - A)$$
$$= -\cos A, \tan (180^\circ - A) = -\tan A$$

$$\sin (270^\circ - A) = -\cos A, \cos (270^\circ - A)$$
$$= -\sin A, \tan (270^\circ - A) = +\cot A$$

$$\sin (360^\circ - A) = -\sin A, \cos (360^\circ - A)$$
$$= +\cos A, \tan (360^\circ - A) = -\tan A$$

24. Using a diagram interpret the meaning of the following and check your results by putting $\sin (-A) = \sin (0 - A)$, etc.:

$$\sin (-A) = -\sin A \quad \cos (-A) = \cos A \quad \tan (-A) = -\tan A$$

25.
$$\begin{aligned} \sin 130^\circ &= \sin (2n \times 90^\circ - 50^\circ) \text{ [where } n = 1] \\ &= \sin (180^\circ - 50^\circ) \\ &= \sin 50^\circ \end{aligned}$$

Alternatively:

$$\begin{aligned} \sin 130^\circ &= \sin (\overline{2n + 1} \times 90^\circ + 40^\circ) \text{ [where } n = 0] \\ &= \cos 40^\circ \\ &= \sin 50^\circ \end{aligned}$$

In the same way find by two alternative methods the values of:

tan 210°	sin 230°	cos 300°
tan 120°	sin 150°	cos 100°

26. Solve the following quadratic equations by the graphical method and check by p. 277:

$$\text{(i) } 5x^2 + 2x = 7$$
$$\text{(ii) } 8x^2 - 2x - 3 = 0$$
$$\text{(iii) } 5x^2 - x - 6 = 0$$

27. Graph sin x and tan x for $x = 0°, 30°, 45°, 60°, 90°$. From your graph read off the approximate values of:

$$\sin 15° \quad \sin 35° \quad \sin 75°$$
$$\tan 15° \quad \tan 35° \quad \tan 75°$$

and compare with the tables.

28. Draw a graph of $y = x^3$ and from your graph construct a table of the cubes and cube roots of the numbers 1 to 20.

29. Draw graphs of:

$$2^x \quad 1{\cdot}5^x \quad 3^x \quad 1{\cdot}1^x$$

From the graphs read off the values of:

$$2^{3\cdot5} \quad (1{\cdot}5)^{1\cdot5} \quad 3^{2\cdot5} \quad (1{\cdot}1)^{0\cdot5}$$

30. Draw the graphs of:

$$y = x^4 \quad y = x^5 \quad y = x^6 \quad y = x^7$$

31. Draw the graphs of the hyperbolas:

$$xy = 4 \quad \textit{and} \quad x^2 - y^2 = 8$$

32. Make a figure to show that the equation of the sphere, when the centre is at the origin, is:

$$x^2 + y^2 + z^2 = r^2$$

33. Find, using the tables, the sine, cosine, and tangent of the following angles: $-20°$, $-108°$, $400°$, $-500°$.

34. Make a graph to connect the amount (y) to which £100 or \$100 grows in x years at 2, $2\frac{1}{2}$, 3, $3\frac{1}{2}$, 4, 5 per cent: (*a*) at simple interest; (*b*) at compound interest.

Things to Memorize

1. Equation of the Circle:

$$r^2 = (x - a)^2 + (y - b)^2$$

2. Equation of the Straight Line:

$$y = (\tan A)x + b$$

3. Equation of the Parabola:

$$y = ax^2$$

4. Equation of the Ellipse:

$$\frac{x^2}{M^2} + \frac{y^2}{m^2} = 1 \quad or \quad r = \frac{m^2}{M(1 + e \,.\, \cos a)}$$

5. Equation of a Hyperbola:

$$y = kx^{-1}$$

6.* $\sin(-A) = -\sin A$ | $\cos(-A) = \cos A$

$\sin(90° - A) = \cos A$ | $\cos(90° - A) = +\sin A$

$\sin(90° + A) = \cos A$ | $\cos(90° + A) = -\sin A$

$\sin(180° - A) = \sin A$ | $\cos(180° - A) = -\cos A$

$\sin(180° + A) = -\sin A$ | $\cos(180° + A) = -\cos A$

$\sin(270° + A) = -\cos A$ | $\cos(270° + A) = \sin A$

7. $\sin(2n\pi + a) = \sin a$

$\cos(2n\pi \pm a) = \cos a$

$\tan(n\pi + a) = \tan a$

$$\sin\left[\frac{2n+1}{2}\pi \pm a\right] = (-1)^n \cos a, \text{ etc.}$$

8. Difference Formulae:

$$\sin(A - B) = \sin A \,.\, \cos B - \cos A \,.\, \sin B$$
$$\cos(A - B) = \cos A \,.\, \cos B + \sin A \,.\, \sin B$$

9. Inverse Functions

$$\sin^{-1} b = A \equiv \sin A = b$$
$$\cos^{-1} b = A \equiv \cos A = b$$
$$\tan^{-1} b = A \equiv \tan A = b$$

10. Relation between Polar and Cartesian Co-ordinates:

$$r = \sqrt{x^2 + y^2} \quad and \quad a = \tan^{-1}(y \div x)$$

* It is very important to be familiar with the values of general angles, but they are most easily recalled, not by memorizing the formulae but by visualizing a diagram.

CHAPTER 9

LOGARITHMS AND THE SEARCH FOR SERIES

IN THE domain of mathematical studies, the century of the Pilgrim Fathers witnessed three major innovations which bore abundant fruit in the century of the American Revolution. One has been the topic of our last chapter. One which will be the topic of our next was in large measure the offspring of the new geometry which could accommodate the passage of time. Another had its origin in the need for rapid computation with theoretical consequences which no one could have foreseen when Napier, a Scotsman, and Bürgi, a Swiss, respectively and independently published in 1614 and 1620 the first account of what the former called *logarithms*. The explanations offered by both inventors were obscure, and we need not here add to our difficulties by attempting to understand what Briggs (1624) an English professor of Geometry and Vlacq (1628) a Dutch bookseller made so much more intelligible to their contemporaries. The initial intention was essentially practical. To quote the 1631 edition of *Logarithmall Arithmetike* by Briggs, who first compiled the tables now used: 'Logarithmes are numbers invented for the more easie working of questions in arithmetike and geometrie . . . by them all troublesome multiplications and divisions are avoided and performed only by addition instead of multiplication and by subtraction instead of division. The curious and laborious extraction of roots are also performed with great ease. . . . In a word, all questions not only in Arithmetike and Geometrie but in Astronomie also are thereby most plainely and easily answered. . . .'

Two needs and two independent lines of reasoning initially contributed to the recognition of adding numbers which one can tabulate 'for ever' in Napier's own words as a means of sidestepping lengthy processes of multiplication. The first arose in connection with the preparation of trigonometrical tables for use in navigation. The second was closely connected with the laborious calculations involved in reckoning compound interest upon investments.

At the end of the century preceding the work of Napier, Denmark had achieved pre-eminence in the domain of astronomy through the important researches of Tycho Brahe. In this setting two Danish mathematicians, Wittich (1584) and Clavius, whose treatise *de Astrolabio* came out in 1593, suggested use of trigonometrical tables to shorten the process of multiplication. With this end in view, we recall the *sum* and *difference* formulae for *sines*, i.e.:

$$\sin(A + B) = \sin A \,.\, \cos B + \cos A \,.\, \sin B$$
$$\sin(A - B) = \sin A \,.\, \cos B - \cos A \,.\, \sin B$$
$$\therefore \quad \sin(A + B) + \sin(A - B) = 2 \sin A \cos B$$

or

$$\sin A \cos B = \tfrac{1}{2} \sin(A + B) + \tfrac{1}{2} \sin(A - B)$$

This equation can be used to substitute the socialized information of the sine and cosine tables for the individual labour of multiplying two numbers. Thus, to multiply:

$$0{\cdot}17365 \times 0{\cdot}99027$$

we look up the tables and find:

$$\sin 10^\circ = 0{\cdot}17365$$
$$\cos 8^\circ = 0{\cdot}99027$$

Our formula tells that:

$$\sin 10^\circ \,.\, \cos 8^\circ = \tfrac{1}{2}(\sin 18^\circ + \sin 2^\circ)$$

The tables tell us that:

$$\sin 18^\circ = 0{\cdot}30902$$
$$\sin 2^\circ = 0{\cdot}03490$$
$$\sin 18^\circ + \sin 2^\circ = 0{\cdot}34392$$
$$\tfrac{1}{2}(\sin 18^\circ + \sin 2^\circ) = 0{\cdot}17196$$

Thus, correct to five decimal places:

$$0{\cdot}17365 \times 0{\cdot}99027 = 0{\cdot}17196$$

You will see that this is correct to the fifth place by multiplying out as follows:*

* The inaccuracy involved simply depends on the tables used. In this case five-figure tables were used, and these do not assure complete accuracy beyond the fourth place. To get seven figures correct we should need eight-figure tables.

$$\begin{array}{r} 0{\cdot}17365 \\ 0{\cdot}99027 \\ \hline 0{\cdot}156285 \\ 156285 \\ 34730 \\ 121555 \\ \hline 0{\cdot}17196\ldots \end{array}$$

Alternatively, one might use the cosine formulae:

$$\cos(A-B) = \cos A \,.\, \cos B + \sin A \,.\, \sin B$$
$$\cos(A+B) = \cos A \,.\, \cos B - \sin A \,.\, \sin B$$
$$\therefore\ \tfrac{1}{2}\cos(A-B) + \tfrac{1}{2}\cos(A+B) = \cos A \,.\, \cos B$$

Such devices were a dead end, but are relevant to Napier's intention and may have prompted his train of thought. The tables Napier published under his own name were indeed logarithms of *sines*, and had no relevance to computations other than trigonometrical.

However, it rarely if ever happens in the history of science that a great discovery is made singly. The same social context which demanded quicker methods for calculating the position of the stars in the heavens called for quicker ways of calculating the wealth which accumulated through voyages which could not have been made without the use of astronomy to find the ship's position at sea. One line which led to the discovery of logarithms was the preparation of tables for calculating interest.

The calculation of compound interest is a practical application of the use of geometric series. If r is the rate of interest per pound invested, £1 in one year grows to £$(1 + r)$. For instance, if r is 5 per cent ($\frac{5}{100}$), £1 grows to £1·05. At the end of the second year every £1 invested at the end of the first year will be worth £1·05. If no interest is paid on the first year £1·05 will be invested at the beginning of the second year for every £1 invested at the beginning of the first year. So the original £1 will have grown to £1·05 × 1·05 or £$(1{\cdot}05)^2$ at the end of the second year. Similarly, it will be £$(1{\cdot}05)^3$ at the end of the third year. So we may tabulate the rate of growth of £1 in the following way:

At the end of	0	1	2	3	4	years
	1	$(1 + r)$	$(1 + r)^2$	$(1 + r)^3$	$(1 + r)^4$	

The series above is an arithmetic and the series below a geometric series. If we wish to calculate compound interest to quarter years we must extend the table, using *fractional* powers thus:

$$\begin{array}{cccccccc} 0 & \frac{1}{4} & \frac{1}{2} & \frac{3}{4} & 1 & 1\frac{1}{4} & 1\frac{1}{2} & 1\frac{3}{4}\text{, etc.} \\ 1 & (1+r)^{\frac{1}{4}} & (1+r)^{\frac{1}{2}} & (1+r)^{\frac{3}{4}} & (1+r) & (1+r)^{\frac{5}{4}} & (1+r)^{\frac{3}{2}} & (1+r)^{\frac{7}{4}} \end{array}$$

To get the value of £156, which has been accumulating compound interest at 3 per cent for $2\frac{3}{4}$ years, all we have to do is to multiply thus:

$$£156 \times (1{\cdot}03)^{2\frac{3}{4}} = £156 \times (1{\cdot}03)^{\frac{11}{4}}$$

Stevinus, to whom we have already referred on more than one occasion, published tables like this for calculations in commercial arithmetic; and before his time, Stifel had recognized four simple rules which hold good when we place side by side terms of corresponding rank of an arithmetic and geometric series, viz.:

(*a*) Addition of terms in the arithmetic series corresponds with multiplication of the terms in the geometric series.

(*b*) Subtraction of terms in the arithmetic series corresponds with division of terms in the geometric series.

(*c*) Multiplication of a term in the arithmetic series by a constant corresponds with raising a term in the geometric series to a given power.

(*d*) Division of a term in the arithmetic series by a constant corresponds with extracting a given root of a term in the geometric series.

Having got so far, it seems a great pity that Luther's convert spent so much time in arithmetical calculations to prove that Pope Leo X was the Beast of the Apocalypse, instead of undertaking the socially useful task of compiling tables like those of Bürgi or Briggs.

When Joost Bürgi published his *Tables of Arithmetical and Geometric Progressions* in 1620 he was indeed implementing a programme hinted at by several sixteenth-century writers on commercial arithmetic. The geometric series Bürgi tabulated was $(1{\cdot}0001)^n$, or, as we should now say, the base of his logarithms was 1·0001. Why he may have chosen this number we shall see later. Although Bürgi's table was mentioned by Kepler as a useful device for astronomical calculations, its origin is not directly connected with the need for ready-reckoning in navigation like

Napier's logarithms of sines. It simply carried the use of the compound-interest tables of Stevinus a stage farther.

The pairing of terms of a geometric series and of its generating arithmetic series, i.e. the natural numbers, in the way suggested by Stifel does indeed introduce us to the construction of a table of logarithms and to its uses by the simplest possible route. Below are a few terms of the geometric series 2^x (Fig. 144) and corresponding natural numbers:

1	2	3	4	5	6	7
2^1	2^2	2^3	2^4	2^5	2^6	2^7
2	4	8	16	32	64	128

Following Napier we shall now call the numbers in the upper or arithmetic series *logarithms* and the numbers in the bottom or geometric series *antilogarithms*. The principle of Archimedes is that if we want to multiply any two numbers in the bottom series we add the corresponding numbers in the top series and look for the corresponding number in the bottom series. One way of writing this rule is;

$$a^m \times a^n = a^{m+n}$$

e.g.

$$2^3 \times 2^4 = 2^7$$
$$(8 \times 16 = 128)$$

The operator '*log*' written in front of a number means, 'Look up in the table the power to which a has to be raised to give the number.' '*Antilog*' written in front of a number means, 'Look up in the table the value of the base (a) when raised to the power represented by the number.' Thus, if:

$$p = a^m$$
$$m = \log_a p$$
$$p = \text{antilog}_a\, m$$

We can thus write the rule for multiplication in an alternative form, putting:

$$q = a^n \text{ so that } n = \log_a q$$
$$p \times q = a^{m+n} \text{ so that } m + n = \log_a (p \times q)$$
$$\text{or} \quad p \times q = \text{antilog}_a\,(m + n)$$
$$= \text{antilog}_a\,(\log_a p + \log_a q)$$

The particular numerical example given in the new symbols would read:

$$8 \times 16 = \text{antilog}_2 (\log_2 8 + \log_2 16)$$
$$= \text{antilog}_2 (3 + 4)$$
$$= \text{antilog}_2 7$$

The last step means, 'Look up the number in the bottom row of antilogarithms corresponding with 7 in the top row of logarithms.' On looking it up, we find that it is $128 = 2^7$.

In Chapter 2, we have seen that we can extend our abacus for base b into the negative domain by labelling columns as:

$$b^0 = 1 \quad ; \quad b^{-1} = \frac{1}{b} \quad ; \quad b^{-2} = \frac{1}{b^2} \quad ; \quad b^{-3} = \frac{1}{b^3}, \text{ etc.}$$

We can thus write down the following logarithms and antilogarithms based on the geometric progression 3^n:

Log	-3	-2	-1	0	1	2	3	4
	3^{-3}	3^{-2}	3^{-1}	3^0	3^1	3^2	3^3	3^4
Antilog	$0{\cdot}\dot{0}3\dot{7}$	$0{\cdot}\dot{1}$	$0{\cdot}\dot{3}$	1	3	9	27	81

Needless to say, a table of which all the entries of $\log_b N$ refer only to whole number values of N, negative or positive, would be of little practical use. We now therefore recall earlier remarks (Chapter 6, p. 247) on the meaning of a^n when n is a fraction. If the rule of Archimedes ($a^m \times a^n = a^{m+n}$) holds good:

$$\sqrt[2]{a^1} = \sqrt[2]{a^{\frac{1}{2}+\frac{1}{2}}} = \sqrt[2]{a^{\frac{1}{2}} \times a^{\frac{1}{2}}} = a^{\frac{1}{2}}$$

i.e. $\quad a^{\frac{1}{2}} = \sqrt[2]{a}$

Similarly, $\quad \sqrt[3]{a^1} = \sqrt[3]{a^{\frac{1}{3}+\frac{1}{3}+\frac{1}{3}}} = \sqrt[3]{a^{\frac{1}{3}} \times a^{\frac{1}{3}} \times a^{\frac{1}{3}}} = a^{\frac{1}{3}}$

i.e. $\quad a^{\frac{1}{3}} = \sqrt[3]{a}$

So in general $\quad a^{\frac{1}{n}} = \sqrt[n]{a}$

Thus $3^{2{\cdot}5}$ means:

$$3^{2+\frac{1}{2}} = 3^2 \times 3^{\frac{1}{2}} = 9\sqrt{3}$$

Similarly, $2^{\frac{4}{3}}$ means:

$$2^{1+\frac{1}{3}} = 2^1 \times 2^{\frac{1}{3}} = 2\sqrt[3]{2}$$

A quantity like $2^{\frac{4}{3}}$ or $3^{\frac{5}{2}}$, or in general $a^{\frac{m}{n}}$, is translatable in another way, if we use the fact that:

$$\sqrt[2]{a} \times \sqrt[2]{b} = \sqrt[2]{ab}$$

or $\quad \sqrt[3]{a} \times \sqrt[3]{b} = \sqrt[3]{ab} \quad$ etc.

Thus we may put:

$$3^{2\cdot5} = 3^{\frac{5}{2}} = 3^2 \times \sqrt{3} = \sqrt{3^4}\,\sqrt{3} = \sqrt{3^5}$$

$$2^{\frac{4}{3}} = 2\sqrt[3]{2} = \sqrt[3]{2^3} \times \sqrt[3]{2} = \sqrt[3]{2^4}$$

Hence the rule is:

$$a^{\frac{p}{q}} = \sqrt[q]{a^p}$$

The two rules which we have given for fractional and negative powers were first stated by Oresmus in a book called *Algorismus Proportionum*, published about AD 1350. It took the human race a thousand years to bridge the gulf between the rule of Archimedes and the next stage in the evolution of the logarithm table. You need not be discouraged if it takes you a few hours or days to get accustomed to the use of a fractional or negative power.

With these two rules in mind, we can now extend a table of logarithms as far as we like. Thus for logarithms based on the geometric series 2^n, we can draw up a table which is correct to three decimal places as below:

$n = \log_2 N$	$N = (\text{antilog}_2\, n)$	
0	1	1·000
0·5	$\sqrt{2}$	1·414
1·0	2	2·000
1·5	$\sqrt{2^3}$	2·828
2·0	4	4·000
2·5	$\sqrt{2^5}$	5·657
3·0	8	8·000
3·5	$\sqrt{2^7}$	11·314
4·0	16	16·000
etc.	etc.	

We can continue to make the interval in the left-hand column of such a table by the same procedure, e.g. between $\log_2 0$ and $\log_2 1\cdot0$, we have:

$$\text{antilog}_2\,(0\cdot125) = 2^{\frac{1}{8}} = \sqrt[8]{2}$$

$$\text{antilog}_2\,(0\cdot25) = 2^{\frac{1}{4}} = \sqrt[4]{2}$$

$$\text{antilog}_2\,(0\cdot375) = 2^{\frac{3}{8}} = \sqrt[8]{8}$$

$$\text{antilog}_2\,(0\cdot5) = 2^{\frac{1}{2}} = \sqrt{2}$$

$$\text{antilog}_2\,(0\cdot625) = 2^{\frac{5}{8}} = \sqrt[8]{32}$$

$$\text{antilog}_2\,(0\cdot75) = 2^{\frac{3}{4}} = \sqrt[4]{8}$$

$$\text{antilog}_2\,(0\cdot875) = 2^{\frac{7}{8}} = \sqrt[8]{128}$$

How useful the table is depends on how small we make the interval (in the above, 0·125) of the *log* column.

Once we have got such a table we can use it to multiply numbers as follows. Suppose we wish to multiply 2·828 by 5·657. The table tells us that:

$$\log_2 2{\cdot}288 = 1{\cdot}5 \text{ or } 2^{1\cdot 5} = 2{\cdot}828$$
$$\log_2 5{\cdot}657 = 2{\cdot}5 \text{ or } 2^{2\cdot 5} = 5{\cdot}657$$

The rule of Archimedes tells us that:

$$2{\cdot}828 \times 5{\cdot}657 = 2^{1\cdot 5} \times 2^{2\cdot 5} = 2^{1\cdot 5 + 2\cdot 5} = 2^4$$

So the number we are looking for is the number whose logarithm is 4, i.e.:

$$\begin{aligned} \text{antilog}_2\, 4 &= \text{antilog}_2\, (1{\cdot}5 + 2{\cdot}5) \\ &= \text{antilog}_2\, (\log 2{\cdot}828 + \log 5{\cdot}657) \end{aligned}$$

The table shows that $\text{antilog}_2\, 4$ is 16. To check this multiply out:

$$\begin{array}{l} 2{\cdot}828 \\ 5{\cdot}657 \\ \hline 14{\cdot}140 \\ 1{\cdot}6968 \\ 14140 \\ 19796 \\ \hline 15{\cdot}997996 = 16 \text{ to four significant figures} \end{array}$$

The discrepancy is 2 in 16,000, an error of little more than 1 in 10,000. Of course, a better result could be obtained by using a table in which the figures given are correct to five, seven, nine, or more decimal places. The rule for multiplication with logarithms will now be clear: 'To multiply two numbers find these numbers in the column of antilogs, add the corresponding numbers in the column of logs, and find the number corresponding with the result in the column of antilogs.'

The foregoing remarks illustrate the most primitive method of making a table of logarithms and antilogarithms, i.e. by successive extraction of square roots. There is, in fact, an algorithm for this based on the binomial theorem; and we can adapt with the same end in view the method of representing a square root by a continued fraction (p. 104). In the last resort, we can invoke an iterative procedure (p. 390). The labour entailed in the construction of the tables of Briggs and Vlacq is one of the most astonishing monuments of human industry. Briggs published his first table of 1,000 entries in 1617. In his *Arithmetica Logarithmica* of 1624 there were 40,000. To do so, states J. F. Scott, he had to calculate without any mechanical aids successive square roots up to fifty-four times and cited results correct to thirty decimal places. Vlacq's tables listed 100,000 entries correct to ten places.

To Briggs we owe the immense convenience which accrues from the choice for the computation of logarithms to the same base as that of our numeral system. This choice greatly simplifies their use for a reason which we shall now examine. Let us first notice a few values of $log_{10} N$ when N is a whole number as below:

Log	−2	−1	0	1	2	3	4,	etc.
Antilog	0·01	0·1	1	10	100	1,000	10,000	

Thus $\log_{10} 1 = 0$, $\log_{10} 10 = 1$, and $\log_{10} \sqrt{10} = \log_{10} 10^{\frac{1}{2}} = 0{\cdot}50$. The square root of 10 (to three decimals) is 3·162. So we can write

$$\log_{10} 3{\cdot}162 = 0{\cdot}500$$

Now $$31{\cdot}62 = 3{\cdot}162 \times 10$$

Hence $$\begin{aligned}\log_{10} 31{\cdot}62 &= \log_{10} (3{\cdot}162 \times 10)\\ &= \log_{10} 3{\cdot}162 + \log_{10} 10\\ &= 0{\cdot}500 + 1\\ &= 1{\cdot}5\end{aligned}$$

Likewise $$\begin{aligned}\log_{10} 316{\cdot}2 &= \log_{10} (3{\cdot}162 \times 100)\\ &= \log_{10} 3{\cdot}162 + \log_{10} 100\\ &= 0{\cdot}500 + 2\\ &= 2{\cdot}5\end{aligned}$$

In short, shifting the decimal place in a number does not alter the value to the right of the decimal place in its logarithm, and the number to the left of the decimal place can be written down by common sense. Since $10^0 = 1$ and $10^1 = 10$, the number to the left of the decimal place in the logarithm is 0 for all numbers between 1 and 10. Since $10^1 = 10$ and $10^2 = 100$, it is 1 for numbers between 10 and 100. Since $10^2 = 100$ and $10^3 = 1{,}000$, it is 2 for numbers between 100 and 1,000. Similarly, 1 put in front of the fractional part of a number whose antilogarithm we are looking for means, 'Multiply the antilogarithm of the fractional part by 10'; 2 in front means, 'Multiply the antilogarithm of the fractional part by 100', and so on. This means that if we have the logarithms of all numbers between 1 and 10 in suitable intervals we have all we need for multiplying. For instance, we may want to multiply 1·536 by 77. The tables tell us that:

$$\log_{10} 1{\cdot}536 = 0{\cdot}1864$$
$$\log_{10} 7{\cdot}7 = 0{\cdot}8865$$

Hence $$\log_{10} 77 = 1{\cdot}8865$$

So $$\begin{aligned} 1{\cdot}536 \times 77 &= \text{antilog}_{10}\,(0{\cdot}1864 + 1{\cdot}8865) \\ &= \text{antilog}_{10}\,2{\cdot}0729 \\ &= 100 \times \text{antilog}_{10}\,0{\cdot}0729 \\ &= 100 \times 1{\cdot}183 \\ &= 118{\cdot}3 \end{aligned}$$

The result, as you will see by multiplying, is correct to four significant figures (first decimal place), and that is all you can expect since we have only used four-figure tables.

So far we have illustrated the use of a table of logarithms only for performing the operation of multiplication. Let us now consider its use for that of division and for the extraction of roots.

If $$10^n = N, \quad \log_{10} N = n$$
$$10^m = M, \quad \log_{10} M = m$$
$$\therefore \frac{N}{M} = 10^{n-m} \text{ and } \log_{10}\frac{N}{M} = n - m$$
$$\therefore \log_{10}\frac{N}{M} = \log_{10} N - \log_{10} M$$

or $$\frac{N}{M} = \text{antilog}_{10}\,(\log_{10} N - \log_{10} M)$$

Suppose we want to find 20 ÷ 5. We put:

$$20 \div 5 = \text{antilog}_{10}\,(\log_{10} 20 - \log_{10} 5)$$

The crude table (four figures) at the end of this book cites:

$$\log_{10} 20 = 1{\cdot}3010 \quad ; \quad \log_{10} 5 = 0{\cdot}6990$$
$$\therefore \log_{10} 20 - \log_{10} 5 = 0{\cdot}6020$$
$$\text{antilog}_{10}\,(0{\cdot}6020) = 4$$

The extraction of a root depends on the rule:

$$\sqrt[n]{10} = 10^{\frac{1}{n}}$$

So if

$$N = 10^m$$
$$m = \log_{10} N$$

Likewise

$$\sqrt[n]{N} = N^{\frac{1}{n}} = 10^{\frac{m}{n}}$$

$$\log_{10} \sqrt[n]{N} = \frac{m}{n}$$

$$\log_{10} \sqrt[n]{N} = \frac{1}{n} \log_{10} N$$

For illustrative purposes, let us find the cube root of 8. We put:

$$3 \sqrt{8} = \text{antilog}_{10} (\tfrac{1}{3} \log_{10} 8)$$

Our four-figure table cites log_{10} 8 = 0·9031, so that:

$$\tfrac{1}{3} \log_{10} 8 = 0{\cdot}3010$$
$$\text{antilog}_{10} (0{\cdot}3010) = 2$$

The recipe for the converse operation of raising N to the power m follows at once:

$$N^m = \text{antilog}_{10} (m \,.\, \log_{10} N)$$

For instance:

$$2^{12} = \text{antilog}_{10} (12 \log_{10} 2) = \text{antilog}_{10} 12(0{\cdot}3010)$$
$$= \text{antilog}_{10} (3{\cdot}6120)$$

Our four-figure table cites $antilog_{10}$ 0·6120 = 0·4093, so that $antilog_{10}$ 3·6120 = 4093. The correct value is 4096, and the error (less than one in a thousand) is due to the fact that our table is correct only to the fourth decimal place.

When speaking of logarithms to the base 10, one refers to the part in front of the decimal place, e.g. 3 in 3·6120 above, as its *characteristic*. While we can always place correctly by inspection the decimal place in an operation of multiplication or division, and can therefore neglect the characteristic when performing one or the other, it is indispensable when we wish to compute N^m or $\sqrt[m]{N}$. The characteristic will, of course, be negative if $N < 1$, e.g. −1 for 0·2, −2 for 0·02, and in general $-n$ if there are $n - 1$ zeros immediately after the decimal point.

Now tables of logarithms cite only positive values, for instance, $log_{10}2$ = 0·3010, so that:

$$\log_{10}(0{\cdot}002) = \log_{10}(2 \times 0{\cdot}001) = \log_{10}(2 \times 10^{-3}) =$$
$$= \log_{10}2 + \log 10^{-3} = +0{\cdot}3010 - 3$$
$$\therefore \log_{10}(0{\cdot}002) = -2{\cdot}6990$$

Let us suppose that we wish to find $\sqrt{0{\cdot}002}$, i.e.:

$$antilog_{10}(\tfrac{1}{2}\log_{10} 0{\cdot}002) = antilog\,(-1{\cdot}3495)$$
$$= antilog\,(-2 + 0{\cdot}6505)$$

As cited in our four-figure tables, $antilog_{10}(0{\cdot}6505) = 4{\cdot}472$ and the characteristic -2 means that there is one zero immediately behind the decimal place, so that:

$$\sqrt{0{\cdot}002} = 0{\cdot}04472$$

(Direct multiplication gives $(0{\cdot}04472)^2 = {\cdot}0019998884$ an error of about one in 20,000 due to the use of only four figures.)

We can never know how grateful the shadowy author of the Apocalypse would have been to the Scots laird of Merchiston who publicized his alleged intentions fifteen hundred years later in Lowland Scots dialect as the *Plaine Discovery of the Whole Revelation of St John set down in Two Treatises*, and dedicated his discovery (1593) to James VI of Scotland (afterwards James I of England). Napier wrote his much shorter *Canonis Descriptio*, setting forth his invention of Logarithms in Latin, and nothing Edward Wright (1617) his translator accomplished in the attempt to render it in lucid English could have conceivably earned the title of a *Plaine Discovery*. Indeed, it is difficult to believe that the author was wholly clear about his own intentions until Briggs entered into correspondence with him. The outcome – logarithms to the base 10 – was certainly remote from the author's original undertaking, and the major credit of the invention is properly due as much to the immense labour Briggs undertook as to his ingenious exploitation of what might otherwise have remained a curiosity of historical interest. Even so, the major achievement of Briggs in its own setting was an essentially practical contribution to the art of computation with ostensibly little promise of a challenge to mathematical ingenuity in other domains.

Series for Making Tables. The first tables of logarithms contained inaccuracies which were noticed and corrected from time to time. The labour expended in constructing them was stupendous. Not unnaturally it stimulated the search for more congenial methods of calculating them. This search gave a new impetus to the study of what mathematicians call *infinite series*. At the beginning of this book we used the recurring decimal $0{\cdot}\dot{1}$ to illustrate a series which never grows beyond a certain limiting value however many terms we continue to add. Later we were able to show that any geometric series with a fractional base less than unity has the convenient characteristic of choking off in this way. The invention of logarithms was followed by the discovery of a large family of series which do the same thing. Foremost among these was the Binomial Theorem $(1 \pm x)^n$ when n is negative or fractional and $0 < x < 1$.

Before discussing this, it may protect the author from unnecessary correspondence to state the truth about one of those historical tags on all fours with the ludicrous assertion that a document called Magna Carta which a circus of half-literate Baronial gangsters forced an English monarch to sign is the Keystone of British and/or United States democracy. It is almost as widely current that Newton discovered the binomial theorem at the age of sixteen – or thereabouts. The truth is that the Persians and Chinese had known the Binomial Theorem for the index n as a positive whole number several centuries before the birth of Newton. It is also true that Newton did surmise, and seemingly before his contemporaries at a time when logarithms had familiarized them with the use of negative and fractional indices, that the theorem can hold good when n is neither positive – nor a whole number. As we shall now do, he recognized the usefulness of this extension as an empirical fact. He certainly did not offer a proof which would satisfy the punctilious requirements of a nineteenth- or twentieth-century mathematician; and we must defer a satisfactory proof in that sense till we can invoke the use of the infinitesimal calculus to generate infinite series that converge.

The series $(1 \pm x)^n$ consists of $n + 1$ terms when n is positive and a whole number, being expressible in the form:

$$(1 + x)^n = 1 + nx + \frac{n(n-1)}{2!}x^2 + \frac{n(n-1)(n-2)}{3!}x^3 \ldots$$

$$(1 - x)^n = 1 - nx + \frac{n(n-1)}{2!}x^2 - \frac{n(n-1)(n-2)}{3!}x^3 \ldots$$

When n is a positive whole number the coefficients of the last two terms are $n! \div (n-1)! = n$ and $n! \div n! = 1$. So the series terminates at the term of rank n if we label the initial term $1^n . x^0 = 1$ as the term of rank 0. When n is a fraction (or is negative) the coefficients of the form $n^{(r)} \div r!$ in the symbolism of Chapter 4 (p. 197) do not terminate. The expression obtained is an unending series which is meaningful in the sense that it is *convergent*, i.e. that it has a finite sum if $0 < x < 1$ (n negative) or $0 < x \leqslant 1$ (n fractional and positive). This means that we can extract any root of $(1 \pm x)$, e.g. the tenth root of $\frac{3}{4} = (1 - \frac{1}{4})^{0 \cdot 1}$, when $(1 \pm x)$ is not greater than 2, or less than zero.

The importance of this depends on the fact that a very simple rule makes it possible to convert logarithms from one base to another. It is therefore economical to calculate them to a base

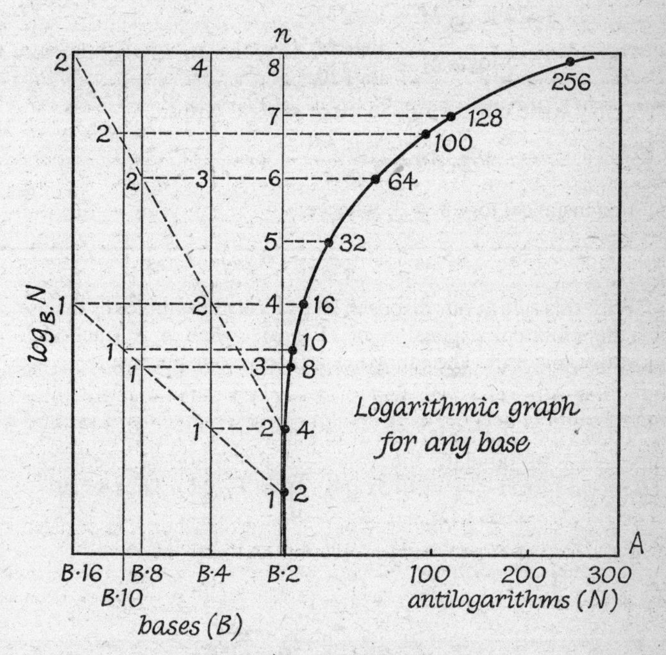

Fig. 156. The Logarithmic Curve

If a and b are bases of two logarithms of a number N, $\log_b N = \log_b a \,.\, \log_a N$. Since $\log_b a$ is a numerically fixed quantity, the change from one base to another is merely a matter of the scale of the Y-axis when $y = \log_b N$ and $x = N = \text{antilog}_b\, y$. We can thus pair off our graduation of the Y-axis for one base with that of the Y-axis of any other, and read off $\log_b N$ accordingly.

for which we can extract roots with as little effort as possible and then to convert them to a base, i.e. the base 10 of so-called common logarithms, more convenient for everyday usage. The trick depends on the fact that $(n^a)^b = n^{ab}$, e.g. $8^3 = (2^3)^3 = 2^9$, as you can see by multiplying. Thus the rule is as follows. Suppose we have calculated a table of logarithms and antilogarithms to the base 2. On looking over the table we find $\log_2 10 = 3{\cdot}322$, or in other words $2^{3\cdot 322} = 10$. So if:

$$M = 10^m, \text{ i.e. } \quad m = \log_{10} M$$

$$M = 2^{3\cdot 322m}, \text{ i.e. } \quad 3{\cdot}322m = \log_2 M$$

$$\therefore \ 3{\cdot}322 \log_{10} M = \log_2 M$$

or
$$\log_{10} M = \frac{\log_2 M}{\log_2 10}$$

We may write out the rule in more general terms:

$$\log_a M = \frac{\log_b M}{\log_b a}$$

For instance, $\log_2 8 = 3$, whence:

$$\log_{10} 8 = \frac{3}{3{\cdot}322} = 0{\cdot}903$$

With this rule at our disposal, let us look at examples of the use of the binomial expansion of $(1 \pm b)^n$ when n is fractional, negative, or both. The following will serve our purpose:

$$\sqrt{\tfrac{3}{4}} = (1 - \tfrac{1}{4})^{\frac{1}{2}} \quad \textit{and} \quad \sqrt{2} = (\tfrac{1}{2})^{-\frac{1}{2}} = (1 - \tfrac{1}{2})^{-\frac{1}{2}}$$

To use the binomial theorem to get a series for $\sqrt{\frac{3}{4}}$ and $\sqrt{2}$ we need first to tabulate the values of the binomial coefficients ($B_r = n_{(r)}$) for $n = \frac{1}{2} = (0{\cdot}5)$ and $-\frac{1}{2} = (-0{\cdot}5)$. Thus when:

$$n = \tfrac{1}{2} = B_1$$

$$B_2 = \frac{n(n-1)}{2!} = \frac{\frac{1}{2}(\frac{1}{2} - 1)}{2}$$

$$= -\tfrac{1}{8}$$

$$= -0{\cdot}125$$

When
$$n = -\tfrac{1}{2} = B_1$$

$$B_2 = \frac{n(n-1)}{2} = \frac{-\frac{1}{2}(-\frac{1}{2} - 1)}{2}$$

$$= +\tfrac{3}{8}$$

$$= +0{\cdot}375$$

So we have:

	$n = \frac{1}{2}$	$(n = -\frac{1}{2})$
$B_1 = n$	+0·5	−0·5
$B_2 = \dfrac{n(n-1)}{2!}$	−0·125	+0·375
$B_3 = \dfrac{n(n-1)(n-2)}{3!}$	+0·0625	−0·3125
$B_4 = \dfrac{n(n-1)(n-2)(n-3)}{4!}$	−0·0390625	+0·2734375
B_5	+0·02734375	−0·24609375
B_6	−0·0205078125	+0·2255859375
B_7	+0·01611328125	−0·20947265625
B_8	−0·013092041016	+0·196380615234
B_9	+0·010910034180	−0·185470581054
B_{10}	−0·009273529053	+0·176197052001
B_{11}	+0·008008956909	−0·168188095092
B_{12}	−0·007007837295	+0·161180257797

To get $\sqrt{\frac{3}{4}}$ we want $(1 + b)^n$, where $b = -\frac{1}{4}$ and $n = \frac{1}{2}$. So the binomial series is:

$$1 - 0{\cdot}5(\tfrac{1}{4}) - 0{\cdot}125(\tfrac{1}{16}) - (0{\cdot}0625(\tfrac{1}{64}) - 0{\cdot}0390625(\tfrac{1}{256}) - 0{\cdot}02734375(\tfrac{1}{1024}) \ldots$$

If we take the first two terms of this series we have:

$$1 - \tfrac{1}{8} = 0{\cdot}875$$

We can tabulate the values we get by taking the sum of the first few terms thus:

Terms	Sum
1	1
2	0·875
3	0·8671875
4	0·8662109375
5	0·866058349609375
6	0·866031646728515625
7	0·866026639983544922

However many terms we take, the sum of this series never grows smaller than 0·866025, which is the value of $\sqrt{\frac{3}{4}}$ correct to six decimals. We only need to take the sum of the first seven terms of the series to get $\sqrt{\frac{3}{4}}$ for making a five-figure table of sines or cosines (see table in Chapter 5, p. 215).

The series for $\sqrt{2}$ does not choke off so quickly. By now you will probably remember that the correct value to four figures is 1·414. The binomial series is obtained by putting $b = -\frac{1}{2}$ and $n = -\frac{1}{2}$ in the expression:

$$(1 - b)^n = 1 + (-0{\cdot}5)(-\tfrac{1}{2}) + (0{\cdot}375)(-\tfrac{1}{2})^2 + (-0{\cdot}3125)(-\tfrac{1}{2})^3 \ldots$$

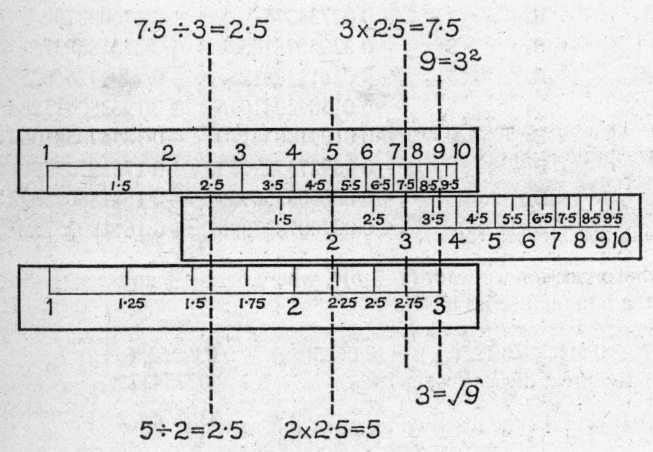

Fig. 157. The Slide Rule

The logarithm of any number (Fig. 156) on one scale differs only from that of the same number by a fixed multiple. Therefore the scale spaced in terms of any base is irrelevant to the log of the number which labels its division. Of this inexpensively procurable slide rule, the spacings of the movable middle scale like those of the one above are identical. We know that $\log (ab) = \log a + \log b$ and $\log (a \div b) = \log a - \log b$. So the setting of mark 1 on the middle against mark N on the upper allows one to read off N times x on the top, or conversely $N \div x$ from x on the upper to the setting of mark 1 on the lower. For the slide rule shown, units on the bottom scale are exactly half as in the top one. Since $\log a^2 = 2 \log a$ and $\log\sqrt{a} = \frac{1}{2} \log a$, one can read off either N^2 or $\sqrt{N}$ by setting the rider in alignment with N on one or other scale.

Tabulating as in the preceding illustration the sum of the first few terms, and giving the answer correct to four figures only, we have:

Terms	Sum
1	1
2	1·250
3	1·344
4	1·383
5	1·400
6	1·408
7	1·411
8	1·413
9	1·414

This series never grows as large as 1·4143 however many terms we go on adding.

As an additional check on the use of the binomial series for negative powers the following result will be used later to get an unlimited series for π. We may write in the form: $(1 + x)^{-1}$ the expression:

$$\frac{1}{1 + x}$$

By direct division we get:

$$\begin{array}{l} 1 + x)\ 1\ (1 - x + x^2 - x^3 + x^4 - x^5 \ldots \\ \quad\quad 1 + x \\ \hline \quad\quad\quad - x \\ \quad\quad\quad - x - x^2 \\ \hline \quad\quad\quad\quad\quad x^2 \\ \quad\quad\quad\quad\quad x^2 + x^3 \\ \hline \quad\quad\quad\quad\quad\quad - x^3 \\ \quad\quad\quad\quad\quad\quad - x^3 - x^4 \\ \hline \quad\quad\quad\quad\quad\quad\quad\quad x^4 \\ \quad\quad\quad\quad\quad\quad\quad\quad x^4 + x^5 \\ \hline \quad\quad\quad\quad\quad\quad\quad\quad\quad - x^5 \ldots \end{array}$$

By using the binomial series we get:

$$(1 + x)^{-1} = 1 + (-1)x + \frac{(-1)(-1-1)x^2}{2} + \frac{(-1)(-1-1)(-1-2)x^3}{3.2} + \frac{(-1)(-1-1)(-1-2)(-1-3)x^4}{4.3.2} \ldots$$

$$= 1 - x + x^2 - x^3 + x^4 \ldots$$

The result is therefore the same as the one which we get by direct division.

The reader will naturally ask how we know when such series do eventually choke off. The first mathematicians who used series of unlimited length did not bother themselves to find a satisfactory test to decide when a series of unlimited length chokes off and when it does not. They were content to use them because they found that they led to useful results. It is very comforting to recall the curious mistakes which some of the most eminent mathematicians of the seventeenth century made before such tests were discovered. A series which puzzled Leibnitz, whose immense contributions to mathematics will be dealt with later was the foregoing series for $(1 + x)^{-1}$ when $x = 1$. It then becomes

$$1 - 1 + 1 - 1 + 1 - 1 \ldots, \text{etc.}$$

This series is not convergent. The sum of an even number of terms is zero, and the sum of an odd number of terms is unity. On the other hand, we shall now see that it is convergent when $x < 1$. We may then say: if $p > 1$ and $x = \frac{1}{p}$ the series becomes:

$$\left(1 + \frac{1}{p}\right)^{-1} = 1 - \frac{1}{p} + \frac{1}{p^2} - \frac{1}{p^3} + \frac{1}{p^4} - \frac{1}{p^5} + \frac{1}{p^6} - \frac{1}{p^7} + \frac{1}{p^8} \ldots$$

$$= 1 - \left(\frac{1}{p} - \frac{1}{p^2}\right) - \left(\frac{1}{p^3} - \frac{1}{p^4}\right) - \left(\frac{1}{p^5} - \frac{1}{p^6}\right) - \left(\frac{1}{p^7} - \frac{1}{p^8}\right) \ldots$$

$$= 1 - \frac{p-1}{p^2} - \frac{p-1}{p^4} - \frac{p-1}{p^6} - \frac{p-1}{p^8} \ldots$$

$$= 1 - (p-1)\left(\frac{1}{p^2} + \frac{1}{p^4} + \frac{1}{p^6} + \frac{1}{p^8}\right) \ldots$$

$$= 1 - (p-1)S$$

We may here write as elsewhere when dealing with a geometric series:

$$S = \frac{1}{p^2} + \frac{1}{p^4} + \frac{1}{p^6} + \frac{1}{p^8} \ldots$$

$$\therefore\ p^2 . S = 1 + \frac{1}{p^2} + \frac{1}{p^4} + \frac{1}{p^6} + \frac{1}{p^8} \ldots$$

$$\therefore\ (1 - p^2) S = -1 \quad \textit{and} \quad S = \frac{1}{p^2 - 1}$$

$$\therefore\quad \left(1 + \frac{1}{p}\right)^{-1} = 1 - \frac{p - 1}{p^2 - 1} = 1 - \frac{1}{p + 1} = \frac{p}{p + 1}$$

For instance, if $p = 5$:

$$\left(1 + \frac{1}{5}\right)^{-1} = \left(\frac{6}{5}\right)^{-1} = \frac{5}{6} = \frac{p}{p + 1}$$

This example illustrates what we may call the *yardstick method* of testing convergence. We know that the sum of any geometric series $a^0 + a^1 + a^2 + a^3 \ldots$, etc., is convergent, if $a < 1$, since we may write, as above:

$$S = 1 + a^1 + a^2 + a^3 + a^4 \ldots$$

$$aS = \quad a^1 + a^2 + a^3 + a^4 \ldots$$

$$\therefore\ S - aS = S(1 - a) = 1 \quad \textit{and} \quad S = \frac{1}{1 - a}$$

Now it may be possible to show that every term of another series after the rth is less than the corresponding term of a convergent geometric series, so that the tail of the former (i.e. all terms after the rth) is less than the known sum of a series which does converge. When $x < 1$, the series for $(1 + x)^n$ for negative and/or fractional values of n does in fact stand up to this test. As an exercise, the reader may test how rapidly $(1{\cdot}0001)^5$, in which 1·0001 is Bürgi's base, converges, and how few terms suffice to yield a result which entails a negligible error even when n is a positive integer. Thus the sum of the first two terms is 1·0005, of the first three terms 1·0005001, of the first four terms 1·00050010001. The successive terms of the series are always less than a thousandth of the preceding one. So the result is correct to ten decimals, even if we only take the first three terms of the series.

The Exponential Series. Of all the series whose discovery occurred in the context of emergent use of logarithms and exploration of the properties of the binomial expansion, none plays a more remarkable role in mathematics than the one which we shall now examine. At an early stage in the development of the infinitesimal calculus, it became clear that the following function of x, expressed as an unending series, has a remarkable property which we shall examine in the next chapter:

$$y = 1 + x + \frac{x^2}{2!} + \frac{x^3}{3!} + \frac{x^4}{4!} \ldots, \textit{ ad infinitum}$$

What concerns us here is a property of the number $e = y$, when $x = 1$, i.e.:

$$e = 1 + 1 + \frac{1}{2!} + \frac{1}{3!} + \frac{1}{4!} \ldots, \text{ etc.}$$

As already mentioned (p. 104), this series chokes off rapidly. Its value (p. 428 below) lies between 2·7 and 2·72. Even if we add up only the first five terms, it is 2·702 to three decimal places, and if we add up even the first six, the sum is 2·717 to three places. Here what precisely e is (to whatever order of precision we want it to be) is interesting only because of what e^x means. We can arrive at a formula for $e^2 = e \,.\, e$ by straight-forward multiplication as below:

$$\begin{array}{r}
1 + 1 + \frac{1}{2!} + \frac{1}{3!} + \frac{1}{4!} + \frac{1}{5!} + \frac{1}{6!} \ldots \\
1 + 1 + \frac{1}{2!} + \frac{1}{3!} + \frac{1}{4!} + \frac{1}{5!} + \frac{1}{6!} \ldots \\
\hline
1 + 1 + \frac{1}{2!} + \frac{1}{3!} + \frac{1}{4!} + \frac{1}{5!} + \frac{1}{6!} \ldots \\
1 + 1 + \frac{1}{2!} + \frac{1}{3!} + \frac{1}{4!} + \frac{1}{5!} \ldots \\
\frac{1}{2!} + \frac{1}{2!} + \frac{1}{2!2!} + \frac{1}{2!3!} + \frac{1}{2!4!} \ldots \\
\frac{1}{3!} + \frac{1}{3!} + \frac{1}{3!2!} + \frac{1}{3!3!} \ldots
\end{array}$$

$$\frac{1}{4!} + \frac{1}{4!} + \frac{1}{4!2!} \cdots$$

$$+ \frac{1}{5!} + \frac{1}{5!} \cdots$$

$$+ \frac{1}{6!} \cdots$$

Let us now add the items in each column.

$$1 + 2 + \frac{(1 + 2 + 1)}{2!} + \frac{(1 + 3 + 3 + 1)}{3!}$$

$$+ \frac{1 + 4 + 6 + 4 + 1}{4!} + \frac{(1 + 5 + 10 + 10 + 5 + 1)}{5!}$$

$$+ \frac{(1 + 6 + 15 + 20 + 15 + 6 + 1)}{6!}, \textit{ and so on}$$

More briefly:

$$e^2 = 1 + 2 + \frac{4}{2!} + \frac{8}{3!} + \frac{16}{4!} + \frac{32}{5!} + \frac{64}{6!}, \textit{ and so on}$$

$$= 1 + 2 + \frac{2^2}{2!} + \frac{2^3}{3!} + \frac{2^4}{4!} + \frac{2^5}{5!} + \frac{2^6}{6!}, \textit{ and so on}$$

This suggests a formula for e^n, true at least when n is a whole number, viz.:

$$e^n = 1 + n + \frac{n^2}{2!} + \frac{n^3}{3!} + \frac{n^4}{4!} + \frac{n^5}{5!}, \textit{ and so on}$$

Such a formula is consistent with the terms for e^0, e^1, and e^2. In accordance with the principle of *induction* (p. 108), it must be true for any integer n, if

$$e^{n+1} = 1 + (n + 1) + \frac{(n + 1)^2}{2!} + \frac{(n + 1)^3}{3!} + \frac{(n + 1)^4}{4!}, \text{ etc.}$$

Now $e^{n+1} = e(e^n)$ and if our proposed formula is true, we may obtain e^{n+1} by straight-forward multiplication as below:

$$1 + n + \frac{n^2}{2!} + \frac{n^3}{3!} + \frac{n^4}{4!} + \frac{n^5}{5!} \ldots, \text{ etc.}$$

$$1 + 1 + \frac{1}{2!} + \frac{1}{3!} + \frac{1}{4!} + \frac{1}{5!} \ldots, \text{ etc.}$$

$$1 + n + \frac{n^2}{2!} + \frac{n^3}{3!} + \frac{n^4}{4!} + \frac{n^5}{5!} \ldots, \text{ etc.}$$

$$1 + n + \frac{n^2}{2!} + \frac{n^3}{3!} + \frac{n^4}{4!} \ldots, \text{ etc.}$$

$$\frac{1}{2!} + \frac{n}{2!} + \frac{n^2}{2!2!} + \frac{n^3}{2!3!} \ldots, \text{ etc.}$$

$$+ \frac{1}{3!} + \frac{n}{3!} + \frac{n^2}{3!2!} \ldots, \text{ etc.}$$

$$+ \frac{1}{4!} + \frac{n}{4!} \ldots, \text{ etc.}$$

$$+ \frac{1}{5!} \ldots, \text{ etc.}$$

When we add the terms in each column we obtain:

$$1 + (n + 1) + \frac{n^2 + 2n + 1}{2!} + \frac{n^2 + 3n^2 + 3n + 1}{3!}$$
$$+ \frac{n^4 + 4n^3 + 6n^2 + 4n^2 + 1}{4!}$$
$$+ \frac{n^5 + 5n^4 + 10n^3 + 10n^2 + 5n + 1}{5!}, \text{ etc.}$$
$$= 1 + (n + 1) + \frac{(n + 1)^2}{2!} + \frac{(n + 1)^3}{3!}$$
$$+ \frac{(n + 1)^4}{4!} + \frac{(n + 1)^5}{5!}, \text{ etc.}$$

This completes the Case for the Defence of the so-called *Exponential Theorem* when n in e^n is an integer and positive, i.e.:

$$e^n = 1 + n + \frac{n^2}{2!} + \frac{n^3}{3!} + \frac{n^4}{4!} \ldots, \text{ etc.}$$

By definition $e^{-n} = 1 \div e^n$. We may set this out as an exercise in direct division as abbreviated below to disclose the pattern:

$$1 + n + \frac{n^2}{2!} + \frac{n^3}{3!} + \frac{n^4}{4!} \ldots 1 + 0 + 0 + 0 + 0 \ldots$$

$$1 + n + \frac{n^2}{2!} + \frac{n^3}{3!} + \frac{n^4}{4!} \ldots$$

$$-n - \frac{n^2}{2!} - \frac{n^3}{3!} - \frac{n^4}{4!} \dots$$

$$-n - n^2 - \frac{n^3}{2!} - \frac{n^4}{3!} \dots$$

$$\frac{n^2}{2!} + \frac{2n^3}{3!} + \frac{3n^4}{4!} \dots$$

$$\frac{n^2}{2!} + \frac{n^3}{2!} + \frac{n^4}{2!2!}$$

$$\frac{-n^3}{3!} - \frac{3n^4}{4!}$$

$$\frac{-n^3}{3!} - \frac{n^4}{3!}$$

$$\frac{n^4}{4!}$$

It is thus evident that:

$$e^{-n} = 1 - n + \frac{n^2}{2!} - \frac{n^3}{3!} + \frac{n^4}{4!} - \frac{n^5}{5!}, \text{ etc.}$$

This is consistent with our expression for e^n, being what we should obtain by substitution of $-n$ for n in each term, i.e. our formula for e^n holds good for negative $(-n)$ as well as a positive $(+n)$, when n itself is a whole number. To satisfy ourselves that it holds for any rational fraction $f = (p \div q)$ expressible as the ratio of two whole numbers p and q, it will here suffice to show that we may paint in f and $(1 - f)$ for n in the foregoing expression and obtain by direct multiplication as for n and $(n + 1)$ above:

$$e^f . e^{1-f} = e^1 = e$$

For instance, we may put $f = 0{\cdot}\dot{3}$ and $(1 - f) = 0{\cdot}\dot{6}$, $f = 0{\cdot}4$ and $(1 - f) = 0{\cdot}6$ or $f = 0{\cdot}5 = (1 - f)$. To exhibit the method, we shall here select $f = 0{\cdot}5$ and leave it to the reader to try out as an exercise $f = 0{\cdot}\dot{3}$ or $f = 0{\cdot}4$. If e^f is meaningful in a sense consistent with e^n when n is a whole number:

$$e^{0 \cdot 5} = 1 + 0{\cdot}5 + \frac{(0{\cdot}5)^2}{2!} + \frac{(0{\cdot}5)^3}{3!} + \frac{(0{\cdot}5)^4}{4!}, \text{ etc.}$$

We may lay out the square of $e^{0 \cdot 5}$ for direct multiplication as below:

$$1 + 0{\cdot}05 + \frac{(0{\cdot}5)^2}{2!} + \frac{(0{\cdot}5)^3}{3!} + \frac{(0{\cdot}5)^4}{4!} \ldots, \text{etc.}$$

$$1 + 0{\cdot}05 + \frac{(0{\cdot}5)^2}{2!} + \frac{(0{\cdot}5)^3}{3!} + \frac{(0{\cdot}5)^4}{4!} \ldots, \text{etc.}$$

$$1 + {\cdot}05 + \frac{(0{\cdot}5)^2}{2!} + \frac{(0{\cdot}5)^3}{3!} + \frac{(0{\cdot}5)^4}{4!} \ldots, \text{etc.}$$

$$0{\cdot}05 + (0{\cdot}5)^2 + \frac{(0{\cdot}5)^3}{2!} + \frac{(0{\cdot}5)^4}{3!} \ldots, \text{etc.}$$

$$\frac{(0{\cdot}5)^2}{2!} + \frac{(0{\cdot}5)^3}{2!} + \frac{(0{\cdot}5)^4}{2!3!} \ldots, \text{etc.}$$

$$+ \frac{(0{\cdot}5)^3}{3!} + \frac{(0{\cdot}5)^4}{3!} \ldots, \text{etc.}$$

$$+ \frac{(0{\cdot}5)^4}{3!} \ldots, \text{etc.}$$

Total:

$$1 + 1 + 0{\cdot}25\left(\frac{1}{2!} + \frac{2}{2!} + \frac{1}{2!}\right) + 0{\cdot}125 \left(\frac{1}{3!} + \frac{3}{3!} + \frac{3}{3!} + \frac{1}{3!}\right) + 0{\cdot}0625\left(\frac{1}{4!} + \frac{4}{4!} + \frac{6}{4!} + \frac{4}{4!} + \frac{1}{4!}\right) \ldots, \text{etc.}$$

$$= 1 + 1 + \frac{1}{2!} + \frac{1}{3!} + \frac{1}{4!}, \text{ etc.} = e$$

Needless to say, this example which establishes that e^f satisfies the acid test of consistency, i.e. $e^f \,.\, e^{1-f} = e$, only when $f = 0{\cdot}5$, does not prove the rule that e^f is meaningful in terms of our definition of e^n when f is a different fraction. What it does is to illustrate a pattern for setting out, as an exercise in multiplication, the rule itself. For brevity we may write $f = a$ and $(1 - f) = b$, so that $(a + b) = 1$ in the product e^a and e^b as below:

$$e^a = 1 + a + \frac{a^2}{2!} + \frac{a^3}{3!} + \frac{a^4}{4!} + \frac{a^5}{5!} \ldots$$

$$e^b = 1 + b + \frac{b^2}{2!} + \frac{b^3}{3!} + \frac{b^4}{4!} + \frac{b^5}{5!} \ldots$$

$$
\begin{array}{cccccccccccc}
1 + a + & \frac{a^2}{2!} & + & \frac{a^3}{3!} & + & \frac{a^4}{4!} & + & \frac{a^5}{5!} & \cdots \\
b + ab + & \frac{a^2b}{2!} & + & \frac{a^3b}{3!} & + & \frac{a^4b}{4!} & \cdots \\
\frac{b^2}{2!} + & \frac{ab^2}{2!} & + & \frac{a^2b^2}{(2!)^2} & + & \frac{a^3b^2}{2!3!} & \cdots \\
& \frac{b^3}{3!} & + & \frac{ab^3}{3!} & + & \frac{a^2b^2}{3!2!} & \cdots \\
& & & \frac{b^4}{4!} & + & \frac{ab^4}{4!} & \cdots \\
& & & & & \frac{b^5}{5!} & \cdots
\end{array}
$$

The column totals are:

$$
1 + (a + b) + \frac{(a^2 + 2ab + b^2)}{2!} \frac{(a^3 + 3a^2b + 3ab^2 + b^3)}{3!} + \frac{(a^4 + 4a^3b + 6a^2b^2 + 4ab^3 + b^4)}{4!}, \text{ etc.}
$$

$$
= 1 + (a + b) + \frac{(a + b)^2}{2!} + \frac{(a + b)^3}{3!} + \frac{(a + b)^4}{4!}, \text{ etc.}
$$

If $b = 1 - a$, $(a + b) = 1$, and the foregoing series is:

$$
1 + 1 + \frac{1}{2!} + \frac{1}{3!} + \frac{1}{4!} \ldots, \text{ etc.}
$$

If, as is usual, we speak of e^n as a *discrete* function of n, if n is an integer, and of e^x, as a *continuous* function of x for all so-called real values of x, we also need to prove that e^x is meaningful when x is not a so-called *rational* fraction f as defined above.

However, we have seen that a robot mathematician would find it difficult to distinguish the rational from the so-called irrational, and that the distinction has little relevance to the way we handle numbers when we ask electronic brains to do the job for which we should otherwise have to rely on our own grey matter. So we shall henceforth speak of $y = e^x$ as the *exponential function*, and the only useful question, relevant to our present aim, is in what circumstances does the series, which we represent by e^x, choke off?

Let us first consider why e itself is a namable number to any required level of precision. We shall label its initial term as t_0, since $e^0 = 1$ is obtainable by painting in $x = 0$ in e^x throughout. We need then consider only:

$$t_{11} + t_{12} + t_{13}, \text{ etc.} = + \frac{1}{11!} + \frac{1}{12!} + \frac{1}{13!}, \text{ etc.}$$

$$= \frac{1}{11(10!)} + \frac{1}{12\,.\,11(10!)} + \frac{1}{13\,.\,12\,.\,11\,.\,(10!)}, \text{ etc.}$$

$$= \frac{1}{10!}\left(\frac{1}{11} + \frac{1}{12\,.\,11} + \frac{1}{13\,.\,12\,.\,11} \ldots, \text{ etc.}\right)$$

This series is clearly less than:

$$= \frac{1}{10!}\left(\frac{1}{10} + \frac{1}{100} + \frac{1}{1000} \ldots, \text{ etc.}\right)$$

It is therefore less than:

$$\frac{1}{10!}\,(0{\cdot}1111\ldots)$$

To see what this means in hard cash, let us look at the sum of the first ten terms, i.e. from rank 0 to rank 9:

$1 + 1$	$= 2{\cdot}000\ 000\ 0$	
$1 \div 2!$	$= 0{\cdot}500\ 000\ 0$	
$1 \div 3!$	$= 0{\cdot}166\ 666\ 7$	(correct to seventh decimal place)
$1 \div 4!$	$= 0{\cdot}041\ 666\ 7$	” ” ” ” ”
$1 \div 5!$	$= 0{\cdot}008\ 333\ 3$	” ” ” ” ”
$1 \div 6!$	$= 0{\cdot}001\ 388\ 9$	” ” ” ” ”
$1 \div 7!$	$= 0{\cdot}000\ 198\ 4$	” ” ” ” ”
$1 \div 8!$	$= 0{\cdot}000\ 024\ 8$	” ” ” ” ”
$1 \div 9!$	$= 0{\cdot}000\ 002\ 8$	” ” ” ” ”
Total:	$2{\cdot}718\ 281\ 6$	

We have seen that the total of all the terms after $t_{10} = (1 \div 10!)$ is less than $(0{\cdot}\dot{1})t_{10}$, i.e. less than $(0{\cdot}0\dot{1})t_9$, so that the *remainder* is of the order 0·0000003; and we may assume that:

$$2{\cdot}718.281 < e < 2{\cdot}718282$$

Clearly, if x is positive, e^x is less than e if x is less than 1, and e^{-x} is less than e^x. So it remains to concern ourselves only with the possibility that e^x is necessarily convergent (i.e. chokes off like the infinite series represented by $0 \cdot \dot{1}$) when $x > 1$. This will be true, if we can define a rank beyond which the sum of all the remaining terms is less than a namable number such as $0 \cdot \dot{1}$. To explore this possibility, we may start by examining the ratio of any term (t_r) of rank r to its successor as below:

$$t_{r+1} = \frac{x^{r+1}}{(r+1)!} = \frac{x \, . \, x^r}{(r+1) \, . \, r!} = \frac{x}{r+1} \cdot t_r$$

Similarly:

$$t_{r+2} = \frac{x}{r+2} \cdot t_{r+1} = \frac{x}{r+1} \cdot \frac{x}{r+2} \cdot t_r$$

$$t_{r+3} = \frac{x}{r+3} \cdot t_{r+2} = \frac{x}{r+1} \cdot \frac{x}{r+2} \cdot \frac{x}{r+3} \cdot t_r$$

We may thus write for the remainder after the term of rank r:

$$= t_r\left(\frac{x}{r+1} + \frac{x^2}{(r+1)(r+2)} + \frac{x^3}{(r+1)(r+2)(r+3)} \ldots, \text{etc.}\right)$$

If y is a positive integer greater than x, every term after the first in the series for e^y is greater than the corresponding one in that of e^x, so that $e^x < e^y$ and the series for e^x converges if that for e^y does so. Now successive terms of e^y begin to diminish only when $r = y$, so that $t_{r+1} \div t_r = r \div (r+1)$. Thus we may write as the tail of the series for e^y for values of $r \geqslant y$:

$$t_r\left(\frac{r}{r+1} + \frac{r^2}{(r+1)(r+2)} + \frac{r^3}{(r+1)(r+2)(r+3)} \ldots\right)$$

Successive terms in this series are less than corresponding ones of the series below, and the series above must converge if the following converges:

$$t_r\left(\frac{r}{r+1} + \frac{r^2}{(r+1)^2} + \frac{r^3}{(r+1)^3} \ldots, \text{etc.}\right)$$

$$= \frac{r \, . \, t_r}{r+1}\left(1 + \frac{r}{(r+1)} + \frac{r^2}{(r+1)^2} + \frac{r^3}{(r+1)^3} \ldots, \text{etc.}\right)$$

Let us now denote the sequence in brackets as S:

$$S = 1 + \frac{r}{(r+1)} + \frac{r^2}{(r+1)^2} + \frac{r^3}{(r+1)^3} \cdots$$

$$\frac{r}{(r+1)} \cdot S = \cdots \frac{r}{(r+1)} + \frac{r^2}{(r+1)^2} + \frac{r^3}{(r+1)^3} \cdots$$

$$\therefore\ S\left(1 - \frac{r}{r+1}\right) = 1, \quad \textit{so that} \quad S = (r+1)$$

For the tail of the series under discussion, we may thus write

$$t_r\left(\frac{r}{r+1} + \frac{r^2}{(r+1)(r+2)} + \frac{r^3}{(r+1)(r+2)(r+3)} \cdots\right) < r \,.\, t_r$$

As an illustration, let us consider the case when $x = 3$, i.e.:

$$e^3 = \left(1 + 3 + \frac{3^2}{2!} + \frac{3^3}{3!}\right) + \frac{3^4}{4!} + \frac{3^5}{5!} + \frac{3^6}{6!} \cdots$$

Here the tail beginning with the term of rank $(x + 1) = 4$ cannot be greater than $x \,.\, t_x = 3 \,.\, 3^3 \div 3! = 13{\cdot}5$. Since the sum of the preceding terms is 13, we see that $e^3 < 13 + 13{\cdot}5$, i.e. $13 < e^3 < 26{\cdot}5$.

Let us now consider the result of summing the first ten terms (rank 0 to rank 9), so that the remainder is:

$$\frac{3^{10}}{10!} + \frac{3^{11}}{11!} + \frac{3^{12}}{12!} \ldots, \text{etc.}$$

$$= \frac{3^{10}}{10!}\left(1 + \frac{3}{11} + \frac{3^2}{12 \,.\, 11} + \frac{3^{13}}{13 \,.\, 12 \,.\, 11} \ldots, \text{etc.}\right)$$

From this we see that the sum of the terms in the tail beyond that of rank 9 is less than:

$$\frac{3^{10}}{10!}\left(1 + \frac{3}{11} + \frac{3^2}{11^2} + \frac{3^3}{11^3} \cdots\right)$$

In this expression, we may write:

$$S_t = 1 + \frac{3}{11} + \frac{3^2}{11^2} + \frac{3^3}{11^3} \cdots$$

$$\frac{3}{11} \cdot S_t = \frac{3}{11} + \frac{3^2}{11^2} + \frac{3^3}{11^3} \cdots$$

$$\frac{8}{11} \cdot S_t = 1 \quad and \quad S_t = \frac{11}{8}$$

Our remainder is now therefore less than:

$$\frac{3^{10}}{10!} \times \frac{11}{8} = \frac{3^6 . 11}{358400} = \frac{8019}{358400} \simeq 0{\cdot}0\dot{2}$$

By summing the first ten terms (rank 0 to rank 9), we thus arrive at a result correct to one decimal place and, since there are two figures in front of the decimal point, correct to three significant figures.

If, however, $x < 1$ convergence is very rapid. For instance, if $x = \frac{1}{5}$:

$$e^x = 1 + 0{\cdot}2 + \frac{(0{\cdot}2)^2}{2!} + \frac{(0{\cdot}2)^3}{3!} + \frac{(0{\cdot}2)^4}{4!} + \frac{(0{\cdot}2)^5}{5!} \cdots$$
$$= 1 + 0{\cdot}2 + 0{\cdot}02 + 0{\cdot}001\dot{3} + 0{\cdot}0000\dot{6} + 0{\cdot}000002\dot{6}$$

The totals for the sum of terms up to rank 0, 1, 2 . . . 5 are:

0	1	2	3	4	5
1	1·2	1·22	1·221$\dot{3}$	1·2213$\dot{9}$	1·221402$\dot{6}$

The terms of rank 6, 7, etc., are:

$$\frac{1}{5^6 . 6!} + \frac{1}{5^7 . 7!} + \frac{1}{5^8 . 8!} + \frac{1}{5^9 . 9!} \cdots$$
$$= \frac{1}{5^6 . 6!}\left(1 + \frac{1}{5 . 7} + \frac{1}{5^2 . 7 . 8} + \frac{1}{5^3 . 7 . 8 . 9} \cdots\right)$$

The series in brackets converges more rapidly than the geometric series:

$$\left(1 + \frac{1}{5 . 7} + \frac{1}{5^2 . 7^2} + \frac{1}{5^3 . 7^3} \cdots\right) = \frac{35}{34}$$

Thus the sum of terms of rank 6 and thereafter is less than:

$$\frac{1}{5^6 . 6!} \cdot \frac{35}{34} = \frac{7}{5^5 . 720 . 34} = \frac{7}{78499200} < 10^{-7}$$

This means that the value of $e^{0 \cdot 2}$ cannot exceed 1·221402$\dot{6}$ by more than 0·0000001, so that 1·221401$\dot{6}$ $< e^{0 \cdot 2} <$ 1·2214028.

The Use of the Imaginary Number. The most interesting, and at first surprising, thing about the mathematical pronoun e is that it is closely connected with quantities met with in trigonometry. In tabulating logarithms for the sines of angles Napier set out to find how the length of the half chord varies (see Fig. 129) as it moves along the diameter of a circle in steps equivalent to equal strips of the circumference. In a circle of unit radius (p. 211) the half chord is the sine of the angle enclosed by an arc. The practical problem was to connect with the length of the half chord quantities whose addition corresponds to the multiplication of sines. In the language of modern mathematics this was equivalent to calculating the logarithms of sines to the base $e^{-1} = (0{\cdot}368\ldots)$. Napier was not acquainted with the series which we have just given, and the reason why e is connected with the behaviour of sines did not become clear before the discovery of the first really important use of the imaginary i.

De Moivre, who discovered it, was a Huguenot refugee who settled (1685) in England, where he earned a livelihood from an Insurance firm as a pioneer in the field of actuarial mathematics.

Interest in i started as the result of the quest for an algebraic solution of a cubic equation by two sixteenth-century Italian mathematicians Tartaglia and Cardano. No problem need arise, if we can express it as the product of three so-called *real* factors whose product is zero, e.g.:

$$x^3 + 2x^2 - x - 2 = 0 = (x - 1)(x + 1)(x + 2)$$

Here the equation is valid: (*a*) if $(x - 1) = 0$, so that $x = +1$; (*b*) if $(x + 1) = 0$, so that $x = -1$; (*c*) if $(x + 2) = 0$, so that $x = -2$. If we can resolve it into two factors only, a different situation may arise, e.g.:

$$x^3 + 3x^3 + 4x + 2 = 0 = (x + 1)(x^2 + 2x + 2)$$

Here the equation is valid if $(x + 1) = 0$, in which event $x = -1$, and if $x^2 + 2x + 2 = 0$, but the solution of the second factor by the usual method yields two *complex* numbers:

$$x = \frac{-2 \pm \sqrt{4 - 8}}{2} = -1 \pm \tfrac{1}{2}\sqrt{-4} = -1 + i \quad or \quad -1 - i$$

As first developed by Tartaglia and Cardano in the sixteenth century, a general method of solving a cubic equation, without recourse to graphical methods or to iteration, proceeds as illustrated by the following:

$$x^3 + 6x^2 - 6x - 63 = 0$$

The treatment of this equation is *not* essential to an understanding of de Moivre's discovery; and *the reader who skips it will not be at a disadvantage when we come to the latter.* It does however give one an insight into the circumstances which first forced mathematicians to take i seriously.

In dealing with such an equation as last cited by the general algebraic method for the solution of a cubic, our first step (see Chapter 8, p. 389) is to shift the origin on the X-axis to dispose of the term in x^2 as we did when constructing the graph of Fig. 136. We therefore put $X + a = x$, and note that this will vanish if:

$$3aX^2 + 6X^2 = 0, \quad \textit{so that} \quad a = -2 \quad \textit{and} \quad x = X - 2$$

Our equation then becomes:

$$(X^3 - 6X^2 + 12X - 8) + 6(X^2 - 4X + 4) \\ - 6(X - 2) - 63 = 0$$

$$\therefore \ X^3 - 18X - 35 = 0$$

The next step depends on noticing that:

$$(u + v)^3 = u^3 + 3u^2v + 3uv^2 + v^3 = u^3 + v^3 + 3uv(u + v)$$

$$\therefore \ (u + v)^3 - 3uv(u + v) - (u^3 + v^3) = 0$$

If we put $X = (u + v)$ in the above:

$$X^3 - 3uv\ X - (u^3 + v^3) = 0 = X^3 - 18X - 35$$

We therefore see that $-3uv = -18$ and $-(u^3 + v^3) = -35$, so that $uv = 6$, $u^3v^3 = 216$, and $(u^3 + v^3)^2 = u^6 + 2u^3v^3 + v^6 = 35^2 = 1225$. We can therefore write:

$$u^6 + 2u^3v^3 + v^6 = 1225$$

$$4u^3v^3 = 864$$

$$\therefore \ u^6 - 2u^3v^3 + v^6 = (1225 - 864)$$

$$\therefore \ (u^3 - v^3)^2 = 19^2$$

Thus we now have:

$$u^3 + v^3 = 35 \textit{ and } u^3 - v^3 = 19, \textit{ so that}$$

$$2u^3 = 54 \text{ and } -2v^3 = 16$$

$$\therefore \ u^3 = 27 \text{ and } v^3 = 8, \text{ when } u = 3,\ v = 2 \text{ and } X = (u + v) = 5$$

Since $x = X - 2$, $x = (5 - 2) = 3$ is a solution. As it happens, $(x - 3)$ is a factor of the original expression equated to zero, the other being $x^2 + 9x + 21$. If we equate this to zero, we find that the equation yields two complex solutions:

$$\tfrac{1}{2}(-9 + i\sqrt{3}) \quad \textit{and} \quad \tfrac{1}{2}(-9 - i\sqrt{3})$$

At first sight, this method seems to break down when all three so-called *roots* (i.e. solutions) are real, as is true of:

$$x^3 - 6x^2 + 11x - 6 = 0 = (x - 1)(x - 2)(x - 3)$$

If we here eliminate the square term in the usual way by writing $x = X + 2$, this equation becomes:

$$X^3 - X = 0$$

If therefore $X = (u + v)$, the foregoing procedure gives:

$$X^3 - 3uv\,X - (u^3 + v^3) = 0 = X^3 - X$$
$$\therefore\ 3uv = 1 \textit{ and } u^3 + v^3 = 0$$
$$\therefore\ uv = \tfrac{1}{3} \textit{ and } 4u^3v^3 = \tfrac{4}{27}$$

We may now proceed as before:

$$u^6 + 2u^3v^3 + v^6 = 0$$
$$4u^3v^3 = \tfrac{4}{27}$$
$$\therefore\ u^3 - v^3 = \sqrt{(-4 \div 27)} = \frac{2i}{3\sqrt{3}}$$

Whence we obtain:

$$u^3 = \frac{i}{3\sqrt{3}} \quad \text{and} \quad v^3 = \frac{-i}{3\sqrt{3}}$$

Now direct multiplication shows that on substitution of $i^2 = -1$:

$$(i + \sqrt{3})^3 = i^3 + 3i^2\sqrt{3} + 3i(\sqrt{3})^2 + 3\sqrt{3}$$
$$= -i - 3\sqrt{3} + 9i + 3\sqrt{3} = 8i$$
$$(i - \sqrt{3})^3 = i^3 - 3i^2\sqrt{3} + 3i(\sqrt{3})^2 - 3\sqrt{3}$$
$$= -i + 3\sqrt{3} + 9i - 3\sqrt{3} = 8i$$

Hence:

$$\frac{(i \pm \sqrt{3})^3}{(2\sqrt{3})^3} = \frac{8i}{8(\sqrt{3})^3} = \frac{i}{3\sqrt{3}} = u^3 \quad \textit{and} \quad u = \frac{(i \pm \sqrt{3})}{2\sqrt{3}}$$
$$\frac{-(i \pm \sqrt{3})^3}{(2\sqrt{3})^3} = \frac{8i}{8(\sqrt{3})^3} = \frac{-i}{3\sqrt{3}} = v^3 \quad \textit{and} \quad v = \frac{-(i \pm \sqrt{3})}{2\sqrt{3}}$$

Owing to the ambivalence of the sign of $\sqrt{3}$, we can obtain three different results by adding u and v, as the table below:

	$v = \dfrac{-i + \sqrt{3}}{2\sqrt{3}}$	$v = \dfrac{-i - \sqrt{3}}{2\sqrt{3}}$
$u = \dfrac{i + \sqrt{3}}{2\sqrt{3}}$	$u + v = +1$	$u + v = 0$
$u = \dfrac{i - \sqrt{3}}{2\sqrt{3}}$	$u + v = 0$	$u + v = -1$

Thus $X = \pm 1$ or 0. Since $x = X + 2$, $x = 1, 2$, or 3 as already obtained by factorization more easily. The value of the exercise is this: we can solve the case by the general method when all the solutions are real if we treat i according to the same rules as we invoke when we write $(-1)^2 = 1$, i.e. by substituting -1 for i^2 in all our calculations.

Before we turn to the discovery of de Moivre, let us therefore be explicit about the meaning we shall attach to successive powers of i:

$$
\begin{aligned}
i &= \sqrt{-1} \\
i^2 &= (\sqrt{-1})^2 = -1 \\
i^3 &= (\sqrt{-1})^2 \times i = -i \\
i^4 &= (\sqrt{-1})^3 \times i = -i^2 = +1 \\
i^5 &= (\sqrt{-1})^4 \times i = +i \\
&\text{etc.}
\end{aligned}
$$

so we have:

i	i^2	i^3	i^4	i^5	i^6	i^7	i^8	$i^9 \ldots$
$+i$	-1	$-i$	$+1$	$+i$	-1	$-i$	$+1$	$+i \ldots$

De Moivre's Theorem. Let us now deal with a remarkable discovery in three stages on the two assumptions that we can:

(*a*) apply the ordinary rules of multiplication and substitute everywhere as above: $i^2 = -1$; $i^3 = -i$; $i^4 = +1$; $i^5 = +i$, etc.;

(*b*) treat only complex numbers as equal when both the *real* part of one is equal to the real part of the other, and the *imaginary* part of one is equal to the imaginary part of the other.

We shall first assume that n is a whole number and positive in the expression:

$$(\cos a + i \sin a)^n$$

By direct multiplication, we then see that:

$$(\cos a + i\,.\sin a)^2 = \cos^2 a + 2i\,.\sin a \cos a + i^2 \sin^2 a$$
$$= \cos^2 a - \sin^2 a + 2i\,.\sin a\,.\cos a$$

Whence, by the addition formulae (p. 369) for *cos* $(A + B)$ and *sin* $(A + B)$:

$$(\cos a + i\,.\sin a)^2 = \cos 2a + i \sin 2a$$

This suggests a rule that we can put to the test of *induction* (p. 108), already used several times in this book. Accordingly, we assume provisionally that:

$$(\cos a + i\,.\sin a)^n = \cos na + i\,.\sin na$$

If this is true:

$$(\cos a + i\,.\sin a)^{n+1} = (\cos a + i\,.\sin a)^n\,.(\cos a + i\,.\sin \text{a})$$
$$= (\cos na + i\,.\sin na)(\cos a + i\,.\sin a)$$
$$= \cos na\,.\cos a + i\,.\cos na\,.\sin a$$
$$+ i\,.\sin na\,.\cos a + i^2\,.\sin na\,.\sin a$$
$$= (\cos na\,.\cos a - \sin na\,.\sin a)$$
$$+ i(\cos na\,.\sin a + \sin na\,.\cos a)$$

By the addition formulae, as above, we therefore derive:

$$(\cos a + i \sin a)^{n+1} = \cos (na + a) + i\,.\sin (na + a)$$
$$= \cos (n + 1)\, a + i\,.\sin (n + 1)a$$

Since we know that our guess is correct when $n = 1$ or 2, it must therefore be true if $n = (2 + 1) = 3$ and if true for $n = 3$, for $n = (3 + 1) = 4$ and so on. The reader should be able to complete the proof for the theorem when n is a whole number and positive, i.e. show that:

$$(\cos a - i \sin a)^n = \cos na - i \sin na$$

If n is a whole number but negative, we may write:

$$(\cos a + i \,.\, \sin a)^{-n} = \frac{1}{(\cos a + i \sin a)^n} = \frac{1}{\cos na + i \,.\, \sin na}$$

$$= \frac{1}{\cos na + i \,.\, \sin na} \times \frac{\cos na - i \,.\, \sin na}{\cos na - i \,.\, \sin na}$$

$$= \frac{\cos na - i \,.\, \sin na}{\cos^2 na - i^2 \sin^2 na} = \frac{\cos na - i \sin na}{\cos^2 na + \sin^2 na}$$

$$= \cos na - i \sin na$$

We reach the same result if we substitute $-n$ for n in the original formula, bearing in mind that $cos\ (-a) = cos\ (2\pi - a) = cos\ a$ and $sin\ (-a) = sin\ (2\pi - a) = (-sin\ a)$. We may therefore write:

$$(\cos a + i \sin a)^{-n} = \cos (-n \,.\, a) + i \sin (-n \,.\, a)$$
$$= \cos \,.\, na - i \,.\, \sin \,.\, na$$

If we can now show that the rule holds good when n is the reciprocal of a whole number q and therefore a rational fraction, we need not bother about proving that it is true if $n = p \div q$ is also a rational fraction. We may work backwards thus:

$$\left(\cos \frac{a}{q} + i \,.\, \sin \frac{a}{q}\right)^q = \cos a + i \,.\, \sin a$$

$$\left(\cos \frac{a}{q} + i \sin \frac{a}{q}\right)^{q/q} = (\cos a + i \,.\, \sin a)^{\frac{1}{q}}$$

$$\therefore\ (\cos a + i \sin a)^{1/q} = \cos \frac{a}{q} + i \,.\, \sin \frac{a}{q}$$

This is true, but not the whole truth. If $n = 0, 1, 2, 3, 4$, etc. (see p. 369), $cos\ a$ or $sin\ a = cos$ or $sin\ (a + 2n\pi)$, so that:

$$\cos a + i \,.\, \sin a = \cos (a + 2n\pi) + i \,.\, \sin (a + 2n\pi)$$

$$(\cos a + i \sin a)^{1/q} = \cos \frac{a + 2n\pi}{q} + i \,.\, \sin \frac{a + 2n\pi}{q}$$

One thing about this extension of the rule is not necessarily self-evident at first sight. To recognize what it is, let us consider the cube root of ($cos\ a + i \,.\, sin\ a$), i.e. what the rule means when $q = 3$, so that:

$$(\cos a + i \,.\, \sin a)^{1/3} = \cos \left(\frac{a + 2n\pi}{3}\right) + i \,.\, \sin \frac{a + 2n\pi}{3}$$

By inserting $n = 0, 1, 2, 3, 4, 5, 6$, etc.:

$$\frac{a + 2n\pi}{3} = \frac{a}{3}; \frac{a + 2\pi}{3}; \frac{a + 4\pi}{3}; \frac{a + 6\pi}{3}; \frac{a + 8\pi}{3}; \frac{a + 10\pi}{3}$$

$$= \frac{a}{3}; \frac{a + 2\pi}{3}; \frac{a + 4\pi}{3}; \frac{a}{3} + 2\pi; \frac{a + 2\pi}{3} + 2\pi;$$

$$\frac{a + 4\pi}{3} + 2\pi$$

If $q = 3$ in the expression last obtained, the cube root of $\cos a + i \sin a$ turns out to have three different values, each of which repeats itself periodically. As the reader may ascertain for other values, the qth root of $\cos a + i \sin a$ has q different values, which likewise repeat themselves. We may be more willing to test whether this makes sense, i.e. whether it is *consistent* with our definition of complex numbers, when we have satisfied ourselves that our new rule can indeed deliver the goods in the domain of what we customarily mean by numbers in the world's work. To do so, we may first use it to derive a formula we can establish by another and more familiar method, i.e.:

$$\cos 3a = \cos (a + 2a) = \cos a \cos 2a - \sin a . \sin 2a$$

Since $\cos 2a = \cos (a + a) = \cos^2 a - \sin^2 a$:

$$\cos 3a = \cos^3 a - 3 \sin^2 a \cos a$$

We may also write this as:

$$\begin{aligned} \cos 3a &= \cos^3 a - 3 \cos a(1 - \cos^2 a) \\ &= \cos^3 a - 3 \cos a + 3 \cos^3 a \\ &= 4 \cos^3 a - 3 \cos a \end{aligned}$$

One result which we used to illustrate the truth of de Moivre's rule can naturally be obtained by applying it. Thus if:

$$x = \cos a + i \sin a$$

$$x^{-1} = (\cos a + i \sin a)^{-1}$$

$$\therefore \frac{1}{x} = \cos a - i \sin a$$

$$\therefore x + \frac{1}{x} = 2 \cos a$$

and $$x - \frac{1}{x} = 2i \sin a$$

Likewise we may put $x^n = \cos na + i \sin na$

$$\frac{1}{x^n} = \cos na - i \sin na$$

$$\therefore\ x^n + \frac{1}{x^n} = 2 \cos na$$

and $$x^n - \frac{1}{x^n} = 2i \sin na$$

We thus see that if

$$x + \frac{1}{x} = 2 \cos a$$

$$x^n + \frac{1}{x^n} = 2 \cos na$$

If we wish to find $\cos 3a$, knowing $\cos a$, we therefore put:

$$2 \cos 3a = x^3 + \frac{1}{x^3}$$

$$\therefore\ 2 \cos a = x + \frac{1}{x}$$

$$\therefore\ (2 \cos a)^3 = \left(x + \frac{1}{x}\right)^3$$

$$\therefore\ 8 \cos^3 a = x^3 + 3x^2 . \frac{1}{x} + 3x . \frac{1}{x^2} + \frac{1}{x^3}$$

$$= \left(x^3 + \frac{1}{x^3}\right) + 3\left(x + \frac{1}{x}\right)$$

$$= 2 \cos 3a + 6 \cos a$$

$$\therefore\ 4 \cos^3 a = \cos 3a + 3 \cos a$$

or $$\cos 3a = 4 \cos^3 a - 3 \cos a$$

If you still wish to convince yourself that *imaginary* quantities may be used for calculation, you can now put:

$$(2 \cos a)^6 = \left(x + \frac{1}{x}\right)^6$$

$$64 \cos^6 a = x^6 + 6x^5 . \frac{1}{x} + 15x^4 . \frac{1}{x^2} + 20x^3 . \frac{1}{x^3} + 15x^2 . \frac{1}{x^4} + 6x . \frac{1}{x^5} + \frac{1}{x^6}$$

$$= \left(x^6 + \frac{1}{x^6}\right) + 6\left(x^4 + \frac{1}{x^4}\right) + 15\left(x^2 + \frac{1}{x^2}\right) + 20$$

$$= 2 \cos 6a + 12 \cos 4a + 30 \cos 2a + 20$$

$$\cos 6a = 32 \cos^6 a - 6 \cos 4a - 15 \cos 2a - 10$$

You can check this from two values of cos a which you know already, i.e. (a) cos 90° = 0; (b) cos 60° = ½. So:

(a) $\cos 540° = 32 \cos^6 90° - 6 \cos 360° - 15 \cos 180° - 10$

$\therefore \cos (360 + 180)° = 0 - 6(1) - 15(-1) - 10$

$\therefore \cos 180° = -1$

(b) $\cos 360° = 32 \cos^6 60° - 6 \cos 240° - 15 \cos 120° - 10$

$= 32 \cos^6 60° - 6 \cos (180 + 60)° - 15 \cos (180 - 60)° - 10$

$= 32(\frac{1}{64}) + 6(\frac{1}{2}) + 15(\frac{1}{2}) - 10$

$= 1$

Having now reason to hope that we may make $\sqrt{-1}$ or i do the sort of work which we can check by actual calculations, we shall now use it to derive the two series from which we can obtain the sine or cosine of any angle to any degree of accuracy required. We have seen how the study of trigonometry in connection with the preparation of astronomical tables for navigation prompted the search for quick methods of calculation, and how the discovery of logarithms led on to the study of unlimited series such as the exponential series. A crowning achievement which followed the Great Navigations was the further discovery that the exponential series has a simple relation to the theorem of de Moivre.

To see the connection between the exponential series and the theorem of de Moivre, first put:

$$x = \cos 1 + i \sin 1$$

This is the same as making $a = 1$ in the original expression of de Moivre's rule, and if we now use a like n for *any* number:

$$x^a = \cos a + i \sin a \quad . \quad . \quad . \quad . \quad . \quad . \quad \text{(i)}$$

and $$x^{-a} = \cos a - i \sin a$$

As before, we may put $x^a - x^{-a} = 2i \sin a$

We can also represent x as some power of e, thus:

$$x = e^y$$

or $$y = \log_e x \quad . \quad . \quad . \quad . \quad . \quad . \quad . \quad . \qquad \text{(ii)}$$

Using the exponential series, we have:

$$x^a = e^{ay} = 1 + ay + \frac{a^2y^2}{2!} + \frac{a^3y^3}{3!} + \frac{a^4y^4}{4!} \ldots$$

$$x^{-a} = e^{-ay} = 1 - ay + \frac{a^2y^2}{2!} - \frac{a^3y^3}{3!} + \frac{a^4y^4}{4!} \ldots$$

Subtracting the lower from the upper, we get:

$$x^a - x^{-a} = 2ay + 2\frac{a^3y^3}{3!} + 2\frac{a^5y^5}{5!} + 2\frac{a^7y^7}{7!} \ldots$$

We may also write this as:

$$2i \sin a = 2ya + 2\frac{y^3a^3}{3!} + 2\frac{y^5a^5}{5!} \ldots$$

$$\therefore \quad \frac{i \sin a}{a} = y + \frac{y^3a^2}{3!} + \frac{y^5a^4}{5!} \ldots \qquad \text{(iii)}$$

Since a is any angle, this equation is true when a is so small that we can neglect any term multiplied by a, i.e.:

$$y + \frac{y^3a^2}{3!} + \frac{y^5a^4}{5!} + \frac{y^7a^6}{7!} \ldots = y$$

We have also seen that if a stands for *radians* (see Chapter 6, p. 215) when a is very small:

$$\frac{\sin a}{a} = 1$$

$$\therefore \quad \frac{i \sin a}{a} = i$$

So when a is very small one side of equation (iii) reduces to i and the other side to y. We thus find that:

$$i = y$$

$$\therefore \quad x = e^i$$

$$\therefore \quad x^a = e^{ia}$$

If we write in the value of x^a in (i), this means that if a is measured in radians:

$$e^{ia} = \cos a + i \sin a$$

But since $i^2 = -1$, $i^3 = -i$, $i^4 = +1$, etc.:

$$e^{ia} = 1 + ia - \frac{a^2}{2!} - \frac{ia^3}{3!} + \frac{a^4}{4!} + \frac{ia^5}{5!} - \frac{a^6}{6!} \ldots$$

$$\therefore \cos a + i \sin a = \left(1 - \frac{a^2}{2!} + \frac{a^4}{4!} - \frac{a^6}{6!} \ldots\right) + i\left(a - \frac{a^3}{3!} + \frac{a^5}{5!} \ldots\right)$$

To take the next step in our stride, we must eschew the error of mixing up numbers which stand for one thing with numbers which stand for another. So we recall that x and ix stand (Fig. 153) for measurements in directions at right angles. If we say that we have a apples and p pears, we mean that we have 30 apples and 15 pears when $a = 30$ and $p = 15$. Consequently, we must also say that $p = \cos a$ and $q = \sin a$, if:

$$\cos a + i \sin a = p + iq$$

We now do the same thing with the equation:

$$\cos a + i \sin a = \left(1 - \frac{a^2}{2!} + \frac{a^4}{4!} \ldots\right) + i\left(a - \frac{a^3}{3!} + \frac{a^3}{5!} \ldots\right)$$

From which we conclude that if a is measured in *radians*:

$$\cos a = 1 - \frac{a^2}{2!} + \frac{a^4}{4!} - \frac{a^6}{6!} + \frac{a^8}{8!} \ldots$$

$$\sin a = a - \frac{a^3}{3!} + \frac{a^5}{5!} - \frac{a^7}{7!} \ldots$$

This means that you can calculate the cosine or sine of 1, 0·5, 0·1, etc., radians directly by substituting 1, 0·5, 0·1, etc., for a in the unlimited series given above. You will see that they must choke off quickly if a is less than 1. To assure yourself that this result is trustworthy turn back to Chapter 3, p. 147, and you will find that by using Euclid's geometry we obtained the values:

$$\cos 15° = 0{\cdot}966$$

$$\sin 15° = 0{\cdot}259$$

To use the series given we have to convert 15° into radians, thus:

$$15^\circ = \frac{1}{6}(90^\circ) = \frac{1}{6}\left(\frac{\pi}{2}\right) \text{ radians}$$

If we take $\pi = 3{\cdot}1416$, $15^\circ = 0{\cdot}2618$ radians.
So putting $a = 0{\cdot}2618$, we have:

$$a^2 = 0{\cdot}0685$$
$$a^3 = 0{\cdot}0179$$
$$a^4 = 0{\cdot}005$$
$$\text{etc.}$$

Thus $$\cos 15^\circ = 1 - \frac{0{\cdot}0685}{2} + \frac{0{\cdot}005}{24} \ldots$$
$$\sin 15^\circ = 0{\cdot}2618 - \frac{0{\cdot}0179}{6} \ldots$$

These series choke off very rapidly. We need take in only the first 3 terms to get the values cited in our three-figure table (p. 148):

$$\cos 15^\circ = 0{\cdot}9659;\ \sin 15^\circ = 0{\cdot}2589$$

Without invoking complex numbers in our next chapter, we shall derive the same convergent series for *sin a* and *cos a* on the same assumption: that our unit of a is the radian. We can therefore, in turn, derive de Moivre's theorem by the relation:

$$e^{ix} = \cos x + i\,.\sin x\ ;\ e^{inx} = \cos .\, nx + i\,.\sin .\, nx$$

The function of x denoted by e^{ix} plays a very important role in the solution of many physical problems, such as the damped oscillation of a weight suspended from a spring or of electrical potential during discharge of a condenser. Here we can merely whet the appetite of the reader with some museum pieces which have no such claim to utility. Since $\cos \pi = -1$ and $\sin \pi = 0$:

$$\cos \pi + i \sin \pi = -1 \quad \textit{so that} \quad e^{i\pi} = -1$$

Though of no manifest usefulness, this extraordinary and unexpected relation between two such important non-rational numbers as e and π, opens a window into a vast territory of number lore, wholly uncharted in Newton's boyhood. By the time he died mathematicians of the Western World, for the first time in history, had begun to map it. By the end of the same century, i.e. the eighteenth, and throughout the next one, they were beginning to see that the pioneers had taken for granted many assumptions which are difficult to justify as rules in a game of which the rules are clearly on record in advance. None the less, the territory claimed had yielded much that was usefully new to the stock-in-trade of the physicist.

Roots of Unity. Another museum piece which early attracted the attention of mathematicians is the use of complex numbers to solve the equations:

$$x^n + 1 = 0 \quad \textit{and} \quad x^n - 1 = 0$$

Needless to say, the solutions are:

$$x = \sqrt[n]{-1} \quad \text{and} \quad x = \sqrt[n]{+1}$$

We have already seen that any number has three cube roots. More generally, the number of the *n*th roots is *n*, none, one, or more than one of which may be real. De Moivre's theorem provides us with a method of finding the *n*th root of unity for any value of *n*. This is of no immediate practical value; but it does give us confidence in applying the rules of real algebra when dealing with Complex Numbers. That is to say, the outcome lands us in no inconsistencies. Let us consider the value of $\sqrt[4]{-1}$.

Since *cos* $\pi = -1$ and *sin* $\pi = 0$, we may write:

$-1 = \cos \pi + i \,.\, \sin \pi$ *and*

$$(-1)^{\frac{1}{4}} = \cos \frac{\pi + 2n\pi}{4} + i \,.\, \sin \frac{\pi + 2n\pi}{4}$$

If we paint in $n = 0, 1, 2, 3$, etc., this leads, after ringing the changes on the same, to the following values of $(\pi + 2n\pi) \div 4$:

$$\frac{\pi}{4}; \frac{3\pi}{4}; \frac{5\pi}{4}; \frac{7\pi}{4} \ldots$$

In the foregoing expression for the fourth roots of -1:

$$\frac{\pi}{4} \text{ radians } (= 45°) \quad \textit{and} \quad \cos 45° = \frac{1}{\sqrt{2}} = \sin 45°$$

$$\frac{3\pi}{4} = \frac{\pi}{2} + \frac{\pi}{4}, \quad \textit{so that} \quad \cos \frac{3\pi}{4} = -\frac{1}{\sqrt{2}} \quad \textit{and} \quad \sin \frac{3\pi}{4} = +\frac{1}{\sqrt{2}}$$

$$\frac{5\pi}{4} = \pi + \frac{\pi}{4}. \quad \textit{so that} \quad \cos \frac{5\pi}{4} = -\frac{1}{\sqrt{2}} \quad \textit{and} \quad \sin \frac{5\pi}{4} = -\frac{1}{\sqrt{2}}$$

$$\frac{7\pi}{4} = \pi + \frac{3\pi}{4}, \text{ so that } \cos\frac{7\pi}{4} = +\frac{1}{\sqrt{2}} \text{ and } \sin\frac{7\pi}{4} = -\frac{1}{\sqrt{2}}$$

Thus the four fourth roots of -1 are:

$$\frac{1}{\sqrt{2}} + \frac{i}{\sqrt{2}};\ \frac{-1}{\sqrt{2}} + \frac{i}{\sqrt{2}};\ \frac{-1}{\sqrt{2}} - \frac{i}{\sqrt{2}};\ \frac{1}{\sqrt{2}} - \frac{i}{\sqrt{2}}$$

These we may write more compactly as:

$$\frac{\pm 1}{\sqrt{2}}(1 + i) \quad \text{and} \quad \frac{\pm(1 - i)}{\sqrt{2}}$$

If we play the hand in the usual way $(\sqrt{2})^4 = 4$ and:

$$(1 + i)^4 = 1 + 4i + 6i^2 + 4i^3 + i^4$$
$$= 1 + 4i - 6 - 4i + 1 = -4$$
$$(1 - i)^4 = 1 - 4i + 6i^2 - 4i^3 + i^4$$
$$= 1 - 4i - 6 + 4i + 1 = -4$$

Hence we find that:

$$\left(\frac{\pm(1 + i)}{\sqrt{2}}\right)^4 = -1 \quad \text{and} \quad \left(\frac{\pm(1 - i)}{\sqrt{2}}\right)^4 = -1$$

Hyperbolic Functions. Here may be a proper place to introduce the so-called hyperbolic functions. We first recall that we may may write:

$$e^{ix} = 1 + ix - \frac{x^2}{2!} - \frac{ix^3}{3!} + \frac{x^4}{4!} + \frac{ix^5}{5!} - \frac{x^6}{6!}, \text{ etc.}$$
$$e^{-ix} = 1 - ix - \frac{x^2}{2!} + \frac{ix^3}{3!} + \frac{x^4}{4!} - \frac{ix^5}{5!} - \frac{x^6}{6!}, \text{ etc.}$$

When by addition or subtraction we obtain:

$$e^{ix} + e^{-ix} = 2\left(1 - \frac{x^2}{2!} + \frac{x^4}{4!} - \frac{x^6}{6!} \cdots\right)$$
$$e^{ix} - e^{-ix} = 2i\left(x - \frac{x^3}{3!} + \frac{x^5}{5!} - \frac{x^7}{7!} \cdots\right)$$

Consequently, we may write (in radian measure):

$$\cos x = \frac{e^{ix} + e^{-ix}}{2} \quad \text{and} \quad \sin x = \frac{e^{ix} - e^{-ix}}{2i}$$

If we eliminate i in the above, we obtain two functions of x which have (as we might guess) several properties in common with the customarily trigonometrical (so-called *circular*) functions *cos x* and *sin x*. We speak of one as the *hyperbolic cosine* of x (written as *cosh . x*), and of the other as the *hyperbolic sine* of x (written as *sinh . x*). By definition then:

$$\cosh . x = \frac{e^x + e^{-x}}{2} \quad \textit{and} \quad \sinh . x = \frac{e^x - e^{-x}}{2}$$

Among other similarities between the two classes of functions, the expressions on the left below are all obtainable by direct multiplication, addition, etc.:

Hyperbolic	Circular
$\sinh . 0 = 0$; $\cosh 0 = 1$	$\sin 0 = 0$; $\cos 0 = 1$
$(\cosh . x)^2 - (\sinh . x)^2 = 1$	$\cos^2 x + \sin^2 x = 1$
$\sinh (x + y)$	$\sin (x + y)$
$= \sinh x . \cosh y + \cosh x . \sinh y$	$= \sin x . \cos y + \cos x . \sin y$
$\cosh (x + y)$	$\cos (x + y)$
$= \cosh x . \cosh y + \sinh x . \sinh y$	$= \cos x . \cos y - \sin x . \sin y$
$\cosh . 2x = 2(\cosh x)^2 - 1$	$\cos 2x = 2 \cos^2 x - 1$
$\sinh . 2x = 2 \sinh x . \cosh x$	$\sin 2x = 2 \sin x . \cos x$

Both *sinh . x* and *cosh . x* turn out to have useful properties in the domain of the infinitesimal calculus, but $y = \cosh x$ (if we tack on a suitable numerical constant, so that $y = K \cosh x$) describes a curve which embodies a physical principle. If we substitute $-x$ for x in $e^x + e^{-x}$, it becomes $e^{-x} + e^x$, which is identical. Hence, the curve of *cosh x* is symmetrical about the Y-axis becoming increasingly large as x increases on either side of it. Since $e^x + e^{-x} = 2$ when $x = 0$, so that $\cosh . 0 = 1$, the curve cuts the Y-axis at $y = 1$, $x = 0$, where the function has its least value. The name for the curve is the *catenary* (from the Latin for *chain*). It describes how a chain or cord attached by its extremities at the same horizontal level sags under its own weight in the middle. If K in $y = K . \cosh x$ is larger, the sag will be larger.

Complex Numbers as Vector Labels. In physical science, it is convenient to distinguish between two classes of measurements. Some such as temperature, mass, and speed involve only one system of units. We speak of them as *scalars*. Others such as force, acceleration, and velocity (i.e. speed referable to displacement along a fixed axis) involve a *magnitude* and a *direction*, the latter expressible in terms of angular relation to a base line. We speak of such composite measurements as *vectors*.

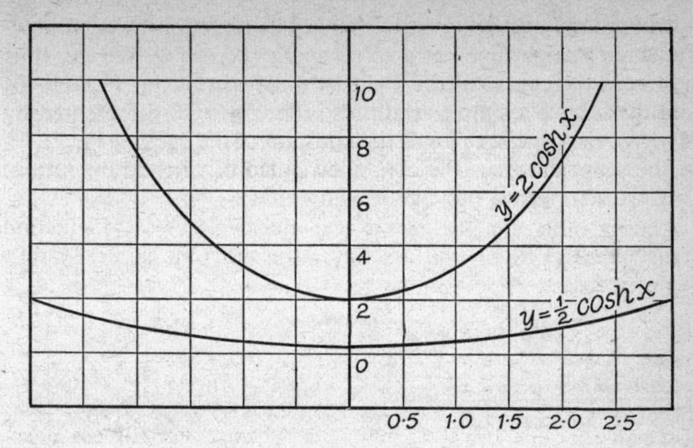

Fig. 158. The Catenary

This figure exhibits the equation of a chain suspended at its ends at the same vertical level.

A visual representation of the *magnitude* of a 2-dimensional vector, as in the left-hand triangle of Fig. 159, is the hypotenuse (OP) of a triangle, whose base is $x_1 = a$ and height $y_1 = d$ if 0 is the origin of a Cartesian grid. Here the angle A which specifies the *direction* of the vector is expressible alternatively (see Fig. 129) as:

$$A = \tan^{-1}\left(\frac{y_1}{x_1}\right) \quad or \quad A = \tan^{-1}\left(\frac{d}{a}\right)$$

The magnitude of the same vector is alternatively expressible as:

$$M = \sqrt{x_1^2 + y_1^2} \quad or \quad M = \sqrt{a^2 + d^2}$$

The last two sets of formulae describe the relation between the length of a journey as the crow flies and the length of its easterly and northerly bearings. We may label such a vector alternatively by writing:

$$V \equiv (x, y) \quad or \quad V \equiv a + di$$

There is no reason to prefer one or the other labels when we are talking about a single vector. When we are speaking of the composition of vectors, as when a person goes from O to Q in Fig. 159 and changes his bearings, the representation of a vector by a complex number is at least an economy of space.

In Fig. 159 our flying crow changes its direction twice, following three straight-line paths in reaching Q from O. We can thus represent the course of the flight by three vectors V_1, V_2, and V_3, and these have a simple relation to the vector V_r represented by OQ. We may speak of V_r as the addition of the vectors V_1, V_2, V_3 in the sense that it is the combined outcome of their operation. In Cartesian terms, our figures show that:

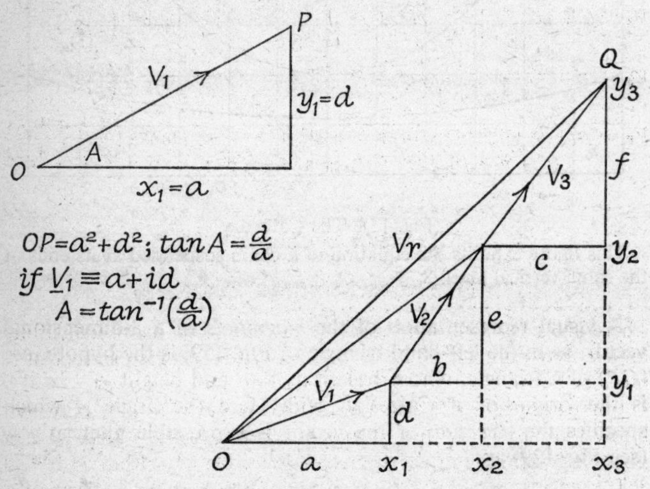

Fig. 159. Vectors as Complex Numbers

Vector	Magnitude	Direction
V_1	$\sqrt{x_1^2 + y_1^2}$	$\tan^{-1}\left(\frac{y_1}{x_1}\right)$
V_2	$\sqrt{(x_2 - x_1)^2 + (y_2 - y_1)^2}$	$\tan^{-1}\left(\frac{y_2 - y_1}{x_2 - x_1}\right)$
V_3	$\sqrt{(x_3 - x_2)^2 + (y_3 - y_2)^2}$	$\tan^{-1}\left(\frac{y_3 - y_2}{x_3 - x_2}\right)$
V_r	$\sqrt{x_3^2 + y_3^2}$	$\tan^{-1}\left(\frac{y_3}{x_3}\right)$

This way of representing the composition of V_r makes explicit the separate contributions of its component vectors (V_1, V_2, V_3) neither to its magnitude nor to its direction. On the other hand, the use of complex numbers as vector labels in what follows, does so in a way which dispenses with the need for any pictorial representation:

	Vector	Magnitude	Direction
V_1	$a + di$	$\sqrt{a^2 + d^2}$	$\tan^{-1}\left(\frac{d}{a}\right)$
V_2	$b + ei$	$\sqrt{b^2 + e^2}$	$\tan^{-1}\left(\frac{e}{b}\right)$
V_3	$c + fi$	$\sqrt{c^2 + f^2}$	$\tan^{-1}\left(\frac{f}{c}\right)$
$V_r =$	$(a + b + c) + (d + e + f)i$	$\sqrt{(a + b + c)^2 + (d + e + f)^2}$	$\tan^{-1}\left(\frac{d + e + f}{a + b + c}\right)$

Exercises on Chapter 9

How to use Logarithm Tables

There are a few details of method to note when using logarithm tables. Any number whatever can be written as the product of a number between 1 and 10 and 10 raised to some power.

For example, 9,876 can be written as $9{\cdot}876 \times 10^3$. Now we know that the logarithm of any number between 1 and 10 is a positive fraction, and we can write down the logarithm of 10^3 as 3. It is easy to see that the logarithm of any number consists of a whole number which can be written down by inspection and a fractional part which is the same for all numbers having the same digits in the same order.

Thus

$$\log 9{\cdot}876 = 0{\cdot}9946$$
$$\log 98{\cdot}76 = \log 10 + \log 9{\cdot}876 = 1{\cdot}9946$$
$$\log 987{\cdot}6 = \log 10^2 + \log 9{\cdot}876 = 2{\cdot}9946$$

etc.

and

$$\log 0{\cdot}9876 = \log 10^{-1} \times \log 9{\cdot}876 = -1 + 0{\cdot}9946$$
$$\log 0{\cdot}09876 = \log 10^{-2} \times \log 9{\cdot}876 = -2 + 0{\cdot}9946$$

The last two logarithms are written as follows:

$$\log 0{\cdot}9876 = \bar{1}{\cdot}9946$$
$$\log 0{\cdot}09876 = \bar{2}{\cdot}9946$$

This is simply a device for making calculations easier. For example:

$$\begin{aligned}\log \frac{182{\cdot}3}{0{\cdot}021} &= \log 182{\cdot}3 - \log 0{\cdot}021 \\ &= 2{\cdot}2608 - (\bar{2}{\cdot}3222) \\ &= 2 + 0{\cdot}2608 - (-2 + 0{\cdot}3222) \\ &= 2 + 0{\cdot}2608 + 2 - 0{\cdot}3222 \\ &= 4 - 0{\cdot}0614 \\ &= 3{\cdot}9386 \\ \therefore \frac{182{\cdot}3}{0{\cdot}021} &= 8682\end{aligned}$$

The positive fractional part of the logarithm is called the *mantissa*, and the integral part, which may be positive or negative, is called the *characteristic*.

To find the logarithm of 9·876 in the table the procedure is as follows. At the left-hand side of the table is a column of double figures beginning with 10. We look down this till we find 98. Looking at the top of the page we see columns headed by the numbers 0 to 9, so that we look along the line beginning 98 till we come to the column headed 7. The number given in this place is 9,943, and this is the mantissa of the logarithm of 9·87. To take into account the last figure, 6, we look at the right-hand of the table, where there is a set of columns with the numbers 1 to 9 at the top. The numbers in these columns indicate what has to be added to the logarithm to take account of the fourth figure in the given number. In the present case we have found the right mantissa for 9·870, and we want to know how much to add on for 9·876. Looking along the line from 98 we see at the extreme right hand under the column headed 6 the number 3. So that the required mantissa is 0·9943 + 0·0003, i.e. 0·9946.

When the logarithm is known, the number of which it is the logarithm can be found by reversing the above process. For example, we can find the number whose logarithm is 2·6276. Considering first the mantissa 0·6276, we can find from either a table of logarithms or a table of antilogarithms that the digits corresponding with it are 4,242. The characteristic 2 tells us that the number lies between 100 and 1,000. Therefore the number required is 424·2.

1. Multiply the following numbers:

(*a*) by using the formula:

$$\sin A \cos B = \tfrac{1}{2}\sin(A + B) + \tfrac{1}{2}\sin(A - B)$$

(*b*) by using the formula:

$$\cos A \cos B = \tfrac{1}{2}\cos(A + B) + \tfrac{1}{2}\cos(A - B)$$

(*c*) by using logarithm tables:

(i) $2{\cdot}738 \times 1504$ (iii) $5{\cdot}412 \times 368$

(ii) $8{\cdot}726 \times 3471$ (iv) $2{\cdot}1505 \times 46{\cdot}12$

Check your results by ordinary multiplication.

2. Calculate the following by logarithms:

$(78{\cdot}91)^2$ $(1{\cdot}003)^3$

$\sqrt{68990{\cdot}3} \div 0{\cdot}0271$ $\sqrt[3]{0{\cdot}0731}$

$9{\cdot}437 \div 484$

$$\frac{\sqrt{273} \times (1{\cdot}1)^3}{0{\cdot}48}$$

3. Find the tenth root of 1,024, the eighth root of 6,561, and the eighth root of $25\frac{161}{256}$ by logarithms, and check your results by multiplication.

4. Plot the curve $y = \log_{10}x$

5. Do not forget that logarithms are a device for doing multiplication and division rapidly. There is no simple way of finding the value of $\log(a + b)$, so that if, for example, you want to calculate $\sqrt{1{\cdot}01} - \sqrt[3]{1{\cdot}01}$, each term must be calculated separately.

Find the value of:

(i) $(29{\cdot}91)^3 + (48{\cdot}24)^3$

(ii) $\sqrt[4]{1001} - \sqrt[3]{101}$

(iii) $$\frac{(0{\cdot}4573)^2}{(0{\cdot}5436)^2 - (0{\cdot}3276)^2}$$

6. Calculate the following, using log tables:

(i) Find the compound interest on £1,000 in 6 years at 4 per cent per annum.

(ii) How long will it take a sum of money to double itself at 10 per cent per annum.

(iii) Find the compound interest on £400 for $5\frac{1}{2}$ years at $3\frac{1}{2}$ per cent per annum, payable half-yearly.

7. Taking e as 2·718, calculate the following:

$$\log_e 1{\cdot}001 \qquad \log_e \sqrt{2}, \qquad \log_e 3789$$

8. In an experiment the following values were obtained for two variables x and y:

x	1·70	2·24	2·89	4·08	5·63	6·80
y	320	411	491	671	903	1,050
x	8·42	12·4	16·3	19·0	24·3	
y	1,270	1,780	2,250	2,520	3,180	

Plot two graphs, one showing x and y, and the other log x and log y. From the second show that the relation between log x and log y is approximately described by the equation:

$$\log y = 0{\cdot}876 \log x + 2{\cdot}299$$

From this write down an equation connecting x and y.

9. Write down the binomial expansion of $(1 + 0{\cdot}05)^{-4}$. Find its value by taking the first five terms. Show that the error involved is less than 0·0000163.

10. Using the 'infinite' series for sin a and cos a, find the values of sin 1° and cos 1°. How many terms do you need to get the values of sin 1° and cos 1° given in four-figure tables?

11. Using the 'infinite' series for sin na and cos na and the values just obtained for sin 1° and cos 1°, make a table of the sines and cosines of 1°, 2°, 3°, 4°, 5°, and compare with the values given in the tables.

Things to Memorize

$$\sin A \cos B = \tfrac{1}{2} \sin (A + B) + \tfrac{1}{2} \sin (A - B)$$
$$\cos A \cos B = \tfrac{1}{2} \cos (A + B) + \tfrac{1}{2} \cos (A - B)$$
$$\log_{10} 100 = 2$$
$$\log_{10} 10 = 1$$
$$\log_{10} 1 = 0$$
$$\log_{10} 0{\cdot}1 = -1$$
$$\log_{10} 0{\cdot}01 = -2$$

$$e = 1 + 1 + \frac{1}{2!} + \frac{1}{3!} + \frac{1}{4!} + \frac{1}{5!} + \ldots$$

$$(\cos a + i \sin a)^n = \cos na + i \sin na$$

$$\sin 3A = 3 \sin A - 4 \sin^3 A$$

$$\cos 3A = 4 \cos^3 A - 3 \cos A$$

$$\sin a \text{ (radians)} = a - \frac{a^3}{3!} + \frac{a^5}{5!} - \frac{a^7}{7!} \ldots$$

$$\cos a \text{ (radians)} = 1 - \frac{a^2}{2!} + \frac{a^4}{4!} + \frac{a^6}{6!} \ldots$$

CHAPTER 10

THE CALCULUS OF NEWTON AND LEIBNITZ

From the Geometry of Motion to the Algebra of Growth

THE WORD *calculus* used in medicine for a stone in the bladder (cf. French *cailloux* for precious stones) is the Latin equivalent of pebble, and its derivative *calculate* recalls the most primitive type of abacus – a tray lined with sand, having grooves for pebbles corresponding to the beads on its more dignified descendant, the counting frame. As used by mathematicians, calculus can properly stand for any group of allied techniques which are aids to calculation. For instance, the vanishing triangle technique of Chapter 6 is part of the *Calculus of Finite Differences*. When used without any qualification, one customarily refers to the *infinitesimal* calculus with its two major branches respectively referred to as the *differential* and the *integral* Calculus.

The reference to the abacus in the last paragraph is not inappropriate. Just as the acquisition of the Hindu–Arabic numerals made it possible to prescribe recipes for calculation – the algorithms – to dispense with the need for the mechanical aid of the abacus, the infinitesimal calculus makes it possible to dispense with pictorial representation when dealing with what were formerly geometrical problems. As such, however, it was an offshoot of the new geometry of motion initiated by Fermat (1636) and Descartes; and Fermat's treatment of a problem which might not seem at first sight to hold out much promise of a vast new territory of mathematical inventiveness or, for that matter, much prospect of useful application, is the most convenient springboard for our theme in this chapter.

Fermat seems to have been the first person to tackle the problem of locating the tangent of a curve, and to tackle it in terms which presuppose both the fixed framework of reference imposed by the new geometry of motion and the novel, if only nascent, concept of an algebraic function relating contour to measurement within such a framework. Two generations after the publication of the first announcement of the new geometry of motion, his successors were exploiting and extending the algebraic symbolism which Descartes enlisted in a more fruitful partnership of algebra and geometry. Consequently, it is possible to trace to earlier sources, what emerges as a system with a single unifying principle in the writings of Newton and Leibnizt. The respective contributions of these two became the *casus belli* of a controversy with disreputable overtones of national prestige. It does little credit to contestants of either camp, and the outcome was highly detrimental to the progress of mathematics in the land of Newton's birth. The issue still has partisans, and its reverberations influenced mathematical thinking long after both Newton and Leibnitz were dead. On that account, it is worthy of brief comment. Unlike the story of Frankie and Johnnie, it has a moral, at least for folks over-proud of their compatriots.

Though a treatise on the topic credibly written by Newton in 1671 did not appear in print (1736) till after his death, it is clear that Newton's views about what we now call *differentiation* and *integration* had taken shape in what he himself called the *method of fluxions* before Leibnitz visited England in 1673. Leibnitz did not claim to have developed his own views before 1674, and did not complete before 1676 an unpublished manuscript with the title *Calculus Tangentium Differentialis*. His major contribution appeared in print ten years later and after he had been in correspondence with Newton about the work of the latter. That Leibnitz picked up many of Newton's more inspired than intelligible tricks second-hand from Newton's friends both by personal contact when in England and in subsequent correspondence is therefore possible, though not proven. We should, however, weigh against the possibility that the debt of Leibnitz to Newton was considerable, a consideration highly relevant to the subsequent progress – or lack of progress – of mathematics in Britain. As we shall later see, Newton used a now defunct notation which is obscure and not adaptable to so wide a range of applications as that of Leibnitz. If his contemporaries were *au fait* with his intentions, it is strange that he relied wholly on Euclidean geometry in the final exposition of his mechanical principles, as set forth in the *Principia* (1687).

It is equally strange that those of his contemporaries most articulate in the advancement of Newton's claims to priority, did little, if anything, to make use of, or to extend, his methods. During the century after the publication of the *Principia*, only two British mathematicians made any outstanding contribution to the common cause. Colin Maclaurin, a Scotsman in the succession of Newton's contemporary Gregory, whose treatment of *finite differences* is a mentionable milestone in the genesis of the Newton–Leibnitz Calculus, did not circulate around the Anglo-London axis in the Oxford–Cambridge orbit. Brook Taylor, born in what is now Greater London, developed Gregory's ideas before he lapsed like Newton, but twenty years before the death of the latter, into a senility of theological disputation and obscurantism. James Stirling, who publicized the important contribution usually attributed to Maclaurin three years after Newton's death was, like Maclaurin, a Scot, and had improved his mind by sojourn abroad in French circles where the notation of Leibnitz was beginning to reap a rich harvest of endeavour.

For a century after the publication of Newton's *Principia*, the infinitesimal calculus was the focus of a flowering of mathematical inventiveness associated especially with the names of the Bernoullis and Euler in Switzerland and later in France with those of Lagrange, Laplace, and Legendre who died respectively in 1813, 1827, and 1833. During the whole of this period no British-born names, other than Taylor, Maclaurin and Sterling, are mentionable. Whether this is because of continental failure

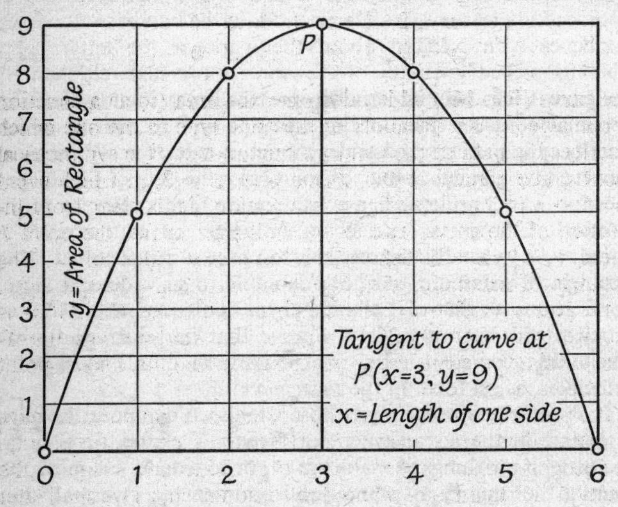

Fig. 160. Graphical Representation of the Maximum Value of the Function $y = 6x - x^2$

to give full credit to Newton, or whether it is because Newton failed to make clear to his British-born contemporaries what he was getting at, are alternatives on which it is now too late to pass judgement with profit to posterity. Let us therefore get back to where Fermat, as much as anyone, should have credit to blazing a trail the end of which he could not conceivably have foreseen.

From Fermat to Newton. Fermat approached an issue with no prevision of its mechanical application. Hence, we may rightly recapture his viewpoint by considering a geometrical problem with no special relevance to motion: *what relation is there between adjacent sides of a rectangle of fixed boundary when its area is greatest?* To present the issue in a form suitable for graphical treatment, let us call the length of adjacent sides a and x, the boundary $B = 2(a + x)$, so that $a = \frac{1}{2}(B - 2x)$ and if y is the area:

$$y = ax = \tfrac{1}{2}x(B - 2x) = \tfrac{1}{2}\,Bx - x^2$$

Let us then suppose that the perimeter is 12 units of length, so that $a = 6 - x$ and $y = 6x - x^2$. Accordingly, we can tabulate x and y as below:

$$\begin{array}{lccccccc} x = & 0 & 1 & 2 & 3 & 4 & 5 & 6 \\ y = & 0 & 5 & 8 & 9 & 8 & 5 & 0 \end{array}$$

The curve (Fig. 160) which describes the area (y) as a function of one side (x) is a parabola of the same type as the one which describes the path of the Galilean cannon-ball. It is symmetrical about a line parallel to the Y-axis when $x = 3$, in which event $a = 6 - x = 3$ and the figure is a square. As is clear from inspection of the curve, and as we shall later prove, the point P where $y = 9$, $x = 3$ defines the *maximum* value of y. The rectangle of maximum area has therefore equal adjacent sides. If we plot $y = x^2 - 6x$, the shape of the curve (Fig. 163) is identical with that of Fig. 160, but it lies wholly below the X-axis within the same range of x-values and its *lowest* point (*minimum* y) is at $x = 3$, $y = -9$.

The problem which Fermat raised, and one for which he gave an essentially correct answer, is this: in such a situation how do we recognize which point P locates the *maximum* value of the function? At this stage, we need not distinguish, as we shall later do, between whether P locates a maximum or a minimum. Either way, we notice that the *tangent at P* (i.e. straight line grazing the curve at *P*) is *parallel to the X-axis*. To say this is to say that:

(*a*) the *tangent* to the curve at a *turning point* (maximum or minimum) inclines at an angle $A = 0$ to the *X*-axis;

(*b*) the curve slopes upwards towards the right before and upwards towards the left after it (maximum) or *vice versa* (minimum).

More precisely, the tangent to the curve on either side of such a point *P* makes with the *X*-axis an angle *numerically* greater than zero on either side of it. This prompts us to ask: what meaning can we give to the *slope of the tangent* to any point on a curve?

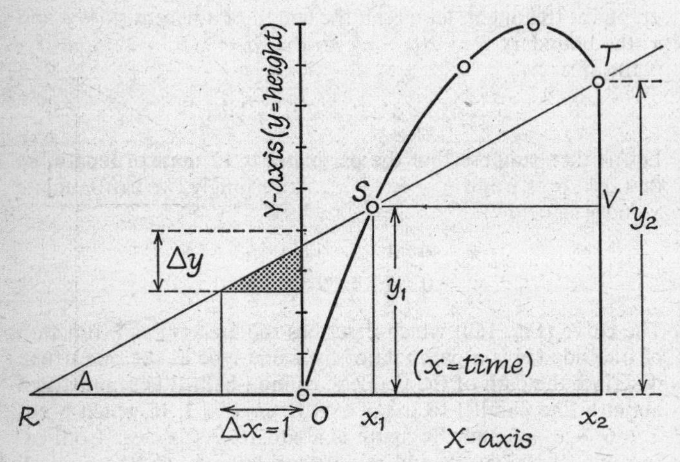

Fig. 161. Graphical Representation of Mean Speed

Since all Newton's thinking was in terms of motion or *fluxion*, as he would say, let us now examine what we mean by a *slope* on a graph in terms of a moving body. Fig. 161 shows part of a curve of much the same shape as that of Fig. 160, i.e. a parabola which describes (with suitable choice of units) the relation between height (y) and time (x) of the cannon-ball trajectory. The straight line *RST* cuts the curve at *S* (co-ordinates x_1, y_1) and *T* (x_2, y_2). In the triangle *STV*, the angle $TSV = A$, so that:

$$\tan TSV = \frac{y_2 - y_1}{x_2 - x_1} = \tan A$$

If the moving body proceeded from S to T at constant vertical speed, the straight line ST would represent its course. Its vertical speed is the ratio of height traversed $(y_2 - y_1)$ to time taken $(x_2 - x_1)$. Thus we may interpret *tan A* as the mean vertical speed in the passage from S to T. In the dotted triangle which is similar to the triangle STV, $\tan A = \Delta y \div \Delta x$ and $\Delta x = 1$. This is another way of saying that *tan A* is the increase of height *per unit interval* of time between time x_1 and time x_2. In more general terms, the tangent of the angle (A) made by a line cutting the curve $y = f(x)$ at two points S (x_1, y_1) and T (x_2, y_2) is the average *rate of change of y per unit change of x* over the range x_1 to x_2 on the x-Axis.

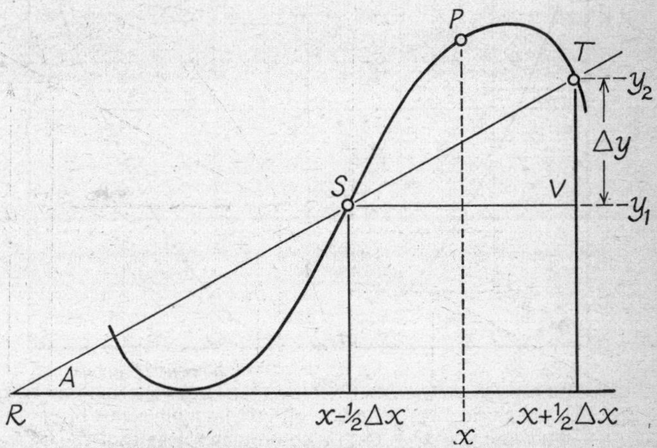

Fig. 162. Measuring the Slope of a Curve

Let us now consider a point (P) located between S and T on a curve, as in Fig. 162. We may label the co-ordinates of S, P, and T so that the x-co-ordinate of P is midway between those of S and T, if we write:

co-ordinates of		S	$x - \frac{1}{2}\Delta x;\ y_1$
,,	,,	P	$x;\ y$
,,	,,	T	$x + \frac{1}{2}\Delta x;\ y_2$

We then see that:

$$\tan A = \frac{TV}{SV} = \frac{y_2 - y_1}{(x + \frac{1}{2}\Delta x) - (x - \frac{1}{2}\Delta x)} = \frac{\Delta y}{\Delta x}$$

If we make Δx sufficiently small, so that *S* and *T* coincide with *P*, the line *RST* in Fig. 162 will coincide with the tangent to the curve at *P*, the angle *A* will be the inclination of the tangent at *P* to the *X*-axis, and tan *A* will represent the rate of change of *y* per unit change of *x* at the point *P*.

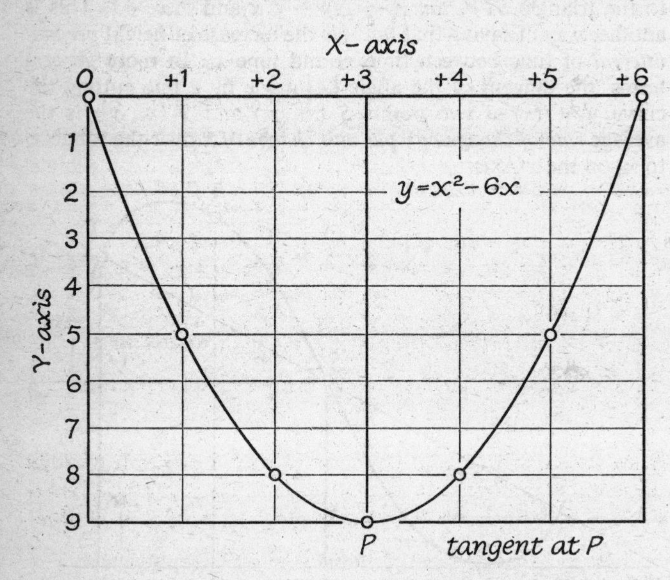

Fig. 163. Graphical Representation of the Minimum Value of the Function $y = x^2 - 6x$; (Y co-ordinates negative)

On this understanding let us return to the function $y = 6x - x^2$ corresponding to the curve of Fig. 160. Here:

$$y_2 = 6(x + \tfrac{1}{2}\Delta x) - (x + \tfrac{1}{2}\Delta x)^2$$
$$= 6x + 3\Delta x - x^2 - x \,.\, \Delta x - \tfrac{1}{4}(\Delta x)^2$$

$$y_1 = 6(x - \tfrac{1}{2}\Delta x) - (x - \tfrac{1}{2}\Delta x)^2$$
$$= 6x - 3\Delta x - x^2 + x \,.\, \Delta x - \tfrac{1}{4}(\Delta x)^2$$

$$\therefore \quad y_2 - y_1 = \Delta y = 6\Delta x - 2x \,.\, \Delta x$$

$$\therefore \quad \frac{\Delta y}{\Delta x} = 6 - 2x = \tan A$$

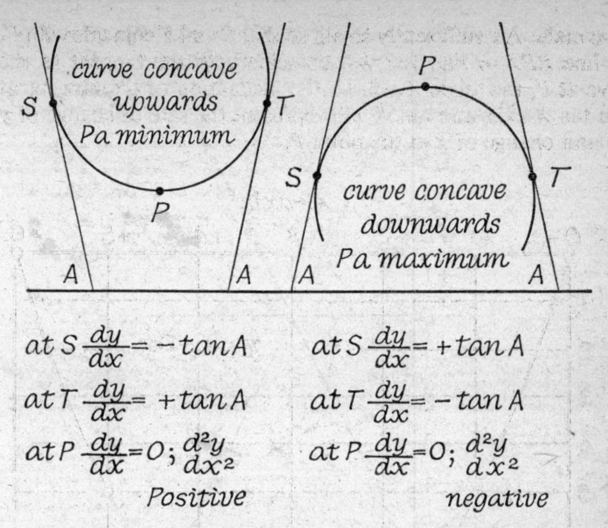

Fig. 164. Changing Slope in the neighbourhood of Turning Points

When the point P lies on the curve in such a position that the tangent at P is parallel to the X-axis as in Fig. 160, the angle A it makes with the latter is zero, and $\tan A = \tan 0 = 0$, so that:

$$6 - 2x = 0 \quad and \quad x = 3$$

Thus, there is a turning-point of the curve shown in Fig. 160, when $x = 3$ and $y = 18 - 9 = 9$. The reader who repeats the argument for the function $y = x^2 - 6x$ (Fig. 163) will find that $\tan A = 0$ when $x = 3$ and $y = -9$, so that the turning-point corresponds to the *lowest* point on the curve. To answer Fermat's problem fully, we must therefore ask how we distinguish between a turning-point at which the value of y is a *maximum*, and one at which y is a *minimum*. We first remind ourselves (Fig. 164) that:

(i) in the neighbourhood of a maximum the slope of the tangent is first upwards to the right (*tan A* positive), then upwards to the left (*tan A* negative);

(ii) in the neighbourhood of a minimum, the slope of the tangent changes from negative to positive.

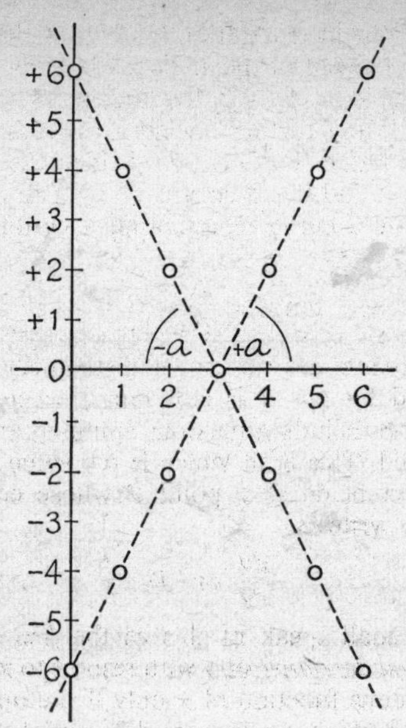

Fig. 165. Graphical Representation of the First Derivatives of Functions $y = 6x - x^2$ and $y = x^2 - 6x$

To keep our feet firmly on the ground, let us again look at the problem, as Newton did, in terms of time and distance. If A is the slope of the tangent at P when y is a measure of distance, x is a measure of time, *tan A* is the *speed at P*. If we plot tan A against x, we are then showing how the speed increases or diminishes with time, so that $\triangle \tan A \div \triangle x$ is a measure of *acceleration at any point P* whose co-ordinates are x and $y = \tan A$. Let us compare the values of *tan A* for the two curves shown in Figs. 160 and 163 from this point of view:

$x =$	0	+1	+2	+3	+4	+5	+6
$\tan A = 2x - 6$	−6	−4	−2	0	+2	+4	+6
$\tan A = 6 - 2x$	+6	+4	+2	0	−2	−4	−6

As we see in Fig. 165, the Cartesian representation of each of these functions is a straight line inclined to the X-axis at *numerically* the same angle (a) one (*tan* $A = 6 - 2x$) to the right downwards, and two (*tan* $A = 2x - 6$) to the right upwards. In accordance with our interpretation (p. 374) of the equation of a straight line, the tangent of the angle a when $tan\ A = 6 - 2x$ is -2, and when $tan\ A = 2x - 6$, the tangent of a is $+2$.

Before drawing any further conclusions from the foregoing, let us refine our terms. Initially, we defined the tangent of the angle A between it and the X-axis made by a straight line which cuts at two points a curve representing a continuous function $y = f(x)$ as:

$$\tan A = \frac{\Delta y}{\Delta x}$$

When tne two points are no longer distinguishably apart, the ratio represented by $tan\ A$ is still finite, except in particular regions of a curve including maxima, minima, and a third case not yet mentioned. This limit which is the value of $tan\ A$ when the two points coincide at a point P whose co-ordinates are x and y, we now write as:

$$\frac{dy}{dx} \text{ or } D_x \, . \, y$$

Henceforth, we shall speak of this as the *first differential co-efficient* or the *first derivative* of y with respect to x, and we speak of y as a continuous function of x only if the operation (called *differentiation*) by which we derive a differential coefficient leads to a unique function of x (e.g. $6 - 2x$ when $y = 6x - x^2$). If therefore $f(x)$ and $F(x)$ stand for different functions of x, we may write:

$$y = f(x) \text{ and } \frac{dy}{dx} = F(x)$$

Now we can repeat the same operation on $F(x)$, and we then speak of the outcome as the *second* differential coefficient or derivative of $y = f(x)$, denoted thus:

$$\frac{d\,F(x)}{dx} \equiv \frac{d^2y}{dx^2} \equiv D_x^{\,2} \, . \, y$$

Needless to say – though not relevant in this context – we can perform the operation again and again, with the same convention:

$$\frac{d^2F(x)}{dx^2} \equiv \frac{d^3y}{dx^3} \equiv D_x^{\,3} \, . \, y \text{ and } \frac{d^3F(x)}{dx^3} \equiv \frac{d^4y}{dx^4} \equiv D_x^{\,4} \, . \, y$$

If we can differentiate $y = f(x)$ twice, we have at our disposal the means of completely solving the problem Fermat was first

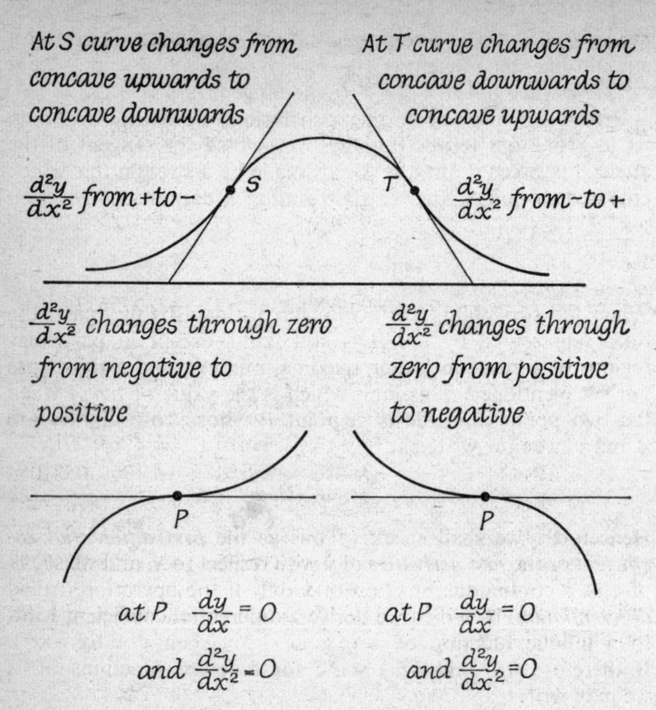

Fig. 166. Points of Inflexion

to tackle. To do so, it will be useful to distinguish between a *turning-point* such as P in Fig. 164 and a *point of inflexion* such as S or T in the upper half and P in the lower half of Fig. 166. A point of inflexion is a point where the curvature changes from concave upwards to concave downwards or vice versa. In a region where a curve is concave downwards (Fig. 164), *tan A* is changing from positive to negative through zero, and the second derivative is therefore negative. In a region where a curve is concave upwards, the reverse is true, and the second derivative is positive. In the region of a point of inflexion, the second derivative thus changes through zero from positive to negative or *vice versa*. We may thus say:

$$P \text{ is a maximum if } \frac{dy}{dx} = 0 \text{ and } \frac{d^2y}{dx^2} \text{ is negative}$$

$$P \text{ is a minimum if } \frac{dy}{dx} = 0 \textit{ and } \frac{d^2y}{dx^2} \text{ is positive}$$

We recognize a third situation in the lower half of Fig. 166. Here P is a point of inflexion so situated that the tangent to the curve is parallel to the X-axis (*tan* $A = 0$) in which event:

$$\frac{dy}{dx} = 0 = \frac{d^2y}{dx^2}$$

At a point of inflexion such as S and T in the upper half of Fig. 166, the second derivative is zero and the first is not.

There is no more to say about Fermat's problem when our concern is wholly with finite values of x; but a fourth situation arises when a curve is *asymptotic* (p. 355) to the X- on one side or both sides of the X-axis. Thus, the first derivative of the hyperbola (p. 354) $y = 16x^{-1}$ is $-16x^{-2}$; and this differs less and less from zero as x becomes immeasurably large. The right-hand limb of the curve then becomes more and more *asymptotic*, i.e. nearly parallel, to the X-axis as y approaches more and more closely to zero.

Newton's Approach. To understand Newton's approach in adapting the device of a *differential triangle* (*STV* in Fig. 161), as propounded by his teacher Isaac Barrow, it is necessary to appreciate a distinction between crude *speed* and *velocity* as physicists now use the terms. Newton's predecessors thought of force in *statical* terms, e.g. the balancing of weights. Newton defined force in *dynamical* terms suggested by Galileo's work on the motion of a ball down a smooth inclined plane, i.e. in terms of *acceleration imparted.*

In everyday life, we think of acceleration in terms of faster motion, i.e. an increase (or decrease) of the speed, i.e. distance–time ratio; but if we are to take stock of expenditure of effort necessary to change the *direction* of motion without change of crude speed, i.e. distance–time ratio in the path traversed, we have to define acceleration in terms which also take stock of a difference between distance traversed in a particular interval of time when the track is and is not a straight line. Accordingly, we distinguish between crude speed defined as distance traversed in unit time in the actual path of a moving object and *velocity* defined as distance traversed in unit time with reference to a straight line.

If a body is moving in a straight line, its crude speed and velocity with reference to that line are identical; and if we then

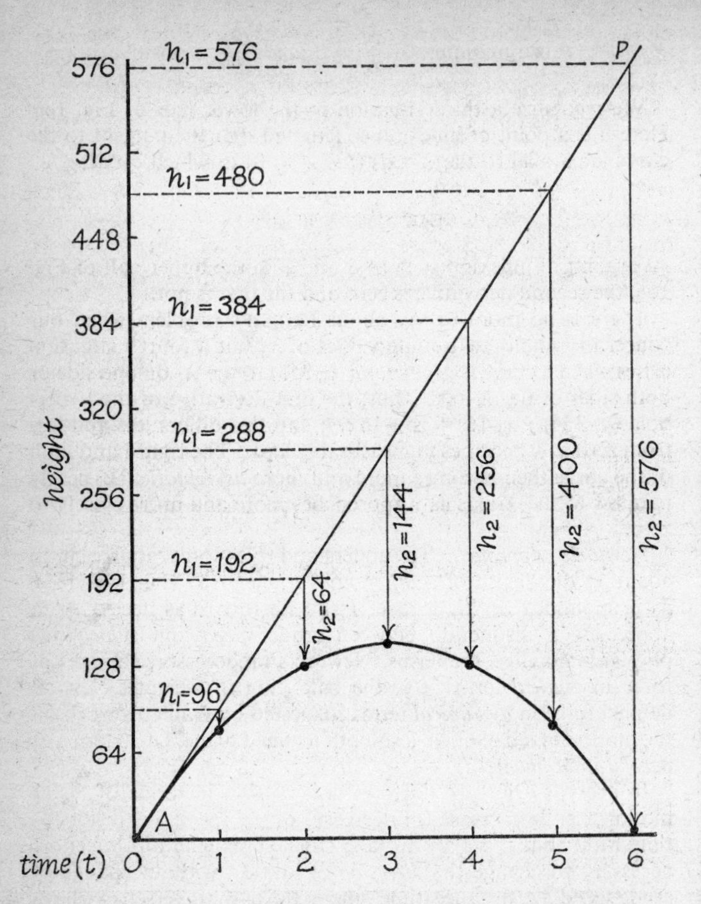

Fig. 167. Gallilean Interpretation of Cannon-ball Trajectory

For convenience of draughtsmanship in this figure, the inclination (*A*) of the line of fire (*OP*) to the ground is approximately $56\frac{1}{4}°$ and the muzzle itself is at ground level. In the absence of any force other than that of the explosion, the missile would move in the straight line *OP* at a constant speed approximately $115\frac{1}{2}$ feet per second, corresponding to a vertical velocity of 96 feet per second *upwards*. Owing to gravitation it falls gaining speed *downwards* at the rate of 32 feet per second

dissect its motion into two components at right angles (e.g. vertical and horizontal, northerly and easterly), the distance traversed in unit time will be constant in either direction, if constant along the actual path (Fig. 167). When a body is moving with fixed speed along a curved path, the distance traversed in unit time in any fixed direction (e.g. northerly or southerly, vertical or horizontal) is not constant. If we define *acceleration* as increase (*positive*) or decrease (*negative*) of *velocity*, we can therefore speak of the acceleration of a body moving in a curved path only in a particular direction, e.g. vertically in the height–time graph of the cannon-ball trajectory.

In a graphical representation of distance (l) and time (t), we can correctly represent either speed or velocity by the first derivative of l with respect to t. Hence, if s is speed along the actual path, v velocity in a particular direction, it is necessary only to specify whether we measure l in one way or the other to write intelligibly:

$$\frac{dl}{dt} = s \text{ (} l \text{ measured along the path traversed)}$$

$$\frac{dl}{dt} = v \text{ (} l \text{ measured with reference to a fixed direction)}$$

When we represent the acceleration (a) of a body by the second derivative, we have to remember that:

$$a = \frac{dv}{dt}, \quad \textit{whence} \quad \frac{d^2l}{dt^2} = a \quad \textit{only if} \quad \frac{dl}{dt} = v$$

The Galilean Cannon-ball Trajectory. We are now ready to look at a problem which Tartaglia (Fig. 126), of cubic equation fame, failed to solve for lack of experimental data established later by his compatriot Galileo. Newton and his contemporaries owed to Galileo the recognition that we can regard the motion of

---

[*Caption continued*]

in each second. We may tabulate the relevant data with respect to distance fallen (h_2) in virtue of gravitation as below:

time (t)	0	. . .	1	. . .	2	. . .	3	. . .	4	. . .	5	. . .	6
velocity (at t)	0	. . .	32	. . .	64	. . .	96	. . .	128	. . .	160	. . .	192
mean velocity in the interval		16		48		80		112		114		176	
h_2	0	. . .	16	. . .	64	. . .	144	. . .	256	. . .	400	. . .	576

a missile as the composite outcome of two independent components: (*a*) its initial and fixed muzzle velocity in a straight line upwards at an angle A to the horizontal; (*b*) its downward mean velocity at a fixed gain per second. This fixed gain is its so-called *acceleration under gravity* denoted by g and approximately equal (at sea level) to 32 feet per second in a second.

To fit these together we have to convert the muzzle velocity (v_m) into its vertical equivalent (V) in the same straight line as the downward pull of gravity. Fig. 167 shows what would be the height after an interval of time t in the absence of any force

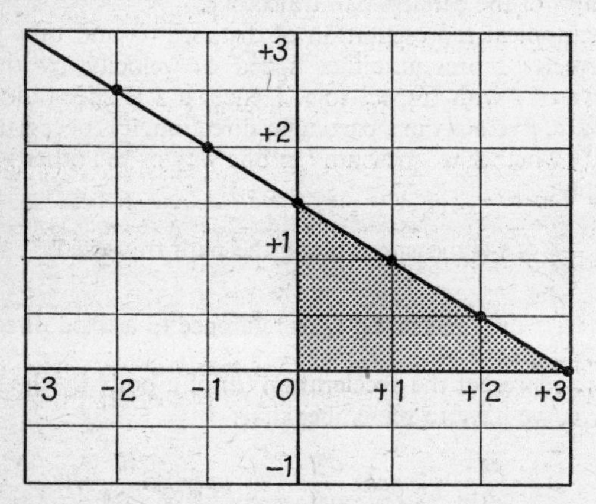

Fig. 168. Downward Acceleration of the Cannon-ball

Along the x axis the units represent time in seconds. Along the y axis the units represent velocity, i.e. distance per second, measured vertically upwards. One y unit corresponds with 64 feet per second. The line slopes from right to left upwards and the sign of the gradient is therefore negative. That is to say, the ball loses speed upwards, i.e. gains speed towards the earth. The gradient as seen from the shaded area is

$$\frac{(0 - 1{\cdot}5) \text{ units of } y}{(3 - 0) \text{ units of } x} \quad \text{or} \; -\frac{1{\cdot}5 \times 64 \text{ feet per second}}{3 \text{ seconds}}$$

i.e. – 32 feet per second per second

the acceleration of the cannon-ball downwards. For the data of Fig. 167:

$$v = \frac{3}{2} - \frac{x}{2}$$

changing the direction of motion or speed along its path in a straight line, i.e.:

$$h_1 = Vt = v_m \,.\, \sin A \,.\, t$$

If v_0 is the initial velocity of a body falling under gravity and v_t its velocity after an interval of time t, its mean velocity is $v = \frac{1}{2}(v_0 + v_t)$. If it gains g feet per second in every second:

$$v_t = v_0 + gt \;\; and \;\; \tfrac{1}{2}(v_0 + v_t) = v_0 + \tfrac{1}{2}gt = v$$

Starting from rest ($v_0 = 0$), its mean velocity downward is therefore $\frac{1}{2}gt$ and since its mean velocity is the ratio of the distance h_2 covered to the time interval:

$$v = \frac{h_2}{t}, \;\; so\ that \;\; h_2 = v \,.\, t = \tfrac{1}{2}gt^2$$

While the missile has been *gaining* height $h_1 = v_m \sin A \,.\, t$ upwards (*positive*), it has also been *losing* height $h_2 = \frac{1}{2}gt^2$ downwards (*negative*) in the same interval t. So its actual height (h) in terms of time is (Fig. 167):

$$h = v_m \,.\, \sin A \,.\, t - \tfrac{1}{2}gt^2$$

The equation of the trajectory of a cannon-ball, shown in Fig. 132 (p. 350), cites its height (in units of 64 feet) and time (unit 1 second) as:

$$h = \frac{3}{2}t - \frac{t^2}{4}$$

If the unit of h is 1 foot this becomes:

$$h = 96t - 16t^2$$

If the angle the barrel of the cannon makes with the ground is 60°, since $sin\ 60° = \frac{1}{2}\sqrt{3}$:

$$\frac{2(96)}{\sqrt{3}} = v_m \simeq 110 \text{ feet per second}$$

This is the muzzle velocity in the line of fire (60° to the ground), and the vertical velocity is:

$$\frac{\Delta h}{\Delta t} = \frac{96(t + \frac{1}{2}\Delta t) - 96(t - \frac{1}{2}\Delta t)}{\Delta t} - \frac{16(t + \frac{1}{2}\Delta t)^2 - 16(t - \frac{1}{2}\Delta t)^2}{\Delta t}$$

$$= 96 - 32t = 96 \text{ when } t = 0$$

The acceleration is obtainable thus:

$$\Delta v = [96 - 16\,(t + \tfrac{1}{2}\Delta t)^2] - [96 - 16\,(t - \tfrac{1}{2}\Delta t)^2]$$
$$= -32\Delta \text{t}$$
$$\frac{\Delta y}{\Delta t} = -32$$

We may therefore write:

$$\frac{d^2h}{dt^2} = -32$$

That is to say, the vertical acceleration (Fig. 168) is negative (i.e. downwards) and independent of t. Thus, the cannon-ball is gaining speed in that direction at a fixed rate of 32 ($=g$) feet per second per second. The path has a turning point when:

$$\frac{dh}{dt} = 0 = 96 - 32t, \quad \textit{so that} \quad t = 3$$

When $t = 3$,

$$h = 96t - 16t^2 = 288 - 144 = 144 \text{ (feet)}$$

This is equivalent to $2\frac{1}{4}$ units of 64 feet as mapped in Fig. 132, and since the second derivative is negative (-32) the turning-point is a maximum, as Fig. 132 also shows.

Before we exhaust interest in the cannon-ball trajectory from the viewpoint of Newton's *fluxions*, we may profitably pause to examine the differential coefficient of functions other than the one we have dealt with so far.

Differentiation. The examples hitherto studied have been of the type $y = Ax + Bx^2$ or $y = A + Cx$ in which A, B, C are numerical constants not necessarily positive. Let us now consider the form Ax^n in which n is a positive integer. If $n = 3$, we may write with respect to 2 points S and T whose co-ordinates are $x - \frac{1}{2}\Delta x$, y_1 and $x + \frac{1}{2}\Delta x$, y_2:

$$y_2 = A(x + \tfrac{1}{2}\Delta x)^3 = Ax^3 + \tfrac{3}{2}Ax^2 \,.\, \Delta x$$
$$+ \tfrac{3}{4}Ax \,.\, (\Delta x)^2 + \tfrac{1}{8}A(\Delta x)^3$$
$$y_1 = (Ax - \tfrac{1}{2}\Delta x)^3 = Ax^3 - \tfrac{3}{2}Ax^2 \,.\, \Delta x + \tfrac{3}{4}Ax\,(\Delta x)^2$$
$$- \tfrac{1}{8}A(\Delta x)^3$$
$$\therefore \quad \Delta y = y_2 - y_1 = 3Ax^2 \,.\, \Delta x + \tfrac{1}{4}(\Delta x)^3$$
$$\therefore \quad \frac{\Delta y}{\Delta x} = 3Ax \,.\, {}^2 + \tfrac{1}{4}(\Delta x)^2$$

As we have seen, a straight line cutting these two points inclines to the x-axis at angle A such that $\tan A$ is equivalent to the expression on the right. While the two points become nearer, Δx (and *a fortiori* its square) becomes smaller and smaller in comparison with $3Ax^2$. When the gap is immeasurably small, so that the two points coalesce at a single point P, Δx vanishes and the line inclined to the X-axis at an angle whose tangent is $3Ax$ grazes the curve at P. We may then write:

$$\frac{dy}{dx} = 3Ax^2 \quad \textit{if} \quad y = Ax^3$$

If $y = Ax^4$, we may write:

$$y^2 = A(x + \tfrac{1}{2}\Delta x)^4 = Ax^4 + 2Ax^3 . \Delta x + \tfrac{3}{2}Ax^2(\Delta x)^2 + \tfrac{1}{2}Ax(\Delta x)^3 + \tfrac{1}{16}A(\Delta x)^4$$

$$y_1 = A(x - \tfrac{1}{2}\Delta x)^4 = Ax^4 - 2Ax^3 . \Delta x + \tfrac{3}{2}Ax^2(\Delta x)^2 - \tfrac{1}{2}Ax(\Delta x)^3 + \tfrac{1}{16}A(\Delta x)^4$$

$$\therefore \quad \Delta y = 4Ax^3 . \Delta x + Ax(\Delta x)^3$$

$$\therefore \quad \frac{\Delta y}{\Delta x} = 4Ax^3 + Ax(\Delta x)^2$$

In the limit, i.e. when Δx becomes immeasurably small, so that $(x\Delta x)^2$ is negligible, we may write:

$$\frac{dy}{dx} = 4Ax^3$$

More generally, if $y = Ax^n$ (n a positive integer):

$$\Delta y = A\left(n . x^{n-1}\Delta x + \frac{n(n-1)(n-2)}{4(3!)}x^{n-3}(\Delta x)^3 + \frac{n(n-1)(n-2)(n-3)}{16(5!)}x^{n-5}, \text{ etc.}\right)$$

$$\therefore \quad \frac{\Delta y}{\Delta x} = A\left(nx^{n-1} + \frac{n(n-1)(n-2)}{4(3!)}x^{n-3}(\Delta x)^2 + \frac{n(n-1)(n-2)(n-3)(n-4)}{16(5!)}x^{n-5}(\Delta x)^4, \text{ etc.}\right)$$

$$\therefore \quad \frac{dy}{dx} = A . nx^{n-1}$$

The same result is obtainable in another way. By direct division we find:

$$\frac{a^n - b^n}{a - b} = a^{n-1} + a^{n-2}b + a^{n-3}b^2 \ldots + a \, . \, b^{n-2} + b^{n-1}$$

The number of terms on the right is n. As b approaches a, the expression on the right approaches more and more closely to:

$$a^{n-1} + a^{n-1} + a^{n-1} \ldots a^{n-1} = na^{n-1}$$

If we write $a = (x + \frac{1}{2}\Delta x)$ and $b = (x - \frac{1}{2}\Delta x)$, so that $a - b = \Delta x$:

$$\frac{a^n - b^n}{a - b} = \frac{(x + \frac{1}{2}\Delta x)^n - (x - \frac{1}{2}\Delta x)^n}{\Delta x}$$

This approaches more and more closely as we reduce Δx to

$$na^{n-1} = n(x + \tfrac{1}{2}\Delta x)^{n-1}$$

$$= n(x^{n-1} + \frac{n-1}{2} x^{n-2} \Delta x + \frac{(n-1)(n-2)}{4(2!)} \ldots, \text{etc.})$$

In the limit, when Δx is negligible $na^{n-1} = nx^{n-1}$.

Before leaving this function we may note that if $y = x^n$:

$$\frac{dy}{dx} = nx^{n-1}; \frac{d^2y}{dx^2} = n(n-1)x^{n-2}, \frac{d^3y}{dx^3} = n(n-1)(n-2)x^{n-3}, \text{ etc.}$$

More generally:

$$\frac{d^ry}{dx^n} = n(n-1)(n-2) \ldots (n-r+1) \, . \, x^{n-r}$$

Let us now consider the functions $y = sin\ x$ and $y = cos\ x$. If $y = sin\ x$ we then write:

$$\frac{dy}{dx} = \frac{sin\,(x + \frac{1}{2}\Delta x) - sin\,(x - \frac{1}{2}\Delta x)}{\Delta x} \text{ when } \Delta x \text{ vanishes}$$

By the sum and difference formulae (pp. 224–225):

$$\sin(x + \tfrac{1}{2}\Delta x) = \sin x \, . \cos(\tfrac{1}{2}\Delta x) + \cos x \, . \sin(\tfrac{1}{2}\Delta x)$$

$$\sin(x - \tfrac{1}{2}\Delta x) = \sin x \, . \cos(\tfrac{1}{2}\Delta x) - \cos x \sin(\tfrac{1}{2}\Delta x)$$

$$\therefore \quad \Delta y = 2 \cos x \, . \sin(\tfrac{1}{2}\Delta x)$$

$$\therefore \quad \frac{\Delta y}{\Delta x} = \cos x \, . \frac{\sin \frac{1}{2}\Delta x}{\frac{1}{2}\Delta x}$$

If we measure a in radians (p. 215) we have found that $\sin a \div a$ approaches nearer and nearer to 1 as a becomes smaller, so that in the limit, when $\frac{1}{2}\Delta x$ becomes immeasurably small:

$$\frac{\sin(\frac{1}{2}\Delta x)}{\frac{1}{2}\Delta x} = 1$$

$$\therefore \quad \frac{dy}{dx} = \cos x \quad \textit{if} \quad y = \sin x$$

If $y = \cos x$, we may write:

$$\cos(x + \tfrac{1}{2}\Delta x) = \cos x \,.\, \cos(\tfrac{1}{2}\Delta x) - \sin x \,.\, \sin(\tfrac{1}{2}\Delta x)$$

$$\cos(x - \tfrac{1}{2}\Delta x) = \cos x \,.\, \cos(\tfrac{1}{2}\Delta x) + \sin x \,.\, \sin(\tfrac{1}{2}\Delta x)$$

$$\therefore \quad \Delta y = -2 \sin x \,.\, \sin(\tfrac{1}{2}\Delta x)$$

$$\therefore \quad \frac{\Delta y}{\Delta x} = -\sin x \,.\, \frac{\sin(\frac{1}{2}\Delta x)}{\frac{1}{2}\Delta x}$$

In the limit, when we measure x in radians, the right-hand factor is unity as above, so that:

$$\frac{dy}{dx} = -\sin x \quad \textit{if} \quad y = \cos x$$

It is vitally important to remember that the formulae for $y = \sin x$ $y = \cos x$ hold good only if we measure x in *radians*.

Methods of Differentiation. To find the differential coefficient of some functions, we need to know certain rules, which we shall now explore:

Rule 1. If C is any numerical constant its derivative is zero. This is merely another way of saying that the rate of change of a constant is zero by definition; but we can tie it up with the derivative of one function already studied, viz.:

$$\frac{dCx^n}{dx} = nC \,.\, x^{n-1}, \quad \textit{so that} \quad \frac{dC}{dx} = \frac{dCx^0}{dx} = 0 \,.\, C \,.\, x^{-1} = 0$$

Rule 2. If C is a numerical constant:

$$\frac{d \,.\, Cy}{dx} = C \,.\, \frac{dy}{dx}$$

This follows from the fact that:

$$\Delta(Cy) = Cy_2 - Cy_1 = C(y_2 - y_1) = C \,.\, \Delta y$$

Rule 3. If $y_a, y_b, y_c \ldots$ are functions of x:

$$\frac{d}{dx}(y_a \pm y_b \pm y_c \ldots) = \frac{dy_a}{dx} \pm \frac{dy_b}{dx} \pm \frac{dy_c}{dx} \ldots$$

We have already seen that:

$$\frac{d}{dx}\left(\frac{3}{2}x - x^2\right) = \frac{3}{2} - 2x$$

and
$$\frac{d}{dx}\left(\frac{3x}{2}\right) - \frac{dx^2}{dx} = \frac{3}{2}\frac{dx}{dx} - 2x = \frac{3}{2} - 2x$$

The reader should thus be able to satisfy himself or herself that the general rule as stated above is correct.

Rule 4. If z is a function of y, and y itself is a function of x:

$$\frac{dz}{dx} = \frac{dz}{dy} \cdot \frac{dy}{dx}$$

This merely depends on the fact that the derivative is the limit of the composite ratio: $(\Delta z \,.\, \Delta y) \div (\Delta y \,.\, \Delta x)$:

$$\frac{\Delta z}{\Delta x} = \frac{z_2 - z_1}{x_2 - x_1} = \frac{z_2 - z_1}{y_2 - y_1} \times \frac{y_2 - y_1}{x_2 - x_1} = \frac{\Delta z}{\Delta y} \cdot \frac{\Delta y}{\Delta x}$$

Example (i). We may write x^6 as $(x^3)^2$ or as $(x^2)^3$, and we know that its first derivative is $6x^5$. In accordance with our rule:

$$\frac{d}{dx}(x^3)^2 = \frac{d(x^3)^2}{d(x^3)} \cdot \frac{d(x^3)}{dx} = 2x^3\, 3x^2 = 6x^5$$

$$\frac{d(x^2)^3}{dx} = \frac{d(x^2)^3}{d(x^2)} \cdot \frac{dx^2}{dx} = 3(x^2)^2 \,.\, 2x = 3x^4 \,.\, 2x = 6x^5$$

Example (ii). Remembering that $\sin^2 x$ and $\cos^2 x$ respectively mean $(\sin x)^2$ and $(\cos x)^2$, we may write:

$$\frac{d\sin^2 x}{dx} = \frac{d\sin^2 x}{d\sin x} \cdot \frac{d\sin x}{dx} = 2\sin x \,.\, \cos x$$

$$\frac{d\cos^2 x}{dx} = \frac{d\cos^2 x}{d\cos x} \cdot \frac{d\cos x}{dx} = 2\cos x\,(-\sin x) = -2\sin x \,.\, \cos x$$

$$\therefore\ \frac{d}{dx}(\sin^2 x + \cos^2 x) = 2\sin x \,.\, \cos x - 2\sin x \,.\, \cos x = 0$$

This is as should be since $(\sin^2 x + \cos^2 x) = 1$, whence its derivative is zero.

Rule 5. If u and v are each functions of x and $y = uv$:

$$\frac{dy}{dx} = u \cdot \frac{dv}{dx} + v \cdot \frac{du}{dx}$$

To establish this, we consider two adjacent points whose co-ordinates are: x, uv, and $x + \Delta x$, $(u + \Delta u)(v + \Delta v)$

$$\therefore \quad \Delta y = (u + \Delta u)(v + \Delta v) - uv = u\Delta v + v\Delta u + \Delta u \,.\, \Delta v$$

$$\therefore \quad \frac{\Delta y}{\Delta x} = u \cdot \frac{\Delta v}{\Delta x} + v \cdot \frac{\Delta u}{\Delta x} + \Delta u \cdot \frac{\Delta v}{\Delta x}$$

In the limit, we can neglect the last product which we can write alternatively as:

$$\Delta u \cdot \frac{\Delta v}{\Delta x} \quad \textit{or} \quad \Delta v \cdot \frac{\Delta u}{\Delta x}$$

When Δu and Δv are negligible in relation to u and v we may therefore write:

$$\frac{d(uv)}{dx} = u \cdot \frac{dv}{dx} + v \frac{du}{dx}$$

Example (*i*). We may write $y = x^9$ as $y = x^4 \,.\, x^5$ and we know that the first derivative of y is then $9x^8$. By our new rule (5):

$$\frac{dy}{dx} = x^4 \cdot \frac{dx^5}{dx} + x^5 \cdot \frac{dx^4}{dx}$$

$$= x^4 \,.\, 5x^4 + x^5 \,.\, 4x^3 = 5x^8 + 4x^8 = 9x^8$$

Example (*ii*). The reader will recall the following identities (pp. 146 and 147):

$$\tfrac{1}{2} \sin 2x = \cos x \,.\, \sin x \quad \textit{and} \quad \cos 2x = \cos^2 x - \sin^2 x$$

In accordance with Rule 5:

$$\frac{d}{dx}(\cos x \,.\, \sin x) = \cos x \cdot \frac{d}{dx} \sin x + \sin x \cdot \frac{d}{dx} \cos x$$

$$= \cos x \,.\, \cos x + \sin x(-\sin x)$$

$$= \cos^2 x - \sin^2 x = \cos 2x$$

By recourse to Rule 4 above:

$$\frac{d}{dx}(\cos x \,.\, \sin x) = \tfrac{1}{2} \frac{d}{dx}(\sin 2x) = \tfrac{1}{2} \frac{d(\sin 2x)}{d(2x)} \cdot \frac{d(2x)}{dx}$$

$$= \tfrac{1}{2} \,.\, \cos 2x \,.\, 2 = \cos 2x$$

Rule 6. One may safely leave it to the learner to justify the rule embodied in the limit by the statement:

$$\frac{\Delta x}{\Delta y} = \frac{1}{\frac{\Delta y}{\Delta x}}, \quad \textit{so that} \quad \frac{dx}{dy} = \frac{1}{\frac{dy}{dx}}$$

With the help of the six rules last given we may proceed to enlarge our dictionary of differential coefficients, starting with the function $y = x^n$. So far we have shown that the first derivative is nx^{n-1} only if n is a positive whole number. We shall now consider the more general issue:

(*a*) n is a so-called rational fraction $\frac{p}{q}$;

(*b*) n is a negative integer, or is a negative rational fraction.

Rule 7. If $y = x^{-1}$:

$$\frac{dy}{dx} = -\frac{1}{x^2}$$

In this case, we may write:

$$\Delta y = \frac{1}{x + \frac{1}{2}\Delta x} - \frac{1}{x - \frac{1}{2}\Delta x} = \frac{-\Delta x}{x^2 - \frac{1}{4}(\Delta x)^2}$$

$$\therefore \quad \frac{\Delta y}{\Delta x} = \frac{-1}{x^2 - \frac{1}{4}(\Delta x^2)}$$

In the limit, we may reject $\frac{1}{4}(\Delta x)^2$ as negligible.

Rule 8. From Rule 7 and Rule 6 we may find:

$$\frac{d}{dx}\left(\frac{u}{v}\right) = \frac{d}{dx}(uv^{-1}) = u\frac{dv^{-1}}{dx} + v^{-1}\frac{du}{dx}$$

$$= u \cdot \frac{dv^{-1}}{dv} \cdot \frac{dv}{dx} + \frac{1}{v} \cdot \frac{du}{dx}$$

$$= \frac{1}{v} \cdot \frac{du}{dx} - \frac{u}{v^2} \cdot \frac{dv}{dx}$$

Rule 9. We may write:

$$\frac{\Delta y}{\Delta x} = \frac{1}{\frac{\Delta x}{\Delta y}}, \quad \text{and } \textit{in the limit} \quad \frac{dy}{dx} = \frac{1}{\frac{dx}{dy}}$$

Let $y = x^{\frac{p}{q}}$ and $z = x^{\frac{1}{q}}$ so that $y = z^p$. By Rule 5:

$$\frac{dy}{dx} = \frac{dy}{dz} \cdot \frac{dz}{dx} = pz^{p-1} \cdot \frac{dz}{dx}$$

now if:

$$z = x^{\frac{1}{q}},\ x = z^q$$

and $$\frac{dx}{dz} = qz^{q-1}, \quad \textit{so that} \quad \frac{dz}{dx} = \frac{1}{q \, . \, z^{q-1}}$$

$$\therefore \quad \frac{dy}{dx} = \frac{pz^{p-1}}{qz^{q-1}} = \frac{p}{q} \cdot z^{p-q} = \frac{p}{q} x^{\frac{p-q}{q}}$$

$$\therefore \quad \frac{d}{dx}(x^{\frac{p}{q}}) = \frac{p}{q} \cdot x^{\frac{p}{q}-1}$$

This is equivalent to writing the derivative of x^n as nx^{n-1}, if n is a rational fraction $\frac{p}{q}$.

Let us now consider the case $y = x^{-n}$, in which n may be a rational fraction or an integer. We may then write, in accordance with Rule 7:

$$\frac{dy}{dx} = \frac{d(x^{-n})}{d(x^n)} \cdot \frac{dx^n}{dx} = \frac{-1}{(x^n)^2} \cdot nx^{n-1}$$

$$= \frac{-nx^{n-1}}{x^{2n}} = -nx^{n-1-2n} = -nx^{-n-1}$$

This is again consistent with the formula for the derivative of x^n, when n is a positive integer. We may thus say that if n is positive or negative, an integer or a rational fraction:

$$\frac{d \, . \, x^n}{dx} = nx^{n-1}$$

Let us now consider the function $y = e^x$. We have seen that this is expressible as the series:

$$e^x = 1 + x + \frac{x^2}{2!} + \frac{x^3}{3!} + \frac{x^4}{4!} + \frac{x^5}{5!} \ldots, \text{ etc.}$$

If we apply Rules 1 and 2, we may differentiate the series term by term, so that:

$$\frac{de^x}{dx} = 0 + 1 + \frac{2x}{2!} + \frac{3x^2}{3!} + \frac{4x^3}{4!} + \frac{5x^4}{5!} \ldots, \text{ etc.}$$

$$= 1 + x + \frac{2^2}{2!} + \frac{x^3}{3!} + \frac{x^4}{4!} \ldots, \text{etc.}$$

$$\therefore \frac{de^x}{dx} = e^x$$

If k is any numerical constant positive or negative:

$$\frac{de^{kx}}{dx} = \frac{de^{kx}}{d(kx)} \cdot \frac{d(kx)}{dx} = k \,.\, e^{kx}$$

If $k = -1$, we may therefore write:

$$\frac{de^{-x}}{dx} = -e^{-x}$$

To obtain the derivative of $y = a^x$, we first write $c = log_e\, a$, so that $a = e^c$ and $a^x = e^{cx}$, so that:

$$\frac{da^x}{dx} = \frac{de^{cx}}{dx} = c \,.\, e^{cx} = \log_e a \,.\, a^x$$

We can now differentiate $y = log_e\, x$, so that $x = e^y$ and

$$\frac{dx}{dy} = e^y \quad and \quad \frac{dy}{dx} = \frac{1}{e^y}$$

$$\therefore \frac{d}{dx} log_e\, x = \frac{1}{x}$$

The reader, who turns back to p. 446, need not now find it difficult to show that:

$$\frac{d}{dx} \cosh . x = \sinh . x \quad and \quad \frac{d}{dx} \sinh . x = \cosh . x$$

To differentiate $y = tan\, x$, we invoke Rule 8:

$$\frac{d}{dx} \cdot \tan x = \frac{d}{dx}\left(\frac{\sin x}{\cos x}\right) = \frac{1}{\cos x} \cdot \frac{d \sin x}{dx} - \frac{\sin x}{\cos^2 x} \cdot \frac{d \cos x}{dx}$$

$$= \frac{\cos x}{\cos x} - \frac{\sin x}{\cos^2 x} (-\sin x)$$

$$= 1 + \frac{\sin^2 x}{\cos^2 x} = 1 + \tan^2 x$$

On p. 374 we have defined $y = tan^{-1}\, x$ as meaning $x = tan\, y$, so that:

$$\frac{dx}{dy} = 1 + \tan^2 y = 1 + x^2 \quad and \quad \frac{dy}{dx} = \frac{1}{1 + x^2}$$

If $y = sin^{-1}x$ so that $x = sin\ y$, we may write:

$$\frac{dx}{dy} = \cos y$$

Since $\sin^2 y + \cos^2 y = 1$, $\cos^2 y = 1 - \sin^2 y$ *and* $\cos y = \sqrt{1 - \sin^2 y}$:

$$\therefore \quad \frac{dx}{dy} = \sqrt{1 - \sin^2 y} = \sqrt{1 - x^2} \quad and \quad \frac{dy}{dx} = \frac{1}{\sqrt{1 - x^2}}$$

To proceed further we shall need to have at our disposal a list of the functions we can now differentiate, as below:

y	$\frac{dy}{dx}$	y	$\frac{dy}{dx}$
x^n	nx^{n-1}	$\sin kx$	$k \,.\, \cos x$
e^{kx}	$k \,.\, e^{kx}$	$\cos kx$	$-k \sin x$
a^x	$\log_e a \,.\, a^x$	$\tan x$	$1 + \tan^2 x$
$\log_e x$	$\frac{1}{x}$	$\sin^{-1} x$	$\frac{1}{\sqrt{1-x^2}}$
$\cosh \,.\, x$	$\sinh \,.\, x$	$\tan^{-1} x$	$1 + x^2$

The Inverse Operation. The operation usually called *integration* answers the question: what function $F(x)$ of x yields Z when we differentiate it? Since we write $sin^{-1}x = A$ if $sin\ A = x$, it would be consistent with our use of the negative exponent (upper signal) if we also wrote:

$$F(x) = D_x^{-1} \,.\, z \quad if \quad D_x \,.\, F(x) = z$$

Since we speak of x as the logarithm of y and of y as the *antilogarithm* of x when $e^x = y$, it would thus be appropriate to speak of $F(x)$ in the above as the *anti-derivative* of Z. Following Leibnitz, it is more usual to call it the *indefinite integral* and to adopt a symbol suggesting its use for calculating areas as below:

$$y = \int z.dx \quad if \quad \frac{dy}{dx} = z$$

If we ask what function y of x yields x^n when differentiated we then write our question in the form:

$$y = \int x^n \,.\, dx \quad if \quad \frac{dy}{dx} = x^n$$

To answer the question, we need to have a list of derivatives at our disposal. We know that:

$$\frac{dx^n}{dx} = nx^{n-1}, \textit{ so that } \frac{dx^{n+1}}{dx} = (n+1)x^n \textit{ and } \frac{d}{dx}\left(\frac{x^{n+1}}{n+1}\right) = x^n$$

This does not completely answer our question. If C is a numerical constant:

$$\frac{d}{dx}(y+C) = \frac{dy}{dx}, \quad \textit{so that} \quad \frac{dy}{dx} = x^n \quad \textit{if} \quad y = \frac{x^n}{n+1} + C$$

$$\therefore \quad y = \int x^n \, . \, dx \quad \textit{if} \quad y = \frac{x^{n+1}}{n+1} + C$$

To put the inverse operation to work, let us first see how we could derive the cannon-ball trajectory from Galileo's discovery that heavy bodies fall at sea-level with a constant acceleration of approximately $g = 32$ feet *per second per second* when air resistance is trivial. If h is the vertical height of a body:

$$\frac{d^2h}{dt^2} = \frac{dv}{dt} = -g$$

If we wish to find v, we write therefore:

$$v = \int (-g) \, . \, dt, \quad \textit{meaning} \quad \frac{dv}{dt} = -g$$

The function v of t which satisfies the condition on the right is given by $-gt + C$, so that:

$$v = \int (-g)\, dt = -gt + C$$

Now we know that the missile has an initial velocity upwards. If we write $v = v_0$ when $t = 0$:

$$v_0 = C \quad \textit{and} \quad v = v_0 - gt$$

If we want to know at what height (h) the missile is at any time t, we recall that:

$$\frac{dh}{dt} = v = (v_0 - gt), \quad \textit{so that} \quad h = \int (v_0 - gt) dt$$

We thus wish to find the function h of t which yields $v_0 - gt$ when differentiated. Since v_0 and g are constants, $v_0 t$ yields v_0 and $\frac{1}{2}gt^2$ yields gt. If therefore K is the *integration constant*:

$$h = \int (v_0 - gt) dt = v_0 t - \tfrac{1}{2}gt^2 + K$$

Now $h = 0$ when $t = 0$, so that $K = 0$ in this equation, i.e.:

$$h = v_0t - \tfrac{1}{2}gt^2$$

To work backwards in this way, we have to rely on our familiarity with functions whose derivatives we know, and on a few rules, of which four follow from those cited for determining derivatives. In what follows k, K, and C are numerical constants:

Rule 1 $\displaystyle\int Ky \,.\, dx = K\int y \,.\, dx$

Example $\displaystyle\int Kx^n \,.\, dx = K\int x^n \,.\, dx = \frac{K \,.\, x^{n+1}}{n+1} + C$

Proof $\displaystyle\frac{d}{dx}\left(\frac{Kx^{n+1}}{n+1} + C\right) = \frac{K}{n+1}\left(\frac{dx^{n+1}}{dx} + \frac{d \,.\, c}{dx}\right) = Kx^n$

Rule 2 $\displaystyle\int F(kx) \,.\, dx = \frac{1}{k}\int F(kx) \,.\, d(kx) + C$

Example $\displaystyle\int \cos kx \,.\, dx = \frac{1}{k}\int \cos kx \,.\, d(kx) = \frac{1}{k} \,.\, \sin kx + C$

Proof $\displaystyle\frac{d}{dx}\left(\frac{1}{k}\sin kx + C\right) = \frac{1}{k}\frac{d}{dx}(\sin kx) = \frac{k\cos kx}{k}$

$$= \cos kx$$

Rule 3 $\displaystyle\int (y + K)\, dx = \int y \,.\, dx + Kx + C$

Example $\displaystyle\int \frac{1 + Kx}{x} \,.\, dx = \int\left(\frac{1}{x} + K\right) dx = \log_e x + Kx + C$

Proof $\displaystyle\frac{d}{dx}(\log_e x + Kx + C) = \frac{d}{dx}(\log_e x) + K\frac{dx}{dx}$

$$= \frac{1}{x} + K = \frac{1 + Kx}{x}$$

Rule 4 $\displaystyle\int (y_1 + y_2 + y_3 \ldots)\, dx = \int y_1 \,.\, dx + \int y_2 \,.\, dx$

$$+ \int y_3 \,.\, dx \ldots$$

Example $\int(3x^2 - 4x + 4)dx = \int 3x^2 . dx - 4\int x . dx +$

$$\int 4 . dx = \frac{3x^3}{3} - \frac{4x^2}{2} + 4x + C = x^3 - 2x^2 + 4x + C$$

Proof $\frac{d}{dx}(x^3 - 2x^2 + 4x + C) = \frac{dx^3}{dx} - 2\frac{dx^2}{dx} + 4\frac{dx}{dx}$

$$= 3x^2 - 4x + 4$$

A fifth rule is obtainable from that of differentiation of the product (uv) of two functions of x:

Rule 5 $\int uv . dx = u\int v . dx - \int w . \left(\frac{du}{dx}\right) dx$ *if*

$$w = \int v . dx$$

Example $\int \sin x . \cos x . dx$

$$= \sin x \int \cos x . dx - \int \sin x . \frac{d \sin x}{dx} \cdot dx$$

$$= \sin^2 x - \int \sin x . \cos x . dx.$$

$$\therefore \quad 2\int \sin x . \cos x . dx = \sin^2 x \quad \textit{and} \quad \int \sin x . \cos x . dx = \tfrac{1}{2}\sin^2 x$$

Check $\frac{d}{dx}(\frac{1}{2}\sin^2 x) = \frac{1}{2}\cdot\frac{d \sin^2 x}{d(\sin x)}\cdot\frac{d \sin x}{dx} = \frac{1}{2} . 2 \sin x . \cos x$

$$= \sin x . \cos x$$

To understand the derivation of this rule we need to be clear about what we mean by speaking of the so-called indefinite integral as the *anti-derivative* and of the operation of integration (better called *anti-differentiation*) as the inverse of differentiation. To clarify their relation symbolically, let us leave out the x-function on which our instruction is to operate, as below:

$$\textit{differentiation}\ \frac{d}{dx}(\ldots); \quad \textit{integration} \int(\ldots)\, dx$$

By definition:

$$\frac{d}{dx}(y) = Z \quad \textit{means the same as} \quad \int Z . dx = y$$

In each equation we can perform one or other operation if we do the same to both sides, i.e.:

$$\int\left(\frac{dy}{dx}\right) dx = \int Z \,.\, dx = y \quad and \quad \frac{d}{dx}\left(\int Z \,.\, dx\right) = \frac{dy}{dx} = Z$$

Thus, successive application of both operations in either order leaves the function unchanged. With this in mind, let us consider three functions of x labelled as u, v, and w, defining w as above, i.e.:

$$\frac{dw}{dx} = v, \quad so\ that \quad w = \int v \,.\, dx.$$

By the product rule for differentiation, we may write:

$$\frac{d}{dx}(uw) = u\frac{dw}{dx} + w \cdot \frac{du}{dx} = uv + w \cdot \frac{du}{dx}$$

We may now perform on both sides of the equation the operation we have denoted by the integral sign. We then have:

$$uw = \int uv \,.\, dx + \int w \,.\left(\frac{du}{dx}\right) .\, dx$$

$$\therefore \quad u\int v \,.\, dx = \int uv \,.\, dx + \int w\left(\frac{du}{dx}\right) dx$$

$$\therefore \quad \int uv \,.\, dx = u\int v \,.\, dx - \int w\left(\frac{du}{dx}\right) .\, dx$$

Before we enlist the two main devices which assist us to perform the operation of integration with the help of these rules and with that of our familiarity with derivatives already listed, let us make a table as follows:

Z	$\int Z \,.\, dx$	Z	$\int Z \,.\, dx$
x^n	$\frac{x^{n+1}}{n+1} + C$*	$\sin kx$	$-\frac{1}{k}\cos kx + C$
		$\cos kx$	$\frac{1}{k}\sin kx + C$
e^{kx}	$e^{kx} + C$	$1 + \tan^2 x$	$\tan x + C$
a^x	$\frac{a^x}{\log_e a} + C$	$\frac{1}{\sqrt{1 - x^2}}$	$\sin^{-1} x + C$
$\frac{1}{x}$	$\log_e x + C$	$\frac{1}{1 + x^2}$	$\tan^{-1} x + C$

* When $n = -1$ the integral of x^n is $log\ x + C$.

Given a table of anti-derivatives such as the foregoing, one can deal with functions which the five foregoing rules do not suffice to integrate by one of two main methods: (*a*) trigonometrical substitution; (*b*) series expansion.

Trigonometrical Substitution. As an example of this procedure, let us consider the case:

$$\int \sqrt{1 - x^2} \,.\, dx = y$$

We know that $sin^2\, a + cos^2\, a = 1$, so that:

$$1 - sin^2\, a = cos^2\, a \quad and \quad \sqrt{1 - \sin^2 a} = \cos a$$

$$\frac{d}{da} \sin a = \cos a \quad and \quad d \sin a = \cos a\, da$$

Let us substitute in y above $x = sin\ a$, so that:

$$y = \int \sqrt{1 - \sin^2 a} \,.\, d(\sin a) = \int \cos a(\cos a\, da) = \int \cos^2 a \,.\, da$$

From the addition formula:

$$\cos 2a = \cos^2 a - \sin^2 a = 2 \cos^2 a - 1, \quad so\ that \quad \cos^2 a = \tfrac{1}{2}(1 + \cos 2a)$$

$$\therefore \quad y = \tfrac{1}{2} \int (1 + \cos 2a)\, da = \tfrac{1}{2} \int 1 \,.\, da + \tfrac{1}{2} \int \cos 2a \,.\, da$$

$$\therefore \quad y = \tfrac{1}{2}a + \frac{\sin 2a}{4} + C$$

If $x = sin\ a$ in the above $a = sin^{-1} \,.\, x$, and

$$\sin 2a = 2 \sin a \,.\, \cos a = 2 \sin a \sqrt{1 - \sin^2 a} = 2x\sqrt{1 - x^2}$$

$$\therefore \quad y = \tfrac{1}{2}(\sin^{-1} x + x\sqrt{1 - x^2}) + C$$

Series Integration. It is instructive to examine the exponential function in series form, because it emphasizes the role of the integration constant C. We first note that:

$$\int e^x \,.\, dx = e^x + C, \quad since \quad \frac{d}{dx}(e^x + C) = e^x$$

If we expand e^x in series form, we may write:

$$y = \int \left(1 + x + \frac{x^2}{2!} + \frac{x^3}{3!} + \frac{x^4}{4!} \ldots\right) dx$$

$$\therefore \quad y = (x + \frac{x^2}{2!} + \frac{x^3}{3!} + \frac{x^4}{4!} + \frac{x^5}{5!} \ldots) + C$$

At first sight this seems to be inconsistent with the correct value of the integral. However, it is not so. If we put $y_0 = y$ *when* $x = 0$, we see that $C = y_0$; but when $x = 0$, $e^x = e^0 = 1$. Hence the above result – though not of itself useful – draws attention to the importance of taking the integration constant into account.

The reader may check the method of series integration by recourse to the series for *cos* x and *sin* x given on p. 442 and established by a different method below (p. 507).

The following illustration of series integration leads to an evaluation of π. We know that:

$$\int \frac{1}{1 + x^2} \cdot dx = \tan^{-1} x + C$$

Now we can expand the function under the integral sign by direct division as an infinite series which converges slowly when $x \leqslant 1$, viz.

$$\frac{1}{1 + x^2} = 1 - x^2 + x^4 - x^6 + x^8 - x^{10} \ldots, \text{etc.}$$

$$\therefore \int \frac{1}{1 + x^2} \cdot dx = x - \frac{x^3}{3} + \frac{x^5}{5} - \frac{x^7}{7} + \frac{x^9}{9} - \frac{x^{11}}{11} \ldots, \text{etc.}$$

$$\tan^{-1}(1) + C = 1 - \frac{1}{3} + \frac{1}{5} - \frac{1}{7} + \frac{1}{9} - \frac{1}{11} \ldots, \text{etc.}$$

Now $\tan \frac{\pi}{4} = 1$ in circular measure, so that $\tan^{-1}(1) = \frac{\pi}{4}$. Since $\tan 0 = 0$, $\tan^{-1}(0) = 0$ so that when $x = 0$ in the above series $(\tan^{-1} x + C) = 0$ and $C = 0$. Hence:

$$\frac{\pi}{4} = 1 - \frac{1}{3} + \frac{1}{5} - \frac{1}{7} + \frac{1}{9} - \frac{1}{11} \ldots$$

We may group the terms in pairs in two ways:

$$\frac{\pi}{4} = 1 - \left(\frac{1}{3} - \frac{1}{5}\right) - \left(\frac{1}{7} - \frac{1}{9}\right) - \left(\frac{1}{11} - \frac{1}{13}\right) \ldots \text{etc.}$$

$$= 1 - \frac{2}{15} - \frac{2}{63} - \frac{2}{143} \ldots, \text{etc.}$$

Alternatively:

$$\frac{\pi}{4} = \left(1 - \frac{1}{3}\right) + \left(\frac{1}{5} - \frac{1}{7}\right) + \left(\frac{1}{9} - \frac{1}{11}\right) + \left(\frac{1}{13} - \frac{1}{15}\right) \ldots, \text{ etc.}$$

$$= \frac{2}{3} + \frac{2}{35} + \frac{2}{99} + \frac{2}{195} \ldots, \text{ etc.}$$

We thus see that:

$$\frac{2}{3} < \frac{\pi}{4} < 1, \quad \textit{so that} \quad 2.\dot{6} < \pi < 4$$

The foregoing series chokes off very slowly, and we can get a much more convenient form by using other values of tan a. For instance, tan 30°, i.e.:

$$\tan \frac{\pi}{6} = \frac{1}{\sqrt{3}}$$

Hence if:

$$a = \frac{\pi}{6}$$

$$\frac{\pi}{6} = \frac{1}{\sqrt{3}} - \left(\frac{1}{\sqrt{3}}\right)^3 \cdot \frac{1}{3} + \left(\frac{1}{\sqrt{3}}\right)^5 \cdot \frac{1}{5} - \left(\frac{1}{\sqrt{3}}\right)^7 \cdot \frac{1}{7}$$
$$+ \left(\frac{1}{\sqrt{3}}\right)^9 \cdot \frac{1}{9} - \left(\frac{1}{\sqrt{3}}\right)^{11} \cdot \frac{1}{11}$$

$$= \frac{1}{\sqrt{3}}\left(1 - \frac{1}{9}\right) + \left(\frac{1}{\sqrt{3}}\right)^5\left(\frac{1}{5} - \frac{1}{21}\right) + \left(\frac{1}{\sqrt{3}}\right)^9\left(\frac{1}{9} - \frac{1}{33}\right) \ldots$$

$$= \frac{1}{\sqrt{3}}\left(\frac{8}{9}\right) + \frac{1}{9\sqrt{3}}\left(\frac{16}{105}\right) + \frac{1}{81\sqrt{3}}\left(\frac{24}{297}\right) \ldots$$

$$= \sqrt{3}\left\{\frac{8}{27} + \frac{16}{27 \times 105} + \frac{24}{243 \times 297} \ldots\right\}$$

$$\therefore \pi = 3{\cdot}14 \ldots$$

We can obtain a convergent infinite series for $\log_e (1 + x)$ by the same method. We first note that:

$$\frac{d}{dx} \cdot \log_e (1 + x) = \frac{d \log_e (1 + x)}{d(1 + x)} \cdot \frac{d(1 + x)}{dx} = \frac{1}{1 + x}$$

$$\therefore \quad \int \frac{1}{1 + x} \cdot dx = \log_e (1 + x)$$

By direct division:

$$\frac{1}{1+x} = 1 - x + x^2 - x^3 + x^4 - x^5 \ldots, \text{ etc.}$$

$$\therefore \quad \log_e (1 + x) = x - \frac{x^2}{2} + \frac{x^3}{3} - \frac{x^4}{4} + \frac{x^5}{5} \ldots, + C$$

When $x = 0$, $log_e (1 + x) = log_e 1 = C$. Since $e^0 = 1$, $log_e 1 = 0 = C$

$$\therefore \quad \log_e (1 + x) = x - \frac{x^2}{2} + \frac{x^3}{3} - \frac{x^4}{4} + \frac{x^5}{5} - \frac{x^6}{6} \ldots, \text{ etc.}$$

This series is not convergent if $x > 1$, but if $x = 1$ we can cite an upper and lower limit thus:

$$\log_e 2 = 1 - \frac{1}{2} + \frac{1}{3} - \frac{1}{4} + \frac{1}{5} - \frac{1}{6} \ldots, \text{ etc.}$$

$$= 1 - \left(\frac{1}{6} + \frac{1}{20} + \frac{1}{42} \ldots\right)$$

$$= \frac{1}{2} + \left(\frac{1}{12} + \frac{1}{30} + \frac{1}{56} \ldots\right)$$

$$\therefore \quad \tfrac{1}{2} < \log_e 2 < 1$$

The series converges very slowly; but for small values of x it does so fairly fast, e.g.:

$$\log_e (1 \cdot 25) = 0 \cdot 25 - \frac{(0 \cdot 25)^2}{2} + \frac{(0 \cdot 25)^3}{3} - \frac{(0 \cdot 25)^4}{4} \ldots$$

This boils down to:

$$\frac{1}{4}\left(1 - \frac{1}{8}\right) + \frac{1}{64}\left(\frac{1}{3} - \frac{1}{16}\right) + \frac{1}{1024}\left(\frac{1}{5} - \frac{1}{24}\right) \ldots$$

The Definite Integral. From time immemorial – as early as when Egyptian priestly mathematicians correctly calculated the cubic capacity of a pyramid – there have been attempts to solve problems involving areas and volumes by *ad hoc* geometrical methods referred to by such terms as *exhaustion* and *quadrature*. The evaluation of π by Archimedes as a summation of triangular, and by the Japanese as a summation of rectangular, strips embodies a *modus operandi* which takes on a new significance in the light of a particular meaning we can attach to the *anti-derivative*.

Seemingly, that is why Leibnitz, who grasped this more clearly than his other contemporaries, adopted a vertically elongated S (the initial letter of *summatio*) for the operation thenceforth designated as *integration*.

In the setting of the new geometry, a problem raised by Archimedes in connection with his treatment of the parabola took the form: given a curve whose equation is $y = f(x)$, find the area enclosed by the curve itself above, the X-axis below and two *ordinates* (lines parallel to the Y-axis). In Fig. 169, the ordinates we shall consider are: $y_0 = f(x_0)$ and $y_5 = f(x_5)$. We then divide the intervening segment enclosed by the curve and the X-axis into rectangular strips of equal width:

$$x_1 - x_0 = \Delta x = x_2 - x_1 = x_3 - x_2, \text{ etc.}$$

From the figure we see that the total area (A) under discussion lies between two totals ($A_1 < A < A_0$), such that:

$$A_0 = y_0\Delta x + y_1\Delta x + y_2\Delta x + y_3\Delta x + y_4\Delta x$$

$$A_1 = \qquad y_1\Delta x + y_2\Delta x + y_3\Delta x + y_4\Delta x + y_5\Delta x$$

Now we also see that $x_3 = x_0 + 3\Delta x$ or, more generally, $x_r = x_0 + r\Delta x$. Thus, with recourse to the Greek capital S for summation:

$$A_0 = \sum_{r=0}^{r=4} y_r \,.\, \Delta x = \sum_{r=0}^{r=4} f(x_0 + r \,.\, \Delta x) \,.\, \Delta x$$

$$A_1 = \sum_{r=1}^{r=5} y_r \,.\, \Delta x = \sum_{r=1}^{r=5} f(x_0 + r \,.\, \Delta x) \,.\, \Delta x$$

When we write the area as above, all we convey is that we can make smaller and smaller the limits between which its true value lies by making the strips thinner, i.e. by diminishing the interval Δx. This is only an improvement of our notation, but does not of itself prescribe any general rule of procedure. We notice, however, that we may write each such strip in the form $\Delta A = y \,.\, \Delta x$, so that:

$$\frac{\Delta A}{\Delta x} = y$$

With growing confidence in the algebra of growth, we can write

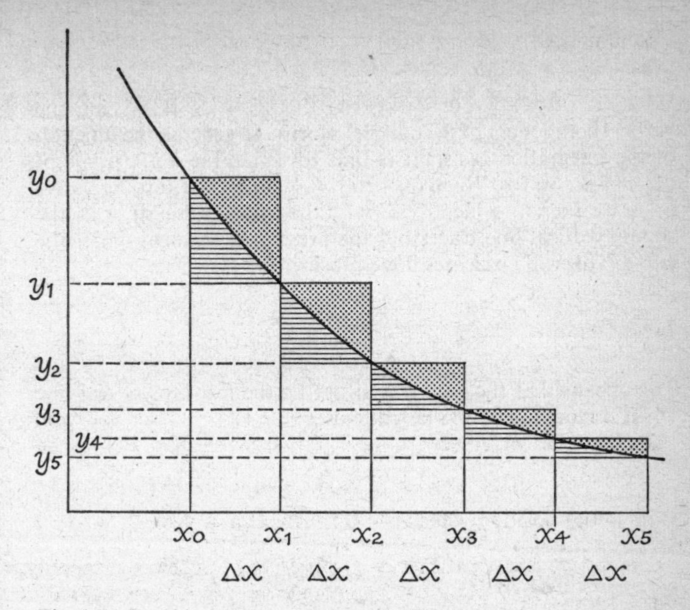

Fig. 169. Graphical Representation of Use of Rectangular Strips to Obtain Approximate Value for Area between Curve and *X*-axis Bounded by Ordinates y_0 and y_5

this as a ratio whose limit, as we diminish Δx, is clearly finite unless $y = 0$, so that:

$$\frac{dA}{dx} = y, \quad \textit{whence} \quad A = \int y \, . \, dx$$

In doing this, we have left the static realm of Greek geometry, and represent A as a *growing* entity which we may picture for the segment bounded in Fig. 169 by y_0 and y_5 as the area traced out by sliding from y_0 to y_5 a straight line of changeable length (e.g. edge of an elastic band) at right angles to the *X*-axis and joining the latter to the curve. So conceived, y is the derivative of A and A is the anti-derivative of y, each expressible as a function of x, but expressed in such terms A is not unique. It must contain a numerical integration constant C; and we have to interpret it as we interpreted C in the cannon-ball problem, when we wrote:

$$\frac{dv}{dt} = -g \quad and \quad v = C - gt$$

When $t = 0$ we wrote $v = v_0$, so that $C = v_0$, this being the vertical component of the muzzle velocity. In general, the function of the integration constant is thus to define the *initial value* of the *anti-derivative.* When (as here) $A = f(x)$ is the anti-derivative of y, we therefore identify C with the initial value of A as the sliding vertical line traces out the prescribed segment under the curve. This will be clearer if we consider the case when $y = x^2$, so that:

$$A = \int x^2 \,.\, dx = \frac{x^3}{3} + C$$

This means that the area traced out by the moving vertical line when it reaches its final position at $x = b$ is:

$$A = \frac{b^3}{3} + C$$

If its initial position when $A = 0$ is at $x = a$

$$A = 0 = \frac{a^3}{3} + C, \quad so\ that \quad C = \frac{-a^3}{3}$$

$$\therefore \quad A = \frac{b^3}{3} - \frac{a^3}{3}$$

It is usual to write this in the form of the so-called *definite* integral:

$$A = \int_a^b x^2 \,.\, dx = \left[\frac{x^3}{3}\right]_{x=a}^{x=b}$$

In the same way, we may treat the hyperbola $y = x^{-1}$ in the top right quadrant. To evaluate the part of the curve between $y = 4$ and $y = 9$, we write:

$$A = \int_4^9 \frac{1}{x} \cdot dx = \Big[\log_e x\Big]_4^9 = \log_e 9 - \log_e 4$$

$$= \log_e \frac{9}{4} \qquad = \log_e 2{\cdot}25$$

The definite integral, with upper and lower limits $x = b$ and $x = a$ respectively, yields the area of a region wholly above or wholly below the X-axis. When the graph of $y = f(x)$ crosses the

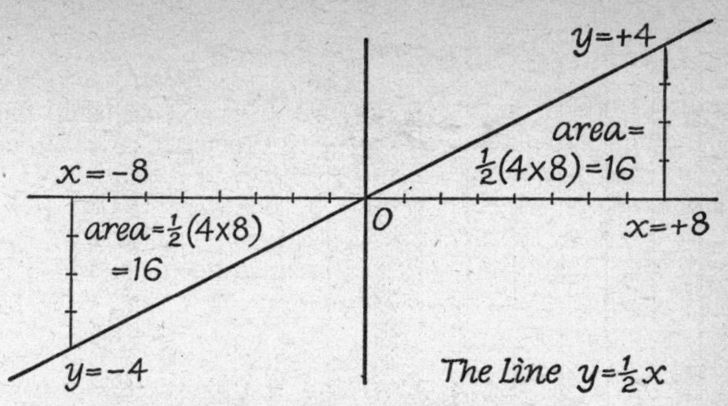

Fig. 170. Areas in Opposite Quadrants

Here the total area between $x = -8, y = -4$ and $x = +8, y = +4$ is $16 + 16 = 32$ units; and

$$\int y \, . \, dx = \tfrac{1}{4}x^2 + C$$

$$\left[\frac{x^2}{4}\right]_{-8}^{+8} = 0; \quad \left[\frac{x^2}{4}\right]_{-8}^{0} = -16; \quad \left[\frac{x^2}{4}\right]_{0}^{+8} = +16$$

Thus the correct area is the *numerical* sum of that of the parts above and below the X-axis regardless of sign.

latter, integration on opposite sides of the X-axis yields values with opposite signs. Thus application of the foregoing rule *as a single operation* will yield a deficient value, if $a < c < b$ and $y = 0$ when $x = c$. The correct procedure is then to integrate separately for regions on opposite sides of the X-axis and add the numerical values obtained *regardless of signs.* Two examples for which the reader should plot graphs will suffice to clarify the procedure.

Let us first consider the function:

$$y = \frac{4x}{5} + 8, \quad \textit{so that} \quad y = 0 \quad \textit{when} \quad x = -10 \quad \textit{and}$$

$$\int_a^b y \, . \, dx = \left[\frac{2x^2}{5} + 8x + C\right]_a^b$$

The region bounded by $a = -20$ to $b = 0$ here consists of two triangles each of base 10 units, height 8 units, and therefore of area 40 units. If we take no stock of the fact that one $(x < -10)$

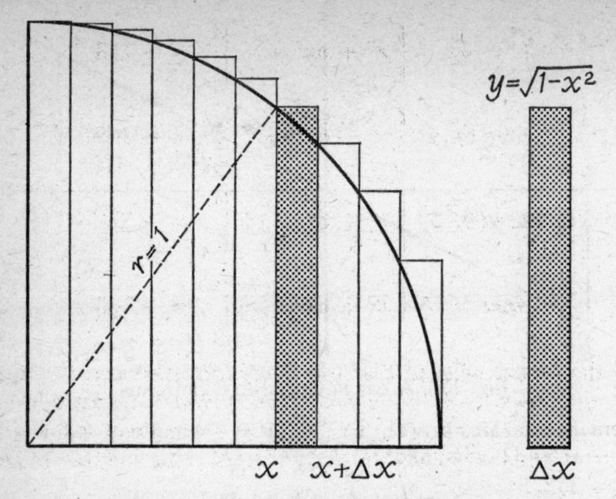

Fig. 171. Area of Circle of Unit Radius

lies below and the other ($x > -10$) above the X-axis, the result we obtain is zero, since:

$$\left[\frac{2x^2}{5} + 8x + C\right]_{-20}^{0} = -\frac{2}{5}(400) + 8(20) = 0$$

We get the correct result $40 + 40 = 80$, if we add the *numerical* value of the integrals:

$$\left[\frac{2x^2}{5} + 8x + C\right]_{-20}^{-10} = -40 \quad \textit{and} \quad \left[\frac{2x^2}{5} + 8x + C\right]_{-10}^{0} = +40$$

Let us also consider a situation in which the whole region from b to a admits of a three-fold split, as is true of the function: $y = 3x^2 + 12x$. The curve crosses the X-axis when $y = 0$, so that:

$$3x^2 + 12x = 0 = x^2 + 4x = x(x + 4)$$

Thus, the curve of this function crosses the X-axis at $x = 0$ and $x = -4$. When $x < -4$ or $x > 0$, the curve lies wholly above the X-axis, but below it in the range $b = 0$, $a = -4$. If we integrate between the limits -8 and $+4$ we therefore embrace three regions (A, B, C), the sign of the middle being negative. Thus

$$\int_{-8}^{4} y \,.\, dx = \Big[x^3 + 6x^2 + C\Big]_{-8}^{4} = +\,288$$

$$\textit{Area of A} \quad \Big[x^3 + 6x^2 + C\Big]_{-8}^{-4} = +\,160$$

$$\textit{Area of B} \quad \Big[x^3 + 6x^2 + C\Big]_{-4}^{0} = -\,32$$

$$\textit{Area of C} \quad \Big[x^3 + 6x^2 + C\Big]_{0}^{4} = +\,160$$

Here the reader will see that the integration between the upper (+4) and lower (−8) limits yields the *algebraic* sum of the three regions, since 288 = (160 − 32 + 160). The correct value is the *numerical* sum (i.e. total regardless of sign). This is: (160 + 32 + 160) = 352.

Charting the Course without Plotting

Since we can intelligently use the definite integral to measure an area only if we can chart our course between the Scylla above the line $y = 0$ and the Charybdis below it, now is a fitting time to discuss blind flying, i.e. forming a mental picture without recourse to a map. As an illustration, let us examine the so-called *normal* function of Fig. 195 in Chapter 12. Its equation is:

$$y = Ae^{-kx^2} = \frac{A}{e^{kx^2}} = A \quad \textit{when} \quad x = 0 \text{ so that } e^{kx^2} = 1$$

We first notice that $(+x)(+x) = x^2 = (-x)(-x)$. This signifies that we can pair off every positive numerical value of x with an equivalent negative numerical value. We can thus infer that one and the same numerical value of x occurs on either side of the Y-axis. Since e^{-kx^2} is positive for all values of x, we must therefore picture the graph of the function as a symmetrical curve lying wholly above the X-axis, if A is also positive.

We next notice that $y = A$ only when $x = 0$. This means that the curve crosses the Y-axis only once. To examine its turning-point and points of inflexion, if any, we apply Rule 4 of p. 474. To do this, we may write $z = kx^2$, so that:

$$\frac{dy}{dx} = \frac{dAe^{-z}}{dz} \cdot \frac{dz}{dx} = -Ae^{-z} \,.\, 2kx = -2Akx \,.\, e^{-kx^2}$$

If the above expression is zero:

$$x \, . \, e^{-kx^2} = 0, \quad \textit{so that} \quad x = 0$$

There is thus a turning-point at $x = 0$ and $y = A$. To get the second derivative, we invoke Rule 5, substituting first $u = -2Akx$ and $v = e^{-kx^2}$, so that:

$$\frac{du}{dx} = -2Ak \quad \textit{and} \quad \frac{dv}{dx} = -2kx \, . \, e^{-kx^2}$$

$$\therefore \quad \frac{d^2y}{dx^2} = -2Ak \, . \, e^{-kx^2} + 4Ak^2x^2 \, . \, e^{-kx^2}$$

When $x = 0$, so that $e^{-kx^2} = 1$:

$$\frac{d^2y}{dx^2} = -2Ak$$

Since the second derivative is negative (p. 464), the turning-point is a maximum. We now equate it to zero, so that:

$$4Ak^2x^2 \, . \, e^{-kx^2} = 2Ak \, . \, e^{-kx^2} \quad \textit{and} \quad 2kx^2 = 1$$

Thus there is a point of inflexion at $x^2 = (2k)^{-1}$, and since we can pair off every positive numerical value with an equivalent value, there is a corresponding point of inflexion at $x = \pm(2k)^{-\frac{1}{2}}$.

It remains to determine the range. To do so, we ask what value y will have when $\pm x$ is numerically as large as conceivable, so that:

$$e^{kx^2} = \infty \quad \textit{and} \quad e^{-kx^2} = \frac{1}{e^{kx^2}} = 0$$

This means that as x becomes numerically larger and larger the curve approaches the X-axis lying approximately parallel thereto. We then say (p. 355) that it is *asymptotic* to the X-axis on either side of the Y-axis.

Without drawing a graph, we have thus disclosed a birds-eye view of the curve which is symmetrical about the Y-axis lying wholly above the X-axis, with a maximum value where it crosses the former at $x = o$, $y = A$. Its limbs are asymptotic to the X-axis, and its curvature changes from concave upwards to concave downwards or vice versa on either side of the Y-axis.

Applications of the Definite Integral. On pp. 484–5, we have learned to use trigonometrical substitution to evaluate the anti-derivative (*indefinite integral*) of $\sqrt{1 - x^2}$ and to evaluate that of $(1 + x^2)^{-1}$ by series integration. Both methods are beset by pitfalls for the unwary when we enlist them to do service for evaluating *definite integrals*, as we may use series integration for computing π to any required level of precision by the procedure we shall now use. Let us recall the formula derived by the trigonometrical trick for:

$$\int \sqrt{1 - x^2} \, . \, dx = \tfrac{1}{2}(\sin^{-1}x + x\sqrt{1 - x^2}) + C$$

Since $\sin 0 = 0$ and $\sin \frac{\pi}{2} = 1$, $\sin^{-1}(0) = 0$ and $\sin^{-1}(1) = \frac{\pi}{2}$, so that the area of a *quadrant* of a circle of unit radius (Fig. 171) is

$$\left[\tfrac{1}{2}(\sin^{-1}x + x\sqrt{1 - x^2}\right]_0^1 = \tfrac{1}{2} \, . \, \frac{\pi}{2} + 0 = \frac{\pi}{4}$$

This agrees with the formula $\pi r^2 = \pi$ for the total area (A) when $r = 1$, A being 4 times that of the quadrant.

At first sight, it may seem that we could proceed in the same way to evaluate π by recourse to the Binomial Theorem as on p. 417, in the code of which:

$$(1 - x^2)^{\frac{1}{2}} = 1 - B_1x^2 + B_2x^4 - B_3x^6 + B_4x^8 \ldots, \text{etc.}$$

Whence series integration would lead us to put:

$$\begin{aligned} &\int (1 - x^2)^{\frac{1}{2}} \, . \, dx \\ &\qquad = x - \frac{B_1 \, . \, x^3}{3} + \frac{B_2 \, . \, x^5}{5} - \frac{B_3 \, . \, x^7}{7} + \frac{B_4 \, . \, x^9}{9} \ldots, \text{etc.} \\ &\qquad = x - \frac{0{\cdot}5 \, . \, x^3}{3} - \frac{0{\cdot}125 \, . \, x^5}{5} - \frac{0{\cdot}0625 \, . \, x^7}{7} \ldots, \text{etc.} \end{aligned}$$

When $x = \frac{1}{2}$, this series evidently converges more rapidly than the one we obtained for $\sqrt{0{\cdot}75}$ on p. 417. To use it rightly, we have however to ask whether the parent series is meaningful. As we shall see later, the Binomial Theorem is valid for $(1 - x)^n$ when n is a fraction, but only if x is numerically less than unity. When $x = 1$ in the above $(1 - x^2)^{\frac{1}{2}} = 0$ and the method of series integration breaks down. That is to say, we cannot use it to derive a series for π by equating $\frac{1}{4}\pi$ to the definite integral:

$$\int_0^1 \sqrt{1 - x^2} \, . \, dx$$

With the aid of a home-made diagram, the reader may however proceed as follows. In a circle (Fig. 171) of unit radius the figure enclosed by the ordinates $x = 0$, $y = 1$ and $x = \frac{1}{2}$, $y = \sqrt{3} \div 2$

is made up of: (i) a segment enclosing an angle of 30° between its two radii, whence of area $\pi \div 12$; (ii) a triangle whose area $\frac{1}{2}xy = \sqrt{3} \div 8$. So we may write:

$$\int_0^{\frac{1}{2}} \sqrt{1 - x^2} \,.\, dx = \frac{\pi}{12} + \frac{\sqrt{3}}{8}$$

or

$$\pi = 12 \int_0^{\frac{1}{2}} \sqrt{1 - x^2} \,.\, dx - \frac{3\sqrt{3}}{2}$$

Recourse to series integration then gives a correct result. It is necessary to use only 3 terms of the integral series to ensure an error not exacting 1 per cent, i.e. a value for π closely corresponding to that of the Rhind papyrus.

It is here important to stress that we have not fully justified the discovery of series for π on pp. 485–6 until we have shown that the series used to derive them are indeed convergent. Here we recall that both the series here invoked and the series generated by integration consist of:

(*a*) successive terms numerically less than their predecessors;
(*b*) successive terms with alternately positive and negative.

When we come to deal with the binomial theorem for negative and/or fractional indices at a later stage (pp. 510–11), we shall see that series which have *both* these properties are always convergent.

Let us now turn to a different class of problems: how to find the volume of a sphere or cone. With this end in view, we imagine

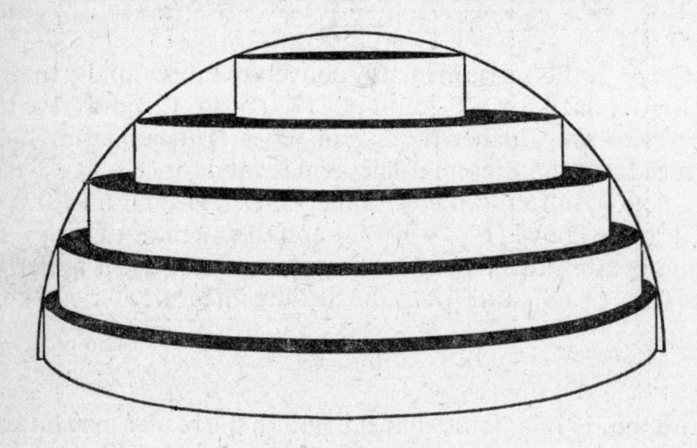

Fig. 172. The Hemisphere Conceived Approximately as a Succession of Cylindrical Slices

that we slice them into measurelessly small layers of cylindrical shape. The volume of a solid figure which has the same cross-section everywhere is the product of the area of the cross-section and the height. Hence the volume of a cylinder is $\pi r^2 h$. To get the volume of the sphere (Figs. 172–3), we therefore proceed as if a series of flat cylindrical slices are placed end to end along the x axis, each flat cylinder being Δx units in *height* when put to stand on its base. The radius of each cylinder will correspond with the y ordinate of the circular cross-section of the sphere. If the radius of the sphere is r, the radius of each cylinder, y, will be given by the equation:

$$y^2 = r^2 - x^2$$

The volume of each slice is:

$$\pi \, . \, y^2 \, . \, \Delta x$$
$$= \pi(r^2 - x^2)\Delta x$$

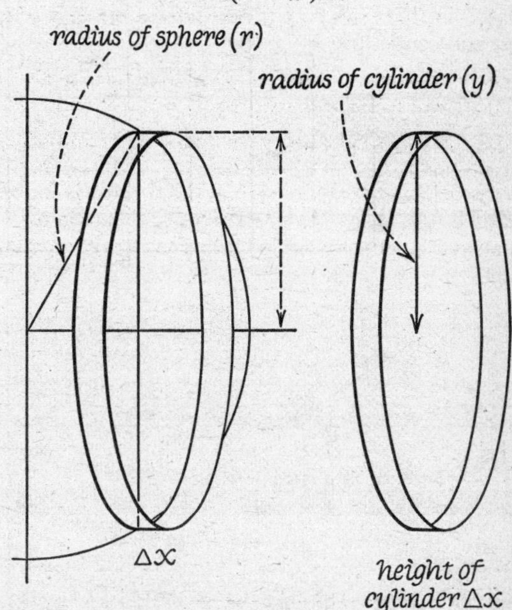

Fig. 173. Using the Integral Calculus to Get the Volume of the Sphere

The volume of the half sphere will be the sum of all the cylinders when Δx becomes indefinitely small, i.e.:

$$\int_0^r \pi(r^2 - x^2)dx$$

For the volume of the half sphere we therefore need to solve the differential equation:

$$\frac{dV}{dx} = \pi r^2 - \pi x^2$$

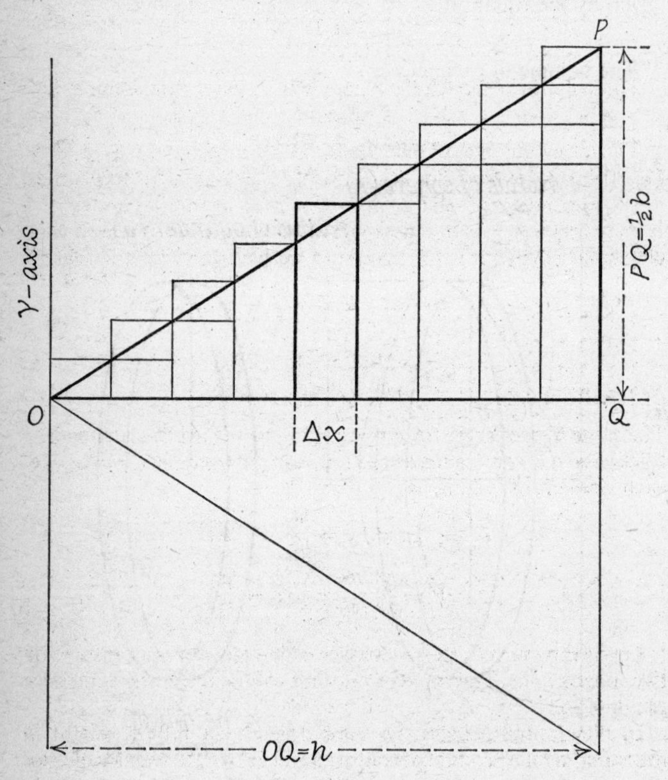

Fig. 174. Using the Integral Calculus to get the Volume of a Cone or Pyramid

The solution of such an equation has been given as:

$$V = \pi r^2 x - \frac{\pi x^3}{3} + c$$

Since we are considering only the volume to the right of $x = 0$, $V = 0$ when $x = 0$, and therefore $c = 0$. Hence the value of the integral is obtained by putting r for x in:

$$\pi r^2 x - \frac{\pi x^3}{3}$$

i.e.

$$V = \pi r^3 - \frac{\pi r^3}{3}$$

$$= \frac{2\pi r^3}{3}$$

The volume of the whole sphere will be twice this volume, i.e.:

$$\tfrac{4}{3}\pi r^3$$

Thus the volume of the earth is approximately $\frac{4}{3} \times \frac{22}{7} \times (4{,}000)^3$ cubic miles. This is roughly 268,000,000,000 cubic miles.

We may deal with the volume of a cone in the same way. If h is its height and b is the diameter of its base, the volume of each elementary cylinder is $\pi y^2 . \Delta x$, and we see from Fig. 174 that:

$$y = \frac{bx}{2h}, \quad \textit{so that} \quad V = \pi \int_0^h \frac{b^2}{4h^2} . x^2 . dx$$

$$\therefore \quad V = \frac{\pi b^2}{4h^2} \cdot \left[\frac{x^3}{3}\right]_0^h = \frac{\pi b^2 h}{12}$$

The same figure serves for a pyramid if we regard the elements of thickness Δx as of square section and of volume $(2y)^2 \Delta x$, i.e. each element of volume is:

$$4\left(\frac{b^2}{4h^2}\right) . x^2 \Delta x = \frac{b^2}{h^2} . x^2 . \Delta x$$

$$\therefore \quad V = \frac{b^2}{h^2} \int_0^h x^2 . dx = \frac{b^2 h}{3}$$

Let us now turn to the mechanics of the Newtonian century for two illustrations. The topic of our first will be the work done by a moving body.

In Newtonian physics the *work* done by a falling weight is measured by the product of the force exerted by the weight and the distance through which it falls ($W = Fl$). If the weight descends through a small distance dl, the small amount of work done is:

$$dW = F \, . \, dl$$

Remembering that Newtonian force is the product of the mass (m) moved and the acceleration produced $\left(\frac{d^2l}{dt^2} \text{ or } \frac{dv}{dt}, \text{ where } v = \frac{dl}{dt} \text{ is the velocity at a given instant}\right)$, we may put:

$$dW = m \cdot \frac{dv}{dt} \cdot dl$$

If we multiply both sides by $dl \div dl\,(= 1)$:

$$dW = m \cdot \frac{dl}{dt} \cdot \frac{dv}{dl} \cdot dl$$

$$= mv \, . \, dv$$

So in starting with a velocity 0 when $l = 0$ and moving till the velocity is V, when the distance fallen is L, the work done is:

$$\int_0^V mv \, . \, dv = \tfrac{1}{2} m V^2$$

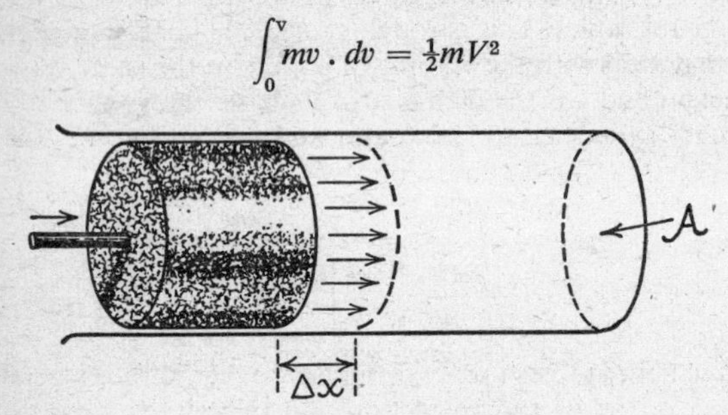

Fig. 175. The Gas Integral

If a vessel of height x has the same area in cross-section (A) throughout like a cylinder its volume is $A \, . \, x$. If the piston is pushed through a distance Δx the decrease in volume of the gas inside the piston is therefore $A \, . \, \Delta x$. Pressure is measured in the mechanics of gases by the force applied to unit area, and Work is measured by the product of the force applied into the distance traversed, i.e.:

$$F = p \, . \, A \quad \text{and} \quad W = F \, . \, D$$

If the piston is pushed through a distance Δx without loss of energy by friction, the work done is therefore $\mathrm{F} \, . \, \Delta x$. This may be written also $p \, . \, A \, , \Delta x$. The small change of volume which is represented by $- A \, . \, \Delta x$ may be written Δv, and the small amount of work done in changing it ΔW, so that:

$$\Delta W = - p \, . \, \Delta v$$

The quantity $\frac{1}{2}mV^2$ is called the Kinetic Energy of m, when it has speed V.

If the weight fell in a vacuum its acceleration would be approximately 32 feet per second per second. The work done by falling through a distance L would therefore be $32mL$, i.e.:

$$32mL = \tfrac{1}{2}mV^2$$

If the weight does not fall in a vacuum, friction with air diminishes its kinetic energy by an amount calculable from the constant usually called the *Mechanical Equivalent of Heat*, but more appropriately called the Thermal Equivalent of Work.

Let us now consider work the *potential energy* of an expanding gas, i.e. what work (W) it could accomplish if it dissipated no energy as heat. The calculation depends on a law discovered by two of Newton's contemporaries, Hooke and Boyle. This states that the volume (v) of a gas is inversely proportional to its pressure (p) at constant temperature, i.e.:

$$v = \frac{K}{p} \quad or \quad p = \frac{K}{v}$$

The legend below Fig. 175 indicates that we may write: $\Delta W = -p \,.\, \Delta v$, so that in a compression from v_1 to v_2 ($>v_1$):

$$W = -\int_{v_1}^{v_2} p \,.\, dv = -K\int_{v_1}^{v_2} \frac{1}{v} \cdot dv$$

$$\therefore \quad W = -K\Big[\log_e v\Big]_{v_1}^{v_2} = -K(\log_e v_2 - \log_e v_1)$$

$$= -K \log \frac{v^2}{v_1}$$

Obituary on the Newton–Leibnitz Quibble. The thrifty reader may well skip the interlude which follows. It will not help anyone to understand the uses of the calculus of Newton and Leibnitz, being merely an attempt to satisfy curiosity about the claims of the contestants to priority. Since a just assessment involves how far the symbols they used did or did not convey their intentions to their contemporaries, what follows invites encouraging comment on inconsistencies of still current mathematical symbols.

Too few people who write about the history of mathematics pay enough attention to what new symbols the beginner finds easy, or otherwise, when everyone except the innovator is still a beginner. We may therefore best attempt to do equal justice to the plaintiff and defendant in the unseemly nationalistic squabble mentioned at the beginning of this chapter, if we first take stock of the gross defects of the symbolism of Newton, then of the still unremedied shortcomings of a notation which the contemporaries of Leibnitz found more easy to decode. In doing so, we shall find one aspect of the controversy more intelligible if we realize three things: (*a*) the economical shorthand we now call algebra was still a nurseling in their time; (*b*) the notion of a *limiting ratio* first mentioned on p. 134 was still novel and half digested; (*c*) neither Newton in Britain nor Leibnitz in Germany anticipated the loopholes which made it possible for contemporary theologians and metaphysicians with no feeling for figures to ridicule plausibly the possibility that the ratio of two vanishing quantities can have a finite value.

In Britain, the bombastic bishop Berkeley, who inaugurated a half century of English cultural decadence by reviving the silliest superstitions of Plato's philosophy, went so far in his attack to declare of Newton's *fluxions* (i.e. differential coefficients): 'what are these . . . velocities of evanescent increments. And what are these same evanescent increments? They are neither finite quantities, nor quantities infinitely small, nor yet nothing. May we not call them the ghosts of departed quantities.' To one who believed so devoutly in ghosts, unholy or otherwise, able also to accommodate a deity simultaneously omniscient, omnipotent, and benevolent, this may make sense. Subsequent generations of mathematicians laboured to find a verbal come-back. To many of us who do not attend divine service regularly, their formulations may be less confusing than to look at the tangent of a graph which gave Fermat a first glimpse of the fact that *tan A* in Fig. 162 remains a measurable quantity when the Δy and Δx of Isaac Barrow's triangle became as ghostly as anything we can conceive either of them to become.

To clarify the advantages of the symbols which Leibnitz used, let us put down side by side the differentiation of x^3:

$$\textit{Leibnitz:}\quad \frac{dy}{dx} = 3x^2 \quad \textit{if} \quad y = x^3$$

$$\textit{Newton:}\quad \dot{y} = 3x^2 \quad \textit{if} \quad y = x^3$$

$$\begin{aligned}\textit{Leibnitz:}\quad dy &= (x + dx)^3 - x^3 \\ &= x^3 + 3x^2 . dx + 3x . (dx)^2 + (dx)^3 - x^3 \\ &= 3x^2 . dx + 3x . (dx)^2 + (dx)^3\end{aligned}$$

$$\textit{Newton:}\quad (x + 0)^3 - x^3 = x^3 + 3x^2(0) + 3x(0)^2 + (0)^3 - x^3$$
$$= 3x^2(0) + 3x(0^2) + (0)^3$$

$$\textit{Leibnitz:}\quad \frac{dy}{dx} = 3x^2 + 3x(dx) + (dx)^2$$

$$\textit{Newton:}\quad \dot{y} = \frac{3x^2(0) + 3x(0)^2 + (0)^3}{0} = 3x^2 + 3x(0) + (0)^3$$

Newton's use of (0) where we might use Δx is somewhat misleading, and his use of $\dot{y}$ is exceptionable on the grounds that it conveys no clue to the independent variable of which y is a function. Since Newton's main preoccupation was with velocities and accelerations, the independent variable which he had mostly in mind was *time*. Thus he wrote:

$$\dot{y} \textit{ for } \frac{dy}{dt} \quad \textit{and} \quad \ddot{y} \textit{ for } \frac{d^2y}{dt^2}$$

In view of the wide range of uses to which we can put the calculus the *dot* notation was an irremediable blemish. The main defect of the notation of Leibnitz was not. In deriving the differential coefficient, he draws no distinction between very small increments Δy or Δx and dy or dx which are the ghosts of departed quantities. In other words, $\Delta y : \Delta x$ of the notation used earlier is the ratio between two identifiable entities, but if we write the derivative as did Leibnitz, we legitimately convey that it is the limit of a ratio $(\Delta y : \Delta x)$ of whose separate components we cannot measure. We can remind ourselves that differentiation is indeed an operation different from arithmetical division, if we write our derivative in the form:

$$\frac{d}{dx}(y) \quad \textit{or better} \quad D_x \, . \, y$$

In passing, we may notice that Leibnitz (following Barrow) in effect represents Δy as $f(x + \Delta x) - f(x)$. In this chapter we have represented it as $f(x + \frac{1}{2}\Delta x) - f(x - \frac{1}{2}\Delta x)$. Since we eventually reject terms in Δx, this makes no difference to the result; but the latter procedure makes it easier to see what is happening in Fig. 162 as the two points S and T coalesce in P.

Though it is in substance identical with that of his rival, Newton's treatment of area was bogged down by bad symbols, or lack of them, even more than his treatment of the tangent, i.e. differentiation; but the symbol introduced by Leibnitz for what one usually calls the indefinite integral, was in one sense a backward step. The definite integral is a particular application of an operation with much wider uses, being the way of solving for z the differential equation in which y is a known function of x, viz.:

$$\frac{dz}{dx} = y \quad or \quad D_x . z = y$$

Just as we write $tan^{-1}x = A$ if $tan\ A = x$, it would bring into focus that so-called integration is differentiation in reverse, if we wrote:

$$D_x^{-1} . y = z \quad if \quad D_x . z = y$$

In the notation of Leibnitz therefore:

$$D_x^{-1} . y = \int y . dx$$

In the same way, we might write the burdensome $log_e\, x$ and *anti-log*$_e\, x$ more compactly as $L_e . x$ and $L_e^{-1}x$ meaning that $L_e^{-1}(a) = b$ if $L_e(b) = a$. Unfortunately, the unwieldy symbols we now use for logarithmic and trigonometrical functions came into use before there was any standardization of symbols. If we write $tan^{-1}x = A$ when $tan\ A = x$, it is grossly misleading to write $(tan\ A)^2$ as $tan^2\ A$; but nothing short of a world conference under the auspices of UNESCO would persuade writers of textbooks to stop using the symbols $tan^2\ A$, $sin^2\ A$, $cos^2\ A$ and *anti-log* . x.

Maclaurin's Theorem: Though we trace its theoretical origin to the problem of maxima and minima on the one hand, and of quadrature on the other, the new tool which Newton and Leibnitz jointly sharpened claimed early triumphs in other fields, more especially in the search for infinite convergent series as a device to expedite tabulation of logarithms, trigonometrical ratios, and other functions. The kingpin of this development was a discovery by Colin Maclaurin (1742), a Scotsman in the following of James Gregory whose treatment of the vanishing triangle we shall now look at in terms relevant to the new ideas.

By turning back to p. 282, the reader will recall that we have labelled u_n as a *discrete* function of n on the assumption that n increases by unit steps. We may define $\Delta u_n = u_{n+1} - u_n$ as the increase of u_n per unit change of n (i.e. from n to $n + 1$); and it would be consistent to write $\Delta n = 1$ for the corresponding change of n, so it is also consistent to write:

$$\Delta u_n \equiv \frac{\Delta u_n}{\Delta n}$$

There are, indeed, several ways in which Gregory's discovery of the basic theorem of what later generations called the *calculus of finite differences*, or more briefly the *finite calculus*, was also a forerunner of the *infinitesimal* calculus of Newton and Leibnitz. This is so because there is no compelling reason why we should restrict ourselves to the conditions $\Delta n = 1$. Suppose we make $\Delta n = \frac{1}{4}$ and $u_n = n^2$ we may then write:

$$\begin{array}{llllllllllll}
n = & 0\ \frac{1}{4} & \frac{1}{2} & \frac{3}{4} & 1 & 1\frac{1}{4} & 1\frac{1}{2} & 1\frac{3}{4} & 2 & 2\frac{1}{4} & 2\frac{1}{2} & \ldots,\text{etc.} \\
u_n = n^2 & 0\ \frac{1}{16} & \frac{4}{16} & \frac{9}{16} & \frac{16}{16} & \frac{25}{16} & \frac{36}{16} & \frac{49}{16} & \frac{64}{16} & \frac{81}{16} & \frac{100}{16} & \ldots,\text{etc.} \\
\Delta u_n & \frac{1}{16} & \frac{3}{16} & \frac{5}{16} & \frac{7}{16} & \frac{9}{16} & \frac{11}{16} & \frac{13}{16} & \frac{15}{16} & \frac{17}{16} & \frac{19}{16} & \ldots,\text{etc.} \\
\frac{\Delta u_n}{\Delta n} & \frac{1}{4} & \frac{3}{4} & \frac{5}{4} & \frac{7}{4} & \frac{9}{4} & \frac{11}{4} & \frac{13}{4} & \frac{15}{4} & \frac{17}{4} & \frac{19}{4} & \ldots,\text{etc.}
\end{array}$$

For this series:

$$\frac{\Delta u_n}{\Delta n} = \frac{(n + \Delta n)^2 - n^2}{\Delta n} = \frac{2n\,.\,\Delta n + (\Delta n)^2}{\Delta n} = 2n + \Delta n$$

$$\left(\textit{Check}:\ \text{if } n = 1\tfrac{1}{4},\quad 2n + \Delta n = 2\left(\frac{5}{4}\right) + \tfrac{1}{4} = \frac{11}{4}\right.$$

$$\left.\text{if } n = 2,\quad 2n + \Delta n = 4 + \tfrac{1}{4} = \frac{17}{4}\right)$$

Clearly, we can therefore make $\Delta u_n \div \Delta n$ approach as closely as we like to the differential coefficient of the corresponding continuous function by making the interval Δn smaller and smaller. Below is an example, which suggests the search for a continuous function whose differential coefficient is itself, as is true of $y = e^x$. If $u_n = 2^n$ and $\Delta n = 1$:

$$\begin{array}{lllllllll}
n & 0 & 1 & 2 & 3 & 4 & 5 & 6 & 7 & \ldots,\text{etc.} \\
u_n = 2^n & 1 & 2 & 4 & 8 & 16 & 32 & 64 & 128 & \ldots,\text{etc.} \\
\frac{\Delta u_n}{\Delta n} = \Delta u_n & 1 & 2 & 4 & 8 & 16 & 32 & 64 & & \ldots,\text{etc.}
\end{array}$$

To bring into sharper focus the trail which Maclaurin blazed, let us rewrite Gregory's series with the foregoing considerations in mind:

$$u_n = u_0 + n \cdot \frac{\Delta u_0}{\Delta n} + \frac{n^{(2)}}{2!} \cdot \frac{\Delta^2 u_0}{(\Delta n)^2} + \frac{n^{(3)}}{3!} \cdot \frac{\Delta^3 u_0}{(\Delta n)^3}$$

$$+ \frac{n^{(4)}}{4!} \frac{\Delta^4 u_0}{(\Delta n)^4} + \frac{h^{(5)}}{5!} \frac{\Delta^5 u_0}{(\Delta n)^5} \ldots, \text{ etc.}$$

In the above, $n^{(r)}$ is meaningful by definition only if $\Delta n = 1$. We shall therefore explore the possibility that we can express the continuous function $y = f(x)$ in a series with substitution of x^r for $x^{(r)}$. We shall expect it to be unending if it is to represent any continuous function, and it will be meaningful only for values of x consistent with its convergence to a finite limit. To exploit these clues we shall assume the possibility of expressing y as an infinite power series (i.e. series whose terms are powers of x and numerical constants) as below:

$$y = A_0 + A_1 x + A_2 x^2 + A_3 x^3 + A_4 x^4 + A_5 x^5 + \ldots$$

$$\frac{dy}{dx} = A_1 + 2A_2 x + 3A_3 x^2 + 4A_4 x^3 + 5A_5 x^4 + \ldots$$

$$\frac{d^2y}{dx^2} = 2A_2 + 3 . 2 . A_3 x + 4 . 3 . A_4 x^2 + 5 . 4 . A_5 x^3 + \ldots$$

$$\frac{d^3y}{dx^3} = 3 . 2\, A_3 + 4 . 3 . 2\, A_4 x + 5 . 4 . 3\, A_5 x^2 + \ldots$$

$$\frac{dy^4}{dx^4} = 4 . 3 . 2\, A_4 + 5 . 4 . 3 . 2 . A_5 x + \ldots$$

$$\frac{d^5y}{dx^5} = 5 . 4 . 3 . 2 . A_5 + \ldots$$

Let us now write $y_0 = y$ when $x = 0$, so that from line 1 above: $y_0 = A_0$. Similarly, let us write:

$$\frac{d}{dx}(0) = \frac{dy}{dx} \text{ when } x = 0$$

$$\frac{d^2}{dx^2}(0) = \frac{d^2y}{dx^2} \text{ when } x = 0$$

$$\frac{d^3}{dx^3}(0) = \frac{d^3y}{dx^2} \text{ when } x = 0$$

and so on

From successive lines above, we see that:

$$\frac{dy}{dx} = A \qquad \text{when } x = 0, \text{ so that } A_1 = \frac{d}{dx}(0)$$

$$\frac{d^2y}{dx^2} = 2A_2 \qquad \text{when } x = 0, \text{ so that } A_2 = \tfrac{1}{2} \cdot \frac{d^2}{dx^2}(0)$$

$$\frac{d^3y}{dx^3} = 3\,.\,2\,.\,A_3 \qquad \text{when } x = 0, \text{ so that } A_3 = \frac{1}{3\,.\,2} \cdot \frac{d^3}{dx^3}(0)$$

$$\frac{d^4y}{dx^4} = 4\,.\,3\,.\,2\,.\,A_4 \qquad \text{when } x = 0, \text{ so that } A_4 = \frac{1}{4\,.\,3\,.\,2} \cdot \frac{d^4}{dx^4}(0)$$

$$\frac{d^5y}{dx^5} = 5\,.\,4\,.\,3\,.\,2\,.\,A_5 \text{ when } x = 0, \text{ so that } A_5 = \frac{1}{5\,.\,4\,.\,3\,.\,2} \cdot \frac{d^5}{dx^5}(0)$$

If we now paint these values of A_0, A_1, etc., into the original power series we derive:

$$y = y_0 + x\,.\,\frac{d}{dx}(0) + \frac{x^2}{2!}\frac{d^2}{dx^2}(0) + \frac{x^3}{3!}\frac{d^3}{dx^3}(0) + \frac{x^4}{4!}\frac{d^4}{dx^4}(0) + \frac{x^5}{5!}\frac{d^5}{dx^5}(0)$$

In the domain of *continuous* functions, this series – that of Maclaurin – corresponds to that of Gregory for *discrete* ones; and the validity of the initial assumption depends only on: (*a*) the function being *differentiable*; (*b*) the derivative series being *convergent*. Let us first try it out on $y = \cos x$ on the assumption that we measure x in radians:

$$y = \cos x; \qquad y_0 = \cos 0 = 1$$

$$\frac{dy}{dx} = -\sin x; \qquad \frac{d(0)}{dx^2} = -\sin 0 = 0$$

$$\frac{d^2y}{d^2x} = -\cos x; \qquad \frac{d^2(0)}{dx^2} = -\cos 0 = -1$$

$$\frac{d^3y}{dx^3} = +\sin x; \qquad \frac{d^3(0)}{dx^3} = +\sin 0 = 0$$

$$\frac{d^4y}{dx^4} = +\cos x; \qquad \frac{d^4(0)}{dx^4} = +\cos 0 = +1$$

It is easy to see how this continues:

$$y = 1 - 0(x) - \frac{x^2}{2!} + \frac{0(x^3)}{3!} + \frac{x^4}{4!} - \frac{0(x^5)}{5!} + \frac{-x^6}{6!} \ldots, \text{etc.}$$

$$\therefore \quad \cos x = 1 - \frac{x^2}{2!} + \frac{x^4}{4!} - \frac{x^6}{6!} + \frac{x^8}{8!} \ldots, \text{etc.}$$

Similarly, we may tabulate for $y = \sin x$:

$$y_0 = \sin 0 = 0 \qquad \frac{d^3(0)}{dx^3} = -\cos 0 = -1$$

$$\frac{d(0)}{dx} = \cos 0 = 1 \qquad \frac{d^4(0)}{dx^4} = +\sin 0 = 0$$

$$\frac{d^2(0)}{dx^2} = -\sin 0 = 0 \qquad \frac{d^5(0)}{dx^5} = \cos 0 = +1$$

and so on

Thus we derive:

$$\sin x = x - \frac{x^3}{3!} + \frac{x^5}{5!} - \frac{x^7}{7!} \ldots, \text{etc.}$$

The series for $y = \cos x$ and $y = \sin x$ are meaningful only if they are convergent for a suitable range of x. All possible *numerical* values of either occur within the range $x = 0$ to $x = \pi/2$. Since $\pi/2 < 2$ radians, successive terms of both series after the initial one are smaller if $x < 2$ than if $x = 2$. So it suffices to show that the series are convergent when $x = 2$.

First let us note that the sum of the series for $\cos x$ must be less than that of the corresponding series of which all the signs are positive, i.e. it suffices to show that the following is convergent within the prescribed range:

$$1 + \frac{x^2}{2!} + \frac{x^4}{4!} + \frac{x^6}{6!} + \frac{x^8}{8!} \ldots$$

If we label the rank of successive terms (t_r) as 0, 1, 2, etc.:

$$t_r = \frac{x^{2r}}{(2r)!}; \quad t_{r+1} = \frac{x^{2r+2}}{(2r+2)!} = \frac{x^2 \cdot x^{2r}}{(2r+2)(2r+1)(2r)!},$$

$$\text{so that} \quad \frac{t_{r+1}}{t_r} = \frac{x^2}{(2r+2)(2r+1)}$$

Thus if $x = 2$ the ratio of the term of rank 4 to that of rank 3 is $4:56 = 1:14$; and that of every subsequent term is less than

1 : 14, so that the sum of the terms after the first two is less than that of the GP.

$$\frac{1}{14} + \frac{1}{14^2} + \frac{1}{14^3}, \text{ etc.}$$

In the same way, the reader can satisfy himself or herself that *sin* 2 is also convergent.

To find the value of sin x and cos x when $x = 1$ Radian (= 57° . 17′ . 45″) we write:

$$\sin 1 = 1 - \frac{1}{6} + \frac{1}{120} - \frac{1}{5040} + \frac{1}{362880} \cdots$$

$$\cos 1 = 1 - \frac{1}{2} + \frac{1}{24} - \frac{1}{720} + \frac{1}{37320} \cdots$$

If we take only the first four terms of each series:

$$\sin 1 = \frac{4241}{5040} = 0{\cdot}84125 \quad (\textit{correct value } 0{\cdot}84135)$$

$$\cos 1 = \frac{389}{720} = 0{\cdot}54028 \quad (\textit{correct value } 0{\cdot}54048)$$

It is thus clear that both series converge very rapidly even for large values of x.

With the aid of Maclaurin's theorem we can show that the binomial theorem holds good when $x < 1$ whether n is positive or negative, a whole number or a rational fraction. This we have shown to be true of the first derivative (nx^{n-1}) of $y = x^n$. Here we need consider only the case when $n = \frac{1}{p}$ and $r > 1$, p being an integer. To recognize the pattern, we need derive only the first four terms:

$$y = (1 + x)^n, \quad \textit{so that} \quad y_0 = 1$$

$$\frac{dy}{dx} = \frac{dy}{d(1 + x)} \cdot \frac{d(1 + x)}{dx} = n(1 + x)^{n-1}, \frac{d\,(0)}{dx} = n = \frac{1}{p}$$

$$\frac{d^2y}{dx^2} = n(n - 1)(1 + x)^{n-2}, \frac{d^2\,(0)}{dx^2} = n(n - 1) = \frac{1}{p}\left(\frac{1}{p} - 1\right)$$

$$\frac{d^3y}{dx^3} = n(n-1)(n-1)(1+x)^{n-3}, \text{ so that } \frac{d^3}{dx^3}(0)$$

$$= n(n-1)(n-2) = \frac{1}{p}\left(\frac{1}{p}-1\right)\left(\frac{1}{p}-2\right)$$

and so on

$$\therefore \quad y = 1 + \frac{1}{p}(x) + \frac{1}{p}\left(\frac{1}{p}-1\right)\frac{x^2}{2!} + \frac{1}{p}\left(\frac{1}{p}-1\right)\left(\frac{1}{p}-2\right)\frac{x^3}{3!}$$

$$+ \frac{1}{p}\left(\frac{1}{p}-1\right)\left(\frac{1}{p}-2\right)\left(\frac{1}{p}-3\right)\cdot\frac{x^4}{4!}\ldots, \text{etc.}$$

If we substitute $\frac{1}{p}$ for n throughout, this series, which is endless, is otherwise identical with the binomial expansion of $(1 + x)^n$ when n is an integer. It is meaningful only if we can show it to be convergent. To establish its convergence when $x \leqslant 1$, we recall expressions for the terms of rank r and $r + 1$ in the form:

$$t_r = \frac{n(n-1)(n-2)\ldots(n-r+1)}{r!}\cdot x^r$$

$$t_{r+1} = \frac{n(n-1)(n-2)\ldots(n-r+1)(n-r)}{(r+1)!}\cdot x^{r+1}$$

$$\therefore \quad \frac{t_{r+1}}{t_r} = \frac{(n-r).x}{r+1} = \frac{1-pr}{p+pr}\cdot x$$

Notice two things about this result:

(*a*) Since $p > 1$ by definition, $(1 - pr)$ is negative unless $r = 0$, i.e. the sign for t_{r+1} is negative if that of t_r is positive, and vice versa except for the first two terms t_0 and t_1, so that from t_1 onwards successive terms have *opposite* signs; (*b*) the quotient $(1 - pr) \div (p + pr)$ is numerically less than unity and t_{r+1} is for all values of r less than t_r, i.e. successive terms *diminish numerically*.

We have now all the clues to satisfy ourselves that our series for $(1 + x)^n$ is meaningful, in the sense that it adds up to nameable result, when $x < 1$ and n is a rational fraction less than unity. Series which consist of diminishing terms tending to zero in the limit do not necessarily choke off if it happens that the signs of successive terms are all positive; but they are necessarily convergent if alternate terms have opposite signs. For instance, of the two series below, (*a*) does not converge, but (*d*) does so:

$$(a)\ 1 + \frac{1}{2} + \frac{1}{3} + \frac{1}{4} + \frac{1}{5} + \frac{1}{6} \ldots$$

$$(b)\ 1 - \frac{1}{2} + \frac{1}{3} - \frac{1}{4} + \frac{1}{5} - \frac{1}{6} \ldots$$

To see that a series such as (*b*) does choke off, we may write the sum as

$$S = a - b + c - d + e - f + g - h \ldots$$
$$= (a - b) + (c - d) + (e - f) + (g - h) \ldots$$

If $a > b > c > d$, etc., each pair in brackets is positive, so that

$$S > (a - b)$$

We may also write:

$$S = a - (b - c) - (d - e) - (f - g) \ldots$$

On the same assumption, each pair in brackets is again positive, so that $S < a$. In short, we may say: $a > S > (a - b)$. In the space-saving symbolism of p. 280, we may write the expansion for $(1 + x)^n$ in the form:

$$1 + nx + n_{(2)}x^2 + n_{(3)}x^3 \ldots \text{etc.} = 1 + S;\ \text{so that}$$
$$S = nx + n_{(2)}x^2 + n_{(3)}x^3\ \text{etc.}$$

When $n = p^{-1}$ and $x < 1$, p being an integer as above, we have seen that successive terms of S diminish without end and have opposite signs so that:

$$nx > S > nx + n_{(2)}\,x^2$$

$$\therefore\ 1 + \frac{x}{p} > (1 + x)^n > 1 + \frac{x}{p} - \frac{(p - 1)x^2}{2p^2}$$

Thus the binomial series derived above satisfies the essential condition that it is convergent. As a numerical illustration, let $x = 0{\cdot}5 = n$, so that $p = 2$ and $(1 + x)^n = \sqrt{1{\cdot}5}$, then

$$\frac{x}{p} = 0{\cdot}25 \text{ and } \frac{(p - 1)\,.\,x^2}{p^2} = 0{\cdot}0625;\ \textit{whence}$$
$$1{\cdot}25 > \sqrt{1{\cdot}5} > 1{\cdot}1875$$

Tables of square roots cite $\sqrt{1{\cdot}5} > 1{\cdot}2247$.

Logarithmic Graph Paper. Graph paper which has the *Y*-axis graduated like a slide rule in intervals corresponding to the logarithms of the numbers which label the division are now widely

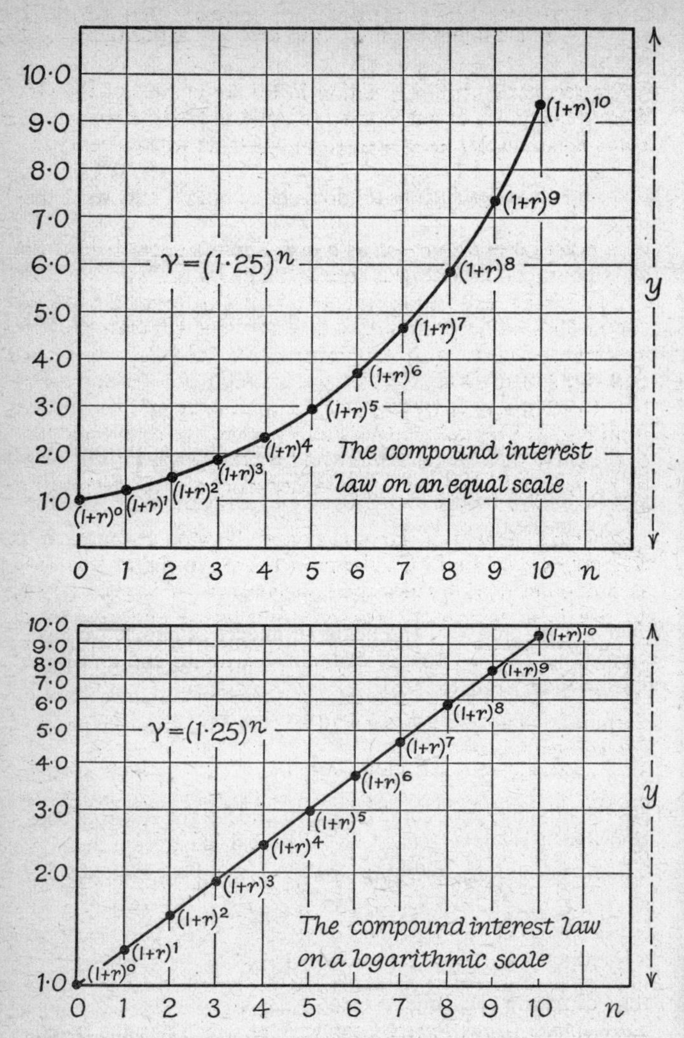

Fig. 176. The Logarithmic Scale Converts an Exponential Curve into a Straight Line

in circulation. The base is immaterial, because logarithms to one base are a simple multiple of logarithms to any other as disclosed by the formula (p. 416):

$$\log_b N = \log_b a \,.\, \log_a N$$

Since the choice of the initial interval (*$\log_a 2$ in Fig. 176*) is arbitrary, the base is thus irrelevant to the ruling of the lines. As we see from Fig. 176, plotting on such a grid converts the exponential curve representing the accumulation of capital at compound interest into a straight line. This is because:

$$\log (1 + r)^n = n \log (1 + r)$$

If therefore we write $x = (1 + r)$, and $y = (1 + r)^n$, $log\ y = n\ log.\ x$. Thus plotting $log\ y$ instead of y on the Y-axis yields a straight line whose slope is n passing through the point $log\ y = 1{\cdot}0$ and $x = 0$.

Of several uses for such paper, the following depends on considerations which the process of differentiation can clarify. Statistics which record declining infant mortality will here serve to illustrate it. When referring to the rate at which a statistic *itself a rate* changes, it is essential to distinguish between two notions. If the infant death rate falls in one year from 150 to 140 *per mille* (i.e. per thousand), the *actual* decrement is 10 per mille or 1 per cent. If it changes from 20 to 10 per mille, the actual decrement is likewise 10 per mille or 1 per cent. Now a decline of 1 per cent from the 20 per mille level if regarded as a percentage of the latter, is a drop to one-half of the previous figure, i.e. a *proportionate* decrement of 50 per cent, whereas a drop of 1 per cent from the 15 per cent (150 per mille) level is a proportionate drop of only one-fifteenth (6·6 per cent). If we record in graphical form statistics of social units for which different levels obtain, we may thus gain totally different impressions from scrutinizing as the alternative co-ordinate to time:

(*a*) the actual rates at the end of each interval;
(*b*) the ratio of the increment or decrement to its initial value in the interval (e.g. year) to which our crude figures are referable.

At the 200 per mille level of infant deaths to be found in some countries a drop to 170 per mille signifies an actual change of 3 per cent which is not outside the range of secular variation in a single year. At the level of 25 per thousand or less (as in most Western European countries), an actual drop of 3 per cent would be an impossible event, since a 2·5 per cent fall could occur only if every live-born infant survived its first year. Evidently, therefore, it is not very profitable to compare the actual decrement of one such statistic with that of another. The more proper procedure when comparing the rate of change of a statistic $x_{a.n}$ with that of another $x_{b.n}$ is to use the *proportionate* decrement or increment of each.

For the time interval $\Delta t = t_{n+1} - t_n$ in successive years, these are by definition:

$$X_{a.n} = \frac{x_{a.n+1} - x_{a.n}}{x_{a.n}} \qquad X_{b.n} = \frac{x_{b.n+1} - x_{b.n}}{x_{b.n}}$$

The corresponding rates of proportionate change which we may thus regard as a proper measure of the tempo of change are identical with the above if $\Delta t = 1$, e.g. if our time unit is one year. For comparative purposes, we may sidestep the labour of calculating $X_{a.n}$, etc., from our crude data $x_{a.n}$, etc., as above by considering the change as continuous. This is not illegitimate, since the crude figures merely record a persistent process at intervals convenient for purposes of observation. On this understanding, the rate of proportionate change of any such statistic is equivalent to the gradient of its logarithm plotted against time. By plotting the crude statistic $x_{a.n}$ etc., on semi-logarithmic paper, we therefore accomplish the end in view without recourse to computing the derivative statistics $X_{a.n}$, etc.

While this is common knowledge, the rationale of the procedure may not be obvious to every reader, and a formal demonstration may be welcome to some. As explained, the actual change $(x_2 - x_1)$ of $x = f(t)$ in a finite interval $(t_2 - t_1)$ corresponds by definition to a proportionate change:

$$y = \frac{x_2 - x_1}{x_1}$$

During a very short interval Δt, we may represent the corresponding change of the rate under consideration as from x to $x + \Delta x$. The proportionate change in the interval is therefore:

$$\Delta y = \frac{(x + \Delta x) - x}{x} = \frac{\Delta x}{x} \quad \textit{and} \quad \frac{\Delta y}{\Delta x} \cdot \frac{\Delta x}{\Delta t} = \frac{1}{x} \cdot \frac{\Delta x}{\Delta t}$$

Thus we may write approximately for the *rate of proportionate change* which is the slope of y plotted against t in the form:

$$\frac{dy}{dt} = \frac{1}{x} \cdot \frac{dx}{dt}$$

Let us now put $z = log\ x$, so that:

$$\frac{dz}{dt} = \frac{d \log x}{dt} = \frac{d \log x}{dx} \cdot \frac{dx}{dt} = \frac{1}{x} \cdot \frac{dx}{dt}$$

Whence we see that:

$$\frac{dy}{dt} = \frac{dz}{dt} = \frac{d \log x}{dt}$$

Thus the rate of change of the logarithm of the crude statistic x is equal to its rate of proportionate change as defined above.

Exercises on Chapter 10

1. Use the series for $\log_e (1 + x)$ to find $\log_e 10$, $\log_e 2$, $\log_e 3$, $\log_e 4$, $\log_e 5$.

Hence, make a table of $\log_{10} 2$, $\log_{10} 3$, $\log_{10} 4$, $\log_{10} 5$.

2. Find π by infinite series correct to three places of decimals by two different methods.

3. Draw very accurately the graph of:

$$y = \tfrac{1}{5}x^2$$

Measure its slope at three points corresponding with three different values of x, and, using a table of tangents, compare your measurements with the values calculated by substituting the appropriate numbers in the differential coefficient $\frac{2x}{5}$.

Find by counting squares the area enclosed by the curve, the x axis, and the y ordinates at the points $x = 5$ and $x = 10$, and compare this with the area calculated from the integral:

$$\int_5^{10} \tfrac{1}{5}x^2 \,.\, dx = \left[\frac{x^3}{15}\right]_5^{10} = \left[\frac{10^3}{15} - \frac{5^3}{15}\right]$$

4. Draw the graph of $y = \sqrt{36 - x^2}$.

Find $\frac{dy}{dx}$ when $x = 1, 2, -2$.

Draw the tangents and compare with the results obtained by differentiation.

5. In the same way draw the graph of $y = \frac{1}{x}$ between $x = 0$ and $x = 4$, and find $\frac{dy}{dx}$ when $x = 1, 2$.

6. Find from first principles (i.e. by applying the methods of pp. 473–9) $\frac{dy}{dx}$ when $y = x + \frac{1}{x}$.

7. Write down $\frac{dy}{dx}$ when

$$y = x^{3 \cdot 6}, \quad 5\sqrt{x}, \quad x^{-1\frac{1}{2}}, \quad \sqrt{x^7}, \quad \frac{3}{\sqrt[3]{x^{-\frac{1}{3}}}}$$

8. What is $\frac{dy}{dx}$ when

$$y = x^5 - 5x^3 + 5x^2 - 4x + 3$$

9. If $pv = k$, k being a constant, show that:

$$\frac{dp}{dv} = -\frac{p}{v}$$

10. Find the turning points of the curve $y = x^3 - 3x$ and plot the curve, showing these points.

11. Find the greatest possible volume of a cylindrical parcel when the length and girth together must not be more than 6 feet.

12. In a dynamo x is the weight of the armature and y the weight of the rest. The cost of running (c) is given by:

$$c = 10x + 3y$$

The power is proportional to xy. If the cost is fixed, find the relation between x and y so as to obtain the maximum power.

13. If the boundary of a rectangle is of fixed length $2L$, one side may be written x, and the adjacent side is $(L - x)$. If the area is a, $a = x(L - x)$. This is a maximum when $\frac{da}{dx} = 0$. Hence, show that the square is the rectangle of fixed perimeter having the greatest area.

Prove that the greatest rectangle that can be inscribed in a circle is a square.

14. Write down the values of y corresponding with the following values of $\frac{dy}{dx}$:

(i) $4x^3$ (ii) $\frac{3}{x}$ (iii) $\frac{x^n}{4}$

(iv) $\sqrt{x}$ (v) $3x^2 + 2x + 1$

15. What values of y correspond with the following values of $\frac{d^2y}{dx^2}$?

(i) $2x$ (ii) 5 (iii) $\sqrt{x}$

16. If $\frac{dp}{dv} = \frac{-700}{v^{2 \cdot 4}}$ and $p = 18{\cdot}95$ when v is 20, express p as a function of v.

17. Write down the values of $\frac{dy}{dx}$ when y has the values

(i) $\cos a^2x$ (ii) $4 \sin 3x$ (iii) $a \sin nx + b \cos nx$

18. Draw a graph of $y = \tan x$ from $x = 0$ to $x = 1{\cdot}2$. Show by the method of p. 474 that when $y = \tan x$, $\frac{dy}{dx} = \sec^2 x$. Verify this from your graph

19. Draw the graph of $y = e^x$, making the y unit one-tenth the x unit. At any point P on the curve, draw PM perpendicular to the axis, meeting it at M. Take a point T one x unit to the left of M.

Show that PT is a tangent to the curve at P.

20. Find by the method of p. 490 the area bounded by $y = 4x + 3$, the x axis, and the y ordinates at the points (i) $x = 4$, $x = 8$, (ii) $x = 2$, $x = 10$, (iii) $x = 5$, $x = 6$.

21. Find the area bounded by $y = 2x^2 + 3x + 1$, $y = 0$, $x = 3$, $x = 7$.

22. Write down the values of the following integrals and check by differentiation before substituting numbers for x:

(i) $\int_2^7 (2x^2)dx$ (ii) $\int_{-1}^1 (ax^2 + bx + c)dx$ (iii) $\int_1^8 \frac{1}{\sqrt[3]{x}}dx$

(iv) $\int_{-3}^5 7dx$ (v) $\int_1^2 \left(x + \frac{1}{x^2}\right)dx$

23. In surveying, the area of a plot bounded by a closed curve is sometimes found by Simpson's rule. The rule is: Divide the area into an even number of strips of equal width by an odd number of ordinates; the area is approximately:

$\frac{1}{3}$. Width of a strip × {sum of extreme ordinates
+ twice sum of other odd ordinates
+ 4 times sum of even ordinates}

Assuming that the curve bounding the area can be described by a curve of the type $y = p + qx + rx^2 + sx^3$, see whether you can justify Simpson's rule as an approximation to the area.

Find the area bounded by $y = 0$, $x = 2$, $x = 10$, and the curve $y = x^4$:

(*a*) by Simpson's rule using three ordinates:
(*b*) by Simpson's rule using nine ordinates;
(*c*) by integration.

24. Find the area bounded by $y = x^3 - 6x^2 + 9x + 5$, the x axis, and the maximum and minimum ordinates.

25. The speed v of a body at the end of t seconds is given by:

$$v = u + at$$

Show that the distance travelled in t seconds is:

$$ut + \tfrac{1}{2}at^2$$

26. Find the work done in the expansion of a quantity of steam at 4,000 lb per square foot pressure from 2 cubic feet to 8 cubic feet. Volume and pressure are connected by the equation:

$$pv^{0\cdot 9} = K \text{ (constant)}$$

27. Find the volume of a cone with radius 5 inches, height 12 inches.

28. In a sphere of radius 12 inches find the volume of a slice contained between two parallel planes distant 3 inches and 6 inches from the centre.

29. Find $\int_0^{\pi} \sin x dx \qquad \int_0^{\frac{\pi}{2}} \cos x dx$

30. Write down:

(i) $\int_0^{\frac{\pi}{2}} \sin 2x dx$ (ii) $\int_{-\frac{\pi}{2}}^{\frac{\pi}{2}} \cos 2x dx$ (iii) $\int_0^{\pi} x \sin x dx$

31. Sometimes a function of x can be recognized as the product of two simpler ones, e.g. $y = x^2 \log x$.

Suppose y can be put in the form uv, where u and v are each simpler functions of x. When u becomes $u + \Delta u$ and v becomes $v + \Delta v$, y becomes $y + \Delta y$; hence, show that $\frac{dy}{dx} = u\frac{dv}{dx} + v\frac{du}{dx}$.

Check this formula by differentiating x^7 and x^5 in the ordinary way, and then by representing them in the form:

$$x^7 = x^5 \times x^2 \quad \text{and} \quad x^5 = x^7 \times x^{-2}$$

In this way differentiate:

(i) $x \sin x$ (ii) $\cos x \tan x$ (iii) $(2x^2 + x + 3)(x + 1)$

first as a product and then by multiplying out.

32. By using the same method as in the previous example, show that if $y = \frac{u}{v}$ when u and v are functions of x, and v is not zero:

$$\frac{dy}{dx} = \frac{v\frac{du}{dx} - u\frac{dv}{dx}}{v^2}$$

Differentiate: (i) $\frac{x}{x + 1}$ (ii) $\frac{\sin x}{x}$ (iii) $\frac{1}{\cos x}$ (iv) $\tan x$

33. Sometimes a function of x (e.g. $\cos^2 x$) can be recognized as a function of a simpler function (in this case $\cos x$) of x. Show that if y is a function of u and u is a simpler function of x, $\frac{dy}{dx} = \frac{dy}{du} \cdot \frac{du}{dx}$.

This can be used to differentiate expressions like the following:

$$y = \cos^2 x$$

$$\frac{dy}{dx} = \frac{d(\cos^2 x)}{d \cos x} \cdot \frac{d \cos x}{dx}$$

$$= 2 \cos x \times (-\sin x)$$

$$= -2 \sin x \cos x$$

Check this formula by using it to differentiate $\log_e x^3$ written as a function of x^3 and $\log_e x^3$ written as $3 \log_e x$.

In this way differentiate:

(i) $\sqrt{\sin x}$ (iii) $\sin (ax + b)$

(ii) $(ax + b)^n$ (iv) $\log_e (ax^2 + bx + c)$

34. If $y = A \cos x + B \sin x$, show that $\frac{d^2y}{dx^2} + y = 0$.

35. Solve the following equations:

(i) $\frac{d^2y}{dx^2} + 4y = 0$ (ii) $\frac{d^2y}{dx^2} - 4y = 0$

Find y in terms of x in each case if $y = 5$ and $\frac{dy}{dx} = 4$ when $x = 0$.

Things to Memorize

1.

y	$\frac{dy}{dx}$
$ax^n + b$	nax^{n-1}
$a^x + b$	$(\log_e a)a^x$
$a \log_e (x + b) + c$	$\frac{a}{x + b}$
$\sin (ax + b)$	$a \cos (ax + b)$
$\cos (ax + b)$	$- a \sin (ax + b)$
e^x	e^x

2.

y	$\int_p^q y \, . \, dx$
$\frac{a}{b + x}$	$a \log_e \frac{(b + q)}{(b + p)}$
ax^n	$\frac{a}{n + 1}(q^{n+1} - p^{n+1})$
$\cos ax$	$\frac{1}{a}(\sin aq - \sin ap)$
e^{cx}	$\frac{1}{c}(e^{cq} - e^{cp})$

Volume of cylinder $= \pi r^2 h$

Volume of sphere $= \frac{4}{3}\pi r^3$

CHAPTER 11

THE ALGEBRA OF THE CHESSBOARD

IN CHAPTER 6, we have seen that the superiority of the Hindu–Arabic number symbols depends solely on two features: (*a*) assigning, in accordance with the use of the abacus, a *position* to every power of the base in an orderly succession; (*b*) using a symbol (*zero*) for the empty column. The ten symbols 0, 1, 2 . . . 8, 9 then suffice to specify any finite number, however large, and any finite fraction; but this is not the only or the chief advantage of exploiting the principle of position. It also makes explicit the operations we can perform with the abacus, and permits us to perform them without the need to handle it.

The representation of a number in the Hindu–Arabic notation relies on the principle of position in one dimension only, i.e. a lay-out exhibiting a straight line power series; but an algorithm, i.e. rule of computation, implicitly employs the same principle in a 2-dimensional set-up. For instance, we perform the first step in the multiplication of 4261 by 315 in accordance with the following chessboard (row and column) lay-out:

3.4	3.2	3.6	3.1	. .	. .
. .	1.4	1.2	1.6	1.1	. .
. .	. .	5.4	5.2	5.6	5.1

Alternatively, we may represent it as:

. .	. .	5.4	5.2	5.6	5.1
. .	1.4	1.2	1.6	1.1	. .
3.4	3.2	3.6	3.1	. .	. .

In such a chessboard scheme, the *column*-position of any product such as 5.4 or 3.1 tells us that it is the coefficient of a particular power of ten, and the *row*-position of any such products records an instruction to assign such a coefficient to its correct column with that end in view. *Matrix* is the technical name for such a grid or chessboard lay-out of symbols to facilitate memorization of a sequence of operations and to economize the use of symbols by assigning a particular meaning to any one of them in virtue of the cell it occupies in the grid. A branch of algebra which relies on this device of cell specification is a matrix algebra. The rules of such an algebra depend solely on the end in view.

In the seventeenth and eighteenth centuries the search for, and the study of the properties of, power series which simplify the preparation of tables of logarithms and trigonometrical ratios reaped a rich harvest from the recognition of the principle of position in one dimension. In the Western World, the invention of algebras which exploit the same principle by recourse to a 2-dimensional grid begins in the nineteenth century; and the earliest to prove its worth is the algebra of *determinants* which is first and foremost a scheme for solving sets of simultaneous equations.

As such, determinants are a programme for mechanically solving sets of equations involving several variables; and as such the Chinese were familiar with the basic principle several centuries before determinants came into use in the Western World. Possibly, this had some connection with their partiality for magic squares, which will indeed provide us later with opportunities for exercise in their evaluation. To understand the use of determinants as a labour-saving device for the solution of linear simultaneous equations involving a considerable number of variables, let us first recall what we do when we solve by *elimination* a pair of equations such as the following: $3x = 5y + 4$ and $4y - 3x = -2$. It will then suffice to show how a *grid* recipe helps us to solve a set of linear equations involving three variables (x, y, z). If the journey seems up-hill and wearisome, the reader may take comfort from the fact that one starts from scratch, or nearly so.

To solve by elimination two linear equations, involving only the two variables x and y, e.g. $3x = 5y + 4$ and $4y - 3x = -2$, our first step is to set them out with corresponding terms in the same position:

$$\begin{aligned} 3x - 5y &= 4 \\ -3x + 4y &= -2 \end{aligned} \qquad \textit{or} \qquad \begin{aligned} 3x - 5y - 4 &= 0 \\ -3x + 4y + 2 &= 0 \end{aligned}$$

The order we choose is immaterial, if we stick to it; and we shall here adopt as our standard pattern for simultaneous linear equations involving two variables:

$$\begin{aligned} ax + by + c &= 0 \\ dx + ey + f &= 0 \end{aligned}$$

The most elementary procedure we can adopt is to reduce such a pair of equations each with two unknown terms to one equation involving only one unknown term. To do so, we either eliminate x by multiplying each term of one equation by the coefficient of x in the other and vice versa, or eliminate y by multiplying each term of one equation by the coefficient of y in the other and vice versa. The steps are as follows:

$$adx + bdy + cd = 0$$
$$adx + aey + af = 0$$
$$\therefore \quad (bd - ae)y = (af - cd)$$
$$\therefore \; -y = \frac{af - cd}{ae - bd}$$

$$aex + bey + ce = 0$$
$$bdx + bey + bf = 0$$
$$\therefore \; (ae - bd)x = (bf - ce)$$
$$\therefore \; x = \frac{bf - ce}{ae - bd} \cdot \quad \text{(i)}$$

Let us now consider how we proceed by elimination, if we have to deal with sets of equations involving more than two variables as below:

$$\begin{aligned} 3x + 5y + z &= 16 \\ x - 2y + 3z &= 6 \\ 2x + 2y + 4z &= 18 \end{aligned}$$

$$\begin{aligned} u + 2v + 3w + z &= 18 \\ 2u + 3v + 4w - 3z &= 8 \\ 3u + 2v - w + 4z &= 16 \\ 4u - 5v + 2w + z &= 4 \end{aligned}$$

We first get rid of one variable in each of any two pairs of equations. In the set on the left, we may choose:

(a) $x - 2y + 3z = 6$ *and* $2x + 2y + 4z = 18$

Hence, by addition: $3x + 7z = 24$:

(b) $2(3x + 5y + z) = 32 = 6x + 10y + 2z$
$5(2x + 2y + 4z) = 90 = 10x + 10y + 20z$

Hence by subtraction: $4x + 18z = 58$. We now have two equations involving x and z only, i.e. $3x + 7z = 24$ and $4x + 18z = 58$, so that $12x + 28z = 96$ and $12x + 54z = 174$, whence, $z = 3$. By substitution for z, $x = 1$. By substituting these values in any of the three original equations (e.g. in the first $3 + 5y + z = 16$) we obtain $y = 2$.

To deal with a set involving the four variables u, v, w, z, as on the right above, we first have to eliminate one of them in each of three pairs to derive three equations in three variables and then proceed as for the three-fold left-hand set to derive values for two of them. The reader who does so as an exercise will begin to realize how formidably laborious the method of elimination becomes, when we have to deal with more than three variables.

This consideration invites us to explore the possibility of a speedier method.

Let us therefore now recall the formulae above, for solution of equations involving only two variables, i.e. for $ax + by + c = 0 = dx + ey + f$:

$$x = \frac{bf - ce}{ae - bd} \quad \textit{and} \quad -y = \frac{af - cd}{ae - bd} \quad .$$

These formulae (i) furnish a computing schema which dispenses with the foregoing steps of elimination of the individual variables; but they suffer from the disadvantage that they are not easily memorizable. To sidestep this, we first note that the denominator of both x and $-y$ is the difference between cross products of coefficients of x and y, and we can visualize this, if we write it in the form:

$$\begin{vmatrix} a & b \\ d & e \end{vmatrix} = (ae - bd)$$

The numerator of x involves the difference between cross products of constants *other than the coefficients of* x, and we can visualize it in accordance with the same convention of subtracting the *right to-left-downwards* from the *left to-right-downwards* diagonal product as before:

$$\begin{vmatrix} b & c \\ e & f \end{vmatrix} = (bf - ce)$$

Similarly, the numerator of $-y$ involves the difference between cross products of constants other than the y-coefficients, and we visualize it accordingly as:

$$\begin{vmatrix} a & c \\ d & f \end{vmatrix} = (af - cd)$$

In this notation, we write:

$$x = \frac{\begin{vmatrix} b & c \\ e & f \end{vmatrix}}{\begin{vmatrix} a & b \\ d & e \end{vmatrix}} \quad ; \quad -y = \frac{\begin{vmatrix} a & c \\ d & f \end{vmatrix}}{\begin{vmatrix} a & b \\ d & e \end{vmatrix}} \qquad \text{(ii)}$$

We here have a more memorizable pattern than the formulae of (i); but we can improve on it by recourse to subscript notation in either of two ways. One is to write our standard pattern in the form:

$$a_1x + b_1y + C_1 = 0$$
$$a_2x + b_2y + C_2 = 0$$

The rule for solution in determinant notation then takes the form:

$$x = \frac{\begin{vmatrix} b_1 & C_1 \\ b_2 & C_2 \end{vmatrix}}{\begin{vmatrix} a_1 & b_1 \\ a_2 & b_2 \end{vmatrix}} \quad ; \quad -y = \frac{\begin{vmatrix} a_1 & C_1 \\ a_2 & C_2 \end{vmatrix}}{\begin{vmatrix} a_1 & b_1 \\ a_2 & b_2 \end{vmatrix}}$$

It is still more economical to write our standard pattern in the form:

$$a_{11}x + a_{12}y + C_1 = 0$$
$$a_{21} + a_{22}y + C_2 = 0$$

The solution then takes the form:

$$x = \frac{\begin{vmatrix} a_{12} & C_1 \\ a_{22} & C_2 \end{vmatrix}}{\begin{vmatrix} a_{11} & a_{12} \\ a_{21} & a_{22} \end{vmatrix}} \quad ;- y = \frac{\begin{vmatrix} a_{11} & C_1 \\ a_{21} & C_2 \end{vmatrix}}{\begin{vmatrix} a_{11} & a_{12} \\ a_{21} & a_{22} \end{vmatrix}}$$

We may now write the determinants themselves in a more compact form by recourse to the generalized symbol *r* for the row:

$$\begin{vmatrix} a_{12} & C_1 \\ a_{22} & C_2 \end{vmatrix} = [a_{ry} \ldots C_r] = a_{12} \,.\, C_2 - a_{22} \,.\, C_1$$

$$\begin{vmatrix} a_{11} & C_1 \\ a_{21} & C_2 \end{vmatrix} = [a_{rx} \ldots C_r] = a_{11} \,.\, C_2 - a_{21} \,.\, C_1$$

$$\begin{vmatrix} a_{11} & a_{12} \\ a_{21} & a_{22} \end{vmatrix} = [a_{rx} \ldots a_{ry}] = a_{11} \,.\, a_{22} - a_{21} \,.\, a_{12}$$

In this more compact notation we then have:

$$a_{11}x + a_{12} + C_1 = 0 \text{ when } \quad x = [a_{ry} \ldots C_r] \div [a_{rx} \ldots a_{ry}]$$
$$a_{21}x + a_{22}y + C_2 = 0 \qquad -y = [a_{rx} \ldots C_r] \div [a_{ry} \ldots a_{rx}]$$

In all this, we have introduced no new rules, merely a new programme as a visual aid to memorizing a method of solution without going through the elementary procedure step by step. The word *determinant* is simply a name for the symmetrical cross product pattern here written in full as a 2×2 set. When we know the numerical value of the cell elements we interpret it, as above, e.g.:

$$\begin{vmatrix} 3 & 4 \\ 2 & 5 \end{vmatrix} = 3(5) - 4(2) = 7$$

Having memorized the pattern, we can proceed to solve a simultaneous equation thus:

$$4x + 5y = 2 \qquad \therefore \quad 4x + 5y - 2 = 0$$
$$3x + 4y = 1 \qquad \therefore \quad 3x + 4y - 1 = 0$$

$$x = \frac{\begin{vmatrix} 5 & -2 \\ 4 & -1 \end{vmatrix}}{\begin{vmatrix} 4 & 5 \\ 3 & 4 \end{vmatrix}} \qquad -y = \frac{\begin{vmatrix} 4 & -2 \\ 3 & -1 \end{vmatrix}}{\begin{vmatrix} 4 & 5 \\ 3 & 4 \end{vmatrix}}$$

$$\therefore \quad x = \frac{5(-1) - 4(-2)}{4(4) - 3(5)} \qquad -y = \frac{4(-1) - 3(-2)}{4(4) - 3(5)}$$

$$\therefore \quad x = \frac{-5 + 8}{16 - 15} = 3 \qquad -y = \frac{-4 + 6}{16 - 15} = 2$$

$$\therefore \quad x = 3 \quad \text{and} \quad y = -2$$

We can make the rule of solution more explicit, if we set out our standard double subscript pattern as a 2×3 matrix:

$$\begin{Vmatrix} a_{11} & a_{12} & C_1 \\ a_{21} & a_{22} & C_2 \end{Vmatrix}$$

From this matrix we can make three determinants of two rows and two columns, viz.:

$$[a_{ry} \ldots C_r] \equiv D(x)$$
$$[a_{rx} \ldots C_r] \equiv D(y)$$
$$[a_{rx} \ldots a_{ry}] \equiv D(C)$$

In this notation:

$$x = D(x) \div D(C) \quad and \quad -y = D(y) \div D(C)$$

We may write this as:

$$\frac{x}{D(x)} = \frac{-y}{D(y)} = \frac{1}{D(C)} \quad . \quad . \quad . \quad . \quad . \quad . \quad \text{(iii)}$$

Practise the use of this formula by making up equations the solutions of which you can check. You may then ask what, if any, benefit we derive from a compact rule such as this for an operation so elementary as the solution of simultaneous equations involving only two unknown quantities. The answer is none, except in so far as we have found a clue to the construction of a *code* for simplifying the laborious solution of simultaneous equations from which it would otherwise be necessary to eliminate successively *many* variables. To follow up our clue, we first note the alternation of signs in (iii); and proceed to write in the same form as (iii) the rule for the solution of a set of linear equations involving three variables, set out in accordance with our standard pattern as follows:

$$ax + by + cz + d = 0$$
$$ex + fy + gz + h = 0$$
$$jx + ky + lz + m = 0$$

The 3 × 4 matrix of the set is:

$$\begin{Vmatrix} a & b & c & d \\ e & f & g & h \\ j & k & l & m \end{Vmatrix} \equiv [a_{rx} \ldots a_{ry} \ldots a_{rz} \ldots C_r]$$

In the notation of (iii) we have:

$$D(x) \equiv [a_{ry} \ldots a_{rz} \ldots C_r]$$

Without prejudging how we are to interpret the meaning of this 3 × 3 determinant, we merely recall at this stage the form it will have, if we can find a consistent code for equations involving any number of variables, i.e.:

$$D(x) \equiv \begin{vmatrix} b & c & d \\ f & g & h \\ k & l & m \end{vmatrix} \equiv [a_{ry} \ldots a_{rz} \ldots C_r]$$

Similarly:

$$D(C) \equiv [a_{rx} \ldots a_{ry} \ldots a_{rz}] \equiv \begin{vmatrix} a & b & c \\ e & f & g \\ j & k & l \end{vmatrix}$$

We now write out the rule of solution we anticipate in accordance with (iii), viz.:

$$\frac{x}{D(x)} = \frac{-y}{D(y)} = \frac{z}{D(z)} = \frac{-1}{D(C)} \quad . \quad . \quad . \quad \text{(iv)}$$

Accordingly, we have:

$$-z = \begin{vmatrix} a & b & d \\ e & f & h \\ j & k & m \end{vmatrix} \div \begin{vmatrix} a & b & c \\ e & f & g \\ j & k & l \end{vmatrix} \quad . \quad . \quad \text{(v)}$$

This pattern, which expresses each variable as a quotient of *third* order (3 × 3) determinants, is useless until we can give the latter a meaning. To do so, we must examine the solution of our three equations by the rule of thumb method. To eliminate x in (i) and (ii), we put

$$aex + bey + cez + de = 0$$
$$aex + afy + agz + ah = 0$$
$$(be - af)y + (ce - ag)z + (de - ah) = 0 \quad . \quad . \quad \text{(vi)}$$

In the same way, we can get from (ii) and (iii):

$$(fj - ek)y + (gj - el)z + (hj - em) = 0 \quad . \quad . \quad . \quad \text{(vii)}$$

We now eliminate **y** from our three standard pattern equations in the usual way obtaining:

$$z = \frac{(be - af)(hj - em) - (fj - ek)(de - ah)}{(fj - ek)(ce - ag) - (be - af)(gj - el)} \text{ . (viii)}$$

$$= \frac{bhj - bem + afm - dfj + edk - ahk}{cfj - eck + agk - bgj + bel - afl}$$

$$= \frac{a(fm - hk) - b(em - hj) + d(ek - fj)}{a(gk - fl) - b(gj - el) + c(fj - ek)} \quad . \quad \text{(ix)}$$

It is now evident that the numerator of (ix) contains the same set of elements as the third-order determinant (v) in the numerator for z expressed above as the quotient of 3×3 determinants; and the denominator of (ix) contains the same set of elements as the third-order determinant (v) in the corresponding denominator. We shall now be in a position to interpret these two third-order determinants, if we rewrite (ix) as follows:

$$-z = \frac{a(fm - hk) - b(em - hj) + d(ek - fj)}{a(fl - gk) - b(el - gj) + c(ek - fj)} \text{ . (x)}$$

By comparing (x) with (v), we now see that our definition of a 3×3 determinant must signify:

$$\begin{vmatrix} a & b & d \\ e & f & h \\ j & k & m \end{vmatrix} = a(fm - hk) - b(em - hj) + d(ek - fj)$$

$$= a\begin{vmatrix} f & h \\ k & m \end{vmatrix} - b\begin{vmatrix} e & h \\ j & m \end{vmatrix} + d\begin{vmatrix} e & f \\ j & k \end{vmatrix} \text{ . (xi)}$$

Likewise:

$$\begin{vmatrix} a & b & c \\ e & f & g \\ j & k & l \end{vmatrix} = a(fl - gk) - b(el - gj) + c(ek - fj)$$

$$= a\begin{vmatrix} f & g \\ k & l \end{vmatrix} - b\begin{vmatrix} e & g \\ j & l \end{vmatrix} + c\begin{vmatrix} e & f \\ j & k \end{vmatrix} \text{ . (xii)}$$

Evidently, the two 3×3 determinants are each reducible to three 2×2 determinants in conformity with one rule, which we may schematize thus:

$$\begin{vmatrix} k_{11} & k_{12} & k_{13} \\ k_{21} & k_{22} & k_{23} \\ k_{31} & k_{32} & k_{33} \end{vmatrix} = k_{11}\begin{vmatrix} k_{22} & k_{23} \\ k_{32} & k_{33} \end{vmatrix} - k_{12}\begin{vmatrix} k_{21} & k_{23} \\ k_{31} & k_{33} \end{vmatrix} + k_{13}\begin{vmatrix} k_{21} & k_{22} \\ k_{31} & k_{32} \end{vmatrix}$$

$$= k_{11}\begin{vmatrix} k_{22} & k_{23} \\ k_{32} & k_{33} \end{vmatrix} - k_{21}\begin{vmatrix} k_{12} & k_{13} \\ k_{32} & k_{33} \end{vmatrix} + k_{31}\begin{vmatrix} k_{12} & k_{13} \\ k_{22} & k_{23} \end{vmatrix}$$

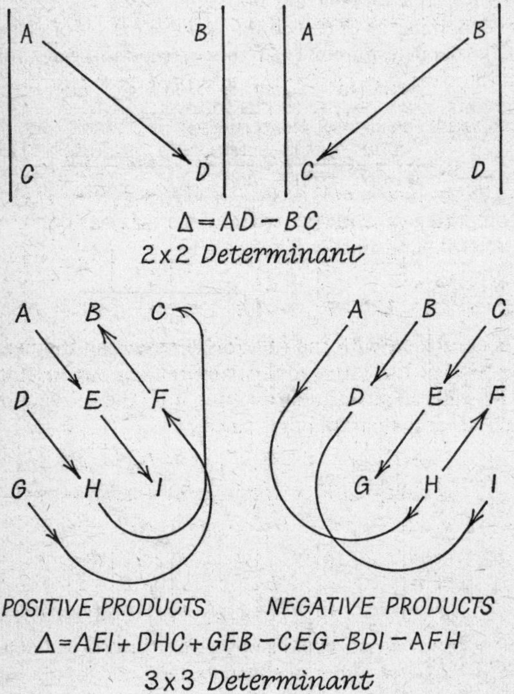
A B C D
A B C D
Δ=AD−BC
2x2 Determinant
A B C D E F G H I
A B C D E F G H I
POSITIVE PRODUCTS
NEGATIVE PRODUCTS
Δ=AEI+DHC+GFB−CEG−BDI−AFH
3x3 Determinant

Fig. 177. The Pattern Common to Determinants of the Second and Third Order

One speaks of the 2×2 determinants in this context as *minors* and of the factors by which we have to multiply them as their *co-factors.*

This interpretation of the third-order (3×3) determinant does not obviously tie up with the evaluation of second-order (2×2) determinant; but it does so if we look at it from a different point of view. The positive term of the 2×2 determinant is the product of the *left to right – downwards* entries, the negative term being that of the *right to left – downwards.* Let us dissect the 3×3 determinant of Fig. 177 on the comparable assumption that we regard as: (*a*) *positive* each oblique three-fold product of entries starting on the left; (*b*) *negative* each oblique three-fold product starting on the right.

We may then assemble our terms as in Fig. 177:

Positive	aei	dhc	gfb
Negative	ceg	fha	idb

We can write the sum of the terms as:

$$\begin{aligned} & aei + dhc + gfb - ceg - fha - idb \\ &= (aei - afh) - (idb - dhc) + (gfb - ceg) \\ &= a(ei - fh) - d(bi - ch) + g(bf - ce) \\ &= a\begin{vmatrix} e & f \\ h & i \end{vmatrix} - d\begin{vmatrix} b & c \\ h & i \end{vmatrix} + g\begin{vmatrix} b & c \\ e & f \end{vmatrix} \end{aligned}$$

This is consistent with the rule for representing the determinant by the left vertical terms and corresponding *minors* (see below). It is also consistent with representing it by the product of the top row terms and their minors, since:

$$\begin{aligned} & aei + dhc + gfb - ceg - fha - idb \\ &= (aei - afh) - (bdi - gfb) + (dhc - ceg) \\ &= a(ei - fh) - b(di - fg) + c(dh - eg) \\ &= a\begin{vmatrix} e & f \\ h & i \end{vmatrix} - b\begin{vmatrix} d & f \\ g & i \end{vmatrix} + c\begin{vmatrix} d & e \\ g & h \end{vmatrix} \end{aligned}$$

To carry out such an expansion of the third-order determinant as the product of three coefficients and corresponding *minors* (lower-order determinants) with alternation of sign, we pick in turn each term of the top row and assign as the corresponding *minor* what is left after removal of both the row and column in which its coefficient occurs. Such will be our interpretation of the numerical value of a third-order determinant, e.g.:

$$\begin{vmatrix} 3 & 4 & 6 \\ 2 & 5 & 1 \\ 0 & 1 & 4 \end{vmatrix} = 3\begin{vmatrix} 5 & 1 \\ 1 & 4 \end{vmatrix} - 4\begin{vmatrix} 2 & 1 \\ 0 & 4 \end{vmatrix} + 6\begin{vmatrix} 2 & 5 \\ 0 & 1 \end{vmatrix}$$

$$= 3(20 - 1) - 4(8 - 0) + 6(2 - 0) = 37$$

The reader will find it a useful exercise to make up equations involving three variables in order to show that the values of x and y obtained by elimination are likewise consistent with (v) above, if we interpret the determinant of the third order in conformity with the rule last stated. The following example illustrates the solution of three equations with numerical constants:

$$2x + 3y + 9 \;= z$$
$$5x - 2y - 4z = 0$$
$$3x + 6y \qquad = -(2z + 4)$$

When arranged in standard form, the 4×3 matrix is:

$$\begin{Vmatrix} 2 & 3 & -1 & 9 \\ 5 & -2 & -4 & 0 \\ 3 & 6 & 2 & 4 \end{Vmatrix}$$

The solution for **x** is given by (iv) as $-\mathbf{x} = D(\mathbf{x}) \div D(C)$ in which:

$$D(x) = \begin{vmatrix} 3 & -1 & 9 \\ -2 & -4 & 0 \\ 6 & 2 & 4 \end{vmatrix}$$

$$= 3\begin{vmatrix} -4 & 0 \\ 2 & 4 \end{vmatrix} + \begin{vmatrix} -2 & 0 \\ 6 & 4 \end{vmatrix} + 9\begin{vmatrix} -2 & -4 \\ 6 & 2 \end{vmatrix}$$

$$= 3(-16) + (-8) + 9(20) = 124$$

$$D(C) = \begin{vmatrix} 2 & 3 & -1 \\ 5 & -2 & -4 \\ 3 & 6 & 2 \end{vmatrix}$$

$$= 2\begin{vmatrix} -2 & -4 \\ 6 & 2 \end{vmatrix} - 3\begin{vmatrix} 5 & -4 \\ 3 & 2 \end{vmatrix} - \begin{vmatrix} 5 & -2 \\ 3 & 6 \end{vmatrix}$$

$$= 2(20) - 3(22) - 36 = -62$$

Thus $-\mathbf{x} = 124 \div (-62)$, so that $\mathbf{x} = 2$. In the same way we get $\mathbf{y} = -3$ and $\mathbf{z} = 4$.

Determinants in General. We may define a fourth-order determinant by analogy with that of the third-order, i.e. as the sum of four 3×3 determinants respectively weighted by the elements of the top row with alternation of signs. Thus one rule for evaluating the fourth-order determinant is:

$$\begin{vmatrix} a & b & c & d \\ e & f & g & h \\ j & k & l & m \\ n & p & q & r \end{vmatrix} = a\begin{vmatrix} f & g & h \\ k & l & m \\ p & q & r \end{vmatrix} - b\begin{vmatrix} e & g & h \\ j & l & m \\ n & q & r \end{vmatrix} + c\begin{vmatrix} e & f & h \\ j & k & m \\ n & p & r \end{vmatrix} - d\begin{vmatrix} e & f & g \\ j & k & l \\ n & p & q \end{vmatrix}$$

Alternatively, we may define it as:

$$a\begin{vmatrix} f & g & h \\ k & l & m \\ p & q & r \end{vmatrix} - e\begin{vmatrix} b & c & d \\ k & l & m \\ p & q & r \end{vmatrix} + j\begin{vmatrix} b & c & d \\ f & g & h \\ p & q & r \end{vmatrix} - n\begin{vmatrix} b & c & d \\ f & g & h \\ k & l & m \end{vmatrix}$$

Either of the above ways of reducing the 4th order determinant to 4 appropriately weighted 3rd order determinants leads to the same result involving 12 four-fold positive and 12 four-fold negative products, *viz*:

Positive:

aflr *agmp* *ahkq* *bemq* *bgjr* *bhln* *cekr* *cfmn* *chjp* *delp* *dfjq* *dgkn*

Negative:

afmq *agkr* *ahlp* *belr* *bgmn* *bhjq* *cemp* *cfjr* *chkn* *dekq* *dfln* *dgjp*

So defined, a 4th order determinant conforms to each of the alternative patterns exhibited for a determinant of the 3rd order in Fig. 178. We have seen that this pattern is consistent with the meaning we have given to a 2nd order determinant; but our definition of a 4th order determinant as above is *not* consistent with an alternative definition of the 3rd order determinant

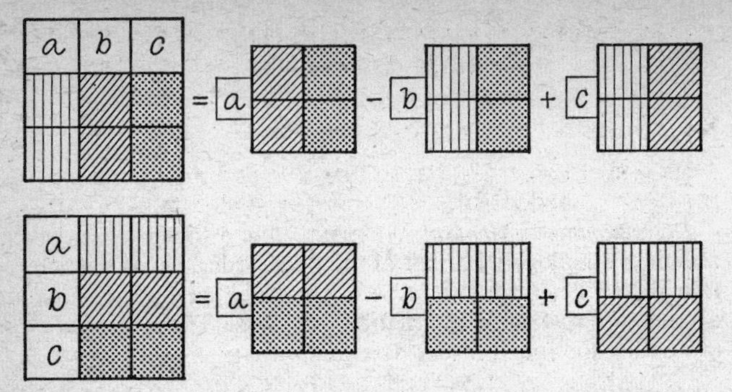

Fig. 178. Visual Mnemonic for expansion of a 3 × 3 Determinant in 2 × 2 Minors

exhibited in Fig. 177. The latter would reduce it to only 4 positive and 4 negative four-fold products. The reason for choosing to define a determinant of the 4th order in terms consistent with Fig. 178 rather than with Fig. 177 depends on the use to which we can put it, i.e., as a device for solving a set of 4 simultaneous linear equations.

It is laborious, but none the less elementary, to show that the solution of a set of four equations involving four variables (x_1, x_2, x_3, x_4) in the standard form:

$a_{r.1}x_1 + a_{r.2}x_2 + a_{r.3}x_3 + a_{r.4}x_4 + C_r = 0$ then conforms to a pattern essentially like (iii) above, viz.:

$$\frac{x_1}{D(x_1)} = \frac{-x_2}{D(x_2)} = \frac{x_3}{D(x_3)} = \frac{-x_4}{D(x_4)} = \frac{1}{D(C)}$$

Each of the four minors in this expansion is amenable to resolution as three determinants of the second order and hence, with due regard to *sign*, the sum of 24 (= 4!) product terms each consisting of four elements. More generally, a determinant of order n is amenable to resolution as: (*a*) the sum of n determinants of order $(n - 1)$ severally weighted by the top row elements; (*b*) the sum of $n!$ products of n elements, half the products being negative and half positive. This statement applies equally to the 2 × 2 determinant which is the sum of two (2!) terms of opposite sign, each the product of two elements. A common law of formation of these n-fold products takes shape if we write the two-fold and three-fold determinants as below:

$$\begin{vmatrix} a_1 & a_2 \\ b_1 & b_2 \end{vmatrix} = a_1 \,|\, b_2 \,| - b_1 \,|\, a_2 \,| = a_1 \,|\, b_2 \,| - a_2 \,|\, b_1 \,|$$

$$= a_1b_2 - a_2b_1$$

$$\begin{vmatrix} a_1 & a_2 & a_3 \\ b_1 & b_2 & b_3 \\ c_1 & c_2 & c_3 \end{vmatrix} = a_1 \begin{vmatrix} b_2 & b_3 \\ c_2 & c_3 \end{vmatrix} - b_1 \begin{vmatrix} a_2 & a_3 \\ c_2 & c_3 \end{vmatrix} + c_1 \begin{vmatrix} a_2 & a_3 \\ b_2 & b_3 \end{vmatrix}$$

$$= a_1 \begin{vmatrix} b_2 & b_3 \\ c_2 & c_3 \end{vmatrix} - a_2 \begin{vmatrix} b_1 & b_3 \\ c_1 & c_3 \end{vmatrix} + a_3 \begin{vmatrix} b_1 & b_2 \\ c_1 & c_2 \end{vmatrix}$$

$$= a_1(b_2c_3 - b_3c_2) - a_2(b_1c_3 - b_3c_1) + a_3(b_1c_2 - b_2c_1)$$

$$= a_1b_2c_3 + a_2b_3c_1 + a_3b_1c_2 - a_1b_3c_2 - a_2b_1c_3 - a_3b_2c_1$$

The whole set of $n!$ products each of n elements includes all possible combinations of n elements subject only to the restriction that elements of the same row or of the same column cannot contribute to one and the same term. The law of signs w.r.t. the ultimate products terms of a determinant breakdown becomes explicit by recourse to the double subscript notation which keeps the distinction between row and column order as clear-cut as possible. When the product elements are arranged as above in the correct row order (*here* a, b, c), the subscripts exhibit the column order. For three-fold products it is noticeable that:

(i) *one* inversion or three inversions are necessary to restore the sequence of the column subscripts of negative terms, viz. 132 to 123, or 321 to 312 to 132 to 123;

(ii) *no* inversion or *two* inversions are necessary to restore the sequence of the column subscripts of *positive* terms, viz. 231 to 213 to 123 and 312 to 132 to 123.

In general, the rule for determinants of any order is that negative products require an odd number of inversions, positive products an even number.

Numerical Evaluation of Determinants. So far, our approach to the notation of determinants has advanced no claim other than it gives us a *conveniently memorizable code* for laying out the computations involved in the solution of sets of linear equations. Of itself, this is no mean advantage, when it is otherwise necessary to steer a course through the repetitive operations of successively eliminating one variable after another of a set of four or more of them. A still greater advantage arises from the possibility of utilizing numerical properties of determinants to speed up the task of computation. We can sum up what are essential properties from this point of view in the following rules, for which it is scarcely necessary to offer formal proof. Testing the rules w.r.t. determinants of orders 2 and 3 should suffice to disclose why they are applicable to determinants of higher order, since the expansion of a 3×3 determinant as three minors of order 2 is on all fours with the expansion of any $n \times n$ determinant as n minors of order $(n - 1)$. To appreciate the utility of each rule *pari passu* will be more helpful for the novice than to set forth a formal demonstration of each.

Rule I. Rotation of the arrays, so that the rth row becomes the rth column and vice versa does not change the value of a determinant:

$$\begin{vmatrix} a_1 & a_2 \\ b_1 & b_2 \end{vmatrix} = (a_1b_2 - a_2b_1) = \begin{vmatrix} a_1 & b_1 \\ a_2 & b_2 \end{vmatrix}$$

The operational value of this rule is evident if we examine the determinant:

$$\begin{vmatrix} 3 & 4 & 5 \\ 0 & 2 & 1 \\ 6 & 6 & 4 \end{vmatrix} = 3\begin{vmatrix} 2 & 1 \\ 6 & 4 \end{vmatrix} - 4\begin{vmatrix} 0 & 1 \\ 6 & 4 \end{vmatrix} + 5\begin{vmatrix} 0 & 2 \\ 6 & 6 \end{vmatrix}$$

By applying the rule of inter-changeability, we can eliminate one minor at one step, since we can write it as:

$$\begin{vmatrix} 3 & 0 & 6 \\ 4 & 2 & 6 \\ 5 & 1 & 4 \end{vmatrix} = 3\begin{vmatrix} 2 & 6 \\ 1 & 4 \end{vmatrix} + 6\begin{vmatrix} 4 & 2 \\ 5 & 1 \end{vmatrix}$$

The rearrangement is unnecessary since the rule implies that we can reduce a determinant in either of two ways:

$$\begin{vmatrix} a_1 & a_2 & a_3 \\ b_1 & b_2 & b_3 \\ c_1 & c_2 & c_3 \end{vmatrix} \equiv a_1\begin{vmatrix} b_2 & b_3 \\ c_2 & c_3 \end{vmatrix} - a_2\begin{vmatrix} b_1 & b_3 \\ c_1 & c_3 \end{vmatrix} + a_3\begin{vmatrix} b_1 & b_2 \\ c_1 & c_2 \end{vmatrix}$$

$$\equiv a_1b_2c_3 + a_2b_3c_1 - a_3b_1c_2 + a_1b_3c_2 - a_2b_1c_3 - a_3b_2c_1$$

$$\equiv a_1\begin{vmatrix} b_2 & b_3 \\ c_2 & c_3 \end{vmatrix} - b_1\begin{vmatrix} a_2 & a_3 \\ c_2 & c_3 \end{vmatrix} + c_1\begin{vmatrix} a_2 & a_3 \\ b_2 & b_3 \end{vmatrix}$$

Rule II. Interchange of a single pair of rows or a single pair of columns reverses the sign of the numerical value of a determinant, e.g.:

$$\begin{vmatrix} a_1 & a_2 \\ b_1 & b_2 \end{vmatrix} \equiv -\begin{vmatrix} a_2 & a_1 \\ b_2 & b_1 \end{vmatrix} \equiv -\begin{vmatrix} b_1 & b_2 \\ a_1 & a_2 \end{vmatrix}$$

The operational value of this rule is evident from the fact that it allows us to get an array containing one or more zero terms into the top row or first column, so that a corresponding number of minors drop out, e.g.:

$$\begin{vmatrix} 2 & 3 & 1 & 4 \\ 4 & 1 & 0 & 3 \\ 3 & 0 & 0 & 6 \\ 1 & 2 & 2 & 4 \end{vmatrix} = -\begin{vmatrix} 2 & 1 & 3 & 4 \\ 4 & 0 & 1 & 3 \\ 3 & 0 & 0 & 6 \\ 1 & 2 & 2 & 4 \end{vmatrix} = +\begin{vmatrix} 1 & 2 & 3 & 4 \\ 0 & 4 & 1 & 3 \\ 0 & 3 & 0 & 6 \\ 2 & 1 & 2 & 4 \end{vmatrix}$$

$$= \begin{vmatrix} 4 & 1 & 3 \\ 3 & 0 & 6 \\ 1 & 2 & 4 \end{vmatrix} - 2\begin{vmatrix} 2 & 3 & 4 \\ 4 & 1 & 3 \\ 3 & 0 & 6 \end{vmatrix} = -2\begin{vmatrix} 3 & 0 & 6 \\ 2 & 3 & 4 \\ 4 & 1 & 3 \end{vmatrix} - \begin{vmatrix} 3 & 0 & 6 \\ 4 & 1 & 3 \\ 1 & 2 & 4 \end{vmatrix}$$

$$= -2(3)\begin{vmatrix} 3 & 4 \\ 1 & 3 \end{vmatrix} - 2(6)\begin{vmatrix} 2 & 3 \\ 4 & 1 \end{vmatrix} - 3\begin{vmatrix} 1 & 3 \\ 2 & 4 \end{vmatrix} - 6\begin{vmatrix} 4 & 1 \\ 1 & 2 \end{vmatrix}$$

$$= -6(9 - 4) - 12(2 - 12) - 3(4 - 6) - 6(8 - 1) = 54$$

Rule III. The numerical value of any determinant is zero if any two rows or any two columns are identical, i.e.:

$$\begin{vmatrix} a & b & c \\ d & e & f \\ a & b & c \end{vmatrix} \equiv 0 \equiv \begin{vmatrix} a & b & b \\ d & e & e \\ g & h & h \end{vmatrix}$$

One value of this rule will be self-evident, when we have examined the meaning of the next one.

Rule IV. Multiplying every element of one row or column by the same factor k is equivalent to a k-fold multiplication of the numerical value of the determinant, e.g.:

$$\begin{vmatrix} ak & bk & ck \\ d & e & f \\ g & h & j \end{vmatrix} \equiv k \begin{vmatrix} a & b & c \\ d & e & f \\ g & h & j \end{vmatrix} \equiv \begin{vmatrix} a & bk & c \\ d & ek & f \\ g & hk & j \end{vmatrix}$$

One use of this rule is to reduce the arithmetical bulk of the elements before further expansion, e.g.:

$$\begin{vmatrix} 25 & 45 & 15 \\ 40 & 24 & 21 \\ 15 & 9 & 21 \end{vmatrix} = (5^2) \begin{vmatrix} 1 & 9 & 3 \\ 8 & 24 & 21 \\ 3 & 9 & 21 \end{vmatrix} = (3^2)(5^2) \begin{vmatrix} 1 & 3 & 1 \\ 8 & 8 & 7 \\ 3 & 3 & 7 \end{vmatrix}$$

Another use depends on Rule III, e.g.:

$$\begin{vmatrix} 25 & 45 & 10 \\ 40 & 24 & 16 \\ 15 & 9 & 6 \end{vmatrix} = 5(2) \begin{vmatrix} 5 & 45 & 5 \\ 8 & 24 & 8 \\ 3 & 9 & 3 \end{vmatrix} = 0$$

Thus we can drop out any minors of an expansion, if corresponding elements of two rows or of two columns are in *the same proportion.*

Rule V. Addition or subtraction of corresponding elements of one row or column from those of any parallel row or column does not change the value of the determinant, i.e.:

$$\begin{vmatrix} a & d & g \\ b & e & h \\ c & f & j \end{vmatrix} \equiv \begin{vmatrix} (a+d) & d & g \\ (b+e) & e & h \\ (c+f) & f & j \end{vmatrix} \equiv \begin{vmatrix} (a-c) & (d-f) & (g-j) \\ b & e & h \\ c & f & j \end{vmatrix}$$

By use of the foregoing rules we can eliminate any minor by rearrangement of arrays to bring a zero element into the top row or first column. By use of the rule last stated we can: (i) introduce new zero terms by making corresponding terms of different parallel arrays identical; (ii) greatly reduce the arithmetical load we have to carry. The following illustration shows every step in a reduction which the practised hand would carry out with much greater economy of space:

$$\begin{vmatrix} 42 & 4 & 39 \\ 13 & 8 & 13 \\ 18 & 10 & 26 \end{vmatrix} = \begin{vmatrix} (42-39) & 4 & 39 \\ (13-13) & 8 & 13 \\ (18-26) & 10 & 26 \end{vmatrix} = \begin{vmatrix} 3 & 4 & 39 \\ 0 & 8 & 13 \\ -8 & 10 & 26 \end{vmatrix}$$

$$= 2(13)\begin{vmatrix} 3 & 2 & 3 \\ 0 & 4 & 1 \\ -8 & 5 & 2 \end{vmatrix} = 26\begin{vmatrix} (3+8) & (2-5) & (3-2) \\ 0 & 4 & 1 \\ -8 & 5 & 2 \end{vmatrix}$$

$$= 26\begin{vmatrix} 11 & -3 & 1 \\ 0 & 4 & 1 \\ (-8-0) & (5-4) & (2-1) \end{vmatrix}$$

$$= 26\begin{vmatrix} 11 & -3 & 1 \\ 0 & 4 & 1 \\ -8 & 1 & 1 \end{vmatrix} = 26\begin{vmatrix} 11 & -3 & 1 \\ 0 & 4 & 1 \\ (-8-0) & (1-4) & 0 \end{vmatrix}$$

$$= 26\begin{vmatrix} 11 & -3 & 1 \\ (0-11) & (4+3) & 0 \\ -8 & -3 & 0 \end{vmatrix}$$

$$= 26\begin{vmatrix} 1 & 11 & -3 \\ 0 & -11 & 7 \\ 0 & -8 & -3 \end{vmatrix} = -26\begin{vmatrix} 11 & 7 \\ 8 & -3 \end{vmatrix}$$

$$= -26(-33-56) = (26)(89) = 2314$$

Rule VI. Addition or subtraction from elements of one array of a fixed multiple of corresponding elements of one (or more) parallel array(s) does not change the value of the determinant. If p and q may each be negative we may write:

$$\begin{vmatrix} a & d & g \\ b & e & h \\ c & f & j \end{vmatrix} \equiv \begin{vmatrix} (a + pd + qg) & d & g \\ (b + pe + qh) & e & h \\ (c + pf + qj) & f & j \end{vmatrix}$$

This one combines Rules IV and V to simplify the task of introducing a *zero* element. If we put $p = -2$ and $q = 4$.

$$= \begin{vmatrix} 6 & 7 & 2 \\ 1 & 5 & 9 \\ 8 & 3 & 4 \end{vmatrix} = \begin{vmatrix} (6 - 14 + 8) & 7 & 2 \\ (1 - 10 + 36) & 5 & 9 \\ (8 - 6 + 16) & 3 & 4 \end{vmatrix}$$

$$= \begin{vmatrix} 0 & 7 & 2 \\ 27 & 5 & 9 \\ 18 & 3 & 4 \end{vmatrix} = -27 \times \begin{vmatrix} 7 & 2 \\ 3 & 4 \end{vmatrix} + 18 \times \begin{vmatrix} 7 & 2 \\ 5 & 9 \end{vmatrix}$$

$$= -27\,(28 - 6) + 18\,(63 - 10) = 360$$

Outline of Proof. Call the determinant for computation D_1. Multiply elements of col. 2 by p and of col. 3 by q. Call the results D_2. By Rule IV: $D_2 = pq \,.\, D_1$. Add elements of col. 2 and of col. 3 in D_2 to corresponding elements of col. 1. Call the result D_3. By Rule V: $D_2 = D_3$. Now divide the elements of col. 2 by p and of col. 3 by q in D_3. Call the result D_4. By Rule IV:

$$D_4 = D_3 \div pq = D_2 \div pq = (pq \,.\, D_1) \div pq = D_1$$

Rule VII. The value of a determinant is zero if all the elements of any row or of any column are zero.

The proof of this is evident, if we transfer the row of zeros to the uppermost tier or the row of columns to the extreme left hand with appropriate change of sign (*Rule II*). Each co-factor of the minors into which we can break up the determinant is then zero, and each term of its expansion as a sum of co-factor minor products vanishes.

Manipulating these rules to advantage depends on practice. Having attained proficiency, one can then proceed to reduce a determinant of any order to one in which all the elements *except* the left-hand top corner (a_{11}) in either the top row or the first column are zeros. Hence, all the minors other than the minor whose co-factor is a_{11} drop out. This reduces the order from n to $(n - 1)$. So successive application of the rules eventually reduces any determinant to one of the second order. The best way to get practice is to *make up* consistent equations for solution by successive elimination and then by determinants. For instance, put $x = 1$, $y = 2$, $z = 3$, so that:

$$2x + 3y + z = 11$$
$$5x + 3y - 2z = 5$$
$$4x + 2y - 3z = -1$$

In the early stages of carrying out the drill, it will be necessary to perform separately several operations which the more experienced player will take in one stride. Other ways of combining the rules may give equally good results, and the student who aims at proficiency should consult Whittaker and Robinson's *Calculus of Observations*. The foregoing account will suffice to show that the method of determinants immeasurably reduces both the labour of solving a large set of simultaneous equations and the danger of making arithmetical slips in doing so.

Geometrical Applications of Determinants. We have hitherto regarded the shorthand of determinants as an economical computing schema for the solution of a set of simultaneous linear equations. Its applications extend far beyond this eminently practical objective including *inter alia* the definition of sundry general principles of co-ordinate geometry. Two examples will suffice to illustrate its geometrical use: (*a*) the condition of *collinearity* of three points in a plane; (*b*) the condition of *concurrence* of three or more lines in a plane, whence also (Fig. 179) the area of a triangle.

Collinearity of three points (p_1, p_2, p_3) signifies that all three are in one and the same straight line. In other words, the line joining p_1 to p_2 and the line joining p_2 to p_3 are inclined to the

Area of Triangle as a Determinant

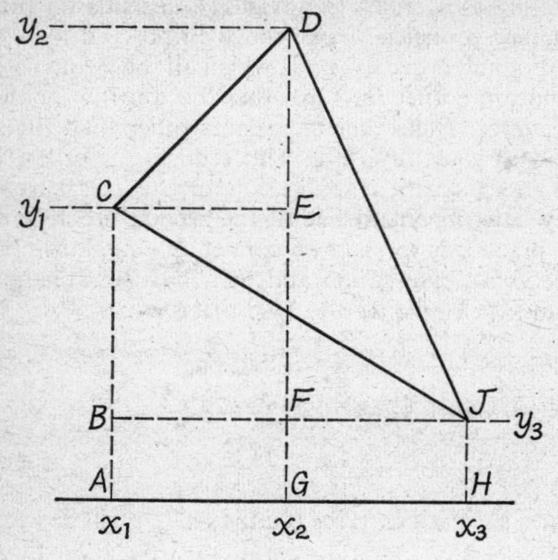

$$
\begin{aligned}
CDJ &= ACDJH - ACJH \\
&= (ACEG + CDE + DFJ + FGJH) - (BCJ + ABJH) \\
&= y_1(x_2 - x_1) + \tfrac{1}{2}(y_2 - y_1)(x_2 - x_1) + \tfrac{1}{2}(y_2 - y_3)(x_3 - x_2) + y_3(x_3 - x_2) \\
&\quad - \tfrac{1}{2}(y_1 - y_3)(x_3 - x_1) - y_3(x_3 - x_1) \\
&= \tfrac{1}{2}(y_1 + y_2)(x_2 - x_1) + \tfrac{1}{2}(y_2 + y_3)(x_3 - x_2) - \tfrac{1}{2}(y_1 + y_3)(x_3 - x_1) \\
\therefore 2(CDJ) &= y_1x_2 - y_1x_1 + y_2x_2 - y_2x_1 + y_2x_3 - y_2x_2 + y_3x_3 - y_3x_2 \\
&\quad - y_1x_3 + y_1x_1 - y_3x_3 + y_3x_1 \\
&= (y_2x_3 - y_3x_2) - (y_1x_3 - y_3x_1) + (y_1x_2 - y_2x_1) \\
&= \begin{vmatrix} y_2 & x_2 \\ y_3 & x_3 \end{vmatrix} - \begin{vmatrix} y_1 & x_1 \\ y_3 & x_3 \end{vmatrix} + \begin{vmatrix} y_1 & x_1 \\ y_2 & x_2 \end{vmatrix} = \begin{vmatrix} 1 & y_1 & x_1 \\ 1 & y_2 & x_2 \\ 1 & y_3 & x_3 \end{vmatrix}
\end{aligned}
$$

Fig. 179. A Geometrical Use of Determinant Symbolism

x-axis at the same angle a. The reader should sketch the appropriate figure in which we may denote the co-ordinates of the three points as (x_1, y_1), (x_2, y_2), (x_3, y_3). In conformity with the sufficient relation that $p_1 p_2$ has the same horizontal inclination $a°$ as $p_2 p_3$, we may write:

$$\frac{y_2 - y_1}{x_2 - x_1} = \tan a = \frac{y_3 - y_2}{x_3 - x_2}$$

$$\therefore \quad (x_3 - x_2)(y_2 - y_1) = (y_3 - y_2)(x_2 - x_1)$$

$$\therefore \quad x_3y_2 - x_3y_1 - x_2y_2 + x_2y_1 = x_2y_3 - x_1y_3 - x_2y_2 + x_1y_2$$

$$\therefore \quad x_3y_2 + x_2y_1 + x_1y_3 - x_3y_1 - x_2y_3 - x_1y_2 = 0$$

$$\therefore \quad x_1(y_3 - y_2) - x_2(y_3 - y_1) + x_3(y_2 - y_1) = 0$$

$$\therefore \quad x_1(y_2 - y_3) - x_2(y_1 - y_3) + x_3(y_1 - y_2) = 0$$

We may write $(y_2 - y_3)$ as a determinant:

$$\begin{vmatrix} y_2 & 1 \\ y_3 & 1 \end{vmatrix}$$

Hence the last equation is equivalent to:

$$x_1 \begin{vmatrix} y_2 & 1 \\ y_3 & 1 \end{vmatrix} - x_2 \begin{vmatrix} y_1 & 1 \\ y_3 & 1 \end{vmatrix} + x_3 \begin{vmatrix} y_1 & 1 \\ y_2 & 1 \end{vmatrix} = 0$$

The three determinants and their co-factors on the left are equivalent to a single determinant of the third order, so that:

$$\begin{vmatrix} x_1 & y_1 & 1 \\ x_2 & y_2 & 1 \\ x_3 & y_3 & 1 \end{vmatrix} = 0$$

The zero value of the determinant on the left thus defines the condition that three points $p_1(x_1, y_1)$, $p_2(x_2, y_2)$, and $p_3(x_3, y_3)$ shall all lie in one straight line.

Example. One way to decide whether (1, 8), (3, 18) and (6, 33) lie in a straight line is thus to evaluate:

$$\begin{vmatrix} 1 & 8 & 1 \\ 3 & 18 & 1 \\ 6 & 33 & 1 \end{vmatrix} = \begin{vmatrix} 0 & 8 & 1 \\ 2 & 18 & 1 \\ 5 & 33 & 1 \end{vmatrix} = -2\begin{vmatrix} 8 & 1 \\ 33 & 1 \end{vmatrix} + 5\begin{vmatrix} 8 & 1 \\ 18 & 1 \end{vmatrix}$$

$$= -2\begin{vmatrix} 8 & 1 \\ 25 & 0 \end{vmatrix} + 5\begin{vmatrix} 8 & 1 \\ 10 & 0 \end{vmatrix} = (50 - 50) = 0$$

We might of course proceed by a more devious route by solving the equation $y = (mx + b)$ for each pair of variables, thus:

$$8 = m + b$$

$$18 = 3m + b$$

$$\therefore \quad 10 = 2m \quad \textit{or} \quad m = 5 \quad \textit{and} \quad b = 3$$

Thus the line on which the points (1, 8) and (3, 18) lie is $y = (5x + 3)$. On inserting the x value of the third point (6, 33), we have $y = 5(6) + 3 = 33$, so that the point whose co-ordinates are $x = 6$, $y = 33$ also lies on the same straight line.

The *concurrence* of three straight lines in a plane signifies they meet in a point. Let us write the three corresponding equations in the form:

$$\text{(i) } y = m_1x + b_1 \quad \textit{or} \quad m_1x - y + b_1 = 0$$

$$\text{(ii) } y = m_2x + b_2 \quad \textit{or} \quad m_2x - y + b_2 = 0$$

$$\text{(iii) } y = m_3x + b_3 \qquad m_3x - y + b_3 = 0$$

In accordance with the graphical method of solving simultaneous equations involving two variables by finding the co-ordinates of the point where the two corresponding straight lines cross, the lines defined by (i) and (ii) intersect at the point whose co-ordinates (x_p, y_p) specify the solution of the two equations, viz.:

$$x_p = \frac{\begin{vmatrix} -1 & b_1 \\ -1 & b_2 \end{vmatrix}}{\begin{vmatrix} m_1 & -1 \\ m_2 & -1 \end{vmatrix}}; \quad -y_p = \frac{\begin{vmatrix} m_1 & b_1 \\ m_2 & b_2 \end{vmatrix}}{\begin{vmatrix} m_1 & -1 \\ m_2 & -1 \end{vmatrix}}$$

If the line defined by the third equation also passes through the point (x_p, y_p):

$$m_3 x_p - y_p + b_3 = 0$$

$$\therefore \quad m_3 \begin{vmatrix} -1 & b_1 \\ -1 & b_2 \end{vmatrix} + \begin{vmatrix} m_1 & b_1 \\ m_2 & b_2 \end{vmatrix} + b_3 \begin{vmatrix} m_1 & -1 \\ m_2 & -1 \end{vmatrix} = 0$$

$$\therefore \quad m_3 \begin{vmatrix} +1 & b_1 \\ +1 & b_2 \end{vmatrix} - \begin{vmatrix} m_1 & b_1 \\ m_2 & b_2 \end{vmatrix} + b_3 \begin{vmatrix} m_1 & +1 \\ m_2 & +1 \end{vmatrix} = 0$$

$$\therefore \quad \begin{vmatrix} m_3 & 1 & b_3 \\ m_1 & 1 & b_1 \\ m_2 & 1 & b_2 \end{vmatrix} = 0 = \begin{vmatrix} m_1 & -1 & b_1 \\ m_2 & -1 & b_2 \\ m_3 & -1 & b_3 \end{vmatrix}$$

Example. Let us ask whether the following three lines are concurrent:

$$y = 5x + 3$$
$$y = 6x + 2$$
$$y = 4x + 4$$

We have to evaluate:

$$\begin{vmatrix} 5 & -1 & 3 \\ 6 & -1 & 2 \\ 4 & -1 & 4 \end{vmatrix} = - \begin{vmatrix} 5 & 1 & 3 \\ 6 & 1 & 2 \\ 4 & 1 & 4 \end{vmatrix} = - \begin{vmatrix} 5 & 1 & 3 \\ 1 & 0 & -1 \\ 4 & 1 & 4 \end{vmatrix}$$

$$= - \begin{vmatrix} 1 & 0 & -1 \\ 1 & 0 & -1 \\ 4 & 1 & 4 \end{vmatrix} = - \begin{vmatrix} 0 & 0 & 0 \\ 1 & 0 & -1 \\ 4 & 1 & 4 \end{vmatrix} = 0$$

Since the value of the appropriate determinant is zero, the lines meet at a point, which is in fact $x = 1$, $y = 9$.

The reader may extend this result to specify the condition of *consistency* w.r.t. 4 simultaneous equations involving three variables in the form $a_m x + b_m y + c_m z + {}_m = 0$, viz.:

$$\begin{vmatrix} a_1 & b_1 & c_1 & d_1 \\ a_2 & b_2 & c_2 & d_2 \\ a_3 & b_3 & c_3 & d_3 \\ a_4 & b_4 & c_4 & d_4 \end{vmatrix} = 0$$

Geometrically, this tells us when four planes intersect at a single point $P = (x_p, y_p, z_p)$. This is necessarily so if they all include the origin, in which case $d_m = 0$, so that the determinant is:

$$\begin{vmatrix} a_1 & b_1 & c_1 & 0 \\ a_2 & b_2 & c_2 & 0 \\ a_3 & b_3 & c_3 & 0 \\ a_4 & b_4 & c_4 & 0 \end{vmatrix}$$

The value of this determinant is necessarily zero in virtue of Rule VII. By recourse to Rule VII we may derive the condition of collinearity for any three points by a route other than the one given above and use the result to establish when four points are co-planar.

Rotation of Axes. The preceding account gives only a very small indication of the wide range of equations for the solution of which determinants do service and of the geometrical relations expressible in grid form. We may now consider a geometrical problem which does not depend on the numerical value of a determinant, or indeed on whether the number of rows and columns is the same. Our task will be to provide an *algebraic pattern* for the co-ordinates of a point after *successive* rotation of the axes, e.g. first through 30°, then through 60°, in a Cartesian grid.

A clue to the pattern depends on a numerical property of determinants, though this turns out to be irrelevant. We shall start by examining the product of two determinants of the second order, viz.

$$\begin{vmatrix} p & q \\ r & s \end{vmatrix} \times \begin{vmatrix} a & b \\ c & d \end{vmatrix} = (ps - qr)(ad - bc)$$

$$= psad - psbc - qrad + qrbc$$

Now let us evaluate the following:

$$\begin{vmatrix} pa + qc & pb + qd \\ ra + sc & rb + sd \end{vmatrix} = (pa + qc)(rb + sd) - (pb + qd)(ra + sc)$$

$$= psad - psbc - qrad + qrbc$$

$$\therefore \begin{vmatrix} p & q \\ r & s \end{vmatrix} \times \begin{vmatrix} a & b \\ c & d \end{vmatrix} = \begin{vmatrix} pa + qc & pb + qd \\ ra + sc & rb + sd \end{vmatrix}$$

To familiarize oneself with this pattern, the reader may try out numerical examples such as:

$$\begin{vmatrix} 3 & 4 \\ 6 & 2 \end{vmatrix} \times \begin{vmatrix} 5 & 1 \\ 10 & 3 \end{vmatrix} = (6 - 24)(15 - 10) = -90$$

$$\begin{vmatrix} 15 + 40 & 3 + 12 \\ 30 + 20 & 6 + 6 \end{vmatrix} = \begin{vmatrix} 55 & 15 \\ 50 & 12 \end{vmatrix} = 15 \times \begin{vmatrix} 11 & 5 \\ 10 & 4 \end{vmatrix}$$

$$= 15(44 - 50) = -90$$

We now recall a recipe for deriving a rotation of 2-dimensional axes through an angle A (Fig. 135) in an anti-clockwise direction:

$$x_2 = \cos A \,.\, x_1 + \sin A \,.\, y_1; \quad y_2 = -\sin A \,.\, x_1 + \cos A \,.\, y_1$$

For brevity, we may write this, with matrices for the coefficients (a, b, c, d) as below on the right:

$$\begin{matrix} x_2 = a \,.\, x_1 + b \,.\, y_1 \\ y_2 = c \,.\, x_1 + d \,.\, y_1 \end{matrix} \qquad \begin{Vmatrix} a & b \\ c & d \end{Vmatrix} = \begin{Vmatrix} \cos A & \sin A \\ -\sin A & \cos A \end{Vmatrix}$$

Similarly, we may write:

$$\begin{matrix} x_3 = p \,.\, x_2 + q \,.\, y_2 \\ y_3 = r \,.\, x_2 + s \,.\, y_3 \end{matrix} \qquad \begin{Vmatrix} p & q \\ r & s \end{Vmatrix} = \begin{Vmatrix} \cos B & \sin B \\ -\sin B & \cos B \end{Vmatrix}$$

By substitution in the second pair from the first, we derive:

$$x_3 = (pa + qc)x_1 + (pb + qd)y_1$$

$$y_3 = (ra + sc)x_1 + (rb + sd)y_1$$

The corresponding matrix is:

$$\begin{Vmatrix} pa + qc & pb + qd \\ ra + sc & rb + sd \end{Vmatrix} = \begin{Vmatrix} p & q \\ r & s \end{Vmatrix} \times \begin{Vmatrix} a & b \\ c & d \end{Vmatrix}$$

The matrix on the left is equivalent to:

$$\begin{Vmatrix} \cos A \,.\, \cos B - \sin A \,.\, \sin B & \cos B \,.\, \sin A + \cos A \,.\, \sin B \\ -\sin B \,.\, \cos A - \sin A \,.\, \cos B & -\sin A \,.\, \sin B + \cos A \,.\, \cos B \end{Vmatrix}$$

By recourse to the addition formulae (Chapter 5) for $\sin (A + B)$ and $\cos (A + B)$, the above reduces to:

$$\begin{Vmatrix} \cos (A + B) & \sin (A + B) \\ -\sin (A + B) & \cos (A + B) \end{Vmatrix}$$

If we regard this as a determinant, its numerical value, regardless of those of A and B, is $cos^2 (A + B) + sin^2 (A + B) = 1$; but this fact is irrelevant to our expressed intention to provide an easily memorizible pattern whereby we may write the equation of the two-fold rotation, i.e.:

$$x_3 = \cos (A + B) \,.\, x_1 + \sin (A + B) \,.\, y_1$$

$$y_3 = -\sin (A + B) x_1 + \cos (A + B) \,.\, y_1$$

The reader can test this in several ways. If $A = 15$ and $B = 30$ the final result is a rotation through $45° = A + B$:

$$\begin{Vmatrix} \cos . 45° & \sin . 45° \\ -\sin . 45° & \cos . 45° \end{Vmatrix} = \begin{Vmatrix} \frac{1}{\sqrt{2}} & \frac{1}{\sqrt{2}} \\ -\frac{1}{\sqrt{2}} & \frac{1}{\sqrt{2}} \end{Vmatrix}$$

Whence, as we have seen (p. 357) already:

$$x_3 = \frac{1}{\sqrt{2}} \,.\, x_1 + \frac{1}{\sqrt{2}} \,.\, y_1$$

$$x_3 = - \frac{1}{\sqrt{2}} \,.\, x_1 + \frac{1}{\sqrt{2}} \,.\, y_1$$

The sum $(A + B) = 90°$, if $A = 45° = B$, or if $A = 15°$, $B = 75°$, or if $A = 30°$, $B = 60°$. The total rotation then implies:

$$\begin{Vmatrix} \cos . 90^\circ & \sin . 90^\circ \\ -\sin . 90^\circ & \cos . 90^\circ \end{Vmatrix} = \begin{Vmatrix} 0 & 1 \\ -1 & 0 \end{Vmatrix}$$

Whence, as we have also seen

$$x_3 = 0 . x_1 + 1 . y_1 = y_1$$

$$y_3 = -1 . x_1 + 0 . y_1 = -y_1$$

The reader who wishes to get more insight into the way a matrix may divulge a pattern for the solution of a geometrical problem with no reliance on visual aids, may find the following exercise instructive, though laborious:

(i) express in the 3 × 3 matrix form the formula for rotating the XY-axis through an angle A_1 and the Z-axis through B_1;

(ii) find the appropriate product matrix for a rotation from A_1 through A_2 to A_3 and B_1 through B_2 to B_3.

Chapter 11

EXERCISES

As exercises in the evaluation of 3 × 3 determinants, we may recall the formula for making nine-cell magic squares given at the beginning of Chapter 4, (p. 173) viz.:

$a + c$	$a + b - c$	$a - b$
$a - b - c$	a	$a + b + c$
$a + b$	$a - b + c$	$a - c$

Below are 3 × 3 magic squares for $a = 5$, $a = 6$, $a = 7$, and $a = 8$ with the numerical value of the corresponding determinant (Δ) cited below each:

$a = 5, b = 3, c = 1$ $\quad \Delta = 360$
$a = 6, b = 3, c = 2$ $\quad \Delta = 270$
$a = 6, b = 3, c = 1$ $\quad \Delta = 432$

. . .

$a = 7, b = 3, c = 2 \qquad \Delta = 315$
$a = 7, b = 4, c = 1 \qquad \Delta = 945$
$a = 7, b = 3, c = 1 \qquad \Delta = 504$

. . .

$a = 8, b = 6, c = 1 \qquad \Delta = 2{,}520$
$a = 8, b = 5, c = 2 \qquad \Delta = 1{,}512$
$a = 8, b = 5, c = 1 \qquad \Delta = 1{,}728$

. . .

$a = 8, b = 4, c = 3 \qquad \Delta = 504$
$a = 8, b = 4, c = 1 \qquad \Delta = 1{,}080$

. . .

$a = 8, b = 3, c = 1 \qquad \Delta = 576$
$a = 8, b = 3, c = 2 \qquad \Delta = 360$

CHAPTER 12

THE ALGEBRA OF CHOICE AND CHANCE

THE RELEVANCE of no branch of mathematics to one or other aspect of the contemporary world's work is more wide open to dispute than is the theory of so-called *probability*. On the other hand, its unsavoury origin is on record. The first impetus came from a situation in which the dissolute nobility of France were competing in a race to ruin at the gaming tables. An algebraic calculus of probability takes its origin from a correspondence between Pascal and Fermat (about AD 1654) over the fortunes and misfortunes of the Chevalier de Méré, a great gambler and by that token *très bon espirit*, but alas (wrote Pascal) *il n'est pas géomètre*. Alas indeed. The Chevalier had made his pile by always betting small favourable odds on getting at least one six in four tosses of a die, and had then lost it by always betting small odds on getting at least one double six in twenty-four double tosses. Albeit useless, the problem out of which the calculus took shape was therefore essentially practical, viz. how to adjust the stakes in a game of chance in accordance with a rule which ensures success if applied consistently regardless of the fortunes of the session. This is the theme song of the later treatise (*Ars Conjectandi*, i.e. art of guesswork) by James Bernoulli (1713) and it was the major pre-occupation of all the writers of what one may call the classical period, i.e. de Moivre, D. Bernoulli, d'Alembert, and Euler.

It may be an accident that Chinese mathematicians and their Japanese pupils had shown an interest in problems of choice before the Chinese art of printing from wood blocks made possible commercial production of playing cards in Europe. This much is certain. Aside from the behaviour of a combination lock, there was little other than games of chance to prompt interest in problems of choice when they began to arouse a tentative interest in Italy a century before Pascal's short treatise on the triangle of binomial coefficients; and it would be more consistent with the attitude of Pascal's immediate successors to the division of the stakes if we were to substitute the word *choosability* for *probability* in a mathematical context. Accordingly, our introduction to the algebraic theory of probability must begin with an account of choice, where we left it off at the end of Chapter 4.

Before we recall what we there learnt let us be clear about a distinction between two kinds of choice or, to use a better term, *sampling*. We can draw five cards from a pack in two ways: (*a*) simultaneously or, as is equivalent, *without replacing* any card before taking another; (*b*) replacing each card before removing another. We shall speak henceforth of (*a*) as *sampling without replacement*, and of (*b*) as *sampling with replacement*. From an algebraic point of view, recording the different ways in which the numbered sectors of a wheel may come to rest against a pointer or the numbered faces of dice may come to rest is sampling *with* replacement.

We have already learned to distinguish between sampling from another point of view: (*a*) by the ingredients alone (*combinations*); (*b*) by the ingredients and by their order when set out in a row (*linear permutations*). The term permutation refers to any classification which includes arrangement, e.g. in a ring, as well as the identity of the constituents of a set of objects; but here we shall use the term only for arrangement in a row. To clarify the distinction between sampling with and sampling without replacement, let us consider the four letters *A*, *B*, *C*, *D*. If our sample consists of three letters, and no letter can occur more than once (sampling *without replacement*), 4 combinations are possible: *ABC*; *ABD*; *ACD*; *BCD*. Each of these admits of 6 arrangements as below. So there are 24 linear permutations:

ABC *ACB* *BAC* *BCA* *CAB* *CBA*
ABD *ADB* *BAD* *BDA* *DAB* *DBA*
ACD *ADC* *CAD* *CDA* *DAC* *DCA*
BCD *BDC* *CBD* *CDB* *DBC* *DCB*

If we sample *with replacement* so that each letter may occur 0, 1, 2, and 3 times in the three-fold sample, the number of combinations is 20, as shown below, and the corresponding number of permutations (in brackets beside each different combination) is 64. As a clue to a discovery the reader may make,* the

* The numbers of combinations of r letters from a set of n when repetition is permissible is (in the symbolism of p. 188, (Chapter 4): $n^{[r]} \div r!$

combinations are classifiable by the first letter below in groups which recall familiar figurate numbers, i.e. A's 10; B's 6; C's 3; D 1:

AAA (1)	ABB (3)	ACC (3)	BBB (1)	BCC (3)	CCC (1)
AAB (3)	ABC (6)	ACD (6)	BBC (3)	BCD (6)	CCD (3)
AAC (3)	ABD (6)	ADD (3)	BBD (3)	BDD (3)	CDD (3)
AAD (3)					DDD (1)

The alert reader may notice that the number of permutations of 4 things taken 3 at a time with replacement is 4^3, being 64. If we write ${}^nR_r$ for the number of permutations with replacement (R because *repetition* is permissible), this suggests that ${}^nR_r = n^r$. This is necessarily so. There are n ways of filling the first place in the row, each of these may occur with any one of n ways of filling the second. Thus there are $n \,.\, n = n^2$ ways of filling the first two places and with each of these n^2 ways of filling the first two places, there are n ways of filling the third, whence $n^2 \,.\, n = n^3$ of filling the first three, and so on. Thus there are six different ways in which we can score 1 toss of a cubical die (1–6 pips), $6^2 = 36$ in which we can score 2 tosses, $6^3 = 216$ in which we can score 3 tosses, $6^4 = 1296$ in which we can score 4 tosses. These numbers correspond to all the ways we can score 1, 2, 3, and 4 cards taken from a set of 6, if we put each card (except the last) back before drawing another. Without replacement, the numbers would be: 6 (1 card); $30 = 6 \,.\, 5$ (2 cards), $120 = 6 \,.\, 5 \,.\, 4$ (3 cards), and $360 = 6 \,.\, 5 \,.\, 4 \,.\, 3$ (4 cards).

The reader should now recall the meaning of $6^{(3)} = 6 \,.\, 5 \,.\, 4$ in the useful symbolism introduced at the end of Chapter 5. If we write ${}^nP_r$ for the number of linear permutations of n things taken r at a time (i.e. r-fold samples from a set of n), we have there seen that ${}^nP_r = n^{(r)}$, and it is easy to remember what the foregoing numerical example illustrates, i.e.:

$$ {}^nR_r = n^r \quad \textit{and} \quad {}^nP_r = n^{(r)} $$

As we have seen in Chapter 5, one writes $n^{(n)}$ in the form $n!$, and can define our terms, so that:

$$ n^0 = 1; \quad n^{(0)} = 1; \quad 0! = 1 $$

In Chapter 5 we have also derived a formula for the number of combinations when sampling is *without replacement*, viz.:

$$^{n}C_r = \frac{n^{(r)}}{r!} = \frac{n!}{r!\,(n-r)!}$$

In the same context, we made the acquaintance with the following convention to economize space when writing binomial coefficients.*

$$n_{(r)} = \frac{n^{(r)}}{r!} = \frac{n!}{r!\,(n-r)!}, \quad \textit{and} \quad n_{(0)} = 1 = n_{(n)}$$

Thus we may write:

$$(1 + x)^n = A_0 \,.\, x^0 + A_1 \,.\, x^1 + A_2 \,.\, x^2 + A_3 \,.\, x^3 \ldots A_n \,.\, x^n$$
$$= n_{(0)} \,.\, x^0 + n_{(1)}x^1 + n_{(2)}x^2 + n_{(3)}x^3 \ldots n_{(n)} \,.\, x^n$$

Choice and Distinguishability. We can give the symbol $n_{(r)}$ a second meaning in terms of sampling without replacement and, as we shall see later, one which has more relevance to the role of the binomial theorem in the theory of probability. It is hence unsatisfactory to write (as above and in many textbooks) binomial coefficients in the form $^{n}C_r$. To understand an alternative meaning of $n_{(r)}$ let us return to our three-fold samples of the four letters *ABCD*. We there took for granted that there are $3(= 3! \div 2!)$ arrangements of samples such as *AAB* (*AAB*, *ABA*, *BAA*) in contradistinction to $6 = 3!$ arrangements of samples such as *ACD* (*ACD*, *ADC*, *CAD*, *CDA*, *DAC*, *DCA*). It is clear that by replacing by *C* one *A* in each of the *AAB* class we have doubled our possibilities, since *AA* is replaceable by *AC* or *CA*. This leads us to a more general rule.

Suppose that there are n letters of which a are *A*'s and $b = (n - a)$ are *B*'s. Let us label the number of arrangements of this set as $^{n}P_{a.b}$. If we replace all but one of the *A*'s by different letters (other than *B*) we can distinguish $a!$ different arrangements corresponding to any one in which the *A*'s occupy a distinguishably different set of places. The number of arrangements is therefore $a!$ times $^{n}P_{a.b}$. If we replace *B* by b different letters, none being the same as those we used to replace the *A*'s, we have made the total number of arrangements $b!$ times as great again, and we have now replaced all our n letters by different ones. If so, the total number of arrangements is then $^{n}P_n = n!$ Thus:

$$a!b!\,.\,^{n}P_{a.b} = {}^{n}P_n, \quad \textit{so that} \quad ^{n}P_{a.b} = \frac{n!}{a!\,(n-a)!} = n_{(a)}$$

* Before going farther, the reader may benefit from working the exercises of the Section A at the end of the chapter.

The reader should be able to generalize the relation further. We may write for the number of permutations of n things when a are indistinguishable members of one *class* A, b indistinguishable of another class B, c *ditto* class C, etc., and $n = a + b + c \ldots$

$$ {}^{n}P_{a.b.c\ldots} = \frac{n!}{a!\,b!\,c!\ldots} $$

This formula, which the reader may practise by working Set B in the exercises at the end of the chapter, is of use in consideration of score sums, e.g. the make up of a score of 14 in a four-fold toss of a die. Here some combinations may have all different elements (e.g. $1 + 3 + 4 + 6$), two elements alike (e.g. $1 + 3 + 5 + 5$) two classes of two alike (e.g. $2 + 2 + 5 + 5$), or one class of 3 alike ($2 + 4 + 4 + 4$).

The Chessboard Diagram. The distinction between two different *factual* meanings of the symbol $n_{(r)}$ is important because early writers on algebraic probability did not always clearly distinguish between whether by all possible ways in which the player could gain a particular score they meant all possible combinations or all possible permutations, least of all did they make explicit why (when consistent) they meant the latter. To

Fig. 180. Equality of Opportunity (Two-fold Choice)

Linear arrangements from a set of four:

With replacement – all pairs . . . . . . $4 . 4 = 4^2$

Without replacement – only block pairs . . . $4 . 3 = 4^{(2)}$

sidestep this source of bewilderment, one can use a visual device which is a powerful weapon in the theory of algebraic probability to build up successively, as in Figs. 180–1, permutations in an instructive way by introducing a new notion: *equality of opportunity to get together* (boy meets boy, boy meets girl, girl meets boy, girl meets girl). These two figures refer to a *universe* of four items, viz.: a pack consisting only of 4 cards: 1 spade, 1 heart, 1 diamond, and 1 club. If we replace each card before picking another, each of the n^2 (here 4^2) ways of extracting a two-fold sample corresponds to one of the pairs of an $n \,.\, n = n^2$ lattice. To each row of the lattice corresponds one item of the universe; and successive pigeon-holes of the same row exhibit the result of picking any one of the n cards after first drawing the one indicated in the left-hand vertical margin. If choice is not repetitive, one such pair drops out of each row, leaving $(n - 1)$ pairs per row and a total of $n(n - 1) = n^{(2)}$ pairs.

In visualizing two-fold samples of either type by this device, the underlying principle is that each card first taken has an *equal opportunity to associate with each remaining card*. We can extend the chessboard method to the representation of linear arrangements of three-fold samples or to larger samples by successive application of the trick employed in Fig. 180. If choice is repetitive, there are n^2 ways of arranging the two items first taken to make up a three-fold sample; and there are n ways of taking the third. Accordingly, we lay out the n^2 ways of extracting a two-fold sample in the left-hand vertical margin and assign each such sample *an equal opportunity* to pair off with each of the n items in the universe reconstituted by replacement of cards extracted before withdrawal. The resulting lattice now has $n^2 \,.\, n = n^3$ pigeon-holes. If there is no replacement the lattice will have only $n^{(2)}$ instead of n rows; and two triplets will drop out of every row, leaving $(n - 2)$ pairs per row or $n^{(2)}\,(n - 2) = n^{(3)}$ in all.

A Master-key Formula. The question to which the formula for ${}^{n}P_{a.b.c}$ is an answer excludes replacement; but the replacement condition is relevant if we ask a question such as the following: in how many ways can I choose from a full pack a sample of 10 cards of which 4 are hearts, 3 are diamonds, 2 are spades, and 1 a club? In more general terms we may formulate the question in this way: from a collection of n things of which a belong to class A, b to class B, c to class C, and so on, in how many ways can I choose r of which u belong to class A, v to class B, w to class C, and so on?

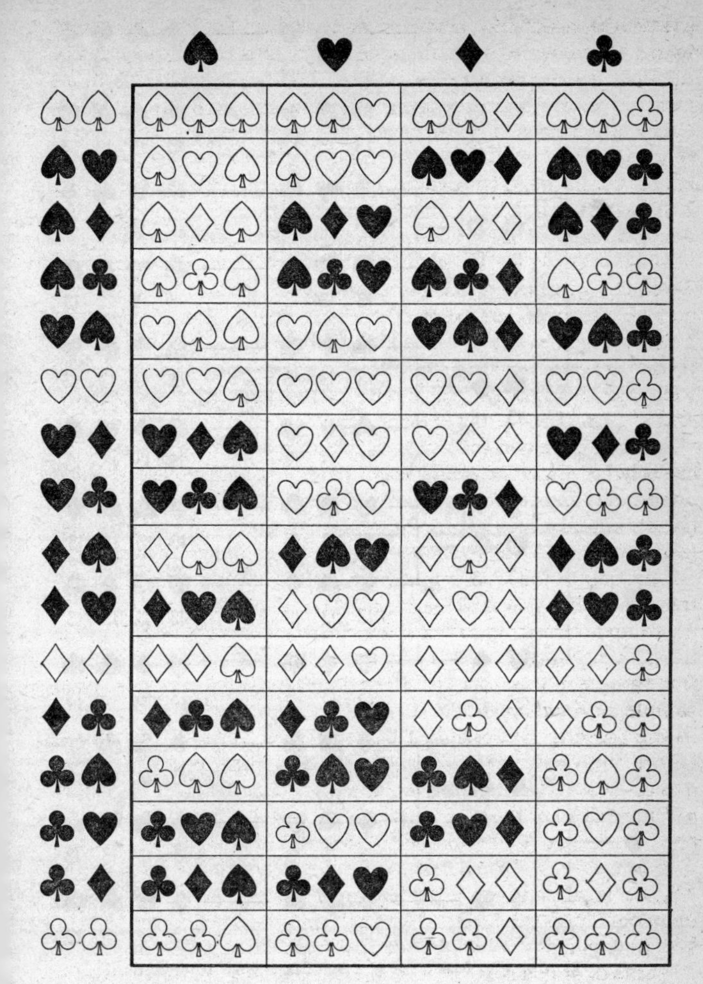

Fig. 181. Equality of Opportunity (Three-fold Choice)

Linear arrangements from a set of four:

With replacement – triplets . . . . . . $4^2 . 4 = 4^3$

Without replacement – only block triplets . . $4 . 3 . 2 = 4^{(3)}$

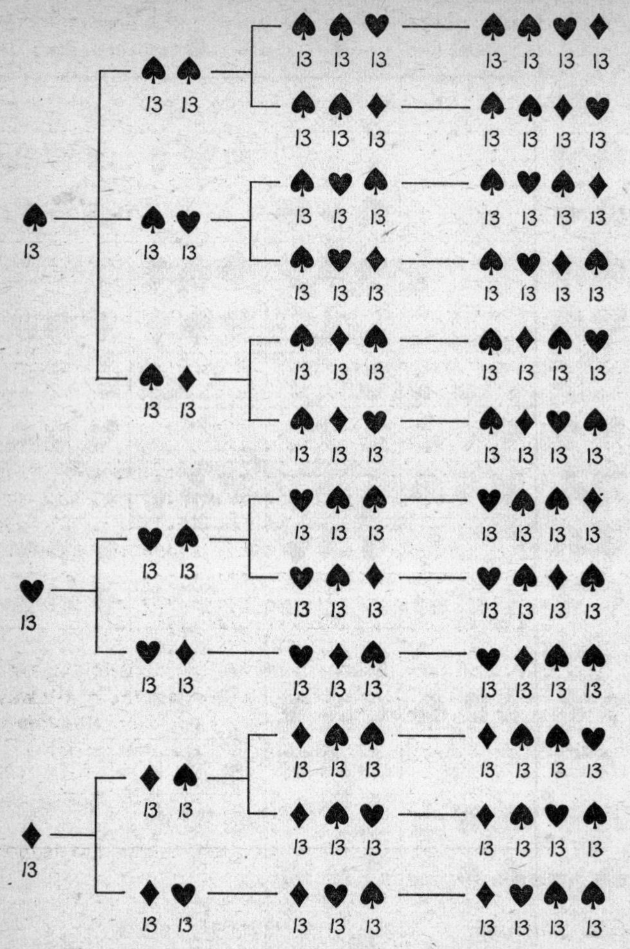

Fig. 182. The Master-key Formula – with replacement

Different permutations of taking from a full pack four cards of which 2 must be spades, 1 a heart, and 1 a diamond when the player replaces his cards and reshuffles before picking out another. The number is:

$$\frac{4!}{2!1!1!} . 13^2 . 13^1 . 13^1$$

Later in this chapter, we shall see that the answer to this question is a master key to the door which separates choice from chance. With the help of a picture (Figs. 182–3) it is not difficult to find it. A pattern emerges from the 3-class case, i.e. when there are only 3 classes, e.g. aces, picture cards, and others. In a full pack there are 4 of the first, 12 of the second and 36 of the other. Let us suppose that we want to know how many row-face-upward arrangements there are of a sample of 9 consisting of 4 aces, 3 picture cards, and 2 others. Now the number of *classes* of permutations of our 9 cards so specified is ${}^9P_{4.3.2}$, and we can dissect each of these classes into its ultimate constituents, if we are clear about whether we are or are not free to replace each of the 8 cards first chosen before choosing another.

If replacement is permissible, each of our ${}^9P_{4.3.2}$ different classes has 4 fixed places allocated to aces, hence ${}^4R_4 = 4^4$ ways of filling them, any one of which is consistent with ${}^{12}\mathrm{R}_3 = 12^3$ ways of filling 3 remaining places allocated to picture cards or $4^4 . 12^3$ ways of filling the places allocated to aces *and* picture cards, each way consistent with filling the remaining 2 places for a card which is neither an ace nor a picture in ${}^{36}\mathrm{R}_2 = 36^2$ ways. Thus the total number of ways of taking a nine-fold sample of 4 aces, 3 picture cards, and 2 others is:

$$4^4\ 12^3\ 36^2\ {}^9P_{4.3.2} = \frac{9!}{4!\,3!\,2!}\,4^4 . 12^3 . 36^2$$

The reader will now be able to retrace the steps of the argument, if there is *no* replacement, so that there are ${}^4P_4 = 4^{(4)}$ ways of filling the places filled by the aces in any one class of permutations. That is to say, the total number of permutations is:

$$4^{(4)} . 12^{(3)} . 36^{(2)}\ {}^9P_{4.3.2} = \frac{9!}{4!\,3!\,2!}\,4^{(4)} . 12^{(3)} . 36^{(2)}$$

Well, the pattern which answers our question stated in its most general terms now comes to life, viz.:

with replacement $$\frac{n!}{u!\,v!\,w!\,\ldots}\,a^u\,b^v\,c^w \ldots$$

without replacement $$\frac{n!}{u!\,v!\,w!\,\ldots}\,a^{(u)}\,b^{(v)}\,c^{(w)} \ldots$$

The two last formulae are going to steer us through a maze of misunderstandings about what the mathematician, in contradistinction to the plain man, means by probability. Often our

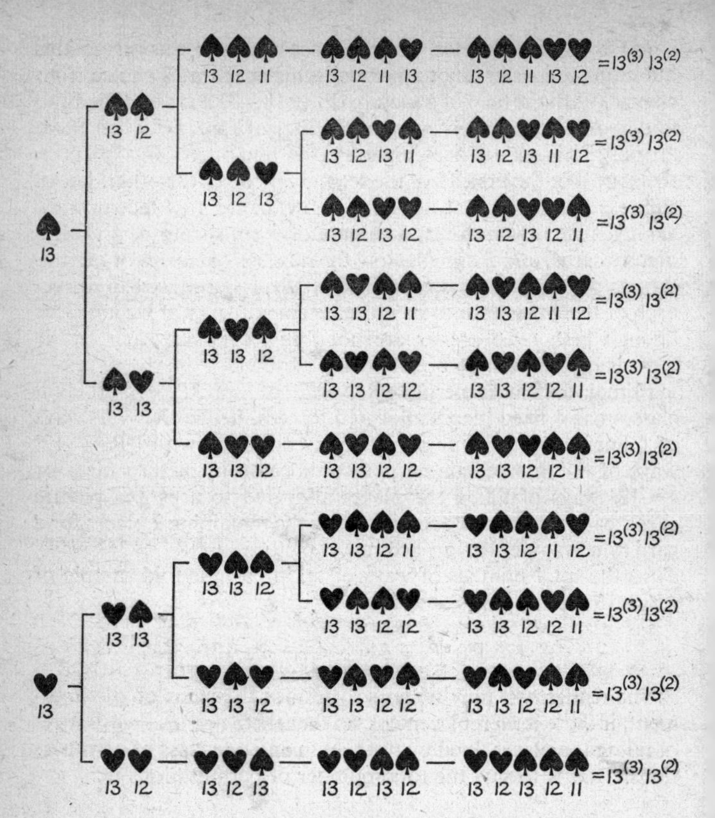

Fig. 183. The Master-key Formula – Without Replacement

Different permutations of five cards taken simultaneously from a full pack when 3 must be spades and 2 hearts. The number is:

$$\frac{5!}{3!\,2!} \cdot 13^{(3)} \cdot 13^{(2)}$$

concern will then be with systems of two classes, in which case we can speak of the choice of a card of one class (e.g. picture cards) as a *success* in contradistinction to the choice of any other card as a *failure*. Our pack of n cards thus consists of s cards of the first class and $f = (n - s)$ of the other. The master key formulae for taking x out of r cards of the class scored as success and hence $r - x$ of the class scored as failures are then:

with replacement $$\frac{r!}{x!\,(r-x)!}\, s^{x} f^{r-x}$$

without replacement $$\frac{r!}{x!\,(r-x)!}\, s^{(x)} f^{(r-x)}$$

There is one tricky point to notice. The master-key formula depends on the assumption that our classification of the pack is exhaustive, as is true of a two-fold classification such as picture cards and others or aces and others. If we are speaking of two-fold samples containing an ace and a picture card taken from a full pack, our classification of the pack itself is exhaustive only if we recognize a third class for 36 cards which are neither aces nor pictures. In this case, the use of our master-key formula requires clarification, as is evident by writing it down as below with the proviso that $n^{(0)} = 1!$:

with replacement $$\frac{2!}{1!\,1!\,0!}\, 4^{1}\, 12^{1}\, 36^{0}$$

without replacement $$\frac{2!}{1!\,1!\,0!}\, 4^{(1)}\, 12^{(1)}\, 36^{(0)}$$

In Chapter 5, we have defined 0! and $n^{(0)}$, so that $0! = 1 = n^{(0)}$; but we can here test the consistency of our definitions with what the theory of choice demands in two ways. First we know that the number of combinations of r out of n things is:

$$^{n}C_{r} = \frac{n!}{r!\,\overline{n-r}!}$$

The number of combinations of n things, if we take all of them is therefore:

$$^{n}C_{n} = \frac{n!}{n!\,\overline{n-n}!} = \frac{n!}{n!\,0!} = \frac{1}{0!}$$

Now there is only one combination of n things taken all at a time, i.e. $^{n}C_{n} = 1$, so that:

$$\frac{1}{0!} = 1 \quad and \quad 0! = 1$$

Our master-key formula demands the same meaning of 0! If we take 1 picture card from a pack, we can do so in 12 ways, and our implicit classification of the pack is two-fold, viz.: picture cards and others. Since replacement is irrelevant to a single choice, the master-key formula of either sort should give the same result, i.e.:

$$\frac{1!}{1!\,0!}\,12^1\,40^0 = 12 = \frac{1!}{1!\,0!}\,12^{(1)}\,40^{(0)}$$

From the left-hand side of this double equation, we see again that $0! = 1$. The right hand makes sense only if $40^{(0)} = 1 = 40^{(0)}$.*

Mathematical Probability. Against the background of the foregoing discussion of choice, we are ready to turn to the algebra of *chance*. This we may treat by asking the following questions:

(*a*) In what sense did the architects of the algebraic theory of probability themselves conceive it to be relevant to the real world?

(*b*) In what terms did they formulate the rules of the calculus and what latent assumptions about the relevance of the rules to reality do their axioms endorse?

(*c*) To what extent does experience confirm the factual relevance of the rules in the native domain of their application?

As regards the first of the above, there is little need to add more to what we learned at the beginning of this chapter about the fortunes and misfortunes of the Chevalier de Mérè. Even de Moivre, who undertook actuarial work for an insurance corporation, called his treatise (1718) on the subject: *Doctrine of Chances: or, A Method of Calculating the Probability of Events in Play*. Thus there is no ambiguity about what the Founding Fathers conceived to be the type specimen of real situations in which a calculus of probability is usefully applicable. At the outset, we should therefore be very clear about what we here mean by a betting rule in a game of chance.

Let us accordingly scrutinize a situation analogous to the dilemma of the Chevalier through the eyes of his own generation. Our task will be to state a rule for division of stakes to ensure success to a gambler who bets on the outcome of taking five cards *simultaneously* (i.e. without replacement) from a well-shuffled pack. We shall assume his bet to be that three of the

* Before going farther the reader may find it helpful to work through Set C of the exercises at the end of this chapter.

cards will be pictures and that two will be aces. In other words, the gambler *asserts* before each game that the hand of five will consist of three picture cards and two aces. Without comment on the rationale of the rule, we shall first illustrate what the operations of the calculus are, and then how we prescribe the rule deemed to be consistent with the outcome. If we approach our problem in the mood of the Founding Fathers of the algebraic theory, we shall reason intuitively as follows:

In a full deck of 52 there are $52^{(5)}$ ways in which the disposition of the cards may occur, this being the number of linear permutations of 52 objects taken 5 at a time without replacement. Out of these $52^{(5)} = 311,853,201$ ways in which the cards might occur, we can derive the number of ways in which the disposition is consistent with the bet by recourse to the *master-key formula* which tells us the number of *recognizably different linear permutations* of 5 things taken from 52 when: (*a*) we regard 4 of the 52 as identical members of class *Q* (*aces*), and 12 as identical members of class *P* (*pictures*); (*b*) the sample consists of 2 members of *Q* and 3 members of *P*. The number is:

$$\frac{5!}{2!\,3!} \cdot 4^{(2)} \,.\, 12^{(3)} = 5_{(2)} \,.\, 4^{(2)} \,.\, 12^{(3)} = 158,400$$

Without here pausing to examine the relation between such usage and what probability signifies in everyday speech, we shall now arbitrarily define in the classical manner the algebraic probability of a *success*, i.e. of the specified five-fold choice, and hence of the truth of the *assertion the gambler proposes to make*, by the ratio $5_{(2)} \,.\, 4^{(2)} \,.\, 12^{(3)} \div 52^{(5)} = 1 : 1,969$.

To get the terms of reference of the operations of the calculus vividly into focus in its initial domain of practice before making them more explicit, we may then postulate *fictitiously* that exactly one in every 1,969 games justifies the gambler's assertion. In this fictitious set-up, we shall suppose that:

(*a*) the gambler A bets that the result will be a success, and agrees to pay on each occasion if wrong a penalty of £*x* to his opponent B;

(*b*) his opponent likewise bets that the result will be a failure, and agrees to pay A a penalty of £*y* if wrong;

(*c*) each gambler adheres to his system throughout a sequence of 1,969 withdrawals.

If $x = 1$ and $y = 1{,}968$, A will part with £1 on 1,968 occasions or £1,968 in all, and B will part with £1,968 on only one occasion. Thus, neither will lose or gain in any completed 1,969-fold sequence. If $x = 1{\cdot}05$, so that A parts with a guinea (21 shillings) whenever wrong and $y = 1{,}968$ as before, A will part with £2,066 8*s* 0*d* in a complete 1,969-fold sequence, and B will gain £98 8*s* 0*d* at his expense. If $x = 1$ but $y = 2{,}000$, A will gain £32 in each such sequence, and B will lose the amount. On the foregoing assumption, therefore, A will always gain if two conditions hold good:

(*i*) A *consistently* follows the rule of asserting that the result of a five-fold withdrawal will be a success;

(*ii*) B agrees to pay a forfeit somewhat greater than £1,968 when A is right, on the understanding that A agrees to pay a forfeit of £1 if wrong.

Needless to say, the Founding Fathers did not postulate that the gambler A would score exactly one success in every 1,969-fold sequence. What they did claim is that the ratio of success to failure, i.e. *of true to false assertion*, would approach closer and closer to 1 : 1,968, if he went on making the same bet consistently in a *sufficiently extended succession of games*. Such then are the odds in favour of success in the idiom of the game. In so far as the classical formulation has any bearing on a calculus of truth and falsehood, it thus refers to the long-run frequency of correct assertion in conformity with *a rule stated in advance*. We are then looking forwards. Nothing we observe on any single occasion entitles us to deviate from adherence to it. Nothing we claim for the usefulness of the calculus confers a numerical valuation on a *retrospective judgement* referable to information derived from observing the outcome of a particular game.

In stating the foregoing rule and its application, we have identified the long-run ratio of the number of successes which *will* occur to the number of failures which *will* also occur with the ratio of the number of different ways in which success *may* occur to the number of different ways in which failure *may* occur. The italicized auxiliaries suffice to show that this calls for justification; but the proponents of the classical theory seem to have been quite satisfied to embrace the identification as a self-evident principle. Even so, we shall see that it does not suffice to prescribe the circumstances in which the calculus actually specifies long-term experience of situations to which they applied it, or even to prescribe any class of situations to which the rule might conceivably be relevant. This will emerge more clearly if we now formulate the definition of the concept of algebraic probability and the rules it endorses more explicitly.

We thus come to the second question we have set ourselves to answer: in what terms did the Founding Fathers formulate the rules of the calculus and what latent assumptions about the relevance of the rules to reality do the axioms condone? As regards games with picture cards, we have anticipated the answer to the first part of the question, viz. if there are b linear permutations of *all* cards which our sample might contain and a linear permutations of the cards consistent with scoring it as a *success*, the probability (p) of success is the ratio $a : b$. Accordingly, we speak of the probability (q) of failure as the ratio $(b - a) : b$, from which we see:

$$q = \frac{b - a}{b} = 1 - \frac{a}{b} = 1 - p$$

$$q = 1 - p \quad \text{and} \quad p = 1 - q$$

In the jargon of the gambler, we then say that the *odds* in favour of success are $a : (b - a)$ and of failure $(b - a) : a$. Thus if $p = 3/7$, so that $q = 4/7$, the odds in favour of success are 3 : 4, and in favour of failure 4 : 3.

When we are speaking of a die, our definition of the number of linear permutations involved leaves open the possibility that the same score, e.g. a 4 can occur 0–5 times in a five-fold sample. Thus the situation is comparable to drawing *successively* 5 cards from a pack, if one replaces each card (and shuffles the pack) before choosing another. If we had so prescribed the previous card game in which our criterion of success was a five-fold draw containing 3 pictures and 2 aces, the master-key formula specifies the probability of success as:

$$\frac{5!}{3!\,2!} \,.\, 4^2 \,.\, 12^3 \div 52^5 = \frac{270}{371293}$$

(Odds *against* success 371023 : 270.)

How to deal with the probability of getting a given total score in r tosses of a die, in terms consistent with our definition, the following example will suffice to show. Let us ask: what is the probability of getting a total score of 25 in the five-fold toss of an ordinary cubical die (1–6 pips on its 6 faces)? The possible combinations consistent with the specification and (in brackets) the number of permutations consistent with the ${}^{n}P_{a.b.c.}$ formula of p. 555 are as below:

66661 (5); 66652 (20); 66643 (20); 66553 (30); 66544 (30); 65554 (20); 55555 (1)

The total number of permutations consistent with the criterion of success is therefore $5 + 20 + 20 + 30 + 30 + 20 + 1 = 126$. The number of all possible five-fold scores is $6^5 = 7776$, so that if p is the probability of success:

$$p = \frac{126}{7776} = \frac{21}{1296} \quad \textit{and} \quad q = \frac{1275}{1296} \text{ (odds \textit{against} success 1275 : 21)}$$

The reader will now appreciate that algebraic probability defined as above is really *proportionate choosability* or, if you prefer it *proportionate possibility*. What it has to do with probability in the usual sense of the term depends on a conclusion illustrated by Figs. 180–1 and 184.

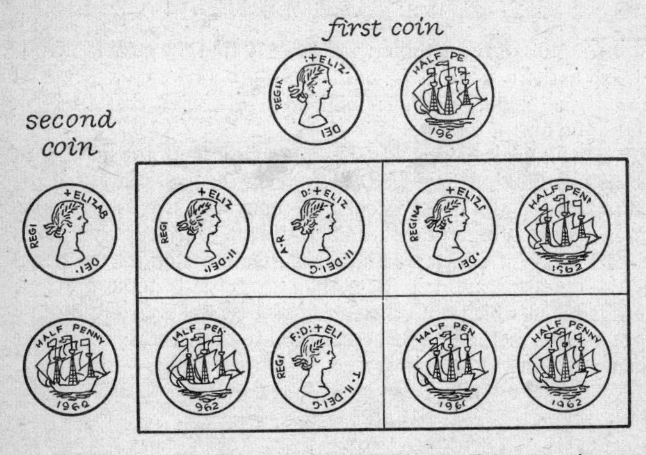

Fig. 184. Tossing Two Coins with Equality of Opportunity

By defining *different ways of choosing* a sample of particular composition in terms of the number of *different permutations* consistent with its structure, we have brought into the picture the notion of *equality of opportunity* for any item first chosen to

Fig. 185. Drawing One Card from Each of Two Packs:
(*a*) of 5 cards (2 spades + 3 clubs) (*b*) of 4 cards (3 spades + 1 club)

associate with any item remaining, and so on. We can recapture Pascal's attack on the problem, if we reason as follows. If I continue to shuffle a pack of cards, I give each card a better opportunity to pair off with any other; and if I go on picking samples (e.g. of three as above) from the pack, the long-run record of my experiment will tally with the consequences of equalizing the opportunity of the cards to do so. Experience of card games gives credibility to this assumption; but it is possible to test its truth. We shall later do so.*

The Three Basic Laws of Probability

Three simple rules sidestep many complications, when we set out to assign the chance of getting a particular result in a game of chance. We shall establish them first on the assumption that choice admits repetition, i.e. on the replacement principle if the model is a card pack or an urn. The behaviour of dice or coins necessarily comes within the framework of this assumption.

The Addition Rule. The foregoing examples illustrate a rule which applies to all problems of choice and chance when the prescribed selection involves the condition EITHER . . . OR. The rule is this: *if two or more selections are* MUTUALLY EXCLUSIVE *the probability of making one or other of them is the sum of their separate probabilities.*

Suppose a two-fold sample consists of a *red* ace and a king. The number of possibilities is then limited to 2 aces and 4 kings. If we replace the first card taken and reshuffle before taking another, the chance of getting such a sample is $\frac{6}{52} = \frac{3}{26}$. The chance of selecting a red ace is $\frac{2}{52}$. That of selecting a king is $\frac{4}{52}$, and our rule states that the chance of selecting *either* a red ace *or* a king of any suit is:

$$\frac{2}{52} + \frac{4}{52} = \frac{6}{52} = \frac{3}{26}$$

The rule of mutually *exclusive* choice applies to any size of sample. For instance, we might thus specify all the choices of the preceding example with their respective proportionate possibilities thus:

1 ace of hearts $\frac{1}{52}$ 1 king of diamonds $\frac{1}{52}$

* Here the reader may try out Set D of the exercises at the end of the chapter.

1 ace of diamonds	$\frac{1}{52}$	1 king of spades	$\frac{1}{52}$
1 king of hearts	$\frac{1}{52}$	1 king of clubs	$\frac{1}{52}$

By the addition rule, the chances of getting a single draw composed of either a red ace or a king is therefore:

$$\frac{1}{52}+\frac{1}{52}+\frac{1}{52}+\frac{1}{52}+\frac{1}{52}+\frac{1}{52}=\frac{6}{52}=\frac{3}{26}$$

The Product Rule. We can visualize the possibilities of a simultaneous draw with the right and left hand from two identical packs, or (what comes to the same thing) a double draw from one pack with replacement of the first card taken before removal of the second, by recourse to the chessboard device for exhibiting the implications of equality of opportunity. In the same way, we may exhibit the result of simultaneously drawing one card from each of two packs of *different* composition as in Figs. 185–6. The composition of the packs in them is as follows:

Left-hand Pack (4)	Right-hand Pack (5)
Ace of spades	Ace of spades
2 of spades	2 of spades
3 of spades	Ace of clubs
Ace of clubs	2 of clubs
. . . .	3 of clubs

There are 4 . 5 (= 20) two-fold permutations, i.e. 20 possibilities consistent with equality of opportunity:

Ace of spades twice	1	2 of spades and 3 of clubs	1
Ace of spades and 2 of spades	2	3 of spades and ace of spades	1
Ace of spades and ace of clubs	2	3 of spades and 2 of spades	1
Ace of spades and 2 of clubs	1	3 of spades and ace of clubs	1
Ace of spades and 3 of clubs	1	3 of spades and 2 of clubs	1
2 of spades twice	1	3 of spades and 3 of clubs	1
2 of spades and ace of clubs	2	Ace of clubs twice	1
2 of spades and 2 of clubs	1	Ace of clubs and 2 of clubs	1
		Ace of clubs and 3 of clubs	1
	—		—
Total	11		9

From this table we may extract the following data w.r.t. pairs of *like suit*:

Fig. 186. The Same as Fig. 185, showing Only the Suit of Each Card Picked

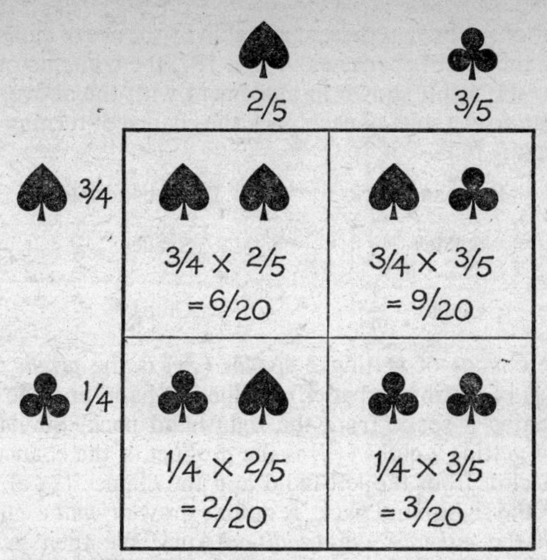

Fig. 187. The Balance Sheet of Figs. 185–6 Illustrating Both the Product Rule and the Addition Rule

(*a*) Two Spades		(*b*) Two Clubs	
Ace of spades *twice*	1	Ace of clubs *twice*	1
Ace of spades and 2 of spades	2	Ace of clubs and 2 of clubs	1
2 of spades *twice*	1	Ace of clubs and 3 of clubs	1
3 of spades and ace of spades	1		
3 of spades and 2 of spades	1		
Total	6	Total	3

Thus $6 + 3 = 9$ possibilities involve a draw of either 2 clubs or 2 spades, and the remaining $20 - 9 = 11$ involve a draw of one club and one spade. The mathematical probability we assign to choice of samples of these three classes are therefore:

2 spades $\quad \frac{6}{20} = \frac{3}{10} = \frac{3}{4} \times \frac{2}{5}$

2 clubs $\quad \frac{3}{20} = \frac{1}{4} \times \frac{3}{5}$

a club and a spade $\quad \frac{11}{20} = \left(\frac{3}{4} \times \frac{3}{5}\right) + \left(\frac{1}{4} \times \frac{2}{5}\right)$

If our concern is only whether we draw one or other of these combinations, we can represent any club by the ace of clubs or any spade by the ace of spades as in Fig. 187, the contents of which reduce to the result shown in agreement with the above. Let us now assign to the suits of each pack the chance of turning up at a single draw:

Left-hand Pack		Right-hand Pack	
Spades	$\frac{3}{4}$	Spades	$\frac{2}{5}$
Clubs	$\frac{1}{4}$	Clubs	$\frac{3}{5}$

Thus the chance of getting 2 spades ($\frac{3}{10}$) is the *product* of the chance ($\frac{3}{4}$) of getting a spade from the left-hand and the chance ($\frac{2}{5}$) of getting a spade from the right-hand pack. Similarly the chance of getting 2 clubs ($\frac{3}{20}$) is the product of the chance ($\frac{1}{4}$) of getting a club from the left-hand and the chance ($\frac{3}{5}$) of getting one from the right-hand pack. A double draw of *unlike* cards also brings in the *either . . . or* (*addition*) rule. We then have two *exclusive* possibilities, left spade–right club or right spade–left club. By the product rule these are respectively ($\frac{3}{4} \times \frac{3}{5}$) $= \frac{9}{20}$ and ($\frac{1}{4} \times \frac{2}{5}$) $= \frac{2}{20}$. The sum is $\frac{11}{20}$ as shown above.

The reader may here find it helpful to construct tables similar to the above for the following (1–10) pairs of incomplete packs of cards all of the *same* suit. Such tables provide all the data for testing out the product and addition rules, by assigning the chance of getting a two-fold sample of one or other sort (*a–d*) below:

(*a*) Double aces.
(*b*) Doubles other than aces.
(*c*) Mixed pairs.
(*d*) Double even numbers.

1. L: A 2 3 4
   R: A 2 3 4
2. L: A 2
   R: A 2 3 4 5
3. L: A 2 3 4
   R: A 2 3
4. L: A 2 3
   R: A 2 3
5. L: A 2 3 4 5
   R: A 2 3 4
6. L: A 2
   R: A 2 3 4
7. L: A 2 3 4 5 6
   R: A 2 3 4
8. L: A 2 3 4 5
   R: A 2 3 4 5
9. L: A 2 3 4
   R: A 2 3 4 5 6 7
10. L: A 2 3 4 5 6
    R: A 2 3 4 5 6

The Subtraction Rule. If our classification of proportionate possibilities, i.e. chances, is *exhaustive*, the total must equal *unity*, as in our last example, viz.:

2 spades	2 clubs	Both suits	Total
$\frac{6}{20}$	$\frac{3}{20}$	$\frac{11}{20}$	$\frac{20}{20} = 1$

If we classify our selections in 2 ways according as they have or have *not* a particular characteristic, the probability p of having the characteristic and q, that of *not* having it, therefore add up to 1, i.e. $(p + q) = 1$. Hence:

$$p = 1 - q \quad and \quad q = 1 - p$$

For example, the respective chances of getting and not getting a *red royal* (K *or* Q) card at a single draw from a full pack are $\frac{4}{52}$ and $(1 - \frac{4}{52}) = \frac{48}{52}$ or $\frac{1}{13}$ and $\frac{12}{13}$. It is often possible to short-circuit work involved in solving problems of sampling by using this *subtraction rule*, more especially if they call for a determination of the chance of getting *at least one* combination of a particular category, as when we ask: *what is the chance of getting* AT LEAST ONE *red card in a single draw from each of two full packs?*

Here, the important point to grasp is that we must *either* get no red card in the combined draw *or* at least one. In other words, this two-fold classification is exhaustive, i.e. complete in itself. Hence the chances we assign to the two classes of combinations so defined must add up to unity. Now the chance that a card will be black in a single draw from one pack is $\frac{1}{2}$, which is of course the chance that it will *not* be red. The chance (p) that both cards of a simultaneous draw from two full packs will be black is (by the product rule):

$$\tfrac{1}{2} \times \tfrac{1}{2} = \tfrac{1}{4}$$

Hence the proportionate possibility (q) that both cards will *not* be black is:

$$1 - q = 1 - \tfrac{1}{4} = \tfrac{3}{4}$$

Since the condition that both cards will not be black is another way of stating that *at least one* will be red, this is the required answer.

It is just as simple to answer the question: what is the chance of getting at least one heart in eight successive draws with replacement? By the product rule, the chance of getting 8 cards which are *not* hearts is $(\frac{3}{4})^8$. Thus the chance of getting at least one heart is:

$$1 - (\tfrac{3}{4})^8 = 1 - \frac{6561}{65536} = \frac{58975}{65536}$$

The respective probabilities of getting at least one heart and of getting none at all are thus 58975 : 6561 or roughly 9 : 1. We customarily express this saying the odds are roughly 9 : 1 in favour of getting at least one heart in the eight-fold draw.

The occasion which prompted the celebrated controversy between Pascal and Fermat furnishes a neat illustration of our negation rule for the *at least one* type of problem. The Chevalier de Mérè had made a fortune by betting small favourable odds on getting at least one six in 4 tosses of a die, and lost it by betting small favourable odds on getting a double six in 24 double tosses. The subtraction rule gives the explanation.

Chance of getting at least one six in 4 tosses:

$1 - (\frac{5}{6})^4 = 0{\cdot}517$ (*more* likely than not).

Chance of getting a double six in 24 double tosses:

$1 - (\frac{35}{36})^{24} = 0{\cdot}491$ (*less* likely than not).*

The Meaning of Independence. It is common to state the product rule as follows: the probability of a combination of two or more *independent* events is the product of their individual probabilities. The key word in this statement is *independent*. A simultaneous draw from two card packs or a simultaneous toss of two dice are independent events; and the same is true of successive tosses of one and the same die. It is not necessarily true of *successive* draws from the same card pack. The result of tossing a die a second time is not dependent on what happened the first time, because the number of faces remains the same; but a similar statement applies to a card pack *only* if we replace the card previously drawn before picking on another. It cannot be true of a simultaneous two-fold draw.

Let us now therefore see how we can apply our three fundamental rules to sampling without replacement. We shall suppose that our pack consists of 8 different cards, classified by suit as follows:

Clubs 1; *Hearts* 3; *Diamonds* 4

* Before going further the reader may usefully try Sets E–G of the exercises at the end of the chapter.

Let us call this pack, as initially constituted, *pack* A. If we draw a card from it, it must be a club, a heart or a diamond. We are thus left with one of 3 *residual* 7-card packs respectively composed of:

	Clubs	Hearts	Diamonds
B	0	3	4
C	1	2	4
D	1	3	3

Thus the chances of drawing a club, heart, or diamond from the original (A) and the residual (B–D) packs are:

	Clubs	Hearts	Diamonds
A	$\frac{1}{8}$	$\frac{3}{8}$	$\frac{1}{2}$
B	0	$\frac{3}{7}$	$\frac{4}{7}$
C	$\frac{1}{7}$	$\frac{2}{7}$	$\frac{4}{7}$
D	$\frac{1}{7}$	$\frac{3}{7}$	$\frac{3}{7}$

We can now regard the problem of drawing two cards without replacement in a new way, viz. as that of choosing a card from pack (A) with the left hand and choosing a card from another pack (B, C, *or* D) with the right hand. We can regard the card we draw from A as a lottery ticket which entitles us to draw from one or other of the packs B, C, D, the composition of each of which is fixed. Drawing from one and the same pack A *first* a club and *then*, without replacing it, a heart is thus equivalent to two acts of choice which are truly *independent*:

(i) Choosing a *club* from pack A.

(ii) Choosing a *heart* from pack B.

We may therefore apply the product rule for independent events. The chances of (i) and (ii) are, respectively, $\frac{1}{8}$ and $\frac{3}{7}$. Hence that of the combined choice is $\frac{1}{8} \times \frac{3}{7} = \frac{3}{56}$. In this way we can make a complete balance sheet of probabilities for the combined acts of choice:

Club–*club*	0	Club–*heart*	$\frac{3}{56}$	Club–*diamond*	$\frac{4}{56}$
Heart–*club*	$\frac{3}{56}$	Heart–*heart*	$\frac{6}{56}$	Heart–*diamond*	$\frac{12}{56}$
Diamond–*club*	$\frac{4}{56}$	Diamond–*heart*	$\frac{12}{56}$	Diamond–*diamond*	$\frac{12}{56}$

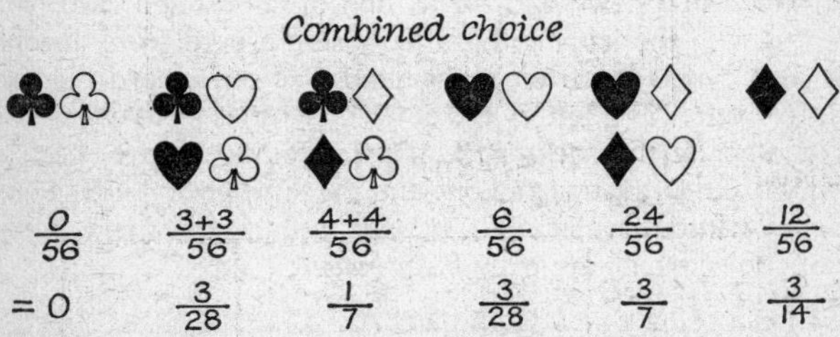

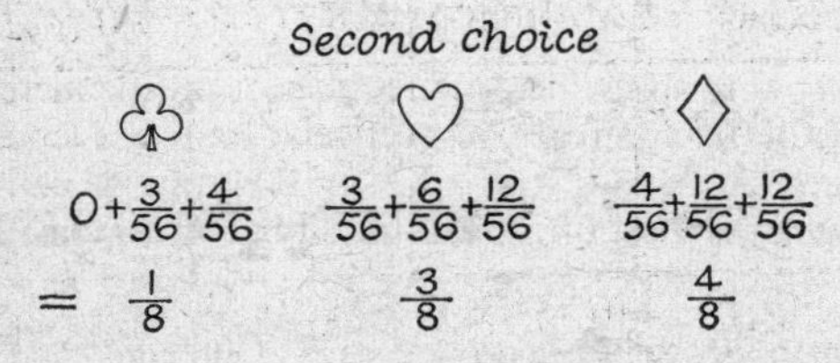

Fig. 188. Staircase Model

For Two-fold choice *without* replacement.

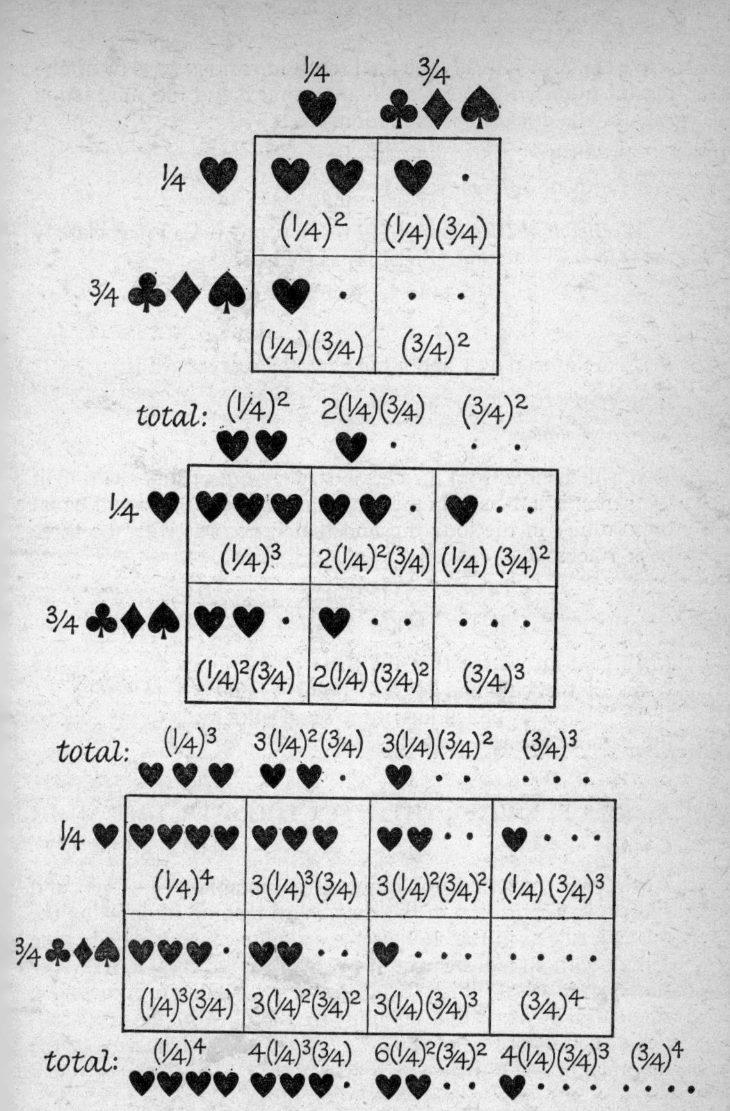

Fig. 189. The Binomial Replacement Distribution

We can also* classify the final result in various ways by applying the addition rule for alternative choice and the subtraction rule for the *at least one* condition, e.g.:

Two cards of the *same* suit $0 + \frac{6}{56} + \frac{12}{56} = \frac{9}{28}$

Cards of *different* suits $1 - \frac{9}{28} = \frac{19}{28}$

The Binomial Distribution. If I toss a penny twice I may classify the four-fold outcome (Fig. 184) as follows:

Result	2 tails	Unlike faces	2 heads
Chance	$\frac{1}{4}$	$\frac{1}{2}$	$\frac{1}{4}$

If I score a head as 1 and a tail as 0, the scoreboard is:

Score	0	1	2
Chance	$\frac{1}{4}$	$\frac{1}{2}$	$\frac{1}{4}$

This is deducible from the chessboard lay-out of Fig. 184 only if the penny is unbiased in the sense that it obeys the law of equal opportunity in the long-run and that it accords with the three basic rules:

TT	TH or HT	HH
$(\frac{1}{2})(\frac{1}{2})$	$(\frac{1}{2})(\frac{1}{2}) + (\frac{1}{2})(\frac{1}{2})$	$(\frac{1}{2})(\frac{1}{2})$

In this set-up we see that the three probabilities are successive terms of the binomial $(\frac{1}{2} + \frac{1}{2})^2 = (\frac{1}{2})^2 + 2(\frac{1}{2})^2 + (\frac{1}{2})^2$. For the three-fold toss, application of the same rules leads to the following balance sheet:

Score	0	1	2	3
Result	TTT	TTH, THT, HTT	THH, HTH, HHT	HHH
Chance	$(\frac{1}{2})^3$	$3(\frac{1}{2})^3$	$3(\frac{1}{2})^3$	$(\frac{1}{2})^3$

Again, we get successive terms of a binomial: $(\frac{1}{2} + \frac{1}{2})^3$; and successive application of the chessboard method of applying the product rule as in Fig. 189 discloses a quite general law when we sample without replacement. If we score our results as 0, 1, 2 . . . r successes in an r-fold trial denoting the probability of success in a single trial by p and that of failure by $q = (1 - p)$, the probabilities of getting scores of 0, 1, 2 . . . r are successive terms of the binomial $(q + p)^r$, i.e.:

* Here the reader may try out Set H of the exercises at the end of this chapter.

Score	0	1	2	etc.
Chance	q^r	$rq^{r-1} \cdot p$	$\frac{r(r-1)q^{r-2}p^2}{2!}$	etc.

Now we could get to this result right away by using our master-key formula. Let us suppose that there are n cards in a pack of which s (successes) are of one sort and $f = (s - n)$ are of another, our classification is then exhaustive. If we replace, the number of all possible arrangements of r cards is ${}^nR_r = n^r$. The number of r-fold samples which contain x successes and $(r - x)$ failures is:

$$\frac{r!}{x!\,(r-x)!}\,s^x f^{r-x}$$

The proportionate contribution of samples so specified to all possible r-fold samples is therefore:

$$\frac{r!}{x!\,(r-x)!}\,\frac{s^x f^{r-x}}{n^r} = \frac{r!}{x!\,(r-x)!}\left(\frac{s}{n}\right)^x\left(\frac{f}{n}\right)^{r-x}$$

In this expression $(s \div n)$ is the proportion (p) of cards classified as successes, and $(f \div n)$ is the proportion (q) of cards classified as failures, since the probability of success at a single trial is simply $(s \div n)$. Thus the chance of getting x successes is:

$$\frac{r!}{x!\,(r-x)!}\cdot q^{r-x}p^x$$

We have already learned (Chapter 6) that this is the term of rank x in the expansion of $(q + p)^r$, when we label the initial term as the term of rank 0. Fig. 190 shows that an analogous law governs sampling with replacement and incidentally illustrates a rule known as *Vandermonde's Theorem*, i.e. the expansion of the binomial in *factorial indices*, e.g.:

$$\begin{aligned}(f+s)^{(2)} &= (f+s)(f+s-1)\\ &= f(f-1) + 2fs + s(s-1)\\ &= f^{(2)} + 2f^{(1)}s^{(1)} + s^{(2)}\end{aligned}$$

For non-replacement the 2-class law of distribution is summarized in the expansion $(f + s)^{(r)} \div n^{(r)}$, so that the probability of x successes (Fig. 190) is:*

$$\frac{r!}{x!\,(r-x)!}\,\frac{f^{(r-x)}s^{(x)}}{n^{(r)}}$$

* Here the reader may try out Set J of exercises at the end of the chapter.

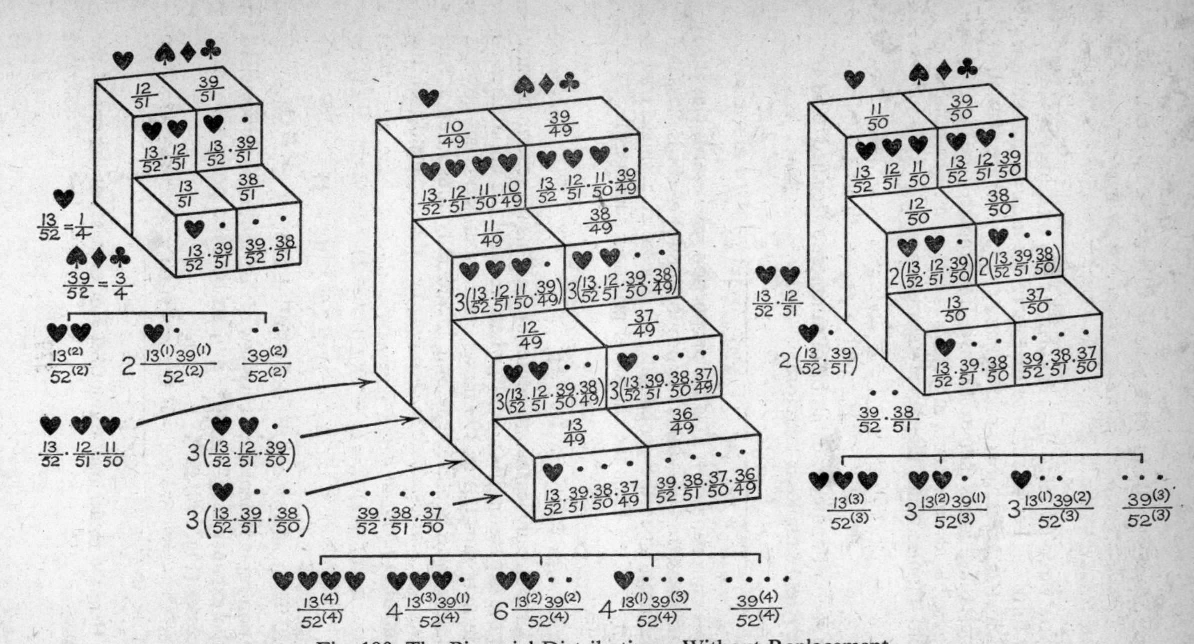

Fig. 190. The Binomial Distribution – Without Replacement

Division of the Stakes. The relevance of the binomial distribution of the division of the stakes depends on a simple application of the *addition rule*. If $p = 1 - q$ is the probability of scoring a success in a single game, we may set this out for a sequence of five games as follows:

Number (*s*) of successes	Probability	Not more than *s* successes
0	q^5	q^5
1	$5pq^4$	$q^5 + 5pq^4$
2	$10p^2q^3$	$q^5 + 5pq^4 + 10p^2q^3$
3	$10p^3q^2$	$q^5 + 5pq^4 + 10p^2q^3 + 10p^3q^2$
4	$5p^4q$	$q^5 + 5pq^4 + 10p^2q^3 + 10p^3q^2 + 5p^4q$
5	p^5	$q^5 + 5pq^4 + 10p^2q^3 + 10p^3q^2 + 5p^4q + p^5$

When $p = q$ and $p^5 = 2^{-5} = q^5$, we may write the above in the form:

Successes . (*s*)	0	1	2	3	4	5
Probability of *s* (×32)	1	5	10	10	5	1
Probability of *not more* than *s* (×32)	1	6	16	26	31	32(= 2^5)

Now let us consider a situation in which each game is the outcome of a three-fold toss of a tetrahedral die numbered with 1–4 pips on one or other of its four triangular faces. We assume that: (*a*) the recorded score is the number of pips on the face it falls on; (*b*) the long run frequency of each face score is equal. By means of a 1 × 1- and 1 × 2-toss chessboard diagram, we may then derive the following distributions for the probabilities (P_1, P_2, P_3) of scores obtainable from 1, 2, and 3 tosses as below:

Scores	1	2	3	4	5	6	7	8	9	10	11	12
$P_1 \times 4$	1	1	1	1	—	—	—	—	—	—	—	—
$P_2 \times 16$	—	1	2	3	4	3	2	1	—	—	—	—
$P_3 \times 64$	—	—	1	3	6	10	12	12	10	6	3	1

From the above we see that the probability of scoring less than 8 in the three-fold toss is:

$$\frac{1}{64}(1 + 3 + 6 + 10 + 12) = \frac{32}{64} = \frac{1}{2}$$

If therefore we count as a *success* a score of 8 or more, the probability of success is $p = 0{\cdot}5 = q$; and for sequences of 6 games and 12 games, respectively, the probabilities of $P_{s.6}$ and $P_{s.12}$ successes are as shown below:

Score (s)	0	1	2	3	4	5	6	7	8	9	10	11	12
$P_{s.6} \times 4096$	64	384	960	1280	960	384	64	—	—	—	—	—	—
$P_{s.12} \times 4096$	1	12	66	220	495	792	924	792	495	220	66	12	1

We now suppose that:

(*a*) a gambler A decides to bet *consistently* that the score in a game (i.e. three-fold toss) will be over 7 (8 or more), and shall speak of such a score as his success:

(*b*) A agrees to pay B \$$y$ from a failure, i.e. if he (A) is wrong (score < 8); and B agrees to pay \$$x$ for a success to A, i.e. if he (A) is right (score > 7).

On these assumptions the net gain (+) or loss (−) of A (in dollars) for a score of s successes is:

6-game sequence: $s \, . \, x - (6 - s)y = s(x + y) - 6y$

12-game sequence: $s \, . \, x - (12 - s)y = s(x + y) - 12y$

For instance, if A receives \$3 for each success and B receives \$2 for each failure, the net gain of A for five successes is: (*a*) +13 in a six-fold sequence; (*b*) +1 in a twelve-fold sequence. We may therefore draw up balance sheets for stakes as follows: (i) $x = 3$, $y = 2$; (ii) $x = 4$, $y = 2$; (iii) $x = 4$, $y = 3$. Table I shows the balance sheet for a six-fold sequence of games, Table II shows one for a twelve-fold sequence. The number of games in the two sequences are small to facilitate computation by straightforward arithmetic; and, above, the probabilities disclosed are for the same denominator ($2^{12} = 4096$) to make comparison easy at a glance.

Next, let us notice that it is possible, though not very likely, for A to lose every game in an n-fold set. If so, he forfeits ny dollars on the foregoing assumption that he has to pay a forfeit of \$$y$ for every failure. If he is a prudent gambler he will propose a division of stakes acceptable to his opponent B on the assumption that he can afford to lose no more than this amount in a prearranged number (n) of games. For instance, if B receives \$2 for each failure of A in a sequence of 6 games he must be prepared to lose \$12, and B receives \$3 he must be prepared to lose \$18.

The probability that he will lose all he can afford to lose is in either case only:

$$\frac{64}{4096} = \frac{1}{64}$$

For stakes $x = 3$, $y = 2$ and $x = 4$, $y = 3$, Table I shows the probability that A will lose at least $2 is:

$$\frac{1408}{4096} = \frac{22}{64}$$

If the stakes are $x = 4$, $y = 2$, the probability that he gains nothing is the same; but he may neither lose nor gain, and this is also the probability that the 6 games end with the *status quo*. He can then play another round of 6 without abandoning his original resolve.

Table I

(All probabilities multiplied by 4,096)

Successes (s) in 6 games	Probability of s	Probability of s or less	A's net gain $x = 3$, $y = 2$	$x = 4$, $y = 2$	$x = 4$, $y = 3$
0	64	64	−12	−12	−18
1	384	448	−7	−6	−11
2	960	1,408	−2	0	−4
3	1,280	2,688	+3	+6	+3
4	960	3,648	+8	+12	+10
5	384	4,032	+13	+18	+17
6	64	4,096	+18	+24	+24

Otherwise, if he plays a sequence of 12 games with the same intention he will need an initial outlay of twice as much, i.e. $24 for stakes 3 : 2 or 4 : 2 and $36 for stakes 4 : 3. However, the chance (Table II) of losing all he is willing (or able) to part with is now very much smaller being: $(1 \div 4096)$. Thus the risk of losing all in a 6-game sequence is 64 times as great as it would be in a 12-game sequence. The same table also shows the probability of making no gain when the stakes are 3 : 2 and 4 : 2 is less in the twelve-fold than in the six-fold sequence, being:

$$\frac{794}{4096} = \frac{397}{2048}$$

We now assume that A had lost all he has to lose in a game sequence, and we bring into the picture a gambler C who requires

TABLE II

(All probabilities multiplied by 4,096)

Successes (s) in 12 games	Probability of s	Probability of s or less	A's net gain $x = 3, y = 2$	A's net gain $x = 4, y = 2$	A's net gain $x = 4, y = 3$
0	1	1	−24	−24	−36
1	12	13	−19	−18	−29
2	66	79	−14	−12	−22
3	220	299	−9	−6	−15
4	495	794	−4	0	−8
5	792	1,586	+1	+6	−1
6	924	2,510	+6	+12	+6
7	792	3,302	+11	+18	+13
8	495	3,797	+16	+24	+20
9	220	4,017	+21	+30	+27
10	66	4,083	+26	+36	+34
11	12	4,095	+31	+43	+41
12	1	4,096	+36	+48	+48

$4 from his opponent if he wins a game but is prepared to pay a forfeit of $3 when he loses one. Thus, the most he can lose in a 12-game sequence is $36 with a probability $\frac{1}{4096}$. We may now explore three situations in which we assume that A has lost all in his first sequence of 6 games:

(i) C has at his disposal $48 and is prepared to lend A $12 to continue his play with B at stakes of 3 : 2 or 4 : 2 as before, while he himself plays with another opponent on the foregoing assumption that he receives $4 from a success and parts with $3 for a failure.

In this event, C can lose all he has at his disposal only if A loses every game in his six-fold sequence ($12) and he himself loses every game in his 12-fold sequence ($36). These are independent events to which the product rule applies. Hence the probability of the joint one is exceedingly small, being:

$$\frac{64}{4096} \times \frac{1}{4096} = \frac{1}{262{,}144}$$

Thus the odds against this calamity are somewhat more than a quarter of a million to one, *i.e.* 262,143 : 1.

(ii) C has at his disposal the same amount and is prepared to lend to A as much as A can lose ($48), if A will pay a $4 forfeit when C wins a game in return for receipt of a $3 forfeit when C loses one in a set of 12.

In this event, C still has at his disposal $12 if A wins every game. The probability of this is the same as the probability that C will win every game being:

$$\frac{1}{4096}$$

(iii) C agrees to lend A sufficient to go on playing if A will pay him 25 per cent of his earnings, C setting the limit to the number of games and A setting the stakes favourably to himself with an opponent B.

To safeguard him from being in the red, C needs to be able to afford only $12 for a 6-game set with stakes $x = 3$ or 4, $y = 2$, and only $18 for stakes $x = 4$, $y = 3$; and his greatest possible gain is $6. If he sets the limit at 12 games, he can safeguard himself against being in the red if able to afford $24 (*stakes* 3 or 4 : 2) or $36 (*stakes* 4 : 3) as a loan to A; and his greatest possible gain is $12. The chances of losing *all* his loan are:

$$\text{6 game set} \quad \frac{1}{64}$$

$$\text{12 game set} \quad \frac{1}{4096}$$

To sum up the lesson of these model situations, we may say:

(i) for equally favourable stakes, the gambler with more capital can continue playing longer with decreasing risk of losing all of it;

(ii) a gambler A with more capital can afford to offer his opponent B less unfavourable stakes;

(iii) a money-lender with sufficient capital can finance a gambler A to go on playing in return for a part share in his gains with negligible risk of total loss and an increasing probability of some gain.

If the reader remarks that B is a mug to agree to a division of the stakes favourable to A, this is merely another way of saying that there would be no bankrupt participants in games of pure chance if there were no mugs.

It is now time to be more clear about the second part of the second question raised at the beginning of our discussion of the algebra of chance: what *latent* assumptions about the relevance of the rules to reality do the axioms of the Founding Fathers endorse?

It is not possible to provide an exhaustive answer to this in the space at our disposal; but we can put the spotlight on a few of the issues. On the assumption that A can find a mug, he can rely on the mathematical treatment of the problem only if two other conditions hold good:

(i) Having agreed *in advance* about the division of the stakes and what constitutes a success for A, the latter must adhere to the agreement regardless of the fortunes of the game. We exclude the possibility of the *Backward Look*, i.e. change of stakes and/or of the criterion of success in the light of previous disappointments.

(ii) Having prescribed that both the claim, i.e. the *assertion* A *makes* as a criterion of success (e.g. $s > 7$ in each game consisting of a three-fold tetrahedral die), we must insist also that:

> (*a*) the physical apparatus (*die*, *wheel*, *card pack*) remains the same throughout the entire sequence;
>
> (*b*) the procedure (i.e. method of shuffling or tossing) must ensure *randomization* before each game, i.e. in the terms previously used, *equality of opportunity for association.*

Clearly therefore, gambling on the outcome of games of skill (e.g. baseball, greyhound, or horse-racing) has nothing to do with the mathematical theory of probability.

The trickiest part of this catalogue of requirements is the meaning of the word *random*. In the opinion of the writer, current manuals which distribute cookery book recipes for statistical tests abound with word-magic about choosing a *random sample*. This expression implies a fundamental misunderstanding. Randomization is a process which does not terminate. We can therefore speak meaningfully of *random sampling* only as a *repetitive* procedure for generating successive samples. How we can prescribe such a procedure is still a controversial issue when we transgress the boundaries of games of chance; but it is at least possible to recognize within their own domain essential features common to that of at least one other class of situations, i.e. errors of observation in performing scientific experiments or repeated measurements of the same natural event.

A game is a game of chance only if we assume conformity to a programme of instructions for shuffling, tossing, or spinning with the ostensible aim of imposing *disorder* on successive observations. Three features are common to such programmes. All impose the condition that the *sensory discrimination of the agent is impotent to influence the outcome*. All embrace the possibility that strict adherence to the programme is consistent with the realizable occurrence of every *conceptually possible outcome* in a single trial. All exclude the possibility that any orderly rhythm of external agencies intervenes to impose a periodicity on successive trials. Whether these three features suffice to specify a randomizing process as such is difficult, if not impossible, to prove; but they confer some plausibility on the comparison between some categories of observational error and the results of tossing a die.

Theory and Practice. We have still to find a satisfactory answer to the third question raised on p. 562, viz. *to what extent does experience confirm the factual relevance of the rules on the original domain of their application?*

At the outset, let us admit that few of us doubt the reality of some factual basis for the faith of the Founding Fathers, and the relevance of their faith to the fate and fortunes of the Chevalier. A few investigations recorded in more recent times will indeed encourage us still to explore hopefully the credentials of a doctrine with so little ostensible relevance to its applications at the most elementary level. All we shall ask of the theory at this stage is that it works, if the gambler goes on long enough. We here interpret this to mean that:

(*a*) over 1,000 is a large enough number of games in the context;

(*b*) as a criterion of what works, we shall be satisfied with a correspondence between theory and observation, if it conforms to the familiar standard set by the tabular content of any current textbook of physical chemistry.

Karl Pearson (*Biometrika* 16) cites the following counts for 3,400 hands (of 13 cards) at whist, with corresponding probabilities calculated in accordance with the distribution defined by the terms of $(13 + 39)^{(13)} \div 52^{(13)}$:

No. of Trumps per hand	No. of hands observed	No. of hands expected
under 3	1,021	1,016
3–4	1,788	1,785
over 4	591	599

Uspensky (*Introduction to Mathematical Probability*) records two experiments of this type. The first is of seven experiments each based on 1,000 games in which the score is a success if a card of each suit occurs in a four-fold simultaneous withdrawal from a pack without picture cards. The probability of success is thus:

$$\frac{4!}{1!\,1!\,1!\,1!}\;\frac{10^{(1)}\,10^{(1)}\,10^{(1)}\,10^{(1)}}{40^{(4)}} = 0{\cdot}1094$$

The outcome for the seven successive 1,000-fold trials (I–VII) was as follows:

I	II	III	IV	V	VI	VII
0·113	0·113	0·103	0·105	0·105	0·118	0·108

A second experiment subsumes 1,000 games in which the score for the five-fold withdrawal from a full pack is the number of different denominations cited below with the corresponding theoretical and observed proportions correct to three places:

	1 + 1 + 1 + 1 + 1	2 + 1 + 1 + 1	2 + 2 + 1
Observed	0·503	0·436	0·045
Expected	0·507	0·423	0·048
	3 + 1 + 1	3 + 2	4 + 1
Observed	0·014	0·002	0·000
Expected	0·021	0·001	0·000

In the domain of die models, results obtained from experiments on the needle problem of the eighteenth-century French naturalist Buffon are both relevant and arresting. The gamester drops a needle of length l on a flat surface ruled with parallel lines at a distance (h) apart, scoring a success if it falls across, and a failure if it falls between them. Theory prescribes that the ratio of success to failure involves π, the probability of success being $2l/h\pi$. If we equate this to the proportionate frequency of success, we may thus be able to give a more or less satisfactory evaluation for π, and its correspondence with the known value will then be a criterion of the adequacy of the calculus. Uspensky (*loc. cit.*) cites two such:

Investigator	No. of throws	Estimate of π	Error
Wolf 1849–53	5,000	3·1596	$<0{\cdot}019$
Smith 1855	3,204	3·1412–3·155	$<0{\cdot}015$

The alternative figures in the second line of the table take doubtful intersections into account. In 1901, Lazzerini, an Italian mathematician (cited by Kastner and Newman, *Mathematics and the Imagination*), carried out an experiment involving 3,408 tosses of the needle, and obtained the value $\pi \simeq 3{\cdot}1415929$, an error of only 0·0000003.

The theory of this experiment need not concern us here; but the nature of the game calls for comment because the model set-up is seemingly more comparable to that of the Chevalier's die than are some situations cited above. In the context elsewhere cited, Uspensky refers to records of Bancroft Brown on experience of American dice in the game of craps. The caster wins if he scores a natural (7 or 11) at the initial double toss, loses if he scores craps (2, 3, or 12) and otherwise has the right to toss his two dice till the score is the same on two successive occasions, in which event he wins, or till the score for the double toss is 7, in which event he loses. Correct to three places, the probability of winning is 0·493 and of scoring craps is 0·111. In 9,900 games recorded by Brown the corresponding frequencies were 0·492 and 0·106. Keynes thus describes the outcome of experiments on die models:

> The earliest recorded experiment was carried out by Buffon, who, assisted by a child tossing a coin into the air, played 2,048 partis of the Petersburg game, in which a coin is thrown successively until the parti is brought to an end by the appearance of heads. The same experiment was repeated by a young pupil of De Morgan's 'for his own satisfaction'. In Buffon's trials there were 1,992 tails to 2,048 heads; in Mr. H.'s (De Morgan's pupil) 2,044 tails to 2,048 heads. . . . Following in this same tradition is the experiment of Jevons, who made 2,048 throws of ten coins at a time, recording the proportion of heads at each throw and the proportion of heads altogether. In the whole number of 20,480 single throws, he obtained heads 10,353 times . . .
>
> . . . Czuber has made calculations based on the lotteries of Prague (2,854 drawings) and Brunn (2,703 drawings) between the years 1754 and 1886, in which the actual results agree very well with theoretical predictions. Fechner employed the lists of the ten State lotteries of Saxony between the years 1843 and 1852. Of a rather more interesting character are Professor Karl Pearson's investigations into the results of Monte Carlo Roulette as recorded in Le Monaco in the course of eight weeks . . . on the hypothesis of the equi-probability of all the compartments throughout the investigation, he found that the

actually recorded proportions of red and black were not unexpected, but that alternations and long runs were so much in excess that . . . a priori odds were at least a thousand millions to one against some of the recorded deviations. Professor Pearson concluded, therefore, that Monte Carlo Roulette is not objectively a game of chance in the sense that the tables on which it is played are absolutely devoid of bias.

The Insurance Illusion. During the latter half of the eighteenth century the government of Royalist France reaped a handsome revenue from a lottery which throws light on the predilection of Laplace for the urn model. Other European governments operated similar state lotteries in the following century. A French citizen of the earlier period could purchase one or more billets numbered from 1 to 90 inclusive. At appointed intervals an official drew randomwise 5 tickets of a complete set of 90. On the announcement of the result, the holder of one or more tickets bearing the same number as a ticket drawn could claim compensation. The holder of one or more tickets with the same winning number could claim 15 times the cost of each, of one or more pairs with two different winning numbers 270 times the cost of each and so on. For each claim so specified the government set its stake to gain in the long run. If N is the number on a single ticket, the probability* that one of 5 taken from 90 will be N is $\frac{5}{90} = p = \frac{1}{18}$. If the holder pays t francs for it, the government thus gains t with probability $\frac{17}{18}$ that N is *not* one of the winning numbers and otherwise loses $(15 - 1)\,t$ with probability $q = \frac{1}{18}$. The long run gain of the government for t francs by sale of tickets is therefore positive, being:

$$\frac{17}{18}t - \frac{14}{18}t = \frac{t}{6}$$

Till abandoned in 1789, the enterprise proved profitable to the regal gambler. Had the monarch chosen to limit the issue of

* Such is Uspensky's account. It implies that the lottery pool is comparable to a very large card pack which contains a very large number of tickets with the same number. When this is so, the chance of choosing *simultaneously* a 5-fold sample does not appreciably differ from the chance of choosing a 5-fold sample *with replacement.*

tickets and to seek coverage against insolvency like gambler A in our parable of how to divide the stakes (p. 582), we may suppose that an insurance company C with more capital than the treasury might also have profited by underwriting possible loss and have covered itself against insolvency by a comparable arrangement with one of the big banking houses (D). At a time when a banking family could (and did) in effect underwrite the risk that Wellington would lose the Battle of Waterloo, the supposition is not unlikely. Such then was the setting in which one of the first surviving British life insurance corporations, the Equitable, came into being (1762). There had been less successful enterprises of the kind for more than two centuries, before which other forms of insurance had emerged as a form of commercial gambling.

The earliest reputable branch of insurance to get under way on a large scale was for loss of ships at sea. In the history of Flemish shipping we can trace it to the beginning of the fourteenth century. By the sixteenth, it was a well-established financial transaction; and Sir Nicholas Bacon, addressing Elizabeth's first Parliament, asked: 'Does not the wise merchant in every adventure give part to have the rest assured?' Earlier writers are by no means so unanimous in associating insurance with the virtue of moral prudence. In its initial stages it was a pure gamble which went hand in hand with less reputable forms of speculation. With these less reputable practices the origins of life insurance are closely associated. Money-lending at high rates of interest to princes, with a serious prospect of repudiation after a term of years, and credit transactions for business in the medieval fairs were not the only basis for the power which finance began to wield in the fourteenth and fifteenth centuries. Side by side with business loans at the fairs flourished the practice of making wagers on the life of an individual or the birth of a child, and a variety of fantastic speculations. There was frequent legislation during the sixteenth century to restrict the activities of the continental bourses and exchanges to credit operations, while prohibiting various kinds of wager insurance on which the ecclesiastical authorities frowned.

A sixteenth-century writer complains that 'a part of the nobles and merchants . . . employ all their available capital in dealing in money . . . the soil remains untilled, trade in commodities is neglected, there is often increase in prices'. The astrologer Kurz, who used the horoscope to prophesy the prices of pepper, ginger, and saffron a fortnight in advance, was 'surrounded with work as a man in the ocean with water'. With no surer basis than astrology the financial transactions which enriched the merchant princes of the medieval centres were, to a large extent, gambling in the most literal sense. At the medieval fairs merchants with capital would lay wagers on the sex of the unborn child, or on the time of a person's death. Examples of these wager insurances, which were the precursors of life insurance, are given by Goris in his study on the southern merchants' colonies. Thus there is an extant contract between Domingo Symon Maiar and his brother Bernardo with two women to whom they agree to pay 30 livres if the offspring is a girl on the undertaking that they receive 48 livres in gratitude for the birth of a son. In 1542 Villalon wrote:

> 'Of late in Flanders a horrible thing has arisen, a kind of cruel tyranny which the merchants there have invented among themselves. They wager on the rate of exchange in the Spanish fairs at Antwerp. They call these wagers parturas according to the former manner of winning money at a birth when a man wagers the child shall be a boy. . . . One wagers that the exchange rate shall be at 2 per cent premium or discount, another at 3 per cent, etc. They promise each other to pay the difference in accordance with the result. This sort of wager seems to me like marine insurance business. . . . For dealing of this kind is only common among merchants who *hold much capital*. . . . By their *great capital* and their tricks they can arrange that in any case they have profit.'

The last sentence anticipates the practical link between a mathematical theory of probability and success in speculation. Numerical devices which had served a long apprenticeship in the practice of magic provided the basis for a theory of mathematical probability when financiers who were gambling in the exchanges required a more certain guidance than the astrologer could give them.

Since the biologist now interprets the birth of a boy and girl on the assumption that fertilization of an ovum by sperms with an X and sperms with a Y chromosome is on all fours with tossing a coin, the merchant with *great capital* here mentioned is legitimately comparable to gambler C in our discussion of the division of stakes in a game of chance. Once the comparison had caught on, it was difficult to dislodge even in circumstances where it is manifestly as irrelevant to the end in view as it is in life insurance. If by *chance of dying* at a particular age, we mean the proportion of people who do so in a particular year, nothing is more certain than that it is different in successive years. Death is therefore comparable to a lottery only if we assume that the probability of success in a single game changes from game to game. If so, the mathematical theory of probability as formulated by the Founding Fathers has no bearing whatever on life policies and annuities.

The prosperity of the *Equitable* with the advice of Dr Richard Price, an English nonconformist divine, signalizes the end of what we may call the classical period of algebraic probability. In the history of mathematics he is memorable for preparing for posthumous publication in the *Transactions of the Royal Society* a memoir by his friend Thomas Bayes. This introduced a new notion, that of *inverse probability*, developed at length by Laplace. According to its proponents, who have never gained universal assent among mathematicians or scientific workers, this prescribes the possibility of assessing with what assurance we can forecast a future event on the basis of past occurrence.

Coming from a *savant* who told Napoleon that God is an unnecessary hypothesis, the terms on which Laplace entered the lists to lend the immense authority of his well-deserved eminence as a pure mathematician and as an astronomer to the pretensions of Richard Price and the guesswork of Thomas Bayes are far less explicable than the sophistries with which the pious Bishop Berkeley (p. 502) put back the clock in the Anglican Universities by denouncing Newton's fluxions as a passport to atheism. The kingpin of the doctrine of inverse probability, as transmitted by Laplace to posterity was transparently a figment of the celestial and unattainable domain of Plato's *universals*. Translated and embalmed in the English language, it is an *infinite hypothetical population* of which the funeral records and baptismal registers of Northampton dated AD 1735 were (or are) a single sample chosen on high at random from the sinful shadow-world of human experience.

We can get to the core of the debate, if we are clear about the implications of a question often asked by beginners: if the result of spinning a coin ten times is a score of 10 heads, what is the probability that the coin is biased? In this context, the word biased means $p > 0{\cdot}5$. The question is inadmissible in terms of the classical doctrine as expounded in this chapter. Probability so defined applies only to situations in which we know, or at least claim to know, in advance the value of p. If $p = 0{\cdot}5$, the product rule prescribes that the probability (as a measure of long-run frequency) is $(0{\cdot}5)^{10} = 1024^{-1}$. Accordingly, classical theory says in effect: if I *rightly* assert that the coin is unbiased, I shall very rarely be wrong in asserting that at least one tail will turn up in a ten-fold toss. The situation is entirely different, if I know that some coins are normal ($p \simeq 0{\cdot}5$) and the remainder minted defectively with a head on both sides ($p = 1$). If I also know the proportion of each in circulation I can solve the problem in the classical manner by regarding it as a game in two stages: (i) picking randomwise and blindfold a coin of one or other sort; (ii) tossing the coin so chosen.

In such a two-stage game, textbooks speak of P_1 the proportion of unbiased, and P_2 that of the two-headed coin as *a priori* probabilities. If we cannot assign a correct value to each, the classical doctrine dismisses the possibility of arriving at a judgement. When we have no such knowledge, Bayes first, with Price, Laplace, and many after him, adopted the principle of *insufficient reason* to the effect that we can put $P_1 = 0{\cdot}5 = P_2$ equal prior probabilities of picking either coin in the above situation. The reader who fails to see how his ignorance of whether a card under the carpet is face up or face down has anything to do with the long run frequency of being one or the other is in very good company. Fewer and fewer mathematicians commit themselves to the principle of insufficient reason which is the corner stone of inverse probability as defined by Laplace; but very few are prepared to jettison the many applications of the theory of probability propounded when it was fashionably acceptable and relevant only if one makes assumptions foreign to the formulation of the classical period.

One example of such irrelevance is glaring. When life insurance was in its infancy there were several attempts to make what one calls a Life Table on the basis of somewhat faulty data. The author of the earliest English one was Halley of comet fame. He based it on statistics of the city of Breslau; and the Royal Society published it in 1693. A Life Table records the proportion of people of y years of age who die in a particular calendar year, whence the requisite data for calculating the mean age at death of each year. As such it is an arithmetical exercise in simple proportion having no necessary connection with, or need to rely on, algebraic probability. Dr Richard Price thought otherwise. Thereafter, the prosperity of the Equitable, with him as guide, philosopher, and friend, encouraged the belief that algebraic probability is the *sine qua non* of the actuary's professional equipment.

During the lifetime of Price, there was no compulsory registration of births and deaths in Britain. So it fell to the actuary to collect such data as best he could. Advised to that effect by Price, the Equitable felicitously based its early premiums on the Northampton Life Table, the author of which grossly under-estimated the population at risk by using the data of Established Church baptismal registers in a township with a large dissenting community. By this error, the firm added a vast sum to its capital assets. By the same error, the British government of that time lost heavily on annuity outpayments to its pensioners. Thereafter, the corporation exercised its prerogative to revise its premiums at its own convenience. If life insurance was to them a game, it was a game in which the winning partner could change the stakes and the dealer could change the composition of the card pack when so disposed.

Clearly therefore, the theory of probability could bear scrutiny henceforth only if developed from axioms other than those which the Founding Fathers, Bernoulli in particular, had made explicit. There have been several subsequent attempts to develop the same algebraic operations (basically the three rules discussed above), from assumptions which have a wider range of relevance to the real world, if justifiable; but the justifiability of the assumptions have failed to commend themselves to all the experts. Consequently, there is still room for uncertainty about what class of situations other than games of chance algebraic probability can rightly take within its scope. We can do full justice to their claims, only if we acquaint ourselves with a geometrical device which brings into the picture the so-called *normal curve*.

Odds and Area. We represent the growth of something *measurable* by a graph on the assumption that it grows smoothly without jumps. Statistics deal largely with whole numbers, and whole numbers do not grow larger in this way. For a reason which we shall soon see more clearly, it is convenient to represent a function of whole numbers by a device called a histogram. A histogram is a visual *fiction*, which hardly needs explanation with pictures such as Figs. 191–4 before us. This shows how y_x the chance of getting a score of 0, 1, 2, etc., depends on x the score itself. We represent y_x a particular value of y corresponding to a particular value of x by the height of a column whose *midpoint* on the base line is x. This is a fiction in the sense that y_x

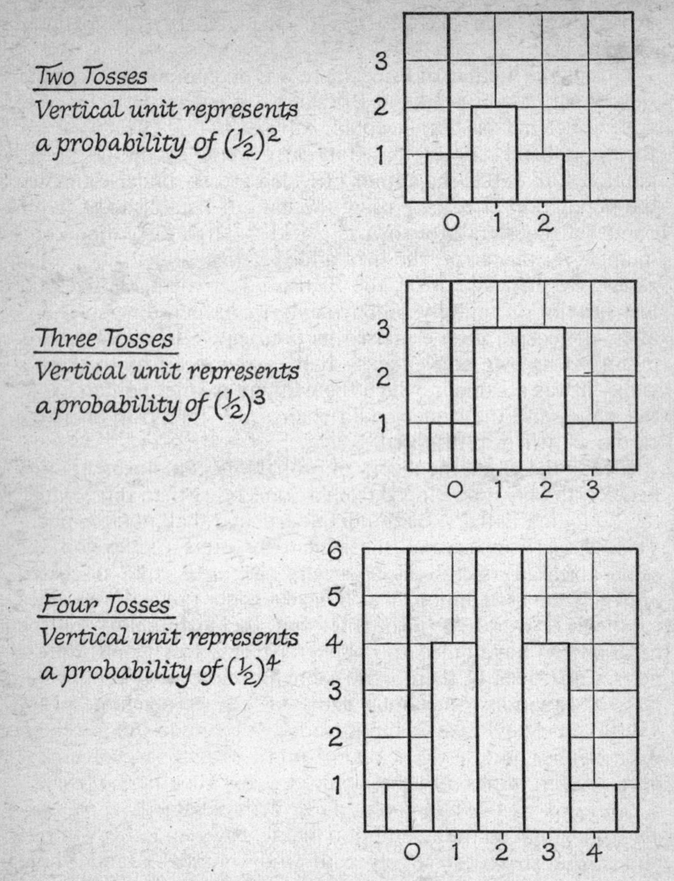

Fig. 191. Histogram of Spinning a Coin ($p = \frac{1}{2} = q$)
In each figure one horizontal unit corresponds to a success (i.e. head)

has no value between x, (e.g. 3 tosses), and $(x + 1)$, e.g. 4 tosses, or $(x - 1)$, e.g. 2 tosses; but the fiction has a use, as we shall now see.

Here the mid-point of each column is x, and the width of each column is $x \pm \frac{1}{2}$, and since successive values of x differ by unity,

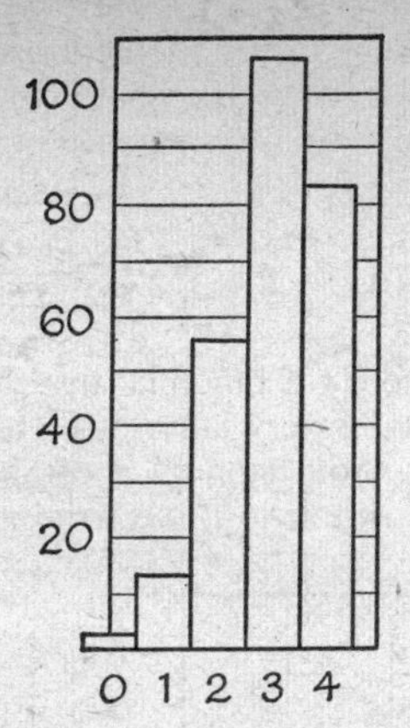

Fig. 192. Histogram of Four-fold Trial ($p = \frac{3}{4}$; $q = \frac{1}{4}$)

Number of red balls drawn in a trial of four represented along the X-axis. Vertical heights correspond to relative frequency, the unit being $(\frac{1}{4})^4$. The player replaces each ball drawn before taking another from the bag.

the width Δx of the column is also unity. That is to say $y_x \,.\, \Delta x = y_x$. Now the chance of getting one or other of any possible value of x is itself unity in accordance with the addition and subtraction rules. When the sample is r-fold we may therefore write the sum of all score values which increase from 0 to r in unit steps in the form:

$$\sum_{x=0}^{x=r} y_x = 1 = \sum_{x=0}^{x=r} y_x \,.\, \Delta x$$

When r is very large a highly characteristic curve (Fig. 195) goes through the mid point at the top of each column. This so-called *normal curve* is asymptotic to the X-axis at both ends. It is therefore impracticable to represent our score values as from 0 to r. Instead we represent them as deviations (X) from the mean score M, as below for the 6-fold toss of a coin for which $M = 3$, and $X = x - M$:

frequency ($y_x \times 2^6$)	1	6	15	20	15	6	1
head score (x)	0	1	2	3	4	5	6
score deviation (X)	−3	−2	−1	0	+1	+2	+3

The frequency terms of the foregoing distribution corresponds to those of the expansion of $(\frac{1}{2} + \frac{1}{2})^6$. If we ask what is the area included by deviations of 2 from the mean on either side of it, the appropriate expression is:

$$\sum_{X=-2}^{X=+2} y_x \, . \, \Delta X \qquad (\Delta x = 1 = \Delta X)$$

The derivation of the normal curve for a binomial distribution involving different values of p and q is very laborious. Here we shall consider only the case (e.g. r-fold spin of a coin) when $p = \frac{1}{2} = q$, and it simplifies our arithmetic if we write $r = 2n$

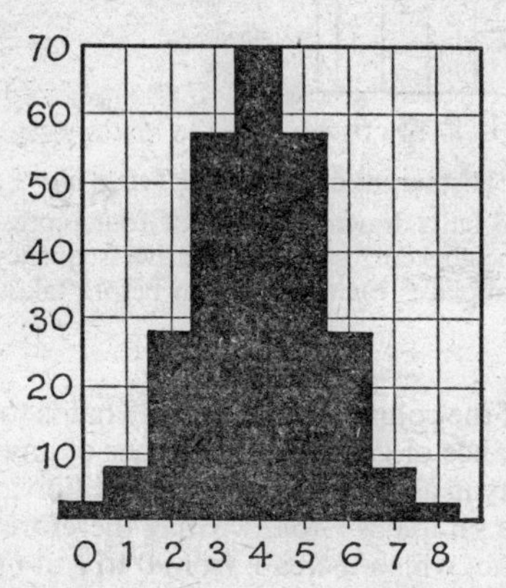

Fig. 193. Probability as the Ratio of Two Areas

If $p = \frac{1}{2} = q$ meaning equal probability that a newly born will be a boy or a girl, we can represent the probability that a family of eight will consist of 0, 1, 2 . . . 8 girls (or boys) as indicated on the horizontal axis. The vertical unit is $(\frac{1}{2})^8$. The area of the first four columns is then the probability that there will not be more than three of the same sex, that of the remaining five that there will be at least four.

in the expression $(\frac{1}{2} + \frac{1}{2})^r$. The mean score is then $M = n$. Without ambiguity, we may use the symbol y_x to signify both the frequency of a score x and of its corresponding score deviation $X = x - n$, so that $x = X + n$. As we have seen

$$y_x = 2^{-2n} \cdot \frac{(2n)!}{(2n - x)!x!} = \frac{2^{-2n} \, . \, (2n)! \; (x + 1)}{(2n - x)! \; (x + 1)!}$$

$$y_{x+1} = 2^{-2n} \cdot \frac{(2n)!}{(2n - x - 1)!(x + 1)!} = \frac{2^{-2n} \cdot (2n)!(2n - x)}{(2n - x)!(x + 1)!}$$

$$\therefore\ y_{x+1} = \frac{2n - x}{x + 1} \cdot y_x$$

$$\therefore\ \Delta y_x = y_{x+1} - y_x = \left(\frac{2n - x}{x + 1} - 1\right) y_x$$

$$\therefore\ \frac{1}{y_x} \cdot \Delta y_x = \frac{2n - 2x - 1}{x + 1}$$

Since $x = n + X$

$$\frac{1}{y_x} \cdot \Delta y_x = \frac{-2X - 1}{n + X + 1} = \frac{-2X}{n + X + 1} - \frac{1}{n + X + 1}$$

Our assumption in seeking a curve cutting very closely the mid-points at the top of each column of the histogram is that n is very large. So the overwhelming majority of values of X will cluster round the mean, i.e. $X = 0$ since $(x - n) = X$. We may thus plausibly explore the possibility of getting a good fit by writing:

$$(n + X + 1) \simeq n \quad \text{and} \quad \frac{1}{n + X + 1} \simeq 0$$

$$\therefore \frac{1}{y_x} \cdot \Delta y_x \simeq \frac{-2X}{n}$$

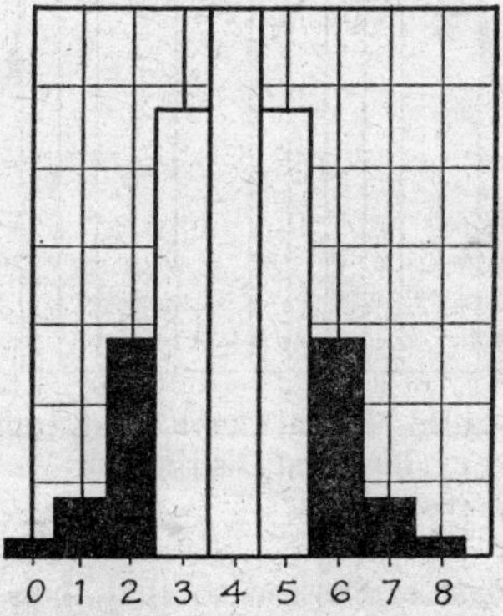

0 1 2 3 4 5 6 7 8

Fig. 194. Probability as the Ratio of Two Areas

Units as in Fig. 193. The unblackened area represents the probability that a family of eight will have not less than three or more than five of the same sex.

Since $\Delta X = 1$:

$$\frac{1}{y_x} \cdot \Delta y_x \simeq \frac{-2}{n} \cdot X \cdot \Delta X$$

When, as we assume, n is very large, we may write:

$$\frac{1}{y} \cdot dy \simeq \frac{-2X}{n} \cdot dX$$

$$\therefore \quad \int \frac{1}{y} dy \simeq \frac{-2}{n} \int X \cdot dX$$

$$\therefore \quad \log_e y = \frac{-X^2}{n} + C$$

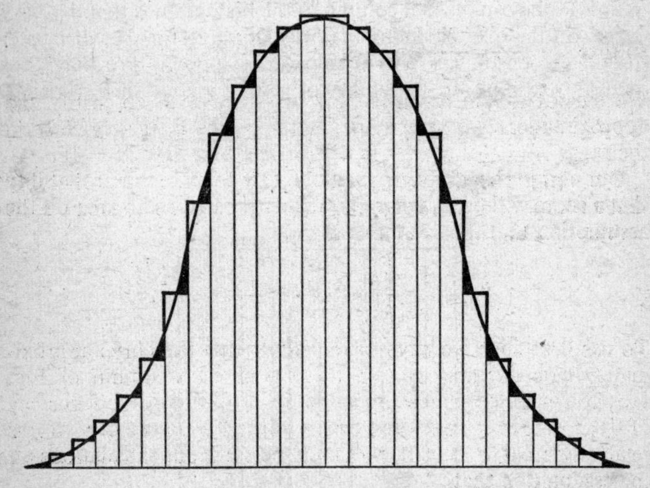

Fig. 195. The so-called Normal Curve as an Approximation to the Binomial Distribution

To tailor the last expression so that X is the independent variable, we may put $C = \log_e A$, so that

$$\log_e y - \log_e A = \frac{-X^2}{n} = \log_e \left(\frac{y}{A}\right)$$

Bearing in mind that $log_e\, a = b$ means $a = e^b$, we may therefore write:

$$\frac{y}{A} = e^{\frac{-X^2}{n}} \quad \text{and} \quad y = A \,.\, e^{\frac{-X^2}{n}}$$

The value of A is obtainable by putting $X = 0$ in which event $y = y_n$, i.e. the ordinate of the mean, so that:

$$y_n = \frac{2^{-2n} \cdot (2n)!}{n!\, n!} = A$$

We have already dealt with the properties of this curve in chapter 10 (page 493). For tabulation of values of the definite integral cited below, it is useful to invoke a formula which gives good values for $n!$ in A, so long as n is larger than 10, i.e.

$$n! \simeq e^{-n} \,.\, n^n \,.\, \sqrt{2\pi n}$$

so that $$A = \frac{2^{-2n} \cdot e^{-2n} \,.\, 2^{2n} \sqrt{4\pi n}}{e^{-2n} \,.\, 2^{2n}\, 2\pi n} = \frac{1}{\sqrt{\pi n}}$$

We cannot say in advance how big n must be to justify the approximations we have made. Actually, the fit is very close in the range $X = \pm 2$ when n is as small as 8 so that $2n = 16$.

Our aim in this derivation is to be able to state the probability that a score will lie in a certain range expressed as an area on the assumption that the total area is unity, i.e.

$$\int_{\infty}^{+\infty} . y \,.\, dx = 1$$

To use it as such we have to remember that the curve approximately cuts the midpoint of the top of each column of Fig. 195, i.e. for the column score value $X = \pm a$ there is a deficiency of $\pm \frac{1}{2}$ in the range enclosed by the particular values of y. If n is relatively small ($n < 100$), we should perform our calculations on the assumption that

$$\sum_{X=-a}^{X=-a} y_x \simeq \int_{-a+\frac{1}{2}}^{a-\frac{1}{2}} y \,.\, dX$$

The reader who hopes to gain an intelligent grasp of how to use such tables should verify the meaning of the half interval correction ($a \pm \frac{1}{2}$) above by drawing the histogram of $(\frac{1}{2} + \frac{1}{2})^{10}$ and the curve which cuts the tops of each column approximately at its midpoint.

To perform the calculation, we can refer to tabulated values of the definite integral on the right. This implies that it is possible to evaluate the indefinite integral, as can be done by series integration (p.584) which shows that the value of the definite integral over its whole range is unity, i.e.:

$$\frac{1}{\sqrt{\pi n}} \int_{\infty}^{+\infty} e^{\frac{-X^2}{n}} . dx = 1$$

Probability and the Real World. Aside from its earliest use in connection with games of chance and its later misuse in connection with life insurance, we may broadly distinguish three domains in which the algebraic calculus of probability has staked a claim. One is the *theory of error*, largely due to Gauss. What prompts the question to which it seeks an answer is the circumstance that successive observations on the same phenomenon, especially ones made with instruments having a large assemblage of cog-wheels as in an astronomical observatory, are never exactly the same. Such an application of the theory of probability seeks to side-step the dilemma which arises when the investigator finds himself confronted with two sets of repeated measurements or counts respectively clustering around different mean values and to some extent overlapping. Here the decision sought is whether the difference between the means may or may not betoken two *true* values rather than one. In common parlance: is an observed difference between two means consistent with errors of observation? We need not dismiss the considerable plausibility of the assumption that the distribution of errors is comparable to what we obtain when we toss a die; but the practical issue is mathematically intractable unless we can justifiably invoke a correct value for p and q.

A second use of the algebraic theory in scientific investigation is wholly above criticism. One may call it a Calculus of Aggregates. This term embraces scientific hypotheses which deal with populations of particles, atoms, molecules, genes, and chromosomes, whose properties the observer can study in large assemblages. Such is the situation when we explore the Brownian movement of microscopic solid particles bombarded by molecules in solution, or when we explore the effect on the proportions of progeny of a particular type when an extra chromosome turns up in the parental stock. In such situations, the investigator can use a card pack, die, roulette wheel, or lottery 'urn' as a model of

interpretation. The justification for this, as for reliance on any other model, depends solely on *whether it works*. In the reflected glory of the spectacular success of scientific hypotheses of this class, e.g. the Kinetic Theory of gases in physical chemistry and the Theory of the Gene in biology, there has however proliferated an overgrowth of statistical theory which involves assumptions which one cannot hope to justify indisputably by whether they do work. One may distinguish between two broad categories: (*a*) techniques for *exploration*; (*b*) tests for the *validity of judgements*. The writer has elsewhere set forth at length the exceptionable assumptions these involve. Here he will content himself with the task of introducing the reader to an example of each category without critical comment.

Correlation. As an example of a *technique of exploration* which invokes algebraic probability, we may instructively examine Spearman's *Rank Coefficient of Correlation*. The end in view is the search for connected characteristics, as if one asks: is ability to do mental arithmetic associated with parental income? If we arrange a class of boys and girls first in the order of the marks obtained in an arithmetical test and find the same or reverse order when we again arrange them according to the annual income of their parents, we should conclude that arithmetical facility and economic prosperity are connected in some way. Such complete correspondence would occur only if the effect of all contributory factors were perfectly standardized or negligible, and in practice we should not be discouraged from drawing a positive conclusion if a few names appeared out of place. Drawing the conclusion that a correspondence exists thus depends on adopting some standard for the amount of displacement which can occur when two such 'arrays' are compared.

A fundamental measure of displacement when we compare two arrays in this way is called *rank gain* or *loss*. Suppose the arithmetic marks of three boys A, B, C are 75, 52, and 39. Then A, B, C respectively have the ranks 1, 2, 3 in descending order of proficiency. If their parents' incomes are £500, £320, and £450 their ranks are 1, 3, 2, the rank of A is the same (1) in each array. The rank of B has decreased by 1 and that of C has increased by 1. The total rank gains and losses must always be the same, and the total number of either can be used as a criterion of correspondence.

To explore such a criterion, let us set out the number of all possible ways of arranging three things, i.e. $3! = 6$, viz.:

A	A	B	B	C	C
B	C	A	C	A	B
C	B	C	A	B	A

If we take any one of these as the standard order there will be a net loss (or gain) of rank when any of the other five are compared with it. If we take the first as the standard the orders of rank are:

1	1	2	2	3	3
2	3	1	3	1	2
3	2	3	1	2	1

The rank gains (positive sign) and losses (negative) are:

..	0	−1	−1	−2	−2
..	−1	+1	−1	+1	0
..	+1	0	+2	+1	+2

This makes the total gains 8 and the losses 8. So when a group of 3 objects are successively arranged in all the 6 possible ways in which they can be arranged, the total rank loss is 8 and the mean for all possible ways is $\frac{8}{6}$. If l is the loss (negative sign) of rank in any compartment of the last table, the mean rank loss for a re-arrangement of n objects when equal value is given to every possible order is therefore

$$\frac{\sum l}{n!} = S$$

If the total rank loss (T) when we compare two arrays does not differ greatly from the mean rank loss when all possible arrangements are given the same value, we have no reason to suspect that there exists any connection between the marking or measurements on which the order of the arrays depends. An index which indicates the correspondence in the following way is called Spearman's rank coefficient:

$$R = 1 - \frac{T}{S}$$

If the total rank loss is equivalent to the mean rank loss $R = 0$, and if the total rank loss is zero $R = 1$. So values of R between 1 and 0 indicate greater or less correspondence. To use this formula we only need to know how to find S for an array composed of a fairly large number of items. The expression is:

$$S = \frac{n^2 - 1}{6}$$

Thus when there are 3 objects, $S = \frac{9 - 1}{6} = \frac{8}{6}$, as we have already found. We can build up the formula for S from the structure of the table:

1	1	.	2	2	.	3	3
2	3	.	1	3	.	1	2
3	2	.	3	1	.	2	1

As indicated before there are $n!$ possible arrangements of n items, and the number of rank gains is the same as the number of rank losses. Looking at the top row you will see that the number of times each of the n numbers occurs is the number of arrangements of the remaining $(n - 1)$ numbers, i.e. each number occurs $(n - 1)!$ times. The highest number (n) can lose rank by 0, 1, 2 on to $(n - 1)$ and each loss occurs $(n - 1)!$ times. The next highest $(n - 1)$ can lose rank 0, 1, 2 on to $(n - 2)$ and each loss occurs $(n - 1)!$ times. So all rank losses can be tabulated thus:

nth number	$(n - 1)!\,[0 + 1 + 2 \ldots + (n - 1)]$
$(n - 1)$th number	$(n - 1)!\,[0 + 1 + 2 \ldots + (n - 2)]$
lowest number but two	$(n - 1)!\,[0 + 1 + 2]$
lowest number but one	$(n - 1)!\,[0 + 1]$
lowest number	$(n - 1)!\,[0]$

The sum of all the items of this table when the addition is rearranged according to the vertical columns is:

$$(n - 1)!\,[n(0) + (n - 1)(1) + (n - 2)(2) + (n - 3)(3) \ldots]$$
$$= (n - 1)!\,[(0 + n + 2n + 3n \ldots) - (0 + 1^2 + 2^2 + 3^2 \ldots)]$$

In each of the two series in this expression there are n terms beginning with 0, hence ending with $n - 1$ and $(n - 1)^2$. So we may rewrite it:

$$(n - 1)!\,[n(1 + 2 + 3 \ldots \overline{n - 1}) - (1^2 + 2^2 + 3^2 \ldots + \overline{n - 1}^2)]$$

We have met the summation of the first n whole numbers and their squares in Figs. 76 and 77. By substituting $(n - 1)$ for n in these expressions we find that the sum of the first $(n - 1)$ numbers and of their squares are respectively $\frac{n(n-1)}{2}$ and $\frac{(n-1)(2n-1)}{6}$.

So we may rewrite the total as:

$$(n-1)!\left\{\frac{n \,.\, n(n-1)}{2} - \frac{n(-1)(2n-1)}{6}\right\}$$

$$= n(n-1)!\left(\frac{n^2-1}{6}\right)$$

$$= n!\left(\frac{n^2-1}{6}\right)$$

This is the total rank loss. To get the mean we have to divide by $n!$ So the mean rank loss is, as stated:

$$\frac{n^2-1}{6}$$

As an illustration of the use of the Spearman coefficient we will suppose that the marks for scripture (i) and parent's income (ii) of 8 boys are as follows:

	(i)	(ii)		(i)	(ii)
A	70	£720	E	60	£250
B	80	£800	F	55	£500
C	21	£750	G	24	£300
D	42	£450	H	30	£200

The ordinal position of the boys on the two scales is

	(i)	(ii)	Rank difference		(i)	(ii)	Rank difference
A	2	3	−1	E	3	7	−4
B	1	1	0	F	4	4	0
C	8	2	+6	G	7	6	+1
D	5	5	0	H	6	8	−2

The total rank gains (or losses) are ± 7. The mean rank loss is $\frac{8^2-1}{6} = \frac{63}{6}$. Substituting in the formula:

$$R = 1 - \frac{7 \times 6}{63} = 0{\cdot}3$$

Thus the ordinal correspondence is 33 per cent as measured by this index.

Detecting a Real Difference. A common type of question which, rightly or wrongly, invokes the algebraic theory of probability to test the *validity of judgements* is one such as this: is vaccination effective against smallpox? It would be unnecessary to appeal to mathematics if it were true both that no vaccinated person ever gets smallpox and that a high proportion of unvaccinated people do. As things are, neither one nor the other is true. We have to base a verdict on the possibility of answering the following question: is the incidence of smallpox *significantly* higher among persons who have not been than among persons who have been vaccinated? The keyword here is *significantly.*

We cannot here attempt to give an exhaustive answer to a question of this sort. All that is possible is to give a lead. The way the statistician tackles the problem is to ask first: is the result of an experiment on a vaccine a fluke? In other words, is it the sort of result we should not regard as very odd in a game of chance? So we make a model. To make our model as elementary as possible we shall assume that we can observe what happens when we do or do not vaccinate the *same* number of people. Of course, this is a simplification of the problem; but it is not difficult to adjust our procedure for solving it with a view to comparison of vaccination statistics of populations which are not of equal size. Within the framework of this limitation, it is easier to get the task of detecting a real difference into focus.

We have found, we suppose, that less people get smallpox if vaccinated; and we ask: is such a difference what would not uncommonly occur in a game of chance, if we took samples of such and such a size from two *identical packs.* Well, the number of people who do and do not get vaccinated is very large, and we have seen that replacement is then irrelevant to the calculation of risks. So we set up a model game of chance regardless of the replacement condition. Our two card packs stand for the populations we compare. The number of hearts in the sample stand for the number of smallpox cases. We seek to find how often the difference between our *heart-scores* would be so much. Figs. 196–197 show how it is possible to find an answer by combining what we have learnt about the binomial replacement distribution and the use of the chessboard trick. What we can do with samples of 4, we can extend to the treatment of larger samples.

If we set up a *null* hypothesis, i.e. a hypothesis to knock down in a situation such as this, we ask ourselves: is the score difference we record in our model situation an occurrence so rare as to make us doubt that the two card packs are in fact identical? We thus assume as the hypothesis we seek to *nullify* that people in general do *not* benefit from vaccination. If our null hypothesis is right,

	• • • • q^4	• • • ♥ $4pq^3$	• • ♥♥ $6p^2q^2$	• ♥♥♥ $4p^3q$	♥♥♥♥ p^4
• • • • q^4	• • • • q^8	• • • ♥ $4pq^7$	• • ♥♥ $6p^2q^6$	• ♥♥♥ $4p^3q^4$	♥♥♥♥ p^4q^4
• • • ♥ $4pq^3$	• • • ♡ $4pq^7$	• • • • $16p^2q^6$	• • • ♥ $24p^3q^5$	• • ♥♥ $16p^4q^4$	• ♥♥♥ $4p^5q^3$
• • ♥♥ $6p^2q^2$	• • ♡♡ $6p^2q^6$	• • • ♡ $24p^3q^5$	• • • • $36p^4q^4$	• • • ♥ $24p^5q^3$	• • ♥♥ $6p^6q^2$
• ♥♥♥ $4p^3q$	• ♡♡♡ $4p^3q^5$	• • ♡♡ $16p^4q^4$	• • • ♡ $24p^5q^3$	• • • • $16p^6q^2$	• • • ♥ $4p^7q$
♥♥♥♥ p^4	♡♡♡♡ p^4q^4	• ♡♡♡ $4p^5q^3$	• • ♡♡ $6p^6q^2$	• • • ♡ $4p^7q$	• • • • p^8

Fig. 196. Heart Score Differences – Pairs of Four

Here hearts count as successes ($p = \frac{1}{4}$). The diagram assumes replacement from identical full packs in accordance with the principle of equal opportunity to associate.

each of our samples – vaccinated or not vaccinated – is like one of two identical card packs. By pooling them we get a larger sample from the same source, and hence a more precise figure for the probability of getting smallpox from the ratio of the sum of smallpox cases in both samples to the total size of the combined sample. This is the p we use to get the marginal distribution $(p + q)^n$ for 2 r-fold samples (r vaccinated and r unvaccinated persons) in applying the chessboard procedure of Figs. 196–7. Part of statistics called the *theory of confidence*, aims at defining how often we may be wrong by acting too literally on this assumption.

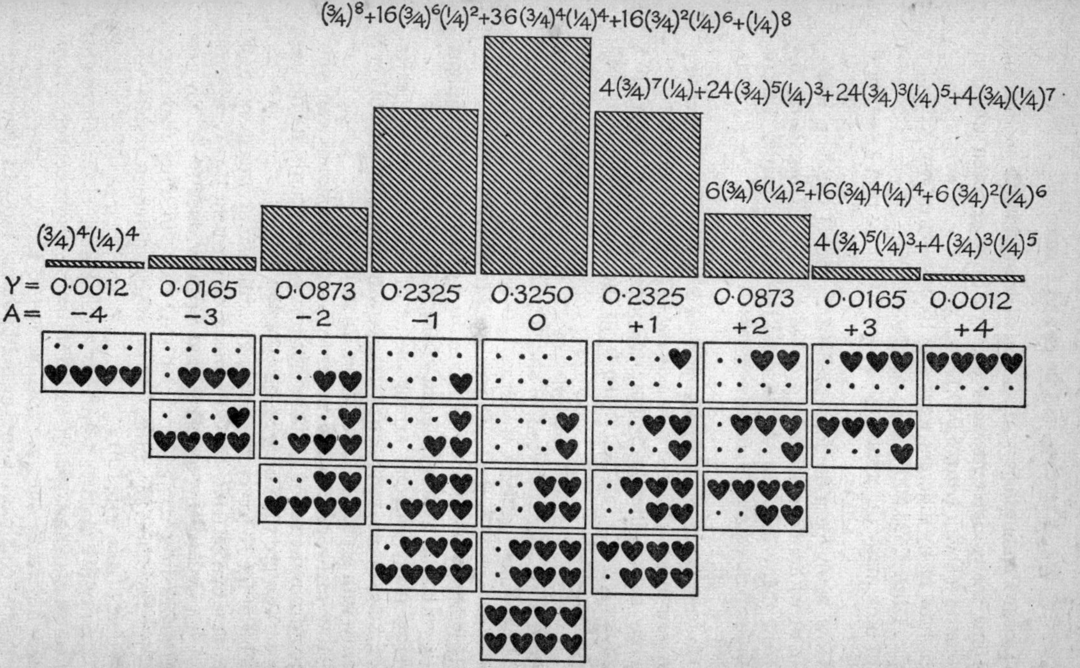

Fig. 197. Histogram of Heart Score Differences – Pairs of four as in Fig. 196

Exercises on Choice and Chance

A. Linear Permutations – All Objects Distinguishable

In how many different ways can we arrange seven different books on a shelf? (*Ans.* 5,040.)

In how many different ways can 10 boys sit on one side of a table? (*Ans.* 3,628,800.)

In how many different ways is it possible to ring a peal of 8 bells? (*Ans.* 40,320.)

How many permutations each of 2 vowels can you make, with or without replacement, from *a*, *e*, *i*, *o*, *u*? (*Ans.* 25 or 20.)

How many different sets of 4 initials can be made, with or without replacement, from the letters *C*, *F*, *H*, *K*, *N*, *P*? (*Ans.* 1,296 *or* 360.)

How many different arrangements of 5 letters without repetition can you make from the whole alphabet? (*Ans.* 7,893,600.)

How many different 5-figure telephone numbers (with all 5 figures different) can you make from the figures 0–9 inclusive? (*Ans.* 30, 240.)

Leaving out 11 and multiples of 11, such as 44, how many numbers are there between 10 and 100? (*Check.*)

How many numbers are there between 100 and 1,000, none containing any figure more than once? (*Check.*)

How many 4-figure numbers are there composed of odd digits only with all 4 figures different? (*Ans.* 120.)

How many different arrangements of 3 of the 4 top hearts (ace, king, queen, jack) can you make with or without repetition? Verify with a pack of cards. (*Ans.* 24 *or* 64.)

How many different arrangements of 5 diamonds can you make from the set of 13 diamonds in a pack of cards, if you use no cards more than once? (*Ans.* 154,440.)

Out of the 16 top cards, how many different arrangements of 4 can you make? (*Ans.* 43,680 *without* replacement.)

A goose, a turkey, a porker, and a rubber duck for booby are the prizes in a competition. Nobody can win more than one. There are 20 competitors. How many different results are possible? (*Ans.* 116,280)

Twenty-five children all sit for a scholarship exam. There are three scholarships, worth £80, £50, and £20 respectively. How many different results are possible? (*Ans.* 13,300.)

Twelve children all run in a 100 yards' race. In how many

different ways can awards be made for the first 4 places? (*Ans.* 11,880.)

B. Linear Permutations – Some Objects Indistinguishable

Find the number of permutations of the letters in the following words:

(*a*) Peripatetic. (*b*) Opprobrious.

(*Ans.* (*a*) 2,494,800, (*b*) 1,663,200.)

Do the same with the following, if you feel strong enough to do so:

(*a*) The chemical compound called: *Dihydrocholesterol.*
(*b*) The town in Anglesey called:
Llanfairpwllgwyngyllgogerychwyrndrobwllllantysiliogogogoch.

A hand of 13 cards has 1 club, 3 diamonds, 4 hearts, and 5 spades. In how many ways can we classify their linear permutations by *suit*? (*Ans.* 360,360.)

In how many different ways can one arrange in a row 6 pennies, 7 sixpences, and 8 shillings? (*Ans.* 349,188,840.)

How many different numbers can one make from the ten figures, 1, 2, 2, 3, 3, 3, 4, 4, 4, 4? (*Ans.* 12,600.)

How many different linear arrangements can one make with 4 oranges, 6 lemons, 8 apples, 10 pears?

A bookshop shelf has 15 books on it. 4 are copies of *Hamlet*; 2 are copies of *Science for the Citizen*; 3 are copies of *The Three Musketeers*; 6 are copies of the *Bible*. How many different arrangements are possible? (*Ans.* 6,306,300.)

In a garden there are 5 rows of peas, 4 rows of beans, 2 rows of lettuces, and 2 rows of radishes. How many different arrangements of the rows are possible? (*Ans.* 540,540.)

C. The Master-Key Formula

Here are some puzzles involving the master-key formula for solution on the assumption of choice: (*a*) with replacement; (*b*) without replacement: How many arrangements in a row face upwards are consistent with the composition of the following classes of samples taken from a full pack:

A two-fold sample containing 2 aces. (*Ans.* 16 *and* 12.)
A three-fold sample containing 1 ace and 2 picture cards.

(*Ans.* 1,728 *and* 1,584.)

A four-fold sample containing 1 red ace and 3 black picture cards. (*Ans.* 1,728 *and* 960.)

A four-fold sample containing 1 heart, 2 diamonds, and 1 club. (*Ans.* 342,732 *and* 316,368.)

A four-fold sample containing 3 red cards and 1 black ace. (*Ans.* 140,608 *and* 124,800.)

D. Sampling Without Replacement

At this point the reader who is new to the topic of the foregoing paragraphs may profitably pause to specify the mathematical probability of the following selections at a single draw from a complete pack:

An ace of spades. (*Ans.* $\frac{1}{52}$.)
A ten of any suit. (*Ans.* $\frac{1}{13}$.)
A red queen. (*Ans.* $\frac{1}{26}$.)
Any black card. (*Ans.* $\frac{1}{2}$.)
A royal card (i.e. *either* a king *or* queen). (*Ans.* $\frac{2}{13}$.)
An ace, king, queen, or knave. (*Ans.* $\frac{4}{13}$.)
A black ace or any royal card. (*Ans.* $\frac{5}{26}$.)
An ace of hearts or a black royal card. (*Ans.* $\frac{5}{52}$.)
A red card with one of the numbers 2–9 inclusive. (*Ans.* $\frac{4}{13}$.)
A red ten or a black knave. (*Ans.* $\frac{1}{13}$.)

E. The Addition Rule

(*a*) A bag contains 3 blue, 5 red, 4 green, and 2 black balls. Find the chance of drawing *one* ball which is:

1. Either a blue or a red ball. (*Ans.* $\frac{4}{7}$.)
2. Either a blue or a green ball. (*Ans.* $\frac{1}{2}$.)
3. A coloured ball. (*Ans.* $\frac{6}{7}$.)
4. Either a black or a blue ball. (*Ans.* $\frac{5}{14}$.)

(*b*) I throw a 6-face die once. What is the chance that the number of pips on the uppermost face is:

5. Either a one or a six. (*Ans.* $\frac{1}{3}$.)
6. An even number. (*Ans.* $\frac{1}{2}$.)
7. Less than 3. (*Ans.* $\frac{1}{3}$.)
8. Less than 5. (*Ans.* $\frac{2}{3}$.)

F. The Product Rule

1. What is the probability of getting 3 aces in a three-fold draw from a full pack: (*a*) with replacement; (*b*) without replacement? (*Ans.* (*a*) $\frac{1}{2197}$; (*b*) $\frac{1}{5525}$.)

2. What are the odds against drawing *no* diamond in a two-fold draw from a full pack: (*a*) with replacement; (*b*) without replacement? (*Ans.* (*a*) 9 : 7; (*b*) 19 : 15.)

3. In a three-fold toss of a cubical die, what is the probability of getting: (*a*) three doubles of *any* denomination; (*b*) three doubles of 5 and 6? (*Ans.* (*a*) $\frac{1}{36}$; (*b*) $\frac{1}{108}$.)

4. In a three-fold toss of a cubical die, what is the probability of getting a total score of: (*a*) 6; (*b*) 10; (*c*) 14? (*Ans.* (*a*) $\frac{5}{108}$; (*b*) $\frac{1}{8}$; (*c*) $\frac{5}{72}$.)

5. If a wheel has 10 equal sectors numbered 1–10, what is the probability of getting a total score of 8 in a four-fold spin? (*Ans.* $\frac{7}{2000}$.)

G. The Subtraction Rule

1. In a simultaneous draw from the two packs of Fig. 193 what is the chance of getting at least one club? (*Ans.* $\frac{7}{10}$.)

2. *Ditto* at least one ace? (*Ans.* $\frac{7}{10}$.)

3. *Ditto* at least one card whose value is less than 6? (*Ans.* $\frac{4}{5}$.)

4. *Ditto* at least one black card? (*Ans.* 1.)

5. In a simultaneous toss of two dice what is the chance that at least one face upturned has at least 3 dots? (*Ans.* $\frac{8}{9}$.)

6. *Ditto* at least 4 dots? (*Ans.* $\frac{3}{4}$.)

7. Two bags each contain 3 yellow, 4 blue, 5 red, 6 green, and 2 black balls. In a simultaneous draw what is the chance of getting at least 1 black ball? (*Ans.* $\frac{19}{100}$.)

8. *Ditto* at least 1 green ball? (*Ans.* $\frac{51}{100}$.)

9. *Ditto* at least 1 red or 1 green ball? (*Ans.* $\frac{319}{400}$.)

10. *Ditto* at least 1 blue or 1 black ball? (*Ans.* $\frac{51}{100}$.)

H. Miscellaneous Problems of Sampling

The reader may complete the balance sheet for practice, and then try by master-key formula and staircase model to check the chance for a double draw from the same pack without replacement as specified below:

1. *Double spade* from pack composed of 1 heart, 3 spades. (*Ans.* $\frac{1}{2}$.)

2. *Spade and diamond* from pack composed of 2 hearts, 3 spades, 1 diamond. (*Ans.* $\frac{1}{5}$.)

3. *Double diamond* from pack composed of 3 hearts, 4 spades, 2 diamonds, 1 club. (*Ans.* $\frac{1}{45}$.)

4. *Heart and spade* from pack composed of 1 heart, 2 spades, 3 clubs. (*Ans.* $\frac{2}{15}$.)

5. *Double club* from pack composed of 2 spades, 3 diamonds, 4 clubs. (*Ans.* $\frac{1}{6}$.)

6. A *red and a black* card from pack composed of 2 hearts, 3 spades. (*Ans.* $\frac{3}{5}$.)

7. *Two red* cards from pack composed of 4 clubs, 3 hearts, 1 diamond. (*Ans.* $\frac{3}{14}$.)

8. *Two black* cards from pack composed of 2 spades, 3 clubs, 4 hearts. (*Ans.* $\frac{5}{18}$.)

9. A *red and a black* card from pack composed of 1 spade, 1 club, 3 diamonds. (*Ans.* $\frac{3}{5}$.)

10. A *red and a black* card from pack composed of 2 clubs, 5 diamonds, 1 heart. (*Ans.* $\frac{3}{7}$.)

J. The Binomial Distribution

For practice, the reader may here try out the following: What are the chances of the following with (if possible) and without replacement?

1. Getting 0, 3, or 5 red cards in a simultaneous single draw from each of 7 full packs? (*Ans.* $\frac{1}{128}$; $\frac{35}{128}$; $\frac{21}{128}$.)

2. Getting 1, 2, or 4 picture cards in a simultaneous draw from each of 5 packs? (*Ans.* $\frac{150000}{371293}$; $\frac{90000}{371293}$; $\frac{4050}{371293}$.)

3. Getting 8, 10, or 12 picture cards in a simultaneous draw of one card from each of 12 full packs?

4. Getting an odd number of royal cards in a simultaneous draw from each of 6 full packs? (*Ans.* $\frac{2354580}{4826809}$.)

5. Getting an even number of aces in a simultaneous draw from each of 6 full packs? (*Ans.* $\frac{313201}{4826809}$.)

6. Getting a heart, 2 spades, 1 club, and 3 diamonds in a seven-fold draw from a full pack, the condition being that each card drawn is replaced before the next draw? (*Ans.* $\frac{420}{16384}$.)

7. Getting 2 red and 4 black balls in a simultaneous single draw from 6 bags each containing 1 red, 2 yellow, 3 green, 4 blue, and 5 black balls? (*Ans.* $\frac{1}{1215}$.)

8. Getting 2 yellow, 1 blue, and 1 black in a four-fold draw from one of the same bags as in 7, on condition that each ball drawn is replaced before the next draw? (*Ans.* $\frac{64}{3375}$.)

9. Getting 3 red, 4 green, 1 blue in an eight-fold draw? (*Ans.* $\frac{224}{6328125}$.)

10. Getting 2 red, 3 green, 5 blue in a ten-fold draw?

Discoveries to Make

1. In contradistinction to ${}^nC_r$, which stands for the number of combinations of r letters from a set of n when one can select no letter more than once, we may denote by ${}^nK_r$ the number one can select if the r-fold sample may contain any of the n letters 0, 1, 2, . . . to n times. If we define $n^{[r]}$ as in Chapter 4 we may write:

$$ {}^nK_r = \frac{n^{[r]}}{r!} $$

Clue: try out for $n = 2$, 3, 4, and 5, arranging the terms in columns by initial letters and recall the formulae for figurate numbers.

2. We may represent the total number of combinations of n things taken without repetition 0, 1, 2, 3, . . . n times as:

$$ \sum_{r=0}^{r=n} {}^nC_r = {}^nC_0 + {}^nC_1 + {}^nC_2 \ldots {}^nC_{n-1} + {}^nC_n = 2^n $$

Clue: expand $(1 + 1)^n$ by the binomial theorem.

3. When repetition is possible, we may write:

$$ \sum_{r=0}^{r=n} {}^nK_r = \frac{(n+1)^{[n]}}{n!} = \frac{(2n)!}{(n!)^2} $$

TABLES

Notes on Using the Tables

I. IN TABLE I are given some of the more useful relations connecting weights and measures. In the metric system the units of length, weight, and capacity are the metre, gramme, and litre. Each of these is sub-divided in the same way. A hundredth part has the prefix centi-, a thousandth part has the prefix milli-, and a thousand times has the prefix kilo-.

III. The use of the difference column has already been explained. Values between those shown in the difference column can be found by proportional parts. For example, we may want the square of 28·756. The table gives the square of 28·75 as 826·5. In this part of the difference column a difference of 1 in the number to be squared corresponds to a difference of 7 in the square. Therefore a difference of 0·6 will correspond approximately to a difference of $7 \times 0{\cdot}6$, i.e., about 4. The square required is thus 826·9. Table III can also be used to find square roots. For example, we may want to find the square root of 123·2. By inspection we can see that the square root of this number lies between 11 and 12. In the tables we see that 1232 occurs twice, first corresponding to the digits 111, and then corresponding to the digits 351. The square root of 123·2 is thus 11·1. If we had wanted the square root of 12·32 it would clearly have been 3·51.

IV. In most tables of sines, etc., the parts of a degree are given in minutes, so that the steps are $6'$, $12'$, etc. The custom of expressing the steps as decimal fractions of a degree is, however, gradually being introduced.

The table of sines can also be used to find cosines by using the formula $\cos A = \sin (90° - A)$, e.g. to find cos 31·5°, look up sin 58·5°.

VI. A table of antilogarithms has been omitted for reasons of economy. Table VI can be used for finding numbers from their logarithms by simply reversing the process of finding the logarithm of a number. For directions for using the table see p. 502 and the note on Table III.

Table I

English Weights and Measures

1,760 yards	= 1 mile
4,840 square yards	= 1 acre
640 acres	= 1 square mile
112 lbs.	= 1 cwt.
20 cwts.	= 1 ton
8 pints	= 1 gallon
1 gallon	= 277 cubic inches
1 cubic foot	= 6·23 gallons

Metric Weights and Measures

10 millimetres	= 1 centimetre
100 centimetres	= 1 metre
1,000 metres	= 1 kilometre
1,000 grammes	= 1 kilogramme
100 centilitres	= 1 litre
1 litre	= 1,000 cubic centimetres

Metric and English Equivalents

1 inch	= 2·54 centimetres
1 lb.	= 454 grammes
1 metre	= 1·09 yards
1 kilometre	= 0·621 mile
1 kilogramme	= 2·20 pounds
1 litre	= 0·22 gallon

Table II

Constants

$\pi = 3{\cdot}1416$ $\log_{10} \pi = 0{\cdot}4971$

1 radian = 57·296 degrees

$e = 2{\cdot}7183$ $\log_{10} e = 0{\cdot}4343$

$\log_e N = 2{\cdot}3026 \log_{10} N$

$\log_{10} N = 0{\cdot}4343 \log_e N$

Earth's mean radius = 3,960 miles = $6{\cdot}371 \times 10^8$ centimetres

g = 32·2 feet per second per second, or 981 centimetres per second per second

1 cubic centimetre of water at 4° C. weighs 1 gramme

Table III

Squares

	0	1	2	3	4	5	6	7	8	9	1	2	3	4	5	6	7	8	9
10	1000	1020	1040	1061	1082	1103	1124	1145	1166	1188	2	4	6	8	10	13	15	17	19
11	1210	1232	1254	1277	1300	1323	1346	1369	1392	1416	2	5	7	9	11	14	16	18	21
12	1440	1464	1488	1513	1538	1563	1588	1613	1638	1664	2	5	7	10	12	15	17	20	22
13	1690	1716	1742	1769	1796	1823	1850	1877	1904	1932	3	5	8	11	13	16	19	22	24
14	1960	1988	2016	2045	2074	2103	2132	2161	2190	2220	3	6	9	12	14	17	20	23	26
15	2250	2280	2310	2341	2372	2403	2434	2465	2496	2528	3	6	9	12	15	19	22	25	28
16	2560	2592	2624	2657	2690	2723	2756	2789	2822	2856	3	7	10	13	16	20	23	26	30
17	2890	2924	2958	2993	3028	3063	3098	3133	3168	3204	3	7	10	14	17	21	24	28	31
18	3240	3276	3312	3349	3386	3423	3460	3497	3534	3572	4	7	11	15	18	22	26	30	33
19	3610	3648	3686	3725	3764	3803	3842	3881	3920	3960	4	8	12	16	19	23	27	31	35
20	4000	4040	4080	4121	4162	4203	4244	4285	4326	4368	4	8	12	16	20	25	29	33	37
21	4410	4452	4494	4537	4580	4623	4666	4709	4752	4796	4	9	13	17	21	26	30	34	39
22	4840	4884	4928	4973	5018	5063	5108	5153	5198	5244	4	9	13	18	22	27	31	36	40
23	5290	5336	5382	5429	5476	5523	5570	5617	5664	5712	5	9	14	19	23	28	33	38	42
24	5760	5808	5856	5905	5954	6003	6052	6101	6150	6200	5	10	15	20	24	29	34	39	44
25	6250	6300	6350	6401	6452	6503	6554	6605	6656	6708	5	10	15	20	25	31	36	41	46
26	6760	6812	6864	6917	6970	7023	7076	7129	7182	7236	5	11	16	21	26	32	37	42	48
27	7290	7344	7398	7453	7508	7563	7618	7673	7728	7784	5	11	16	22	27	33	38	44	49
28	7840	7896	7952	8009	8066	8123	8180	8237	8294	8352	6	11	17	23	28	34	40	46	51
29	8410	8468	8526	8585	8644	8703	8762	8821	8880	8940	6	12	18	24	29	35	41	47	53
30	9000	9060	9120	9181	9242	9303	9364	9425	9486	9548	6	12	18	24	30	37	43	49	55
31	9610	9672	9734	9797	9860	9923	9986				6	13	19	25	31	38	44	50	57
31								1005	1011	1018	1	1	2	3	3	4	5	5	6

	0	1	2	3	4	5	6	7	8	9	1	2	3	4	5	6	7	8	9
32	1024	1030	1037	1043	1050	1056	1063	1069	1076	1082	1	1	2	3	3	4	5	5	6
33	1089	1096	1102	1109	1116	1122	1129	1136	1142	1149	1	1	2	3	3	4	5	5	6
34	1156	1163	1170	1176	1183	1190	1197	1204	1211	1218	1	1	2	3	3	4	5	5	6
35	1225	1232	1239	1246	1253	1260	1267	1274	1282	1289	1	1	2	3	4	4	5	6	6
36	1296	1303	1310	1318	1325	1332	1340	1347	1354	1362	1	1	2	3	4	4	5	6	7
37	1369	1376	1384	1391	1399	1406	1414	1421	1429	1436	1	2	2	3	4	5	5	6	7
38	1444	1452	1459	1467	1475	1482	1490	1498	1505	1513	1	2	2	3	4	5	5	6	7
39	1521	1529	1537	1544	1552	1560	1568	1576	1584	1592	1	2	2	3	4	5	6	6	7
40	1600	1608	1616	1624	1632	1640	1648	1656	1665	1673	1	2	2	3	4	5	6	6	7
41	1681	1689	1697	1706	1714	1722	1731	1739	1747	1756	1	2	2	3	4	5	6	7	7
42	1764	1772	1781	1789	1798	1806	1815	1823	1832	1840	1	2	3	3	4	5	6	7	8
43	1849	1858	1866	1875	1884	1892	1901	1910	1918	1927	1	2	3	3	4	5	6	7	8
44	1936	1945	1954	1962	1971	1980	1989	1998	2007	2016	1	2	3	4	4	5	6	7	8
45	2025	2034	2043	2052	2061	2070	2079	2088	2098	2107	1	2	3	4	5	5	6	7	8
46	2116	2125	2134	2144	2153	2162	2172	2181	2190	2200	1	2	3	4	5	6	7	7	8
47	2209	2218	2228	2237	2247	2256	2266	2275	2285	2294	1	2	3	4	5	6	7	8	9
48	2304	2314	2323	2333	2343	2352	2362	2372	2381	2391	1	2	3	4	5	6	7	8	9
49	2401	2411	2421	2430	2440	2450	2460	2470	2480	2490	1	2	3	4	5	6	7	8	9
50	2500	2510	2520	2530	2540	2550	2560	2570	2581	2591	1	2	3	4	5	6	7	8	9
51	2601	2611	2621	2632	2642	2652	2663	2673	2683	2694	1	2	3	4	5	6	7	8	9
52	2704	2714	2725	2735	2746	2756	2767	2777	2788	2798	1	2	3	4	5	6	7	8	9
53	2809	2820	2830	2841	2852	2862	2873	2884	2894	2905	1	2	3	4	5	6	7	9	10
54	2916	2927	2938	2948	2959	2970	2981	2992	3003	3014	1	2	3	4	5	7	8	9	10

Find the position of the decimal point by inspection.

TABLE III (contd.)

Squares

	0	1	2	3	4	5	6	7	8	9	1	2	3	4	5	6	7	8	9
55	3025	3036	3047	3058	3069	3080	3091	3102	3114	3125	1	2	3	4	6	7	8	9	10
56	3136	3147	3158	3170	3181	3192	3204	3215	3226	3238	1	2	3	5	6	7	8	9	10
57	3249	3260	3272	3283	3295	3306	3318	3329	3341	3352	1	2	3	5	6	7	8	9	10
58	3364	3376	3387	3399	3411	3422	3434	3446	3457	3469	1	2	4	5	6	7	8	9	11
59	3481	3493	3505	3516	3528	3540	3552	3564	3576	3588	1	2	4	5	6	7	8	10	11
60	3600	3612	3624	3636	3648	3660	3672	3684	3697	3709	1	2	4	5	6	7	8	10	11
61	3721	3733	3745	3758	3770	3782	3795	3807	3819	3832	1	2	4	5	6	7	9	10	11
62	3844	3856	3869	3881	3894	3906	3919	3931	3944	3956	1	3	4	5	6	7	9	10	11
63	3969	3982	3994	4007	4020	4032	4045	4058	4070	4083	1	3	4	5	6	8	9	10	11
64	4096	4109	4122	4134	4147	4160	4173	4186	4199	4212	1	3	4	5	6	8	9	10	12
65	4225	4238	4251	4264	4277	4290	4303	4316	4330	4343	1	3	4	5	7	8	9	10	12
66	4356	4369	4382	4396	4409	4422	4436	4449	4462	4476	1	3	4	5	7	8	9	11	12
67	4489	4502	4516	4529	4543	4556	4570	4583	4597	4610	1	3	4	5	7	8	9	11	12
68	4624	4638	4651	4665	4679	4692	4706	4720	4733	4747	1	3	4	5	7	8	10	11	12
69	4761	4775	4789	4802	4816	4830	4844	4858	4872	4886	1	3	4	6	7	8	10	11	13
70	4900	4914	4928	4942	4956	4970	4984	4998	5013	5027	1	3	4	6	7	8	10	11	13
71	5041	5055	5069	5084	5098	5112	5127	5141	5155	5170	1	3	4	6	7	9	10	11	13
72	5184	5198	5213	5227	5242	5256	5271	5285	5300	5314	1	3	4	6	7	9	10	12	13
73	5329	5344	5358	5373	5388	5402	5417	5432	5446	5461	1	3	4	6	7	9	10	12	13
74	5476	5491	5506	5520	5535	5550	5565	5580	5595	5610	1	3	4	6	7	9	10	12	13
75	5625	5640	5655	5670	5685	5700	5715	5730	5746	5761	2	3	5	6	8	9	11	12	14
76	5776	5791	5806	5822	5837	5852	5868	5883	5898	5914	2	3	5	6	8	9	11	12	14

	0	1	2	3	4	5	6	7	8	9	1	2	3	4	5	6	7	8	9
77	5929	5944	5960	5975	5991	6006	6022	6037	6053	6068	2	3	5	6	8	9	11	12	14
78	6084	6100	6115	6131	6147	6162	6178	6194	6209	6225	2	3	5	6	8	9	11	13	14
79	6241	6257	6273	6288	6304	6320	6336	6352	6368	6384	2	3	5	6	8	10	11	13	14
80	6400	6416	6432	6448	6464	6480	6496	6512	6529	6545	2	3	5	6	8	10	11	13	14
81	6561	6577	6593	6610	6626	6642	6659	6675	6691	6708	2	3	5	7	8	10	11	13	15
82	6724	6740	6757	6773	6790	6806	6823	6839	6856	6872	2	3	5	7	8	10	12	13	15
83	6889	6906	6922	6939	6956	6972	6989	7006	7022	7039	2	3	5	7	8	10	12	13	15
84	7056	7073	7090	7106	7123	7140	7157	7174	7191	7208	2	3	5	7	8	10	12	14	15
85	7225	7242	7259	7276	7293	7310	7327	7344	7362	7379	2	3	5	7	9	10	12	14	15
86	7396	7413	7430	7448	7465	7482	7500	7517	7534	7552	2	3	5	7	9	10	12	14	16
87	7569	7586	7604	7621	7639	7656	7674	7691	7709	7726	2	4	5	7	9	10	12	14	16
88	7744	7762	7779	7797	7815	7832	7850	7868	7885	7903	2	4	5	7	9	11	12	14	16
89	7921	7939	7957	7974	7992	8010	8028	8046	8064	8082	2	4	5	7	9	11	13	14	16
90	8100	8118	8136	8154	8172	8190	8208	8226	8245	8263	2	4	5	7	9	11	13	14	16
91	8281	8299	8317	8336	8354	8372	8391	8409	8427	8446	2	4	5	7	9	11	13	15	16
92	8464	8482	8501	8519	8538	8556	8575	8593	8612	8630	2	4	6	7	9	11	13	15	17
93	8649	8668	8686	8705	8724	8742	8761	8780	8798	8817	2	4	6	7	9	11	13	15	17
94	8836	8855	8874	8892	8911	8930	8949	8968	8987	9006	2	4	6	8	9	11	13	15	17
95	9025	9044	9063	9082	9101	9210	9139	9158	9178	9197	2	4	6	8	10	11	13	15	17
96	9216	9235	9254	9274	9293	9312	9332	9351	9370	9390	2	4	6	8	10	12	14	15	17
97	9409	9428	9448	9467	9487	9506	9526	9545	9565	9584	2	4	6	8	10	12	14	16	18
98	9604	9624	9643	9663	9683	9702	9722	9742	9761	9781	2	4	6	8	10	12	14	16	18
99	9801	9821	9841	9860	9880	9900	9920	9940	9960	9980	2	4	6	8	10	12	14	16	18

Find the position of the decimal point by inspection.

Table IV

Sines

	·0°	·1°	·2°	·3°	·4°	·5°	·6°	·7°	·8°	·9°
0°	·0000	0017	0035	0052	0070	0087	0105	0122	0140	0157
1	·0175	0192	0209	0227	0244	0262	0279	0297	0314	0332
2	·0349	0366	0384	0401	0419	0436	0454	0471	0488	0506
3	·0523	0541	0558	0576	0593	0610	0628	0645	0663	0680
4	·0698	0715	0732	0750	0767	0785	0802	0819	0837	0854
5	·0872	0889	0906	0924	0941	0958	0976	0993	1011	1028
6	·1045	1063	1080	1097	1115	1132	1149	1167	1184	1201
7	·1219	1236	1253	1271	1288	1305	1323	1340	1357	1374
8	·1392	1409	1426	1444	1461	1478	1495	1513	1530	1547
9	·1564	1582	1599	1616	1633	1650	1668	1685	1702	1719
10	·1736	1754	1771	1788	1805	1822	1840	1857	1874	1891
11	·1908	1925	1942	1959	1977	1994	2011	2028	2045	2062
12	·2079	2096	2113	2130	2147	2164	2181	2198	2215	2233
13	·2250	2267	2284	2300	2317	2334	2351	2368	2385	2402
14	·2419	2436	2453	2470	2487	2504	2521	2538	2554	2571
15	·2588	2605	2622	2639	2656	2672	2689	2706	2723	2740
16	·2756	2773	2790	2807	2823	2840	2857	2874	2890	2907
17	·2924	2940	2957	2974	2990	3007	3024	3040	3057	3074
18	·3090	3107	3123	3140	3156	3173	3190	3206	3223	3239
19	·3256	3272	3289	3305	3322	3338	3355	3371	3387	3404
20	·3420	3437	3453	3469	3486	3502	3518	3535	3551	3567
21	·3584	3600	3616	3633	3649	3665	3681	3697	3714	3730

	·0°	·1°	·2°	·3°	·4°	·5°	·6°	·7°	·8°	·9°
22	·3746	3762	3778	3795	3811	3827	3843	3859	3875	3891
23	·3907	3923	3939	3955	3971	3987	4003	4019	4035	4051
24	·4067	4083	4099	4115	4131	4147	4163	4179	4195	4210
25	·4226	4242	4258	4274	4289	4305	4321	4337	4352	4368
26	·4384	4399	4415	4431	4446	4462	4478	4493	4509	4524
27	·4540	4555	4571	4586	4602	4617	4633	4648	4664	4679
28	·4695	4710	4726	4741	4756	4772	4787	4802	4818	4833
29	·4848	4863	4879	4894	4909	4924	4939	4955	4970	4985
30	·5000	5015	5030	5045	5060	5075	5090	5105	5120	5135
31	·5150	5165	5180	5195	5210	5225	5240	5255	5270	5284
32	·5299	5314	5329	5344	5358	5373	5388	5402	5417	5432
33	·5446	5461	5476	5490	5505	5519	5534	5548	5563	5577
34	·5592	5606	5621	5635	5650	5664	5678	5693	5707	5721
35	·5736	5750	5764	5779	5793	5807	5821	5835	5850	5864
36	·5878	5892	5906	5920	5934	5948	5962	5976	5990	6004
37	·6018	6032	6046	6060	6074	6088	6101	6115	6129	6143
38	·6157	6170	6184	6198	6211	6225	6239	6252	6266	6280
39	·6293	6307	6320	6334	6347	6361	6374	6388	6401	6414
40	·6428	6441	6455	6468	6481	6494	6508	6521	6534	6547
41	·6561	6574	6587	6600	6613	6626	6639	6652	6665	6678
42	·6691	6704	6717	6730	6743	6756	6769	6782	6794	6807
43	·6820	6833	6845	6858	6871	6884	6896	6909	6921	6934
44	·6947	6959	6972	6984	6997	7009	7022	7034	7046	7059

TABLE IV (contd.)

Sines

	·0°	·1°	·2°	·3°	·4°	·5°	·6°	·7°	·8°	·9°
45°	·7071	7083	7096	7108	7120	7133	7145	7157	7169	7181
46	·7193	7206	7218	7230	7242	7254	7266	7278	7290	7302
47	·7314	7325	7337	7349	7361	7373	7385	7396	7408	7420
48	·7431	7443	7455	7466	7478	7490	7501	7513	7524	7536
49	·7547	7559	7570	7581	7593	7604	7615	7627	7638	7649
50	·7660	7672	7683	7694	7705	7716	7727	7738	7749	7760
51	·7771	7782	7793	7804	7815	7826	7837	7848	7859	7869
52	·7880	7891	7902	7912	7923	7934	7944	7955	7965	7976
53	·7986	7997	8007	8018	8028	8039	8049	8059	8070	8080
54	·8090	8100	8111	8121	8131	8141	8151	8161	8171	8181
55	·8192	8202	8211	8221	8231	8241	8251	8261	8271	8281
56	·8290	8300	8310	8320	8329	8339	8348	8358	8368	8377
57	·8387	8396	8406	8415	8425	8434	8443	8453	8462	8471
58	·8480	8490	8499	8508	8517	8526	8536	8545	8554	8563
59	·8572	8581	8590	8599	8607	8616	8625	8634	8643	8652
60	·8660	8669	8678	8686	8695	8704	8712	8721	8729	8738
61	·8746	8755	8763	8771	8780	8788	8796	8805	8813	8821
62	·8829	8838	8846	8854	8862	8870	8878	8886	8894	8902
63	·8910	8918	8926	8934	8942	8949	8957	8965	8973	8980
64	·8988	8996	9003	9011	9018	9026	9033	9041	9048	9056
65	·9063	9070	9078	9085	9092	9100	9107	9114	9121	9128
66	·9135	9143	9150	9157	9164	9171	9178	9184	9191	9198

	·0°	·1°	·2°	·3°	·4°	·5°	·6°	·7°	·8°	·9°
67	·9205	9212	9219	9225	9232	9239	9245	9252	9259	9265
68	·9272	9278	9285	9291	9298	9304	9311	9317	9323	9330
69	·9336	9342	9348	9354	9361	9367	9373	9379	9385	9391
70	·9397	9403	9409	9415	9421	9426	9432	9438	9444	9449
71	·9455	9461	9466	9472	9478	9483	9489	9494	9500	9505
72	·9511	9516	9521	9527	9532	9537	9542	9548	9553	9558
73	·9563	9568	9573	9578	9583	9588	9593	9598	9603	9608
74	·9613	9617	9622	9627	9632	9636	9641	9646	9650	9655
75	·9659	9664	9668	9673	9677	9681	9686	9690	9694	9699
76	·9703	9707	9711	9715	9720	9724	9728	9732	9736	9740
77	·9744	9748	9751	9755	9759	9763	9767	9770	9774	9778
78	·9781	9785	9789	9792	9796	9799	9803	9806	9810	9813
79	·9816	9820	9823	9826	9829	9833	9836	9839	9842	9845
80	·9848	9851	9854	9857	9860	9863	9866	9869	9871	9874
81	·9877	9880	9882	9885	9888	9890	9893	9895	9898	9900
82	·9903	9905	9907	9910	9912	9914	9917	9919	9921	9923
83	·9925	9928	9930	9932	9934	9936	9938	9940	9942	9943
84	·9945	9947	9949	9951	9952	9954	9956	9957	9959	9960
85	·9962	9963	9965	9966	9968	9969	9971	9972	9973	9974
86	·9976	9977	9978	9979	9980	9981	9982	9983	9984	9985
87	·9986	9987	9988	9989	9990	9990	9991	9992	9993	9993
88	·9994	9995	9995	9996	9996	9997	9997	9997	9998	9998
89	·9998	9999	9999	9999	9999	1·000	1·000	1·000	1·000	1·000

Table V

Tangents

	·0°	·1°	·2°	·3°	·4°	·5°	·6°	·7°	·8°	·9°
0°	0·0000	0017	0035	0052	0070	0087	0105	0122	0140	0157
1	0·0175	0192	0209	0227	0244	0262	0279	0297	0314	0332
2	0·0349	0367	0384	0402	0419	0437	0454	0472	0489	0507
3	0·0524	0542	0559	0577	0594	0612	0629	0647	0664	0682
4	0·0699	0717	0734	0752	0769	0787	0805	0822	0840	0857
5	0·0875	0892	0910	0928	0945	0963	0981	0998	1016	1033
6	0·1051	1069	1086	1104	1122	1139	1157	1175	1192	1210
7	0·1228	1246	1263	1281	1299	1317	1334	1352	1370	1388
8	0·1405	1423	1441	1459	1477	1495	1512	1530	1548	1566
9	0·1584	1602	1620	1638	1655	1673	1691	1709	1727	1745
10	0·1763	1781	1799	1817	1835	1853	1871	1890	1908	1926
11	0·1944	1962	1980	1998	2016	2035	2053	2071	2089	2107
12	0·2126	2144	2162	2180	2199	2217	2235	2254	2272	2290
13	0·2309	2327	2345	2364	2382	2401	2419	2438	2456	2475
14	0·2493	2512	2530	2549	2568	2586	2605	2623	2642	2661
15	0·2679	2698	2717	2736	2754	2773	2792	2811	2830	2849
16	0·2867	2886	2905	2924	2943	2962	2981	3000	3019	3038
17	0·3057	3076	3096	3115	3134	3153	3172	3191	3211	3230
18	0·3249	3269	3288	3307	3327	3346	3365	3385	3404	3424
19	0·3443	3463	3482	3502	3522	3541	3561	3581	3600	3620
20	0·3640	3659	3679	3699	3719	3739	3759	3779	3799	3819
21	0·3839	3859	3879	3899	3919	3939	3959	3979	4000	4020

	·0°	·1°	·2°	·3°	·4°	·5°	·6°	·7°	·8°	·9°
22	0·4040	4061	4081	4101	4122	4142	4163	4183	4204	4224
23	0·4245	4265	4286	4307	4327	4348	4369	4390	4411	4431
24	0·4452	4473	4494	4515	4536	4557	4578	4599	4621	4642
25	0·4663	4684	4706	4727	4748	4770	4791	4813	4834	4856
26	0·4877	4899	4921	4942	4964	4986	5008	5029	5051	5073
27	0·5095	5117	5139	5161	5184	5206	5228	5250	5272	5295
28	0·5317	5340	5362	5384	5407	5430	5452	5475	5498	5520
29	0·5543	5566	5589	5612	5635	5658	5681	5704	5727	5750
30	0·5774	5797	5820	5844	5867	5890	5914	5938	5961	5985
31	0·6009	6032	6056	6080	6104	6128	6152	6176	6200	6224
32	0·6249	6273	6297	6322	6346	6371	6395	6420	6445	6469
33	0·6494	6519	6544	6569	6594	6619	6644	6669	6694	6720
34	0·6745	6771	6796	6822	6847	6873	6899	6924	6950	6976
35	0·7002	7028	7054	7080	7107	7133	7159	7186	7212	7239
36	0·7265	7292	7319	7346	7373	7400	7427	7454	7481	7508
37	0·7536	7563	7590	7618	7646	7673	7701	7729	7757	7785
38	0·7813	7841	7869	7898	7926	7954	7983	8012	8040	8069
39	0·8098	8127	8156	8185	8214	8243	8273	8302	8332	8361
40	0·8391	8421	8451	8481	8511	8541	8571	8601	8632	8662
41	0·8693	8724	8754	8785	8816	8847	8878	8910	8941	8972
42	0·9004	9036	9067	9099	9131	9163	9195	9228	9260	9293
43	0·9325	9358	9391	9424	9457	9490	9523	9556	9590	9623
44	0·9657	9691	9725	9759	9793	9827	9861	9896	9930	9965

TABLE V (contd.)

Tangents

	·0°	·1°	·2°	·3°	·4°	·5°	·6°	·7°	·8°	·9°
45°	1·0000	0035	0070	0105	0141	0176	0212	0247	0283	0319
46	1·0355	0392	0428	0464	0501	0538	0575	0612	0649	0686
47	1·0724	0761	0799	0837	0875	0913	0951	0990	1028	1067
48	1·1106	1145	1184	1224	1263	1303	1343	1383	1423	1463
49	1·1504	1544	1585	1626	1667	1708	1750	1792	1833	1875
50	1·1918	1960	2002	2045	2088	2131	2174	2218	2261	2305
51	1·2349	2393	2437	2482	2527	2572	2617	2662	2708	2753
52	1·2799	2846	2892	2938	2985	3032	3079	3127	3175	3222
53	1·3270	3319	3367	3416	3465	3514	3564	3613	3663	3713
54	1·3764	3814	3865	3916	3968	4019	4071	4124	4176	4229
55	1·4281	4335	4388	4442	4496	4550	4605	4659	4715	4770
56	1·4826	4882	4938	4994	5051	5108	5166	5224	5282	5340
57	1·5399	5458	5517	5577	5637	5697	5757	5818	5880	5941
58	1·6003	6066	6128	6191	6255	6319	6383	6447	6512	6577
59	1·6643	6709	6775	6842	6909	6977	7045	7113	7182	7251
60	1·7321	7391	7461	7532	7603	7675	7747	7620	7893	7966
61	1·8040	8115	8190	8265	8341	8418	8495	8572	8650	8728
62	1·8807	8887	8967	9047	9128	9210	9292	9375	9458	9542
63	1·9626	9711	9797	9883	9970	*0057*	*0145*	*0233*	*0323*	*0413*
64	2·0503	0594	0686	0778	0872	0965	1060	1155	1251	1348
65	2·1445	1543	1642	1742	1842	1943	2045	2148	2251	2355
66	2·2460	2566	2673	2781	2889	2998	3109	3220	3332	3445

	·0°	·1°	·2°	·3°	·4°	·5°	·6°	·7°	·8°	·9°
67	2·3559	3673	3789	3906	4023	4142	4262	4383	4504	4627
68	2·4751	4876	5002	5129	5257	5386	5517	5649	5782	5916
69	2·6051	6187	6325	6464	6605	6746	6889	7034	7179	7326
70	2·7475	7625	7776	7929	8083	8239	8397	8556	8716	8878
71	2·9042	9208	9375	9544	9714	9887	*0061*	*0237*	*0415*	*0595*
72	3·0777	0961	1146	1334	1524	1716	1910	2106	2305	2506
73	3·2709	2914	3122	3332	3544	3759	3977	4197	4420	4646
74	3·4874	5105	5339	5576	5816	6059	6305	6554	6806	7062
75	3·7321	7583	7848	8118	8391	8667	8947	9232	9520	9812
76	4·0108	0408	0713	1022	1335	1653	1976	2303	2635	2972
77	4·3315	3662	4015	4373	4737	5107	5483	5864	6252	6646
78	4·7046	7453	7867	8288	8716	9152	9594	*0045*	*0504*	*0970*
79	5·1446	1929	2422	2924	3435	3955	4486	5026	5578	6140
80	5·671	5·730	5·789	5·850	5·912	5·976	6·041	6·107	6·174	6·243
81	6·314	6·386	6·460	6·535	6·612	6·691	6·772	6·855	6·940	7·026
82	7·115	7·207	7·300	7·396	7·495	7·596	7·700	7·806	7·916	8·028
83	8·144	8·264	8·386	8·513	8·643	8·777	8·915	9·058	9·205	9·357
84	9·51	9·68	9·84	10·02	10·20	10·39	10·58	10·78	10·99	11·20
85	11·43	11·66	11·91	12·16	12·43	12·71	13·00	13·30	13·62	13·95
86	14·30	14·67	15·06	15·46	15·89	16·35	16·83	17·34	17·89	18·46
87	19·08	19·74	20·45	21·20	22·02	22·90	23·86	24·90	26·03	27·27
88	28·64	30·14	31·82	33·69	35·80	38·19	40·92	44·07	47·74	52·08
89	57·29	63·66	71·62	81·85	95·49	114·6	143·2	191·0	286·5	573·0

Where the integer changes, the numbers are italicized.

Table VI

Logarithms (to base 10)

	0	1	2	3	4	5	6	7	8	9	1	2	3	4	5	6	7	8	9
10	·0000	0043	0086	0128	0170	0212	0253	0294	0334	0374	4	8	12	17	21	25	29	33	37
11	·0414	0453	0492	0531	0569	0607	0645	0682	0719	0755	4	8	11	15	19	23	26	30	34
12	·0792	0828	0864	0899	0934	0969	1004	1038	1072	1106	3	7	10	14	17	21	24	28	31
13	·1139	1173	1206	1239	1271	1303	1335	1367	1399	1430	3	6	10	13	16	19	23	26	29
14	·1461	1492	1523	1553	1584	1614	1644	1673	1703	1732	3	6	9	12	15	18	21	24	27
15	·1761	1790	1818	1847	1875	1903	1931	1959	1987	2014	3	6	8	11	14	17	20	22	25
16	·2041	2068	2095	2122	2148	2175	2201	2227	2253	2279	3	5	8	11	13	16	18	21	24
17	·2304	2330	2355	2380	2405	2430	2455	2480	2504	2529	2	5	7	10	12	15	17	20	22
18	·2553	2577	2601	2625	2648	2672	2695	2718	2742	2765	2	5	7	9	12	14	16	19	21
19	·2788	2810	2833	2856	2878	2900	2923	2945	2967	2989	2	4	7	9	11	13	16	18	20
20	·3010	3032	3054	3075	3096	3118	3139	3160	3181	3201	2	4	6	8	11	13	15	17	19
21	·3222	3243	3263	3284	3304	3324	3345	3365	3385	3404	2	4	6	8	10	12	14	16	18
22	·3424	3444	3464	3483	3502	3522	3541	3560	3579	3598	2	4	6	8	10	12	14	15	17
23	·3617	3636	3655	3674	3692	3711	3729	3747	3766	3784	2	4	6	7	9	11	13	15	17
24	·3802	3820	3838	3856	3874	3892	3909	3927	3945	3962	2	4	5	7	9	11	12	14	16
25	·3979	3997	4014	4031	4048	4065	4082	4099	4116	4133	2	3	5	7	9	10	12	14	15
26	·4150	4166	4183	4200	4216	4232	4249	4265	4281	4298	2	3	5	7	8	10	11	13	15
27	·4314	4330	4346	4362	4378	4393	4409	4425	4440	4456	2	3	5	6	8	9	11	13	14
28	·4472	4487	4502	4518	4533	4548	4564	4579	4594	4609	2	3	5	6	8	9	11	12	14
29	·4624	4639	4654	4669	4683	4698	4713	4728	4742	4757	1	3	4	6	7	9	10	12	13
30	·4771	4786	4800	4814	4829	4843	4857	4871	4886	4900	1	3	4	6	7	9	10	11	13
31	·4914	4928	4942	4955	4969	4983	4997	5011	5024	5038	1	3	4	6	7	8	10	11	12

	0	1	2	3	4	5	6	7	8	9	1	2	3	4	5	6	7	8	9
32	·5051	5065	5079	5092	5105	5119	5132	5145	5159	5172	1	3	4	5	7	8	9	11	12
33	·5185	5198	5211	5224	5237	5250	5263	5276	5289	5302	1	3	4	5	6	8	9	10	12
34	·5315	5328	5340	5353	5366	5378	5391	5403	5416	5428	1	3	4	5	6	8	9	10	11
35	·5441	5453	5465	5478	5490	5502	5514	5527	5539	5551	1	2	4	5	6	7	9	10	11
36	·5563	5575	5587	5599	5611	5623	5635	5647	5658	5670	1	2	4	5	6	7	8	10	11
37	·5682	5694	5705	5717	5729	5740	5752	5763	5775	5786	1	2	3	5	6	7	8	9	10
38	·5798	5809	5821	5832	5843	5855	5866	5877	5888	5899	1	2	3	5	6	7	8	9	10
39	·5911	5922	5933	5944	5955	5966	5977	5988	5999	6010	1	2	3	4	5	7	8	9	10
40	·6021	6031	6042	6053	6064	6075	6085	6096	6107	6117	1	2	3	4	5	6	8	9	10
41	·6128	6138	6149	6160	6170	6180	6191	6201	6212	6222	1	2	3	4	5	6	7	8	9
42	·6232	6243	6253	6263	6274	6284	6294	6304	6314	6325	1	2	3	4	5	6	7	8	9
43	·6335	6345	6355	6365	6375	6385	6395	6405	6415	6425	1	2	3	4	5	6	7	8	9
44	·6435	6444	6454	6464	6474	6484	6493	6503	6513	6522	1	2	3	4	5	6	7	8	9
45	·6532	6542	6551	6561	6571	6580	6590	6599	6609	6618	1	2	3	4	5	6	7	8	9
46	·6628	6637	6646	6656	6665	6675	6684	6693	6702	6712	1	2	3	4	5	6	7	7	8
47	·6721	6730	6739	6749	6758	6767	6776	6785	6794	6803	1	2	3	4	5	5	6	7	8
48	·6812	6821	6830	6839	6848	6857	6866	6875	6884	6893	1	2	3	4	4	5	6	7	8
49	·6902	6911	6920	6928	6937	6946	6955	6964	6972	6981	1	2	3	4	4	5	6	7	8
50	·6990	6998	7007	7016	7024	7033	7042	7050	7059	7067	1	2	3	3	4	5	6	7	8
51	·7076	7084	7093	7101	7110	7118	7126	7135	7143	7152	1	2	3	3	4	5	6	7	8
52	·7160	7168	7177	7185	7193	7202	7210	7218	7226	7235	1	2	2	3	4	5	6	7	7
53	·7243	7251	7259	7267	7275	7284	7292	7300	7308	7316	1	2	2	3	4	5	6	6	7
54	·7324	7332	7340	7348	7356	7364	7372	7380	7388	7396	1	2	2	3	4	5	6	6	7

TABLE VI (contd.)

Logarithms (to base 10)

	0	1	2	3	4	5	6	7	8	9	1	2	3	4	5	6	7	8	9
55	·7404	7412	7419	7427	7435	7443	7451	7459	7466	7474	1	2	2	3	4	5	5	6	7
56	·7482	7490	7497	7505	7513	7520	7528	7536	7543	7551	1	2	2	3	4	5	5	6	7
57	·7559	7566	7574	7582	7589	7597	7604	7612	7619	7627	1	2	2	3	4	5	5	6	7
58	·7634	7642	7649	7657	7664	7672	7679	7686	7694	7701	1	1	2	3	4	4	5	6	7
59	·7709	7716	7723	7731	7738	7745	7752	7760	7767	7774	1	1	2	3	4	4	5	6	7
60	·7782	7789	7796	7803	7810	7818	7825	7832	7839	7846	1	1	2	3	4	4	5	6	6
61	·7853	7860	7868	7875	7882	7889	7896	7903	7910	7917	1	1	2	3	4	4	5	6	6
62	·7924	7931	7938	7945	7952	7959	7966	7973	7980	7987	1	1	2	3	3	4	5	6	6
63	·7993	8000	8007	8014	8021	8028	8035	8041	8048	8055	1	1	2	3	3	4	5	5	6
64	·8062	8069	8075	8082	8089	8096	8102	8109	8116	8122	1	1	2	3	3	4	5	5	6
65	·8129	8136	8142	8149	8156	8162	8169	8176	8182	8189	1	1	2	3	3	4	5	5	6
66	·8195	8202	8209	8215	8222	8228	8235	8241	8248	8254	1	1	2	3	3	4	5	5	6
67	·8261	8267	8274	8280	8287	8293	8299	8306	8312	8319	1	1	2	3	3	4	5	5	6
68	·8325	8331	8338	8344	8351	8357	8363	8370	8376	8382	1	1	2	3	3	4	4	5	6
69	·8388	8395	8401	8407	8414	8420	8426	8432	8439	8445	1	1	2	2	3	4	4	5	6
70	·8451	8457	8463	8470	8476	8482	8488	8494	8500	8506	1	1	2	2	3	4	4	5	6
71	·8513	8519	8525	8531	8537	8543	8549	8555	8561	8567	1	1	2	2	3	4	4	5	5
72	·8573	8579	8585	8591	8597	8603	8609	8615	8621	8627	1	1	2	2	3	4	4	5	5
73	·8633	8639	8645	8651	8657	8663	8669	8675	8681	8686	1	1	2	2	3	4	4	5	5
74	·8692	8698	8704	8710	8716	8722	8727	8733	8739	8745	1	1	2	2	3	4	4	5	5
75	·8751	8756	8762	8768	8774	8779	8785	8791	8797	8802	1	1	2	2	3	3	4	5	5
76	·8808	8814	8820	8825	8831	8837	8842	8848	8854	8859	1	1	2	2	3	3	4	5	5

	0	1	2	3	4	5	6	7	8	9	1	2	3	4	5	6	7	8	9
77	·8865	8871	8876	8882	8887	8893	8899	8904	8910	8915	1	1	2	2	3	3	4	4	5
78	·8921	8927	8932	8938	8943	8949	8954	8960	8965	8971	1	1	2	2	3	3	4	4	5
79	·8976	8982	8987	8993	8998	9004	9009	9015	9020	9025	1	1	2	2	3	3	4	4	5
80	·9031	9036	9042	9047	9053	9058	9063	9069	9074	9079	1	1	2	2	3	3	4	4	5
81	·9085	9090	9096	9101	9106	9112	9117	9122	9128	9133	1	1	2	2	3	3	4	4	5
82	·9138	9143	9149	9154	9159	9165	9170	9175	9180	9186	1	1	2	2	3	3	4	4	5
83	·9191	9196	9201	9206	9212	9217	9222	9227	9232	9238	1	1	2	2	3	3	4	4	5
84	·9243	9248	9253	9258	9263	9269	9274	9279	9284	9289	1	1	2	2	3	3	4	4	5
85	·9294	9299	9304	9309	9315	9320	9325	9330	9335	9340	1	1	2	2	3	3	4	4	5
86	·9345	9350	9355	9360	9365	9370	9375	9380	9385	9390	1	1	2	2	3	3	4	4	5
87	·9395	9400	9405	9410	9415	9420	9425	9430	9435	9440	0	1	1	2	2	3	3	4	4
88	·9445	9450	9455	9460	9465	9469	9474	9479	9484	9489	0	1	1	2	2	3	3	4	4
89	·9494	9499	9504	9509	9513	9518	9523	9528	9533	9538	0	1	1	2	2	3	3	4	4
90	·9542	9547	9552	9557	9562	9566	9571	9576	9581	9586	0	1	1	2	2	3	3	4	4
91	·9590	9595	9600	9605	9609	9614	9619	9624	9628	9633	0	1	1	2	2	3	3	4	4
92	·9638	9643	9647	9652	9657	9661	9666	9671	9675	9680	0	1	1	2	2	3	3	4	4
93	·9685	9689	9694	9699	9703	9708	9713	9717	9722	9727	0	1	1	2	2	3	3	4	4
94	·9731	9736	9741	9745	9750	9754	9759	9763	9768	9773	0	1	1	2	2	3	3	4	4
95	·9777	9782	9786	9791	9795	9800	9805	9809	9814	9818	0	1	1	2	2	3	3	4	4
96	·9823	9827	9832	9836	9841	9845	9850	9854	9859	9863	0	1	1	2	2	3	3	4	4
97	·9868	9872	9877	9881	9886	9890	9894	9899	9903	9908	0	1	1	2	2	3	3	4	4
98	·9912	9917	9921	9926	9930	9934	9939	9943	9948	9952	0	1	1	2	2	3	3	4	4
99	·9956	9961	9965	9969	9974	9978	9983	9987	9991	9996	0	1	1	2	2	3	3	3	4

ANSWERS TO SOME OF THE EXERCISES

Remember that some of the numerical answers are only approximate, so that you need not agree with them exactly, but, of course, you should not be far out.

Chapter 2.

Exercise No.

6. (*a*) $x^2 + 3xy + y^2$, (*b*) $6x + 6y + 10z$,
(*c*) $3a^2 + 12a + 12 = 3(a + 2)^2$, (*d*) $2x - 3$,
(*e*) $a^2 - 2ab - b^2$, (*f*) $x^2yz + xy^2z + xyz^2 = xyz(x + y + z)$,
(*g*) $6a^3b^4$, (*h*) $2x^6$, (*i*) $- a^2 - 4x^2$, (*j*) $\frac{1}{2}xy^3$, (*k*) $3ab$, (*l*) $\frac{1}{3}ad$.
8. (*a*) 12, (*b*) 12, (*c*) 14, (*d*) 5, (*e*) 2, (*f*) 6, (*g*) 3, (*h*) $\frac{1}{2}$, (*i*) 3, (*j*) 18, (*k*) 1, (*l*) 5, (*m*) 2, (*n*) $6a$, (*o*) $2a + b$, (*p*) $a - b$.
9. £285, £255. 10. £342, £171, £114.
11. $1\frac{1}{2}$ hours from the time Tom starts. 12. 12.
13. For 6000 miles A requires 212 gallons and B 165 gallons.
14. 1*s* 6*d* per peck.

Chapter 3.

7. $3x + 7y$, $2a - 5b$, $4a - 9b$, $a + 3$, etc.
9. $(x - 1)(x + 1)$, $(a + b - c)(a + b + c)$, $(a + b - c)(a - b + c)$, $(a - b)(a + b)(a^2 + b^2)$, $(9 - x)(9 + x)$, $(a - b - c)(a + b + c)$, $(x + y - 1)(x + y + 1)$, $(x - y)(x + y)(x^2 + y^2)(x^4 + y^4)$, $(a + b - 1)(a + b + 1)$, $(x + y - 2)(x + y + 2)$, $3(2x + 1)$.
10. (i) 90°, (ii) 60°, (iii) 50°, (iv) 10°, (v) 78°. 12. 60°, 49°, $38\frac{1}{2}$°.
13. $46\frac{1}{2}$°, $43\frac{1}{2}$°; $57\frac{1}{2}$°, $32\frac{1}{2}$°; 68°, 22°. 15. $3\sqrt{3} \doteqdot 5{\cdot}2$ feet, 6 feet.
16. 68·2°. 17. $3\sqrt{3} \doteqdot 5{\cdot}2$ feet. 18. 45·14°. 20. 42 yards.

Chapter 4.

1. 5·196, 4·243, 3·464, 4·899, 3·162, 5·477. 2. $\frac{1}{4}\sqrt{7}$, $\frac{1}{3}\sqrt{5}$, $\frac{3}{5}$.
5. $l = f + (n - 1)d$, etc. 7. (i) $2n - 1$, n^2, (ii) $3n - 2$, $\frac{1}{2}n(3n - 1)$, (iii) $5n$, $5n(n + 1)/2$, (iv) $\frac{1}{2}n$, $\frac{1}{4}n(n + 1)$, (v) $4n - 10$, $2n^2 - 8n$, (vi) $-(n - 2)a$, $\frac{1}{2}a(3n - n^2)$, (vi) $\frac{1}{3}(14 - 5n)$, $(23n - 5n^2)/6$; 3, 2. 8. $\frac{1}{2}n(n + 1)$.

9. $7\frac{4}{5}$, $9\frac{3}{5}$, $11\frac{2}{5}$, $13\frac{1}{5}$. 10. $1\frac{1}{2}$, 2, $2\frac{1}{2}$.
11. The difference between successive terms is $(l - f)/(n + 1)$.
14. (i) 2^{n-1}, $2^n - 1$, (ii) $(0{\cdot}9)^n$, $10\{0{\cdot}9 - (0{\cdot}9)^{n+1}\}$,
(iii) $3/2^{n+1}$, $\frac{3}{2}(1 - 1/2^n)$,
(iv) $a^{n-1}x^{6-n}$, $x^{6-n}(x^n - a^n)/(x - a)$, (v) 3^{n-1}, $\frac{1}{2}(3^n - 1)$.
15. 25, 125.
16. 2/9, 4/27, 8/81. 17. The ratio of successive terms is $\left(\frac{l}{f}\right)^{1/(n+1)}$
20. 2/3, 25/99, 791/999. 21. 2, $1\frac{1}{4}$.
22. (*a*) $n^3 - (n - 1)^3 = 3n^2 - 3n + 1$, $n(2n - 1)$, (*b*) $n(3n - 2)$.
23. $4! = 24$, $8! = 40320$, $12! = 479001600$, $16!$
24. 56, 495, 4368. 25. 720. 26. 16, 26.
27. 5040, 720, 48, 24. 28. 15, 21.

Chapter 5.

11. 1·57 miles. 12. About 5,910 yards. 13. 60·9 feet.
14. 13·65 miles. 15. 9 miles.
16. $\sin 2A = 2 \sin A \cos A$,
$\sin 3A = 3 \sin A \cos^2 A - \sin^3 A = 3 \sin A - 4 \sin^3 A$,
$\cos 2A = \cos^2 A - \sin^2 A$,
$\cos 3A = \cos^3 A - 3 \cos A \sin^2 A = 4 \cos^3 A - 3 \cos A$.
22. $QK = 7{,}813$ miles, $KP = 5{,}117$ miles, $PQ = 12{,}962$ miles.
23. $AM = 2{,}996$ miles, $AZ = 4{,}863$ miles, $MZ = 1{,}890$ miles.
25. 4,133 miles. 26. 42·4 miles. 27. $33° 53\frac{1}{2}'$ or $28° 6\frac{1}{2}'$.
30. (*a*) $\frac{3}{4}\{1 - (-3)^n\}$, (*b*) $\frac{1}{6}\{1 - (-2)^{-n}\}$, (*c*) $\frac{27}{20}\{1 - (-\frac{2}{3})^n\}$,
(*d*) $\frac{3}{5}\{1 - (-\frac{2}{3})^n\}$. 31. $-ar^{2n-1}$, ar^{2n}.

Chapter 6.

2. (*a*) $(a + 4)(a + 6)$, (*b*) $(p + 2)(p + 3)$, (*c*) $(x - 1)(x - 2)$,
(*d*) $(m + 1)(m + 3)$, (*e*) $(x - 2)(x - 8)$, (*f*), $(f + 5)(f - 4)$,
(*g*) $(t + 5)(t - 8)$, and so on.
3. (*a*) $(x + 6)(x - 6)$, (*b*) $(3x + 5)(3x - 5)$,
(*c*) $4(x + 5)(x - 5)$, (*d*) $25(2y + 1)(2y - 1)$, etc.
(*i*) $(16t + 13s)(16t - 13s)$, . . . , (*o*)$(\sqrt{3} + x)(\sqrt{3} - x)$,
(*p*) $(\sqrt{2} + \sqrt{3}x)(\sqrt{2} - \sqrt{3}x)$, etc.
4. (*a*) $(x + 3)(3x + 1)$, (*b*) $(2x + 5)(3x + 2)$,
(*c*) $(2p + 1)(3p + 1)$,
(*d*) $(t + 5)(3t + 7)$, . . . , (*f*) $(2q - 1)(3q - 2)$, . . . ,
(*h*) $(4x^2 + 1)(5x^2 - 1)$, (*i*) $(3 + 2x)(5 - 2x)$,
(*j*) $(2n - 3)(3n + 4)$, etc.

5. (a) $(2a + 3x)(3a - x)$, (b) $(3a - 5bc)(5a + 3bc)$,
(c) $(a - 7b)(6a + 5b)$, (d) $(a + b)(2a - 9b)$, etc.
6. (a) and (b) $1/(x + y)$, (c), (d) and (e) $1/(x - y)$, (f) $3x/4z$,
(g) $(x + 1)/(x + 3)$, (h) $(x + 1)/(x + 2)$,
(i) $(x + 1)/(2x + 5)$, (j) $(3x + 7)/(x + 7)$,
(k) $b(a + b) \div a$, (l) $(4a^2 + 2a + 1)/(2a - 1)$,
(m) $(2x - 3)/(x - 2)$.
7. (a) $a(b + c)/bc$, (b) ab, (c) $-2x + 3y$, (d) $a^2/(a - 2b)$,
(e) $5a/4(x + 1)$, (f) $(a + 17b)/12$, (g) $a/4(x + 2y)$,
(h) $y^2(y - x)/(y + x)$.
8. (a) $2x/(x^2 - 1)$, (b) $-2/(x^2 - 1)$, (c) $2b/(a^2 - b^2)$,
(d) $(a^2 + 2ab - b^2)/(a^2 - b^2)$, etc., ($h$) $2/(y - 4)(y - 6)$,
(i) $2x/(x + y)$, . . . , (k) $12/(t - 1)(t + 3)(t - 5)$, etc.
10. (a) 19, (b) $a + 2b$, (c) 12/13, (d) $2\frac{1}{2}$, (e) $-4/3$, (f) 10/7.
11. (a) $4\frac{1}{3}$ miles, (b) 30 miles, (c) 3·59%.
12. (a) 10, -21, (b) 11, -8, (c) $1\frac{1}{4}$, $-1\frac{1}{3}$, (d) 2/3, $-3/8$,
(e) 8, $-17/3$, (f) 1, $-\frac{1}{2}$, (g) a, $b - a$, (h) 2, -4.
13. (a) 5, 6, 7, or -7, -6, -5, (b) $\frac{1}{2}(3 + \sqrt{2081})$ feet,
(c) $2\frac{6}{11}$, $2\frac{17}{44}$ feet.
14. (a) $x = 10$, $y = 2$, (b) 3, 4, (c) 6, 2, (d) 9, 15, (e) 11, 12, 13,
(f) 3, 6, 1.
15. (a) $20\frac{4}{7}$, (b) $-17/64$, (c) $l = 18$, $w = 12$ feet, (d) 25, (e) 30.
16. (i) $n(2n - 1)$, (ii) $\frac{1}{2}n^2(n + 1)$, (iii) $n(3n^2 - 2)$,
(iv) $3n^2 - 3n + 1$, (v) $\frac{1}{2}(3n^2 - 3n + 2)$, (vi) $2n^2 - 2n + 1$.
18. (i) 1·1249, (ii) 0·9039, (iii) 1·5735, (iv) 128·7876.

Chapter 7.

3. Lat. 50° N, Long. $5\frac{3}{8}$° W approximately.
4. 20 h. 28 m., 8.1 pm, $48\frac{1}{2}$° W. 5. 46° N.
7. September 23rd. 8. Thin waning crescent, 5.8 am, 11.8 am.
10. About 7,900. 11. North Devon. 12. About September 24th.
18. 57·7°, 7.33 pm.
19. About 6.40 am, 5.20 pm, at Gizeh, 7 am, 5 pm at New York,
7.28 am, 4.32 pm at London.

Chapter 8.

2. (a) $x^2 + y^2 = 16$, (b) $x^2 + y^2 - 4x - 6y = 3$.
8. (a) 45°, (b) 60°. 9. Parallel and equally spaced. 11. $y = C$.
17. $l = 100t^2$. 19. (a) $x = 0, y = 3$ or $x = 4, y = 0$,
(b) $x = 3, y = 4$, or $x = 4, y = 3$, (c) $x = 2, y = 1$, or
$x = 3, y = 2$ or $x = -3, y = -2$ or $x = -2, y = -1$.
26. (i) 1 or $-1·4$, (ii) $\frac{3}{4}$ or $-\frac{1}{2}$, (iii) 1·2 or -1.

Chapter 9.

1. (i) 4118, (ii) 30288, (iii) 1992, (iv) 99·18.
2. 6227, 9692, 0·01950, 45·82, 1·009, 0·4181. 3. 2, 3, 3/2.
5. (i) 125,930, (ii) 0·968, (iii) 1·111.
6. (i) £265, (ii) between 7 and 8 years, (iii) £84.
7. 0·0010, 0·3466, 8·240.

Chapter 10.

1. Hint:—$\log_e (1 + \frac{1}{80}) = 4 \log_e 3 - 4 \log_e 2 - \log_e 5$, $\log_e (1 - \frac{1}{25}) = 3 \log_e 2 + \log_e 3 - 2 \log_e 5$, and so on.
6. $1 - 1/x^2$. 7. $3{\cdot}6x^{2{\cdot}6}$, $5/2\sqrt{x}$, $-1\frac{1}{2}x^{-2\frac{1}{2}}$, $\frac{7}{2}x^{5/2}$, $\frac{1}{3}x^{-8/9}$.
8. $5x^4 - 15x^2 + 10x - 4$. 10. $x = -1, y = 2$, and $x = 1, y = -2$.
11. $8/\pi$ cubic feet. 12. $10x = 3y = \frac{1}{2}c$.
14. (i) $x^4 + C$, (ii) $3 \log_e x + C$, (iii) $\{\frac{1}{4}x^{n+1}/(n + 1)\} + C$, (iv) $\frac{2}{3}x^{3/2} + C$, (v) $x^3 + x^2 + x + C$.
15. (i) $\frac{1}{3}x^3 + Ax + B$, (ii) $\frac{5}{2}x^2 + Ax + B$, (iii) $\frac{4}{15}x^{5/2} + Ax + B$.
16. $p = 11{\cdot}41 + 500v^{-1{\cdot}4}$.
17. (i) $-a^2 \sin a^2x$, (ii) $12 \cos 3x$, (iii) $na \cos nx - nb \sin nx$.
20. $A = 2x^2 + 3x + C$; (i) 108, (ii) 216, (iii) 25. 21. 380.
22. (i) $223\frac{1}{3}$, (ii) $\frac{2}{3}a + 2c$, (iii) $4\frac{1}{2}$, (iv) 56, (v) 2.
23. (i) $20266\frac{2}{3}$, (ii) $19994\frac{2}{3}$, (iii) 19993·6 exactly.
24. 14. 26. 11,900 ft lb wt. 27. 100π cu in. 28. 369π cu in.
30. (i) 1, (ii) 0, (iii) $[\sin x - x \cos x]_0^\pi = \pi$.
31. (i) $x \cos x + \sin x$, (ii) $\cos x \sec^2 x - \sin x \tan x = \cos x$, (iii) $6x^2 + 6x + 4$.
32. (i) $-1/(x + 1)^2$, (ii) $(x \cos x - \sin x)/x^2$, (iii) $-\sin x/\cos^2 x$, (iv) $\sec^2 x$.
33. (i) $\frac{1}{2} \cos x/\sqrt{\sin x}$, (ii) $na(ax + b)^{n-1}$, (iii) $a \cos (ax + b)$, (iv) $(2ax + b)/(ax^2 + bx + c)$.
35. (i) $y = A \cos 2x + B \sin 2x$, (ii) $y = Ae^{2x} + Be^{-2x}$. Second part: (i) $y = 5 \cos 2x + 2 \sin 2x$ (ii) $y = \frac{1}{2}(7e^{2x} + 3e^{-2x})$.

INDEX

Norman G. Pulsford
Modern Crossword Dictionary £1.50

Norman G. Pulsford, an expert crossword compiler, 'worked his way through numerous dictionaries and reference books, making notes. He also analysed 20,000 published crosswords. The result of his labours is likely to make life more difficult for himself' SUNDAY TELEGRAPH

Harry Maddox
How to Study 70p

Successful study depends not only on ability and industry but on effective methods of working. This invaluable and comprehensive handbook tells you how to obtain the greatest benefit from your studies for the least expenditure of energy and effort. Using the author's methods you can speed up your reading rate, better your ability to memorize and take accurate notes, improve your written English and understand elementary mathematics. Whatever you may be studying, the author will help you to work without supervision and realize your full potential.

Clifford Allen
Passing Examinations 50p

This invaluable book for students of all ages explains the techniques of study which are afforded by modern psychology and advises how these can be applied to both written and oral examinations.

'Refreshingly down to earth' BIRMINGHAM POST

Ronald Ridout and Clifford Witting
The Facts of English £1.00

An invaluable new edition of an established work of reference that gives the linguistic and literary facts of the English language from adverbs to zeugma. Thoroughly revised and expaded by Ronald Ridout and Anthony Hern, this edition has been brought up to date with current cultural and commercial usage.

Over 1,000 entries, fully cross-referenced, explained and illustrated with suitable examples, and ranging from the elementary to the specialized, make this an ideal guide for student and expert alike.

English Proverbs Explained 60p

In this fascinating collection of hundreds of English proverbs still in use the authors are concerned with proverbs as a live force and have selected sayings which are at the same time popular, pithy and wise. Teachers, crossword addicts and the general reader will find much to divert them, and students of English can rejoice in an entertaining guide through a maze of bewildering expressions.

C. L. Barber
The Story of Language 80p

In the first half of this book more general topics such as the nature of language, its origins, the causes of linguistic changes and language families, are gone into – the second half is, in effect, a history of the English language.